Statistics and Data Analysis in Geology

Second Edition

JOHN C. DAVIS

Kansas Geological Survey

JOHN WILEY & SONS

New York • Chichester • Brisbane • Toronto • Singapore

Certain tables and figures in this text are copyright and are reproduced with the permission of the copyright owner. The source of each table or figure is noted in its caption and a complete citation to its source is given in the Suggested Readings. Tables 2.10, 2.22, 2.25, and 2.26 are copyright by John Wiley & Sons Inc. and are reproduced with their permission. Tables 2.11, 2.14, and 2.18 are copyright by Penguin Books Ltd. and are reproduced with their permission. Tables 4.30 and 4.31 are copyright by the American Chemical Society and are reproduced with their permission. Parts of Table 5.6 are copyright by the American Statistical Association and parts are copyright by the American Institute of Biological Sciences; the combined table is reproduced with their permissions. Tables 5.7 and 5.9 are copyright by Academic Press Inc. (London) Ltd. and are reproduced with their permission. Figure 5.24 is copyright by the American Statistical Association and is reproduced with their permission. Figure 5.25 is copyright by Harcourt Brace Jovanovich, Inc. and is reproduced with their permission.

Library of Congress Cataloging in Publication Data:

Davis, John C.
  Statistics and data analysis in geology.

  Includes bibliographies and index.
  1. Geology—Data processing. 2. Geology—
Statistical methods. I. Title.

QE48.8.D38   1986   550'.72   85-12331
ISBN 0-471-08079-9

Printed in the United States of America

10 9 8 7 6 5 4 3

Dedication

My first introduction to mathematical geology came late one evening in 1961, in a dimly lit, seedy bar in Independence, Kansas. Dan Merriam and John Harbaugh, ostensibly arranging a field trip to the carbonate reefs of southeastern Kansas, were sharing with me their enthusiasm for a new computer technique called "trend surface analysis." Dan would swoop his hands first one way and then another, demonstrating how a first-order trend could correspond to the tilt of a basin, and how a second-order trend could represent its bowl-like shape. John attempted to explain how a trend was calculated, scribbling in an obscure mathematical notation on a damp paper napkin. I admit I left somewhat more tipsy than I had arrived, still skeptical of all things mathematical and not at all convinced that a computer would ever find a place in my future as a geologist.

Over the next years of my graduate career, however, I painfully puzzled out for myself what trend surfaces were—and discriminant functions, and even principal components. Slowly I came to share the enthusiasm for things quantitative that Dan and John had tried to show me that night. I returned eventually to Kansas to work under Dan Merriam, and later to succeed him at the Kansas Geological Survey. During those same years I worked on many occasions with John Harbaugh, as a collaborator in research and as a colleague. Computers not only intruded into my life, but slowly and surely grew to be an indispensable part of my professional career.

Perhaps the moral of this story is that young geologists should stay out of dingy bars. But if they do find themselves in such disreputable surroundings, I can wish them no better fortune than to fall in with such companions as I did that night.

This book is dedicated to my good friends and mentors, Daniel F. Merriam and John W. Harbaugh.

Preface

When the first edition of this book was published in 1973, the geologist venturing into the world of computing entered a far different environment than now. It was a world of massive mainframes and centralized facilities operated in batch mode, tended by a priesthood of computer specialists behind glass windows and locked doors. The novice user would timorously approach the sanctum, offering a stack of laboriously punched cards through an opening in a window. The precious cards and data would disappear into the inner recesses and a few days later, *perhaps* a stack of output would appear.

Now most geologists, students as well as professionals, can have immediate access to a computer through a time-sharing terminal attached to a remote mainframe or minicomputer, or have a personal computer sitting on their own desk. The computer has become a routine part of modern student life; graduates and undergraduates alike *expect* to do computing on an almost-daily basis. Seasoned professional geologists, as well as those new to their careers, are turning to the computer in the hope of increasing their productivity and effectiveness.

Unfortunately, easy access to computers does not provide easy knowledge about what to do with them. To most geologists, trend surface analysis is as much a mystery today as it was over 10 years ago. Factor analysis is no less opaque now than in the past, and new, even more exotic and unintelligible methods of data analysis have come to the fore. The need for education in quantitative analysis of geologic data was apparent in 1973, and is equally apparent today. It is for this reason that the second edition of this book has been written.

In response to the many comments I have received on the first edition, and based on my own teaching experiences, the book has been revised for this edition. The basic arrangement of topics is retained, progressing from background information to the analysis of sequences, maps, and multivariate observations. Since most students now take one or more courses in programming and FORTRAN has been supplemented if not replaced by other languages in scientific computing, the chapter on programming has been deleted. The first edition contained numerous listings of FORTRAN programs designed for instructional purposes. These now have been removed from the text, expanded and enhanced to make them suitable for routine use as well as for teaching, and are available in diskette form.

The discussion of elementary probability theory has been expanded, since probability forms the basis for almost all data analysis procedures and a thorough understanding of the concepts of probability is essential. A new section on nonparametric methods has been added because geologic data, particularly that col-

v

lected in the field, often fail to meet the requirements necessary for the proper use of traditional statistical tests.

Eigenvalues and eigenvectors remain difficult topics for students, so their treatment in the section on matrix algebra has been revised and the entire section on factor analysis and other eigenvalue methods has been expanded in the final chapter. The essential interrelations between eigenvalue procedures such as principal components, R- and Q-mode factor analysis, principal coordinates, and correspondence analysis have been made explicit.

Some topics in data analysis have assumed a much greater role in the Earth sciences over the past few years. Regionalized variable theory is now invoked by many investigators to explain the spatial behavior of geologic properties. The semivariogram and the process of kriging play central roles in the methodology of regionalized variable analysis, and are discussed at length. Geophysicists have always recognized the critical importance of spectral analysis, and the utility of these methods is now apparent in many other areas of the Earth sciences, ranging from earthquake prediction to description of the shapes of fossils. The section on Fourier analysis has been expanded to reflect this.

Throughout the text, numerous lesser changes, corrections, and additions have been made in response to the suggestions of readers. The list of those offering comments over the years is too long to recount here, but to all who took the time to write or call about some point, I give my thanks. In addition to those named in the first edition, I would like to specifically thank Dr. Paul Brockington, Dr. Jim Campbell, and Dr. Keith Turner for their suggestions. My reviewers, Dr. Dave Best, Professor Frank Ethridge, and Professor Je-an Fang, made many valuable corrections to the final text.

Numerous contributions were made by colleagues at the Kansas Geological Survey, including Dr. David Collins and Dr. Colin Ferguson, and my collaborator on the first edition, Mr. Robert Sampson. Three of my associates provided extensive assistance during my writing; Dr. Ricardo Olea to the sections on regionalized variables, Dr. Zhou Di to sections on eigenvalue analyses, and Dr. John Doveton who generously donated many of the exercises and examples throughout the text and who helped in all phases of the revision. Finally, I would like to acknowledge the efforts of my research assistant and companion, Ms. Jo Anne DeGraffenreid, who edited, corrected, revised, and organized, and without whose support and encouragement this edition would not have been possible.

Lawrence, KS   John C. Davis

Contents

CHAPTER **1**

Introduction

> "...when you can measure what you are
> speaking about and express it in numbers,
> you know something about it; but when you
> cannot express it in numbers, your
> knowledge is of a meagre and unsatis-
> factory kind; it may be the beginning of
> knowledge, but you have scarcely in your
> thoughts advanced to the state of science,
> whatever the matter may be."
>
> Lord Kelvin

Mathematical methods have been employed by a few geologists since the earliest days of the profession. Mining geologists and engineers, for example, have calculated tonnages and estimated ore tenor from samples for centuries. As Fisher pointed out (1953, p. 3), Lyell's subdivision of the Tertiary on the basis of the relative abundance of modern marine organisms was a statistical procedure. Sedimentary petrologists have regarded grain size and shape measurements as important sources of sedimentological information since the beginning of this century. The hybrid Earth sciences of geochemistry, geophysics, and hydrology require a firm background in mathematics, although the procedures used are primarily derived from the nongeological parent. Similarly, mineralogists and crystallographers utilize the mathematical techniques of physical and analytical chemistry.

Although these topics are of undeniable importance in specialized disciplines, they are not the subject of this book. Since the spread of computers through universities and corporations in the late 1950s, geologists have been increasingly attracted to mathematical methods of data analysis. These methods have been borrowed from all scientific and engineering disciplines and applied to every facet of Earth science; it is these more general techniques that are our concern. Geology itself is responsible for some of the advances in computer science, most notably in the area of graphics, including map and contour plotting. However, our science

1

has benefited more than it has contributed to the exchange of quantitative techniques.

In the United States, petroleum companies are among the largest nongovernment users of computers and also the largest employers of geologists. It is not unexpected that a tremendous interest in geomathematical techniques has developed in these organizations. This interest has spread back into the academic world and has resulted in an increasing emphasis on computer languages and mathematical skills in the training of geologists. Unfortunately, there is no widely developed heritage of mathematical analysis in geology, and establishment of adequate educational programs has proceeded only in scattered institutions and through the efforts of a handful of people.

Many geologists have been caught short in the computer revolution: educated in a tradition that emphasized the qualitative at the expense of the quantitative, they are inadequately prepared in mathematics, unfamiliar with statistics. Even so, members of the profession quickly grasped the potential importance of analytical techniques computers now make so readily available. Many institutions, both commercial and public, have provided an extensive library of computer programs that will implement geomathematic procedures. The temptation is strong, perhaps irresistible, to utilize these programs even though the basis on which they operate may not be clearly understood.

The development and explosive proliferation of personal computers has accelerated these trends. Small companies, consulting groups, and even individual geologists now possess computing power that was restricted to large corporations and universities only a few short years ago. Many geologists, who never imagined they would have either a need for a computer or an opportunity to use one, now find they have one sitting on their own desk! To many of these geologists, the personal computer seems to promise more than word processing and financial calculations, if only they knew how to apply it in their professional work.

This book is designed to alleviate partially the difficulties of geologists who feel that they can gain from a mathematical approach to their research but are inadequately prepared in training or experience. Ideally, of course, these people should receive formal schooling in probability, statistics, numerical analysis, and programming, and then should study under a qualified geomathematician. Such an ideal is unrealistic for all but a few fortunate individuals. Most must make their way as best they can; reading, questioning, and educating themselves by trial and error. The path that these people follow is not an orderly progression through topics laid out as in a curriculum. Typically, they proceed backwards, attracted first to those methods they feel offer the most help in their research, exploration, or operational problems. Later they fill in the gaps in their background and attempt to master the precepts of the techniques they have been applying. This is an unsatisfactory and even dangerous method of education, perhaps comparable to on-the-job training for medical doctors, but one many people seem destined to follow. This book may help organize the self-educational process, guiding the impatient rapidly through the necessary initial steps to the glittering algorithmic Grail that he seeks. Along the way the reader will be exposed to those less glamorous topics that constitute the foundation upon which trend surfaces or factor analyses, for example, are built.

This book also is designed to aid another type of geologists, the student who has taken or is taking courses in statistics and programming. Such curriculum requirements are now almost universal in American and European universities. Unfortunately, these topics typically are taught by persons who have little knowledge of geology or appreciation for the types of problems faced by Earth scientists. The relevance of these studies to his primary field often is obscure to the student. This feeling may be compounded by the absence of mathematical applications in many geology courses. His professors may have received their formal education prior to the current emphasis on quantitative methodology, and consequently are untrained in the quantitative subjects the student is required to master. His instructors may therefore find it difficult to demonstrate the pertinence of these topics. In this book, the student will find not only generalized developments of computational techniques, but also numerous examples of their applications in geology. Of course, it is our hope that both the student and his instructor will find something of interest in this book and that it may help promote the widening common ground we refer to as geomathematics.

About This Book

The reader of a book is entitled to know at the onset where he is going and how the author proposes to take him there. He also needs to know what is expected of him, because the author has made certain assumptions about background, training, interests, and abilities of his audience. This book is about quantitative methods of analysis of geologic data, the area of Earth science which is now being called geomathematics. The orientation is methodological, or "how-to-do-it." Theory is not emphasized for several reasons. Most geologists tend to be pragmatists, and are far more interested in results than in theory. Many useful procedures as of yet have no adequate theoretical background. Those which are theoretically developed often are based on statistical assumptions so restrictive that they are not valid for geologic data. Although elementary probability is discussed and many statistical tests described, the detailed development of geostatistical theory is left to others.

Our emphasis is on operations, as the most complex analytical procedure is built up of a series of relatively simple mathematical manipulations. These operations are most easily expressed in matrix algebra, so we will study this subject. The first edition of this text devoted a chapter to the FORTRAN language. Now, almost all students learn at least one programming language, and FORTRAN is no longer the sole vernacular of scientific computing. Of necessity, all scientific computer languages have provisions for the manipulation of matrices, so our matrix algebra techniques can be implemented on these powerful devices, leaving us free to experiment and consider the application.

The discussion in the following chapters begins with the basic topics of elementary statistics and matrix algebra. Thereafter we will consider the analysis of various types of geologic data, which have arbitrarily been classified into three categories: data in which the sequence of observations is important, data in which the two-dimensional relationships between observations are important, and multivariate data in which order and location of the observations are not considered. The first category contains all classes of problems in which data have been collected along a continuum, either of time or distance. It includes time series problems, analysis of stratigraphic sections, and interpretation of chart recordings such as well logs. The second category includes problems in which geographic locations of samples are important: studies of shape and orientation, mapping, trend-surface analysis, kriging, and similar endeavors. The final category is concerned with clustering, classification, and the examination of interrelations among data sets in which sample locations on a map or traverse are not considered. Paleontologic, mineralogic, and geochemical data often are of this type.

The topics proceed from simple to complex. However, each successive topic is built upon its predecessors, so aspects of multiple regression, covered in Chapter 6, have been discussed in trend analysis (Chapter 5), which has in turn been preceded by curvilinear regression (Chapter 4). The basic mathematical procedure involved has been described under the solution of simultaneous equations (Chapter 3), and the statistical basis of regression has been first discussed in Chapter 2. Other techniques are similarly developed.

The first topic discussed in this book is elementary statistics. The final topic is factor analysis. These two subjects are separated by a wide gulf that would require several years to bridge following a typical course of study. Obviously, we cannot cover this span in a single book without omitting a tremendous amount of material. What has been sacrificed are all but the rudiments of statistical theory associated with each of the techniques, the details of all mathematical operations except those that are absolutely essential, and all the embellishments and refinements that typically are added to the basic procedures. What has been retained are the fundamental mathematical algorithms involved in each analysis, discussions of the relations between quantitative techniques, and simple examples of applications to geologic problems.

FORTRAN program listings are not given in this edition, because many libraries of programs are now available for machines ranging in size from supercomputers to desktop models. These libraries contain programs that are far more sophisticated and flexible than any that could be listed in a book. However, to help introduce you to personal scientific computing, a diskette is included in the back; this contains programs for elementary statistics and the basic operations of matrix algebra. The diskette will run on personal computers such as the IBM-PC® and compatible machines that use the popular MS-DOS® operating system. A complete library of programs for most of the procedures discussed in this book is also available for selected personal computers; information on this is listed in the Appendix.

It is our contention that a quantitative approach to geology can yield a fruitful return to an investigator: perhaps not so much by the proof or demonstration of geological hypotheses, but in the insights gained through the critical examination of phenomena, which is required by a quantitative method. The gathering of data of sufficient quality and quantity to be useful in numerical analysis forces a closer familiarity with the objects of study than might otherwise be obtained. Certainly a paleontologist who has carefully measured hundreds of specimens of an organism has a far greater appreciation of the natural variation in these creatures than a person who has simply examined them. The rigor and objectivity required by a quantitative methodology can compensate in part for insight and experience which otherwise must be gained by many years of work. At the same time, the discipline necessary to perform quantitative research will hasten the growth and maturity of the scientist.

The measurement and analysis of data may lead to interpretations that are not obvious or apparent using other means of research. Multivariate methods, for example, may reveal clusterings of objects that are at variance with accepted classifications, or may show relationships between variables where none was expected. These findings require explanation. Sometimes a plausible explanation cannot be found, but in other instances new theories may be suggested which otherwise would have been neglected.

Perhaps the greatest worth of a quantitative methodology lies not in its ability to demonstrate what is true but rather what is false. These techniques can reveal the insufficiency of data, the tenuousness of assumptions, the paucity of information contained in most geologic studies. When carefully and dispassionately analyzed, many geologic interpretations deteriorate into a collection of guesses and hunches based on very little data, most of which is of a contradictory or inconclusive nature. If geology were an experimental science like chemistry or physics, where observations could be verified by any competent worker, controversy and conflict might disappear. However, we are practitioners of an observational science, and the rigorous application of quantitative methods often reveals us for the imperfect observers that we are. Indeed, a decline into scientific skepticism is one of the dangers that often traps geomathematicians. A suspicious and iconoclastic attitude toward geologic platitudes often marks these workers. However, it must be confessed that such cynicism is often justified. Geologists are trained to see patterns and structure in nature. Geomathematical methods provide the objectivity necessary to avoid creating these patterns when they may exist only in the desire for order.

Geostatistics

All of the techniques of quantitative geology we will discuss in this book can be regarded as statistical procedures, or perhaps ''quasistatistical'' or ''protostatistical'' procedures. Few are sufficiently developed to be used in rigorous tests of statistical hypotheses. Neither is there an adequate body of general theory about the nature of geologic populations, although geology can boast of some original contributions to this area, such as the theory of regionalized variables. However, like statistical tests, geomathematical techniques are based on the premise that information about a phenomenon can be deduced from an examination of a small sample collected from a vastly larger set of potential observations on the phenomenon.

Consider subsurface structure mapping for petroleum exploration: data are derived from scattered drill-holes that pierce successive stratigraphic horizons. The elevation of the top of a horizon measured in one of these holes constitutes a single observation. Obviously, an infinite number of measurements of the top of this horizon could be made if we drilled unlimited numbers of holes. This cannot be done: we are restricted to those holes which actually have been drilled, and perhaps the few additional whose drilling we can authorize. From this data we must deduce as best we can the configuration of the top of the horizon between drill-holes. The problem is analogous to statistical analysis, but unlike the classical statistician we cannot design the pattern of drill-holes or control the manner in which the data were obtained. However, we can use quantitative mapping techniques that are either closely related to statistical procedures, or that rely on novel statistical concepts. Even though traditional forms of statistical tests may be beyond our grasp, the basic underlying concepts are the same.

In contrast, we might consider mine development and production. Mining geologists and engineers have, for years, carefully designed sampling schemes and drilling plans and subjected their observations to statistical analysis. A veritable blizzard of publications has been issued on mine sampling. Several elaborate statistical distributions have been proposed to account for the variation in mine values, providing a theoretical basis for formal statistical tests. Where geologists have control of the means of obtaining samples, they have been quick to exploit the opportunity. Their successes in mine development testify to the power of these methods.

Unfortunately, most geologists must take their observations where they can. Oil well records are too expensive to discard because their locations do not fit into a sample design. Paleontologists must be content with the fossils they can glean from the outcrop; those buried in the subsurface are forever beyond their reach. Samples can be collected from the tops of batholiths in exposures along canyon walls. Samples from the roots of these same bodies are hopelessly deep in the Earth. The problem is seldom too much data in one place. Rather, it is too little data elsewhere. Our observations of the Earth are too precious to lightly discard. We must attempt to wring what knowledge we can from them, recognizing the bias and imperfections of that knowledge.

Many publications have appeared on the design of statistical experiments and sample plans. Notable among these is the geological text by Griffiths (1967), which is in large part concerned with the effect sampling has on the outcome of statistical tests. Although Griffiths' examples are drawn from sedimentary petrology, the methods are equally applicable to other problems in the Earth sciences. The book represents a rigorous, formal approach to the interpretation of geologic phenomena using statistical methods. Griffiths' book, unfortunately now out of print, is especially commended to those who wish to perform experiments in geology, and can exercise strict control over their sampling procedures. In this text, we will concern ourselves with those less tractable situations where the sample design (either by chance or ignorance) is beyond our control. However, be warned that an uncontrolled experiment (i.e., one in which the investigator has no influence over where observations are taken) usually takes us outside the realm of classical statistics. This is the area of "quasistatistics" or "protostatistics," where the assumptions of formal statistics cannot safely be made. Here, the well-developed tests of hypotheses do not exist, and the best we can hope from our procedures is help in what ultimately must be a human decision.

Measurement Systems

A quantitative approach to geology requires something more profound than a headlong rush into the field armed with a personal computer. Because the conclusions reached in a quantitative study will be based at least in part on inferences drawn from measurements, the geologist must be aware of the nature of the number systems in which the measurements are made. Not only must the Earth scientist understand the geological significance of the variables he records, but he must also understand the mathematical significance of the measurement scales he uses. This topic is more complex than it might seem at first glance. Detailed discussions and references can be found in the book edited by Churchman and Ratoosh (1959) and, from a geologist's point of view, in an article by Griffiths (1960).

A measurement is a numerical value assigned to an observation which reflects the magnitude or amount of some characteristic. The manner in which numerical values are assigned determines the scale of measurement, and this in turn determines the type of analyses that can be made of the data. There are four measurement scales, each more rigorously defined than its predecessor, and each containing greater information. The first two are the nominal scale and the ordinal scale, in which observations are simply classified into mutually exclusive categories. The final two scales, the interval and ratio, are those we ordinarily think of as "measurements" because they involve determination of the magnitudes of an attribute.

The *nominal scale* of measurement consists of a classification of observations into mutually exclusive categories of equal rank. These categories may be identified by names, such as "red," "green," or "blue," by symbols such as "A," "B," or "C," or by numbers. However, numbers may be used simply as identifiers. There can be no connotation that 2 is "twice as much" as 1, or that 5 is "greater than" 4. The classification of fossils as to type is an example of nominal measurement. Identification of one fossil as a brachiopod and another as a crinoid implies nothing about the relative importance or magnitude of the two.

The number of observations occurring in each state of a nominal system can be counted, and certain nonparametric tests can be performed on nominal data. A classic example we will consider at length is the occurrence of heads or tails in a coin-flipping experiment. Heads and tails constitute two categories of a nominal scale, and our data will consist of the number of observations that fall into them. A geologic equivalent of this problem consists of the appearance of feldspar and quartz grains along a traverse across a thin section. Quartz and feldspar form mutually exclusive categories that cannot be meaningfully ranked in any way.

Sometimes observations can be ranked in a hierarchy of states. Mohs' hardness scale is a classic example of a ranked or *ordinal scale*. Although the minerals on the scale, which extends from one to ten, increase in hardness with higher rank, the steps between successive states are not equal. The difference in absolute hardness between diamond (rank ten) and corundum (rank nine) is greater than the entire range of hardness from one to nine. Similarly, metamorphic rocks may be ranked along a scale of metamorphic grade, which reflects the intensity of alteration. However, the steps between grades do not represent a uniform progression of temperature and pressure.

As with the nominal scale, a quantitative analysis of ordinal measurements is restricted primarily to counting observations in the various states. However, we also can consider the manner in which different ordinal classes succeed one another. This is done, for example, by determining if states tend to be followed an unusual number of times by greater or lesser states on the ordinal scale.

The *interval scale* is so named because the length of successive intervals is a constant. The most commonly cited example of an interval scale is that of temperature. The increase in temperature between 10 and 20°C is exactly the same as the increase between 110 and 120°C. However, an interval scale has no natural zero, or point where the magnitude is nonexistent. Thus, we can have negative temperatures less than zero. The starting point for the centigrade scale was *arbitrarily set* at a point coinciding with the freezing point of water. To convert from one interval scale to another, we must perform two operations; a multiplication and an addition, to shift the arbitrary origin.

Ratio scales not only have equal increments between steps, but also have a true zero point. Measurements of length are of this type. A shell 2 inches long is twice the length of a shell 1 inch long. A shell with zero length does not exist, because it has no length at all. It is generally agreed that "negative lengths" are not possible. To convert from one ratio scale to another, such as from inches to centimeters, we must perform only the single operation of multiplication.

Ratio scales are the highest form of measurement. All types of mathematical and statistical operations may be performed with them. Although interval scales in theory convey less information than ratio scales, for most purposes, the two can be used in the same manner. Most geologic measurements are done on a ratio scale, because they consist of measures of length, volume, mass, and the like. In subsequent chapters, we will be primarily concerned with the analysis of interval and ratio data. No distinction will be made between the two, and they may occur intermixed in the same problem. An example occurs in trend-surface analysis, where an independent variable may be measured on a ratio scale while the geographic coordinates are on an interval scale, because the coordinate grid has an arbitrary origin.

A False Feeling of Security

Perhaps this chapter should be concluded with a precautionary note. If you pursue the following topics, you will become involved with mathematical methods that have a certain aura of exactitude, that express relationships with apparent precision, and that are implemented on devices which have a popular reputation of infallibility. Computers can be used very effectively as devices for intimidation. The presentation of masses of numbers, all expressed to eight decimal places, overwhelms the minds of many people and numbs their natural skepticism. A geologic report couched in mathematical jargon and filled with computer output usually will bluff all but a few critics, and those who understand and comment often do so in equally obtuse terms. Hence, both the report and criticism pass over the heads of most of the intended audience. The greatest danger, however, is to the researcher himself. If he falls sway to his own computer, he may cease to critically examine his data and his interpretative methods. Hypnotized by numbers, he may be led to the most ludicrous conclusions, totally blind to any reality beyond the computer screen. Keep in mind the little phrase posted on the wall of every computation center: GIGO, which means "Garbage In, Garbage Out."

This chapter began with a quotation, and I will close it with another. The following rhyme was left on my desk by an anonymous critic.

> *What could be cuter*
> *Than to feed a computer*
> *With wrong information*
> *But naïve expectation*
> *To obtain with precision*
> *A Napoleonic decision?*

Major Alexander P. de Seversky

SELECTED READINGS

Churchman, C. W., and P. Ratoosh, eds., 1959, *Measurement: definitions and theories:* John Wiley & Sons, Inc., New York, 274 p.

Fisher, R. A., 1953, The expansion of statistics: *Jour. Royal Statistical Soc.*, series A, **116**, p. 1–6.

Griffiths, J. C., 1960, Some aspects of measurement in the geosciences: *Pennsylvania State Univ. Mineral Industries*, **29**, no. 4, pp. 1, 4, 5, 8.

Griffiths, J. C., 1967, *Scientific method in analysis of sediments:* McGraw-Hill, Inc., New York, 508 p.

CHAPTER 2

Elementary Statistics

Geologists are confined to the outer part of the Earth's crust in their direct observations on our globe. Yet, geologists must attempt to understand the nature of the Earth's core and mantle, and the deeper parts of the crust. Furthermore, the processes that modify the Earth, such as mountain building and continental evolution, are generally beyond the geologist's capabilities for direct manipulation. No other group of scientists, with the exception of astronomers, are more removed from the bulk of their study material and less able to experiment on their subject. Geology, to a major extent, remains a science that is principally concerned with observation. Because geologists depend heavily on observations, particularly observations in which there is a large portion of uncertainty, statistics should play an important role in their research. Although the term "statistics" previously referred simply to the collection of numerical facts such as baseball scores, it has come to include the analysis of data, and especially the uncertainty associated with such data. Statistical problems, whether perceived or not, occur wherever there are elements of chance. Geologists need to be conscious of these problems, and of some of the statistical tools that are available to help solve the problems.

Probability

Although many descriptions and definitions of statistics have been written, it perhaps may be best considered as the determination of the probable from the possible. In any circumstance, there are a variety (sometimes an infinity) of possible outcomes. All these have an associated probability that describes their frequency of occurrence. From an analysis of probabilities associated with events, future behavior or past states of the object or event under study may be estimated.

All of us have an intuitive concept of probability. For example, if asked to guess whether it will rain tomorrow, most of us would reply with some confidence that rain is likely or unlikely, or perhaps in rare circumstances, that it is certain to rain, or certain not to rain. An alternative way of expressing our estimate would be to use a numerical scale, as for example a percentage scale. If we state that the chance of rain tomorrow is 30%, then we imply that the chance of it not raining is 70%.

Scientists usually express probability as an arbitrary number ranging from 0 to

1, or an equivalent percentage ranging from 0 to 100%. If we say that the probability of rain tomorrow is 0, we imply that we are absolutely certain that it will not rain. If, on the other hand, we state that the probability of rain is 1, we are absolutely certain that it will. Probability, expressed in this form, pertains to the likelihood of an event. Absolute certainty is expressed at the ends of this scale, 0 and 1, with different degrees of uncertainty in between. For example, if we rate the probability of rain tomorrow as 1/2 (and therefore of no rain as 1/2), we express our view with a maximum degree of uncertainty; the likelihood of rain is equal to that of no rain. If we rate the probability of rain as 3/4 (1/4 probability of no rain), we express a smaller degree of uncertainty, for we imply that it is three times as likely to rain as it is not to rain.

Our estimates of the likelihood of rain may be based on many different factors, including a subjective "feeling" about the matter. We will use a different approach, however, relying on the past behavior of a phenomenon such as the weather to provide insight into its probable future behavior. This "relative frequency" approach to probability is intuitively appealing to geologists, because the concept is closely akin to uniformitarianism. Other methods of defining and arriving at probabilities may be more appropriate in certain circumstances, but their consideration is beyond our needs at this time. The implications contained in various concepts of probability are discussed in books by von Mises (1981) and Fisher (1973).

The chance of rain is an example of discrete probability; it either will or will not rain. A classic example of discrete probability, used almost universally in statistics texts, pertains to the outcome of the toss of an unbiased coin. A single toss has two outcomes, heads or tails. Each is equally likely, so the probability of obtaining a head is 1/2. This does not imply that every other toss will be a head, but rather that, in the long run, heads will appear one-half of the time. Coin tossing is, then, a clear-cut example of discrete probability. The event has two states and must occupy one or the other; except for the vanishingly small possibility that the coin will land precisely on edge, it must come up either heads or tails.

An interesting series of probabilities can be formed based on coin tossing. If the probability of obtaining heads is 1/2, the probability of obtaining two heads in a row is $1/2 \cdot 1/2 = 1/4$. Perhaps we are interested in knowing the probabilities of obtaining three heads in a row; this will be $1/2 \cdot 1/2 \cdot 1/2 = 1/8$. The logic behind this progression is simple. On the first toss, our chances are 1/2 of obtaining a head. If we do, our chances of obtaining a second head are again 1/2, because the second toss is not dependent in any way on the first. Likewise, the third toss is independent of the two preceding ones, and has an associated probability of 1/2 for heads. So, we have "one-half of one-half of one-half" of a chance of getting all three heads.

Suppose instead that we were interested in the probability of obtaining only one head in three tosses. The possible outcomes, denoting heads as H and tails as T, are:

HHH	HTH	TTT
HHT	THH	[THT]
[HTT]	[TTH]	

Bracketed combinations are those that satisfy our requirements that they contain only one head. Because there are eight possible combinations, the probability of getting only one head in three tosses is $3/8$.

What we have found is the number of possible *combinations* of three things (heads or tails), taken one item at a time. This can be generalized to the number of possible combinations of n items taken r at a time. Symbolically, this is represented as $\begin{pmatrix} n \\ r \end{pmatrix}$.

It can be demonstrated that the number of possible combinations of n items, taken r items at a time, is

$$\frac{n!}{r!(n - r)!} \tag{2.1}$$

The exclamation points stand for *factorial*, and mean that the number preceding the exclamation point is multiplied by the number less one, then by the number less two, and so on:

$$n! = n \cdot (n - 1) \cdot (n - 2) \cdot (n - 3) \cdot \cdots \cdot 1 \tag{2.2}$$

The value of 3! is $3 \cdot 2 \cdot 1 = 6$. In our coin-flipping problem,

$$\begin{pmatrix} 3 \\ 1 \end{pmatrix} = \frac{3!}{1!(3 - 1)!} = \frac{3 \cdot 2 \cdot 1}{1 \cdot (2 \cdot 1)} = \frac{6}{2} = 3$$

That is, there are three possible combinations that will contain one head. By this equation, how many possible combinations are there that contain exactly two heads?

$$\begin{pmatrix} 3 \\ 2 \end{pmatrix} = \frac{3!}{2!(3 - 2)!} = \frac{3 \cdot 2 \cdot 1}{2 \cdot 1(1)} = \frac{6}{2} = 3$$

HHH	[HTH]	TTT
[HHT]	[THH]	THT
HTT	TTH	

These combinations are bracketed above in our collection of possible outcomes. Next, how many possible combinations of three tosses contain exactly three heads?

$$\begin{pmatrix} 3 \\ 3 \end{pmatrix} = \frac{3!}{3!(3 - 3)!} = \frac{3 \cdot 2 \cdot 1}{3 \cdot 2 \cdot 1(1)} = 1$$

Note that 0! is defined as being one, not zero. Finally, the remaining possibility is the number of combinations that contain no heads:

$$\begin{pmatrix} 3 \\ 0 \end{pmatrix} \frac{3!}{0!(3 - 0)!} = \frac{3 \cdot 2 \cdot 1}{1(3 \cdot 2 \cdot 1)} = 1$$

Thus, with three flips of a coin, there is one way we can get no heads, three ways we can get one head, three ways we can get two heads, and one way we can get all heads. This can be shown in the form of a bar graph as in Figure 2.1.

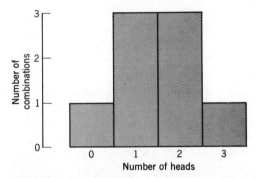

FIGURE 2.1 Bar graph showing the number of different ways to obtain a specified number of heads in three flips of a coin.

We can count the number of total possible combinations, which is eight, and convert the frequencies of occurrence into probabilities. That is, the probability of getting no heads in three flips is one correct combination (TTT) out of eight possible, or 1/8. Our histogram now can be redrawn and expressed in probabilities, giving the *discrete probability distribution* shown in Figure 2.2. The total area under the distribution is 8/8, or 1. We are thus certain of getting *some* combination on the three tosses; the shape of the distribution function describes the likelihood of getting any specific combination. The coin-flipping experiment has four characteristics:

1. There are only two possible outcomes (call them ''success'' and ''failure'') for each trial or flip.
2. Each trial is independent of all others.
3. The probability of a success does not change from trial to trial.
4. The trials are performed a fixed number of times.

The probability distribution that governs experiments such as this is called the *binomial distribution*. Among its geological applications, it may be used to forecast

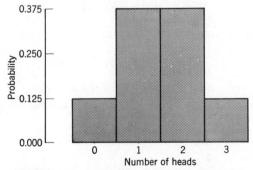

FIGURE 2.2 Discrete distribution giving the probability of obtaining specified numbers of heads in three flips of a coin.

the probability of success in a program of drilling for oil or gas. The four characteristics listed above must be assumed to be true; such assumptions seem most reasonable when applied to "wildcat" exploration in relatively virgin basins. Hence, the binomial distribution often is used to predict the outcomes of drilling programs in frontier areas and offshore concessions.

Under the assumptions of the binomial distribution, each wildcat must be classified as either a discovery ("success") or a dry hole ("failure"). Successive wildcats are presumed to be independent; that is, success or failure of one hole will not influence the outcome of the next hole. (This assumption is difficult to justify in most circumstances, as a discovery usually will affect the selection of subsequent drilling sites. A protracted succession of dry holes will also cause a shift in an exploration program.) The probability of a discovery is assumed to remain unchanged. (This assumption is reasonable at the initiation of exploration, but becomes increasingly tenuous during later phases when a large proportion of the fields in a basin have been discovered.) Finally, the binomial is appropriate when a fixed number of holes will be drilled during an exploratory program, or during a single time period (perhaps a budget cycle) for which the forecast is being made.

The probability p that a wildcat hole will discover oil or gas is estimated using industry-wide success ratios that have been observed during drilling in similar regions, using the success ratio of the particular company making the evaluation, or simply by making a subjective "guess." From p, the binomial model can be developed as it relates to exploratory drilling in the following steps:

1. The probability that a hole will result in a discovery is p.
2. Therefore, the probability that a hole will be dry is $1 - p$.
3. The probability that n successive wildcats will all be dry is

$$P = (1 - p)^n$$

4. The probability that the nth hole drilled will be a discovery but the preceding $(n - 1)$ holes will all be dry is

$$P = (1 - p)^{n-1} p$$

5. The probability of one discovery in a series of n wildcat holes is

$$P = n(1 - p)^{n-1} p$$

since the discovery can occur on *any* of the n wildcats.
6. The probability that $(n - r)$ dry holes will be drilled, followed by r discoveries, is

$$P = (1 - p)^{n-r} p^r$$

7. However, the $(n - r)$ dry holes and the r discoveries may be arranged in $\binom{n}{r}$ combinations, or equivalently in $\dfrac{n!}{(n - r)!r!}$ different ways. So, the

probability that r discoveries will be made in a drilling program of n wildcats is

$$P = \frac{n!}{(n - r)!r!} (1 - p)^{n-r} p^r \qquad (2.3)$$

This is an expression of the binomial distribution, and gives the probability that r successes will occur in n trials, when the probability of success in a single trial is p.

The binomial equation can be solved to determine the probability of occurrence of any particular combination of successes and failures, for any desired number of trials and any specified probability. These probabilities have already been computed and tabulated for many combinations of n, r, and p. Using either the equation or published tables such as those in Hald (1952), many interesting questions can be investigated. For example, suppose we wish to develop the probabilities associated with a five-well exploration program in a virgin basin where the success ratio is anticipated to be about 10%. What is the probability that the entire exploration program will be a total failure, with no discoveries? Such an outcome is called "gambler's ruin" for obvious reasons, and the binomial expression has the terms

$$n = 5$$
$$r = 0$$
$$p = 0.10$$
$$P = \binom{5}{0} \cdot 0.10^0 \cdot (1 - 0.10)^5$$
$$= \frac{5!}{5!0!} \cdot 1 \cdot 0.90^5$$
$$= 1 \cdot 1 \cdot 0.59 = 0.59$$

The probability that no discoveries will result from the exploratory effort is almost 60%.

If only one hole is a discovery, it may pay off the costs of the entire exploration effort. What is the probability that one well will come in during the five-hole exploration campaign?

$$P = \binom{5}{1} \cdot 0.10^1 \cdot (1 - 0.10)^4$$
$$= \frac{5!}{4!1!} \cdot 0.10 \cdot 0.90^4$$
$$= 5 \cdot 0.10 \cdot 0.656 = 0.328$$

Using either the binomial equation or a table of the binomial distribution, the probabilities associated with all possible outcomes of the five-well drilling program can be found. These are shown in Figure 2.3.

Other discrete probability distributions can be developed for those experimental situations where the basic assumptions are different. Suppose, for example, an

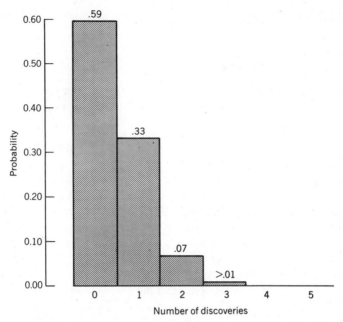

FIGURE 2.3 Discrete distribution giving the probability of making *n* discoveries in a five-well drilling program when the success ratio (probability of a discovery) is 10%.

exploration company is determined to discover two new fields in a virgin basin it is prospecting, and will drill as many holes as required to achieve its goal. We can investigate the probability that it will require 2, 3, 4, . . . , up to *n* exploratory holes before two discoveries are made. The same conditions that govern the binomial distribution may be assumed, except that the number of "trials" is not fixed.

The probability distribution that governs such an experiment is called the *negative binomial,* and its development is very similar to that of the binomial distribution. As in that example, *p* is the probability of a discovery and *r* is the number of "successes" or discovery wells. However, *n*, the number of trials, is not specified. Instead, we wish to find the probability that *x* dry holes will be drilled before *r* discoveries are made. The negative binomial has the form

$$P = \binom{r + x - 1}{x}(1 - p)^x p^r \tag{2.4}$$

Note the similarity between this equation and eq. (2.3); the term $r + x - 1$ appears because the last hole drilled in a sequence must be the *r*th success. Expanding (2.4) gives

$$P = \frac{(r + x - 1)!}{(r - 1)!x!}(1 - p)^x p^r \tag{2.5}$$

If the regional success ratio is assumed to be 10%, the probability that a two-hole exploration program will meet the company's goal of two discoveries can be calculated:

$$P = \frac{(2 + 0 - 1)!}{(2 - 1)!0!} \cdot (1 - 0.10)^0 \cdot 0.10^2$$

$$= \frac{1!}{1!0!} \cdot 0.90^0 \cdot 0.10^2$$

$$= 1 \cdot 1 \cdot 0.01 = 0.01$$

The probabilities attached to other drilling programs having different numbers of holes can be found in a similar way. The possibility that five holes will be required to achieve two successes is

$$P = \frac{(2 + 3 - 1)!}{(2 - 1)!3!} \cdot (1 - 0.10)^3 \cdot 0.10^2$$

$$= \frac{24}{1 \cdot 6} \cdot 0.729 \cdot 0.01 = 0.029$$

The probabilities calculated are low because they relate to the likelihood of obtaining two successes and *exactly* x dry holes. It may be more useful to consider the distribution of the probability that more than x dry holes must be drilled before the goal of r discoveries is achieved. This is found by first calculating the negative binomial distribution in *cumulative form* (which gives the probability that the goal of two successes will be achieved in $(x + r)$ or fewer holes), as shown in Figure 2.4. Each of these probabilities is then subtracted from 1.0 to yield the desired probability distribution (Fig. 2.5). The negative binomial will appear again in

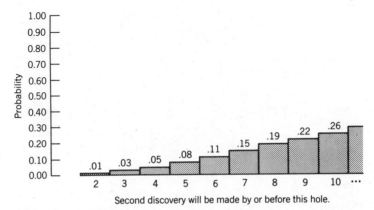

FIGURE 2.4 Discrete distribution giving the cumulative probability that two discoveries will be made by or before a specified hole, when the probability of a discovery is 10%.

Chapter 5, as it constitutes an important model for the distribution of points in space.

There are other discrete probability distributions that apply to experimental situations similar to those appropriate for the binomial. These include the *Poisson distribution,* which can be used instead of the binomial when p, the probability of success, is very small. The Poisson distribution will be discussed in Chapter 4, where it will be applied to the analysis of random events in time (such as earthquakes or volcanic eruptions), and in Chapter 5, where it will serve as a model for objects located randomly in space. The *geometric distribution* is a special case of the negative binomial, appropriate when interest is focused on the number of trials prior to the initial success. The *multinomial distribution* is an extension of the binomial where more than two mutually exclusive outcomes are possible. These topics are extensively developed in most books on probability theory, such as those by Parzen (1960) or Ash (1970).

An important characteristic of all of the discrete probability distributions just discussed is that the probability of success remains constant from trial to trial. Statisticians discuss simple experiments called *sampling with replacement* in which this assumption holds strictly true. A typical experiment would involve an urn filled with red and white balls; if a ball is selected at random, the probability it will be red is equal to the proportion of red balls originally in the urn. If the ball is then returned to the urn, the proportions of the two colors remain unchanged, and the probability of drawing a red ball on a second trial remains unchanged as well. The probability also will remain approximately constant if there are a very large number of balls in the urn, even if those selected are not returned, because their removal causes an infinitesimal change in the proportions among those remaining. This latter condition usually is assumed to prevail in many geological

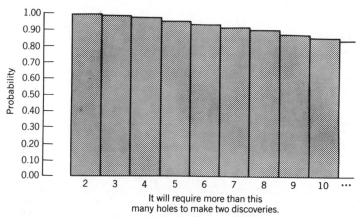

FIGURE 2.5 Discrete distribution giving the probability that more than a specified number of holes must be drilled to make two discoveries, when the probability of a discovery is 10%.

situations where discrete probability distributions are applied. In our example, the "urn" consists of the geologic basin where exploration is occurring, and the red and white balls correspond to undiscovered reservoirs and barren areas. As long as the number of undrilled locations is large, and the number of prospects that have been drilled (and hence "removed from the urn") is small, the assumption of constant probability of discovery seems reasonable. However, if a sampling experiment is performed with a small number of colored balls initially in the urn, and those taken from the urn are not returned, the probabilities obviously change with each draw. Such an experiment is called "sampling without replacement," and is governed by the discrete *hypergeometric distribution*. Geologic problems where its use is appropriate are not common, but McCray (1975) presents an example from geophysical exploration for petroleum.

In some circumstances it is possible to know the size of the population within which discoveries will be made. Suppose an offshore concession contains ten well-defined seismic anomalies that seem to represent structures caused by movement of salt at depth. From experience in nearby offshore tracts, it is known that about 40% of such seismic anomalies prove to be productive structures. Because of budgetary limitations, it is not possible to drill all of the anomalies in the current exploration program. The hypergeometric distribution can be used to estimate the probabilities that specified numbers of discoveries will be made if only some of the known prospects are drilled.

The binomial distribution is not appropriate for this problem because the probability of a discovery changes with each wildcat. If there are four reservoirs distributed among the ten seismic features, the discovery of one reservoir increases the odds against finding another, because there are fewer remaining to be discovered. Conversely, drilling a dry hole on a seismic feature increases the probability that the remaining untested features will prove productive, because one nonproductive feature has been eliminated from the population.

Calculating the hypergeometric probability consists simply of finding all of the possible combinations of producing and dry features within the population, and then enumerating those combinations that yield the desired number of discoveries. The probability of making x discoveries in a drilling program of n holes, when sampling from a population of N prospects of which S are believed to contain reservoirs, is

$$P = \frac{\binom{S}{x}\binom{N-S}{n-x}}{\binom{N}{n}} \qquad (2.6)$$

This is the number of combinations of the reservoirs taken by the number of discoveries, times the number of combinations of barren anomalies taken by the number of dry holes, all divided by the number of combinations of all the prospects taken by the total number of holes in the drilling program.

The hypergeometric probability distribution can be applied to our offshore concession that contains ten seismic anomalies, of which four are likely to be structures containing reservoirs. Unfortunately, we cannot know in advance of drilling *which* four of the ten anomalies will prove productive. If the current season's exploration budget permits the drilling of only three of the prospects, we can determine the probabilities attached to the various possible outcomes.

What is the probability that the drilling program will be a total failure, with no discoveries among the three features tested?

$$P = \frac{\binom{4}{0}\binom{6}{3}}{\binom{10}{3}}$$

$$= \frac{1 \cdot 20}{120} = 0.167$$

The probability of gambler's ruin is approximately 17%.

What is the probability that one discovery will be made?

$$P = \frac{\binom{4}{1}\binom{6}{2}}{\binom{10}{3}}$$

$$= \frac{4 \cdot 15}{120} = 0.50$$

The probability that one discovery will be made is 50%.

A histogram can be prepared which shows the probabilities attached to all possible outcomes in this exploration situation (Fig. 2.6). Note that the probability of *some* success is $(1.00 - 0.17)$, or 83%.

The preceding examples have addressed situations where there are only two possible outcomes: a hole is dry, or oil is discovered. If oil is found, the well cannot be dry, and vice versa. Events in which the occurrence of one outcome precludes the occurrence of the other outcome are said to be *mutually exclusive*. The probability that one event *or* the other happens is the sum of their separate probabilities; that is, p (discovery or dry hole) = p (discovery) + p (dry hole). This is called the *additive rule of probability*.

Events are not necessarily mutually exclusive. For example, we may be drilling an exploratory hole for oil or gas in anticipation of hitting a porous reservoir sandstone in what we have interpreted as an anticlinal structure from seismic data. The two outcomes, *hit porous sandstone* and *drill into an anticline,* are not mutually

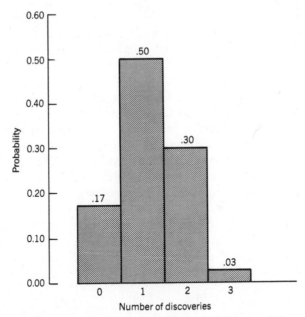

FIGURE 2.6   Discrete distribution giving the probability of *n* discoveries in three holes drilled on ten prospects, when four of the ten prospects contain reservoirs.

exclusive as we hope that both can occur simultaneously. Since the presence of a sandstone is governed by factors that operated at the time of deposition, and since the occurrence of an anticlinal fold is presumed to be related to tectonic conditions at a later time, the two outcomes are unrelated, or *independent*. If two events are not mutually exclusive but *are* independent, the *joint probability* that they will occur simultaneously is the product of their separate probabilities of occurrence. That is, *p* (hit sandstone and drill anticline) = *p* (hit sandstone) × *p* (drill anticline). This is the *multiplicative rule of probability*.

Two events may be related in some way, so that the outcome of one is dependent in part on the outcome of the other. The joint probability of such events is said to be *conditional*. Such events are extremely important in geology, because we may be able to observe one event directly, but the other event is hidden. If the two are conditional, the occurrence of the observable event tells us something about the likely state of the hidden event. For example, the upward movement of magma in chambers beneath a volcano such as Mt. St. Helens in Washington is believed to cause a harmonic tremor, a particular type of earthquake. We cannot directly observe an active magma chamber, but we can observe and record the seismic activity associated with a volcano. If a conditional relationship exists between these two events, the occurrence of harmonic tremors may help predict eruptions. If p (tremor) is the probability that a harmonic tremor occurs and p (eruption) is the probability of a subsequent volcanic eruption, then p (tremor and eruption) $\neq p$ (tremor) $\times p$ (eruption) if the two events have a conditional relationship.

The conditional probability that an eruption will occur, *given* that harmonic tremors have been recorded, is denoted p (eruption|tremor). In this instance the conditional probability of an eruption is greater than the unconditional probability, or p (eruption), which is simply the probability that an eruption will occur without any knowledge of other events. Other conditional probabilities may be lower than the corresponding unconditional probabilities (the probability of finding a fossil, *given* that the terrain is igneous, is much lower than the unconditional probability of finding a fossil). Obviously, geologists exploit conditional probabilities in all phases of their work, whether this is done consciously or not.

The relationship between conditional and unconditional probabilities can be expressed by *Bayes' theorem,* named for Thomas Bayes, an 18th century English clergyman who investigated the manner in which probabilities change as more information becomes available. Bayes' basic equation is:

$$P(A,B) = P(B|A)P(A) \tag{2.7}$$

which states that $P(A,B)$, the *joint probability* that both events A and B occur, is equal to the probability that B will occur given that A has already occurred, times the probability that A will occur. $P(B|A)$ is a *conditional probability* because it expresses the probability that B will occur conditional upon the circumstance that A has already occurred. If events A and B are related (or dependent), the fact that A has already transpired tells us something about the likelihood that B will then occur.

Conversely, it is also true that

$$P(A,B) = P(A|B)P(B)$$

Therefore, the two can be equated, giving

$$P(B|A)P(A) = P(A|B)P(B)$$

which may be rewritten as

$$P(B|A) = \frac{P(A|B)P(B)}{P(A)} \tag{2.8}$$

This is a most useful relationship, because sometimes we know one form of conditional probability but are interested in the other. For example, we may determine that mining districts are often characterized by the presence of abnormal geomagnetic fields. However, we are more interested in the converse, which is the probability that an area will prove to be mineralized, conditional upon the presence of a magnetic anomaly. We can gather estimates of the conditional probability p (anomaly|mineralization) and the unconditional probability p (mineralization) from studies of known mining districts, but it may be more difficult to directly estimate p (mineralization|anomaly) because this would require the examination of geomagnetic anomalies that may not yet have been prospected.

If there is an all-inclusive number of events B_i that are conditionally related to event A, the probability that event A will occur is simply the sum of the conditional probabilities $P(A|B_i)$ times the probabilities that the events B_i occur. That is,

$$P(A) = \sum_{i=1}^{n} P(A|B_i)P(B_i) \tag{2.9}$$

If eq. (2.9) is substituted for $P(A)$ in Bayes' theorem, as given in (2.8), we have the more general equation

$$P(B_i|A) = \frac{P(A|B_i)P(B_i)}{\sum_{i=1}^{n} P(A|B_i)P(B_i)} \tag{2.10}$$

A simple example involving two possible prior events, B_1 and B_2, will illustrate the use of Bayes' theorem.

A fragment of a hitherto unknown species of mosasaur has been found in a stream bed in western Kansas, and a vertebrate paleontologist would like to send a student field party out to search for more complete remains. Unfortunately, the source of the fragment cannot be identified with certainty because the fossil was found below the junction of two dry stream tributaries. The drainage basin of the larger stream contains about 18 square miles, while the basin drained by the smaller stream includes only about 10 square miles. On the basis of just this information alone, we might postulate that the probability that the fragment came from one of the drainage basins is proportional to the area of the basin, or

$$P(B_1) = \frac{18}{28} = 0.64$$

$$P(B_2) = \frac{10}{28} = 0.36$$

However, an examination of a geologic report and map of the region discloses the additional information that about 35% of the outcropping Cretaceous rocks in the larger basin are marine, while almost 80% of the outcropping Cretaceous rocks in the smaller basin are marine. We may therefore postulate the conditional probability that, given a fossil is derived from basin B, it will be a marine fossil, as proportional to the percentage of the Cretaceous outcrop area in the basin that is marine, or for basin B_1

$$P(A|B_1) = 0.35$$

and for basin B_2

$$P(A|B_2) = 0.80$$

Using these probabilities and Bayes' theorem, we can assess the conditional probability that the fossil fragment came from basin B_1, given that the fossil is marine.

$$P(B_1|A) = \frac{P(A|B_1)P(B_1)}{P(A|B_1)P(B_1) + P(A|B_2)P(B_2)}$$

$$= \frac{(0.35)(0.64)}{(0.35)(0.64) + (0.80)(0.36)}$$

$$= 0.44$$

Similarly, the probability that the fossil came from the smaller basin is

$$P(B_2|A) = \frac{(0.80)(0.36)}{(0.35)(0.64) + (0.80)(0.36)}$$

$$= 0.56$$

Fortunately for the students who must search the area, it seems somewhat more likely that the fragment of marine fossil mosasaur came from the smaller basin than from the larger. However, the differences in probability are very small and, of course, depend upon the reasonableness of the assumptions used to estimate the probabilities.

To introduce the next topic we must return briefly to the binomial distribution. Figure 2.2 shows the probability distribution for all possible numbers of heads in three flips of a coin. A similar experiment could be performed that would involve a much larger number of trials. Figure 2.7, for example, gives the probabilities associated with obtaining specified numbers of "successes" (or heads) in ten flips of a coin, and Figure 2.8 shows the probability distribution that describes outcomes from an experiment involving 50 flips of a coin. All of the probabilities were obtained from binomial tables, or could be calculated using the binomial equation.

In each of these experiments, we have enumerated all possible numbers of heads that we could obtain, from zero up to three, to ten, or to fifty. No other combinations of heads and tails can occur. Therefore, the sum of all the probabilities within each experiment must total 1.00, because we are absolutely certain to obtain a result from among those enumerated. We can conveniently represent this by setting the areas underneath histograms in Figures 2.7 and 2.8 equal to 1.00, as was done with the histogram shown in Figure 2.2. If this is done, the increasing number of coin tosses can be accommodated only by making the histogram bars ever more narrow. The histogram becomes increasingly like a smooth and continuous curve. We can imagine an ultimate experiment involving flips of an infinite number of coins, yielding a histogram having an infinite number of bars of infinitesimal width. Then, the histogram would be a continuous curve, and the horizontal axis would represent a *continuous*, rather than discrete, variable.

In the coin-tossing experiment, we are dealing with discrete outcomes—that is, specific combinations of heads and tails. In most experimental work, however, the possible outcomes are not discrete. Rather, there is an infinite continuum of possible

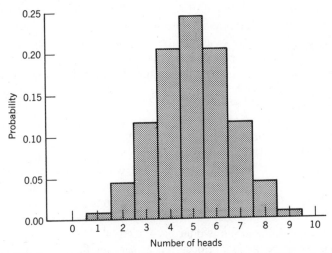

FIGURE 2.7 Discrete distribution giving the probability of obtaining specified numbers of heads in ten flips of a coin.

results that might be obtained. The range of possible outcomes may be finite and in fact quite limited, but within the range the exact result that may appear cannot be predicted. Such events are called *continuous random variables*. Suppose, for example, we measure the length of the hinge line on a brachiopod and find it to be six millimeters long. However, if we perform our measurement using a binocular

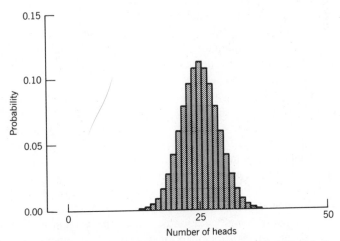

FIGURE 2.8 Discrete distribution giving the probability of obtaining specified numbers of heads in 50 flips of a coin.

microscope, we may obtain a length of 6.2 mm, by using a comparator we may measure 6.23 mm, and with a scanning electron microscope, 6.231 mm. A continuous variate can, in theory, be infinitely refined, which implies that we can always find a difference between two measurements, if we conduct the measurements at a fine enough scale. The corollary of this statement is that every outcome on a continuous scale of measurement is unique, and that the probability of obtaining a specific, exact result must be zero.

If this is true, it would seem impossible to define probability on the basis of relative frequencies of occurrence. However, even though it is impossible to observe a number of outcomes that are, for example, exactly 6.000. . .000 mm, it is entirely feasible to obtain a set of measurements that fall within an interval around this value. Even though the individual measurements are not precisely identical, they are sufficiently close that we can regard them as belonging to the same class. In effect, we divide the continuous scale into discrete segments, and can then count the number of events that occur within each interval. The narrower the class boundaries, the fewer the number of occurrences between them, and the lower the estimates of the probabilities of occurrence.

When dealing with discrete events, we are counting—a process that usually can be done with absolute precision. Continuous variables, however, must be measured by some physical procedure, and these inherently are limited in both their accuracy and precision. Repeated measurements made on the same object will display small differences whose magnitude may reflect both natural variation in the object, variation in the measurement process, and variation inadvertently caused by the person making the measurements. A single, exact, "true" value cannot be determined; rather, we will observe a continuous distribution of possible values. A measurement that has this characteristic is called a *continuous random variable*.

To illustrate the nature of a continuous random variable, we can consider the problem of performing permeability tests on core samples. Permeabilities are determined by measuring the time required to force a certain amount of fluid, under standardized conditions, through a piece of rock. Suppose one test indicates a permeability of 108 md (millidarcies). Is this the "true" permeability of the sample? A second test run on the same specimen may yield a permeability of 93 md, and a third test may register 112 md. The permeability that is recorded on the instruments during any given run is affected by conditions which inevitably vary within the instrument from test to test, vagaries of flow and turbulence that occur within the sample, and inconsistencies in the performance of the test by the operator. No single test can be taken as an absolute measure of the true permeability. The various sources of fluctuation combine to yield a continuously random variable, which we are sampling by making repeated measurements.

Variation induced into measurements by inaccuracy of instrumentation is most apparent when repeated measurements are made on a single object or a test is repeated without change. This variation is called *experimental error*. In contrast, variation may occur between members of a set if measurements or experiments are performed on a series of test objects. This is usually the variation that is of scientific interest. Often, the two types of variations are hopelessly mixed together, or *confounded,* and the experimenter cannot determine what portion of the variability is due to variation between his tests and what is due to error.

Rather than a single piece of rock, suppose we have a sizable length of core taken from a borehole through a sandstone body. We want to determine the permeability of the sandstone, but obviously cannot put 20 ft of core into our permeability apparatus. Instead, we cut small plugs at intervals from the larger core and determine the permeability of each. The variation we see is due in part to differences between the test plugs, but also results from differences in experimental conditions. Devising methods to estimate the magnitude of different sources of variation is one of the major tasks of statistics.

Repeated measurements on large samples drawn from natural populations may produce a characteristic frequency distribution. Most values are clustered around some central value, and the frequency of occurrence declines away from this central point. A graph of the distribution (Fig. 2.9) appears bell-shaped, and is called a *normal* distribution. It frequently is assumed that random variables are normally distributed, and many statistical tests are based on this supposition.

As with all frequency distributions, we may define the total area underneath the normal curve as being equal to 1.00 (or if we wish, as 100%), so we can calculate the probability directly from the curve. You should note the similarity of the bell-shaped continuous curve shown in Figure 2.9 to the histogram in Figure 2.8. However, because there are an infinite number of subdivisions along the horizontal axis, the probability of obtaining one exact, specific event is almost zero. Instead, we consider the probability of obtaining a result within a specified range. This probability is proportional to the area of the frequency curve bounded by these limits. If our specified range is wide, we are more likely to observe an event within them; if the range is extremely narrow, an event is extremely unlikely.

Two terms have been introduced in preceding paragraphs without definition. These are "population" and "sample," two important concepts in statistics. A

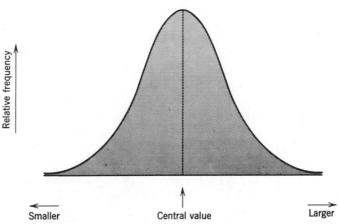

FIGURE 2.9 Plot of the normal frequency distribution.

population consists of a well-defined set (either finite or infinite) of elements. Commonly, these elements are measurements of a specific nature made on items of a specified type. A *sample* is a subset of elements taken from a population. A finite population might consist of all oil wells drilled in Kansas in 1963. An example of an infinite geologic population might be all possible thin sections of the Tensleep Sandstone, or all possible shut-in tests on a well. Note in the latter example that the population includes not only the limited number of tests that have been run, but also all possible tests that could be run. Tests that actually were performed may be regarded as a sample of all potential tests.

If observations with certain characteristics are systematically excluded from the sample, deliberately or inadvertently, it is said to be *biased*. Suppose, for example, we are interested in the porosity of a particular sandstone unit. If we exclude all loose and crumbly rocks from our sample because their porosity is difficult to measure, we will alter the results of the study. It is likely that the range of porosities will be truncated at the high end, biasing the sample toward low values and giving an erroneously low estimate of the variation in porosity within the unit.

Samples should be drawn from populations in a random manner. This means that each item in the population has an equal opportunity to be included in the sample. A random sample will be unbiased, and as the sample size is increased, will provide an increasingly refined picture of the nature of the population. Unfortunately, obtaining a truly random sample may be impractical, as in the situation of sampling a geologic unit that is partially buried. Samples within the unit at depth do not have the same opportunity of being chosen as samples at outcrops. The problems of sampling in such circumstances are complex; some of the references at the end of this chapter discuss the effects of various sampling schemes and the relative merits of different sampling designs. However, many geologic problems involve the analysis of data collected without prior design. The interpretation of subsurface structure from drill-hole data is a prominent example.

Statistics

Distributions have certain characteristics, such as their midpoint; measures indicating the amount of "spread"; and measures of symmetry of the distribution. These characteristics are known as *parameters* if they describe populations, and *statistics* if they refer to samples. Statistics may be used to estimate parameters of parent populations and to test hypotheses about populations.

The most obvious measure of a population or sample is some type of average value. Several measures exist, but only a few are used in practice. The *mode* is the value that occurs with the greatest frequency. In a distribution such as that shown in Figure 2.10, the mode is the highest point on the frequency curve. The *median* is the value midway in the frequency distribution. In Figure 2.10, one-half of the area below the distribution curve is to the right of the median, one-half is to the left. The *mean* is another word for the arithmetic average, and is defined as

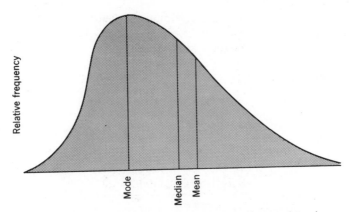

FIGURE 2.10 Relation between measures of central tendency in an asymmetric frequency distribution.

the sum of all observations divided by the number of observations. The *geometric mean* is the *n*th root of the products of the *n* observations, or equivalently, the arithmetic mean of the logarithms of the observations. In asymmetric frequency curves, the median lies between the mean and the mode. In symmetric curves such as the normal distribution, the mean, median, and mode coincide.

Certain symbols traditionally have been assigned to measures of distribution curves. Generally, the symbols for population distributions are Greek letters, and those for sample distributions are Roman. The sample mean, for example, is designated $\overline{X}$, and the population mean is μ (mu). A common objective is to estimate some parameter of a population. A statistic we compute from a sample of the population is used as an estimator of the desired parameter. The use of Greek and Roman symbols serves to emphasize the difference between parameters and the equivalent statistics.

The sample mean has two highly desirable properties that make it more useful as an estimator of the average or central value of a population than either the sample median or mode. First, the sample mean is an *unbiased estimate* of the population mean. A (sample) statistic is an unbiased estimate of the equivalent (population) parameter if the average value of the statistic, from a large series of samples, is equal to the parameter. Second, it can be demonstrated that, for symmetrical distributions such as the normal, the sample means tends to be closer to the population mean than any other unbiased estimate (such as the median) based on the same sample. This is equivalent to saying that sample means have a smaller variance than sample medians, hence they are more *efficient*.

In geochemical analyses, it is common practice to run a series of determinations, or *replicates*, on a single sample. The most nearly correct analytical figure is taken to be the mean value of the determinations. Table 2.1 lists five values for chromium, in parts per million, obtained by spectrographic analysis of a sample of Pennsylvanian shale from southeastern Kansas. Find the mean value of the replicates.

TABLE 2.1 Chromium Content of an Upper Pennsylvanian Shale from Kansas

Replicate	Cr (ppm)
1	205
2	255
3	195
4	220
5	<u>235</u>
TOTAL =	1,110
MEAN =	1,110/5 = 222

Another characteristic of a distribution curve is the spread or dispersion about the mean. Various measures of this property have been suggested, but only two are used to any extent. One is the *variance,* and the other is the square root of the variance, called the *standard deviation.* Variance may be regarded as the average squared deviation of all possible observations from the population mean, and is defined by the equation

$$\sigma^2 = \frac{\sum_{i=1}^{n} (X_i - \mu)^2}{n} \tag{2.11}$$

The variance of a population, σ^2, is given by this equation. The variance of a sample is denoted by the symbol s^2. If the observations $X_1, \ldots, X_n$ are a random sample from a normal distribution, s^2 is an efficient estimate of σ^2.

The reason for using the average of squared deviations may not be obvious. It perhaps seems more logical to define variability as simply the average of deviations from the mean, but a few simple trials will demonstrate that this value will always equal zero. That is,

$$\frac{\sum_{i=1}^{n} (X_i - \bar{X})}{n} = 0 \tag{2.12}$$

Another choice might be the average absolute deviation from the mean, or mean deviation (MD):

$$MD = \frac{\sum_{i=1}^{n} |X_i - \bar{X}|}{n} \tag{2.13}$$

The vertical bars denote that the absolute value (i.e., without sign) of the enclosed quantity is taken. However, it can be demonstrated that this is less efficient than the sample variance. Although not intuitively obvious, the variance has properties that make it far more useful than other measures.

Because variance is the average squared deviation from the mean, its units are the square of the units of the original measurements. A rock, for example, may have feldspar phenocrysts whose longest axes have an average length of 13.2 mm and a variance of 2.0 mm². Many people may find themselves reluctant to regard areas as an appropriate measurement unit for the dispersion of lengths! Fortunately, in most instances where we are concerned with variance it is standardized, or corrected to a form independent of the measurement units. This is a topic discussed in greater detail elsewhere in this chapter.

To provide a statistic that describes dispersion or spread of data around the mean, and is in the units of the measurements of the data, we can calculate the *standard deviation*. This is defined simply as the square root of variance and is symbolically written as σ for the population parameter and s for the sample statistic.

A small standard deviation indicates that observations are clustered tightly around a central value. Conversely, a large standard deviation indicates that values are scattered widely about the mean and the tendency for central clustering is weak. This is illustrated in Figure 2.11, which shows two symmetric frequency curves having different standard deviations. Curve A represents the percentage of oil saturation of core samples from the producing zone of a northeastern Oklahoma oil field. Curve B is the same type of data from a field in west Texas. Mean oil saturation differs in the two fields, but the major difference between the curves reflects the fact that the Texas field has a much greater variation in saturation.

A most useful property of normal distributions is that areas under the curve, within any specified range, can be precisely calculated. For example, slightly over two-thirds (68.27%) of observations will fall within one standard deviation on either side of the mean value. Approximately 95% of all observations are included within the interval from $+2$ to -2 standard deviations, and more than 99% are covered by the interval lying three standard deviations on both sides of the mean. This is illustrated in Figure 2.12.

The distribution of measured oil saturations in cores from the northeastern Oklahoma field (Fig. 2.11, curve A) has a mean of 20.1% saturation and a standard

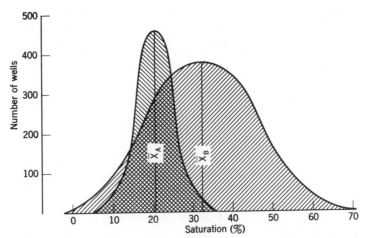

FIGURE 2.11 Frequency distribution of the percentage of oil saturation in an Oklahoma oil field (A) and a Texas oil field (B).

deviation of 4.3%. If we assume that the distribution is normal, we would expect about two-thirds of the cores tested to have oil saturations between about 16 and 24%. Examination of the original data shows that there are 1145 cores having saturations within this range, or about 68% of the data. Only 101 cores, or about 6%, have saturations outside the 2σ range (12–29%).

Persons who have not performed statistical analyses usually find it difficult to obtain a "feel" for the numerical value of a variance or standard deviation. Is a variance of 10 large or small? What is the meaning of a standard deviation of 23? The way to interpret both variance and standard deviation is not to attach a significance to each numerical value, but to compare one variance to another. The sample having the largest variance or standard deviation has the greater spread among the values of the observations, provided all the measurements were made in the same units.

Equation (2.11) is called the definitional equation of variance, but ordinarily it is not used for calculation, involving as it does n subtractions, n multiplications, and n summations. Instead, a computational formula is used which is algebraically equivalent but easier to perform. The equation is

$$s^2 = \frac{\sum_{i=1}^n X_i^2 - n\bar{X}^2}{n - 1} \tag{2.14}$$

or alternately,

$$s^2 = \frac{n \sum_{i=1}^n X_i^2 - (\sum_{i=1}^n X_i)^2}{n(n - 1)} \tag{2.15}$$

On hand calculators, $\sum X_i$ and $\sum X_i^2$ can be found simultaneously, thus reducing the number of operations by n. With computers, (2.15) can be used to find the mean and variance simultaneously, avoiding the necessity of handling the data twice.

To compute variances and standard deviations, we generate intermediate quantities which can be used directly in many techniques we will discuss in following chapters. The *uncorrected sum of squares* is simply $\sum X_i^2$; the *corrected sum of*

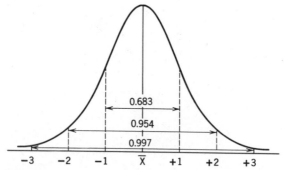

FIGURE 2.12 Areas enclosed by successive standard deviations of the standard normal distribution.

squares (SS) is defined as

$$SS = \sum_{i=1}^{n} (X_i - \bar{X})^2 \tag{2.16}$$

or, in the algebraically equivalent computational form,

$$SS = \sum_{i=1}^{n} X_i^2 - \frac{(\sum_{i=1}^{n} X_i)^2}{n} \tag{2.17}$$

Once this quantity is obtained, the variance can be found by division by $n - 1$.

$$s^2 = \frac{SS}{n-1} = \frac{\sum_{i=1}^{n} X_i^2 - [(\sum_{i=1}^{n} X_i)^2/n]}{n-1} = \frac{n \sum_{i=1}^{n} X_i^2 - (\sum_{i=1}^{n} X_i)^2}{n(n-1)} \tag{2.18}$$

Introduction of the quantity $(n - 1)$, which also appears in (2.14) and (2.15), requires some explanation. Variance is defined as the average squared deviation from the mean. However, when we sample, we do not know the population mean μ but estimate it by the sample mean $\bar{X}$. The sample mean is calculated in a manner that minimizes the squared deviations about it. In other words, the operation $\bar{X} = \Sigma X_i/n$ produces a value $\bar{X}$ for which $\Sigma(X_i - \bar{X})^2$ is the minimum of all possible values that could be selected. Because of this property of the sample mean, it tends to underestimate variance when used in (2.11). That is, $s^2 = \Sigma(X_i - \bar{X})^2/n$ is a biased estimator of $\sigma^2 = \Sigma(X_i - \mu)^2/n$. In order to correct for bias, we reduce the denominator of the variance equation to $n - 1$, producing a larger estimate of the variance.

The computation of these quantities can be illustrated with the geochemical data on chromium in shale from Table 2.1. Rewriting the table to include a column of squares gives Table 2.2.

TABLE 2.2 Sums of Squares and Variance Computations for Data in Table 2.1

X	X²
205	42,025
255	65,025
195	38,025
220	48,400
235	55,225
$\Sigma X_i = 1,110$	$\Sigma X_i^2 = 248,700$

$$(\Sigma X_i)^2 = 1,232,100$$

$$SS = 248,700 - \frac{1,232,100}{5} = 2,280$$

$$s^2 = \frac{2,280}{4} = 570$$

$$s = \sqrt{570} = 23.88$$

TABLE 2.3 Chromium, Nickel, and Vanadium in an Upper Pennsylvanian Shale from Kansas

	Cr (ppm)	Ni (ppm)	V (ppm)
	205	130	180
	255	165	215
	195	100	135
	220	135	200
	235	145	205
TOTALS =	1,110	675	935
MEANS =	222	135	187

Assuming that the analytical values are distributed approximately normally, we would expect about two-thirds of the readings to lie between 198 and 246 ppm. Examination of the table will show that three of the five values, or 60%, do indeed fall in this range.

Note that in the computation of sums of squares for the geochemical data, figures having seven decimal places are created. This tendency to produce extremely large numbers during computation sometimes leads to problems in computers that carry too few significant digits. It also may lead to problems on output if format fields are not large enough to contain the numbers to be printed.

In many geological problems, more than one variable is measured on each observational unit. Examples include a series of measurements made on a collection of corals, results from a sequence of tests run on a group of wells, or sedimentary parameters of a collection of sandstone samples. The data from such problems can conveniently be arranged in an $n \times m$ array, where n is the number of observations and m is the number of variables measured. The complete analyses from which the data for Table 2.1 were taken, for example, contain 17 variables. If we consider three of these, the trace amounts of nickel, vanadium, and chromium, we have the data array shown in Table 2.3.

Each column can be summed and a mean value and standard deviation computed. However, the different variables may not be independent, but rather there may be some form of conditional relationship between them. It is important that we be able to assess the nature and strengths of these conditional relationships, just as it was important to assess the conditional probabilities of occurrences of discrete events.

Joint Variation of Two Variables

Computational procedures used to calculate the variance of a single property can be extended to calculation of a measure of the mutual variability of a pair of properties. This measure, called the *covariance,* is the joint variation of two variables about their common mean. This relation is illustrated in Figure 2.13, which

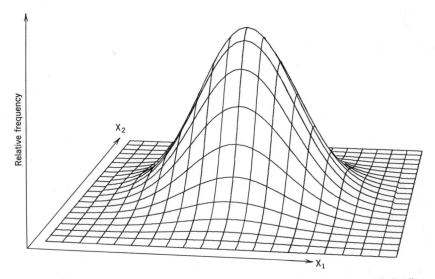

FIGURE 2.13 Joint probability distribution of two independent normal distributions. Both X_1 and X_2 are normally distributed.

shows the form of the probability surface created from two normal curves. Both X_1 and X_2 have probability curves similar to those shown in Figure 2.12. Just as the variance measures the spread of values around the central point as illustrated in Figure 2.12, the covariance measures the distribution of values around a common mean.

To calculate covariance, we first must calculate a quantity analogous to the sum of squares. This is called the *corrected sum of products* (SP) and is defined by

$$SP_{jk} = \sum_{i=1}^{n} (X_{ij} - \overline{X}_j)(X_{ik} - \overline{X}_k) \tag{2.19}$$

In this notation, X_{ij} is the ith measurement of variable j, and X_{ik} is the ith measurement of variable k. The symbol SP_{jk} is the sum of products between variables j and k. In computational form, this becomes

$$SP_{jk} = \sum_{i=1}^{n} (X_{ij}X_{ik}) - \frac{\sum_{i=1}^{n} X_{ij} \sum_{i=1}^{n} X_{ik}}{n} \tag{2.20}$$

The quantity $\Sigma(X_{ij}X_{ik})$ is called the *uncorrected sum of products*. The relationship of SP_{jk} to the sum of squares can easily be seen if we take j and k as being the same. Then,

$$SP_{jj} = \sum_{i=1}^{n} (X_{ij}X_{ij}) - \frac{\sum_{i=1}^{n} X_{ij} \sum_{i=1}^{n} X_{ij}}{n}$$

$$= \frac{\sum_{i=1}^{n} X_{ij}^2 - (\sum_{i=1}^{n} X_{ij})^2}{n} = SS_j \tag{2.21}$$

If we compute sums of products and sums of squares for all possible combinations of our three variables in Table 2.3, we can arrange the results in a 3 × 3 array of the form:

	Cr	Ni	V
Cr	SS_{Cr}	SP_{Cr-Ni}	SP_{Cr-V}
Ni	SP_{Ni-Cr}	SS_{Ni}	SP_{Ni-V}
V	SP_{V-Cr}	SP_{V-Ni}	SS_V

It should be apparent that some of the entries are duplicates; for example, the sum of products for vanadium and nickel is the same as the sum of products for nickel and vanadium. This can be generalized to $SP_{jk} = SP_{kj}$. This feature will be of help to us in subsequent chapters.

Just as variance was calculated by dividing SS by $(n - 1)$, we calculate co-variance by dividing SP by $(n - 1)$.

$$COV_{jk} = \frac{SP_{jk}}{n - 1} = \frac{\sum_{i=1}^n X_{ij}X_{ik} - (\sum_{i=1}^n X_{ij} \sum_{i=1}^n X_{ik}/n)}{n - 1}$$

$$= \frac{n \sum_{i=1}^n X_{ij}X_{ik} - \sum_{i=1}^n X_{ij} \sum_{i=1}^n X_{ik}}{n(n - 1)} \tag{2.22}$$

Returning to the geochemical data in Table 2.3, we now can calculate the covariances between the three elements. Referring to chromium and nickel as X_1 and X_2, respectively, we can calculate the entries in Table 2.4. We now know the variance of X_1 (chromium) and the covariance of X_1 and X_2 (chromium and nickel). We also have all the figures necessary to calculate the variance of X_2 (nickel), following eq. (2.12). The reader should calculate this value, completing the 2 × 2 array shown below.

	X_1	X_2
Chromium (X_1)	570	537.5
Nickel (X_2)	537.5	s_2^2

Three additional quantities remain to be calculated in order to complete our analysis of the geochemical data of Table 2.3. These are the covariances between chromium and vanadium (cov_{13}) and nickel and vanadium (cov_{23}) and the variance of vanadium (s_3^2). Compute the quantity (cov_{13}) following the procedure used in Table 2.4.

Figure 2.14 is a scatter diagram of two variables that are closely related and have a relatively high covariance. The two variables plotted in Figure 2.15 have the same variances as those in Figure 2.14, but are independent of each other and

TABLE 2.4 Computation of Covariance between Chromium (X_1) and Nickel (X_2)

X_1^2	X_1	X_1X_2	X_2	X_2^2
42,025	205	26,650	130	16,900
65,025	255	42,075	165	27,225
38,025	195	19,500	100	10,000
48,400	220	29,700	135	18,225
55,225	235	34,075	145	21,025
$\Sigma X_1^2 = 248,700$	$\Sigma X_1 = 1,110$	$\Sigma X_1X_2 = 152,000$	$\Sigma X_2 = 675$	$\Sigma X_2^2 = 93,375$

$$SP_{1,2} = 152,000 - \frac{(1,110)(675)}{5} = 2,150$$

$$COV_{1,2} = \frac{2,150}{4} = 537.5$$

have relatively low covariance. Interpretation of covariance values must proceed in the same manner as an interpretation of variances; individual values are not too meaningful because they are dependent upon the units of measurement.

In order to estimate the degree of interrelation between variables in a manner not influenced by measurement units, the *correlation coefficient r* is used. Cor-

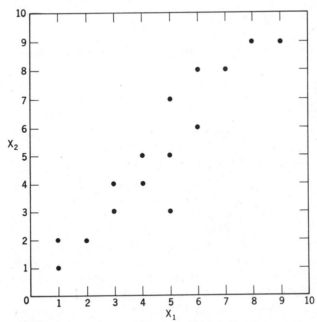

FIGURE 2.14 Scatter diagram of two variables with high covariance. Variance of X_1 = 5.7, variance of X_2 = 7.1, covariance = 5.9.

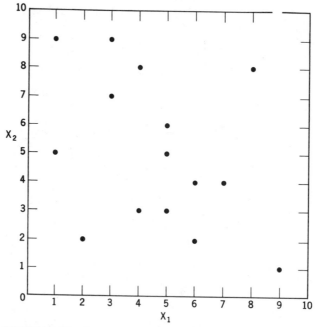

FIGURE 2.15 Scatter diagram of two variables having low covariance. Variances of X_1 and X_2 are the same as in Figure 2.14. Covariance $= -2.3$.

relation is the ratio of the covariance of two variables to the product of their standard deviations:

$$r_{jk} = \frac{\text{cov}_{jk}}{s_j s_k} \tag{2.23}$$

Because the correlation coefficient is a ratio, it is a unitless number. Covariance may equal but cannot exceed the product of the standard deviations of its variables, so correlation ranges from $+1$ to -1. A correlation of $+1$ indicates a perfect direct relationship between two variables; a correlation of -1 indicates that one variable changes inversely with relation to the other. Between the two extremes is a spectrum of less-than-perfect relationships, including zero, which indicates the lack of any sort of linear relationship at all.

In Figure 2.16a, a strong correlation between the variables is evident; the correlation coefficient is almost $+1.00$. A less pronounced correlation is indicated in Figure 2.16b. The correlation coefficient is only $+0.54$. Points on Figure 2.16c were selected from a random number table, so the values of the two variables have no relation to one another. The correlation is near zero. A negative correlation of -0.90 is shown in Figure 2.16d. This illustrates that values of one variable decrease as the other increases. An interesting extreme is shown in Figure 2.16e. One variable is *invariant*, that is, its values do not change. Attempts to calculate the correlation

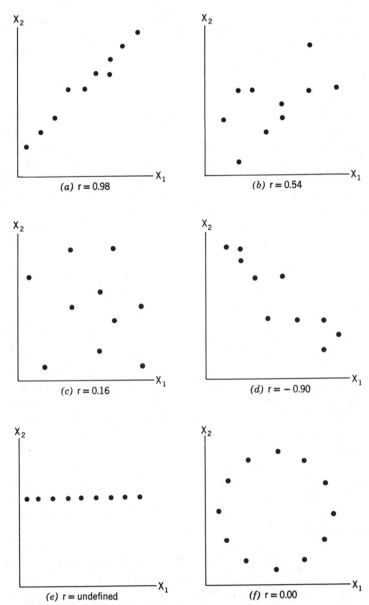

FIGURE 2.16 Scatter diagrams of two variables, showing different correlations between variables.

coefficient will encounter division by zero; the correlation is undefined in this situation. In the example shown in Figure 2.16f, there is an obvious interdependency between the two variables. The observations of X_1 and X_2 lie on a circle, so the relationship between the two variables may be expressed as

$$X_2 = \sqrt{a^2 - X_1^2}$$

assuming the center of the circle lies at the origin. The radius of the circle is equal to the coefficient a. However, if the correlation between X_1 and X_2 is computed, it will be zero. This is because the correlation coefficient is an expression of the *linear* relationship between the two variables, and the circular relationship shown is not linear. There are many possible nonlinear relationships that may exist between two variables; the correlation coefficient is not a satisfactory measure of the strength of such relationships.

In practice, the sample correlation coefficient r commonly is computed by the equation

$$
\begin{aligned}
r_{jk} &= \frac{SP_{jk}}{\sqrt{SS_j \cdot SS_k}} \\[2mm]
&= \frac{\sum_{i=1}^n X_{ij}X_{ik} - (\sum_{i=1}^n X_{ij} \sum_{i=1}^n X_{ik})/n}{\sqrt{\{\sum_{i=1}^n X_{ij}^2 - [(\sum_{i=1}^n X_{ij})^2/n]\}\{\sum_{i=1}^n X_{ik}^2 - [(\sum_{i=1}^n X_{ik})^2/n]\}}}
\end{aligned}
\tag{2.24}
$$

If r measures the linear relationship between two variables, it should be possible to compute the line of dependence between them. This leads into the statistical field of correlation and regression analysis, which is important because most techniques for surface fitting are based on these premises. An extensive discussion of these topics will be deferred until later, and we will content ourselves at this point with the computation of r.

Biological characteristics may be highly correlated within a group of organisms, because measurements are related to the overall size of individuals. For example, consider the data in Table 2.5, giving lengths and widths of shells of the brachiopod, *Composita*. It should be apparent from the observations that there is a strong correspondence between the two measures. Just how strong this correspondence is may be evaluated by the correlation coefficient.

TABLE 2.5 Lengths and Widths of Shells of the Brachiopod, *Composita*

Length (mm)	Width (mm)
18.4	15.4
16.9	15.1
13.6	10.9
11.4	9.7
7.8	7.4
6.3	5.3

TABLE 2.6 Sums of Squares, Cross-Products, and Correlation between Data in Table 2.5

X_1^2	X_1	X_1X_2	X_2	X_2^2
338.56	18.4	283.36	15.4	237.16
285.61	16.9	255.19	15.1	228.01
184.96	13.6	148.24	10.9	118.81
129.96	11.4	110.58	9.7	94.09
60.84	7.8	57.72	7.4	54.76
39.69	6.3	33.39	5.3	28.09
$\Sigma X_1^2 = 1,039.62$	$\Sigma X_1 = 74.4$	$\Sigma X_1X_2 = 888.48$	$\Sigma X_2 = 63.8$	$\Sigma X_2^2 = 760.92$

$$SP_{1,2} = (888.48) - \frac{(74.4)(63.8)}{6} = 97.37$$

$$COV_{1,2} = \frac{97.37}{5} = 19.47$$

$$SS_1 = (1039.62) - \frac{(74.4)^2}{6} = 117.06$$

$$SS_2 = (760.92) - \frac{(63.8)^2}{6} = 82.51$$

$$s_1^2 = \frac{117.06}{5} = 23.41 \qquad s_1 = \sqrt{23.41} = 4.84$$

$$s_2^2 = \frac{82.51}{5} = 16.50 \qquad s_2 = \sqrt{16.50} = 4.06$$

$$r_{1,2} = \frac{19.47}{(4.84)(4.06)} = 0.991$$

To find the correlation between the two measurements, it is necessary to create columns of squares and cross-products. This is done in Table 2.6, where length is X_1 and width is X_2. A correlation of 0.99 is extremely high, and confirms our suspicions that there is a direct relationship between shell length and width in the organisms. Such extremely high correlations are not always found; in fact, it may be a problem to determine if any real correlation exists at all. This is a subject we will discuss again.

Induced Correlations

Some correlations between variables do not reflect the relationships between them, but are induced by an operation or transformation that has been performed on the variables. Two unrelated random variables are expected to have a correlation of zero. However, certain operations on the variables may lead to a correlation other than zero even though there is no linear relationship between them. Correlations that do exist may be changed or even reversed by such operations.

Suppose pebbles are randomly selected from a shingle beach and three orthogonal axes are measured on each pebble. No attempt is made to measure the longest or shortest axes of the pebbles in any particular order. We might suppose that the measurements will be correlated, because a large pebble will most likely have large values for all three axes, and a small pebble will conversely have small measurements for all three axes.

Table 2.7 lists the measurements made on a collection of pebbles, and the correlations between the variables. The data are also shown in the form of scatter diagrams in Figure 2.17. However, if the axes are now defined in conventional terms (that is, the longest axis of a pebble is, by definition, the a-axis; the shortest axis is the c-axis; and the intermediate axis is the b-axis), the ordering causes a change in the correlations (Table 2.8). This is especially apparent on the scatter diagrams, also shown on Figure 2.17, as the definitions force all of the points to plot within the lower 45° segment of the graphs. Because of this, there must always be a positive correlation between any pair of axes, or between the ratios of two axes and the third axis (for example, between b/a versus c).

The most troublesome induced correlations are *spurious negative correlations* that appear in closed data sets. A *closed data set* is one in which all variables measured on an individual add to a fixed total such as 1.00 or 100%, which means the variables are proportions of a whole. Because the sum of the variables is a fixed number, an increase in the proportion of one variable can only occur at the expense of other variables.

In an open data set in which the measurements are not expressed as proportions, two linearly independent variables will have a correlation that is not significantly different from zero. If an open data set is closed by converting the measurements

TABLE 2.7 Axial Lengths, in Centimeters, of Pebbles Collected on a Shingle Beach; Axes are Listed in Order of Measurement

Sample	Axis 1	Axis 2	Axis 3
a	3	7	8
b	16	5	8
c	10	12	9
d	13	5	12
e	14	16	5
f	9	8	14
g	16	13	13
h	6	3	11
i	9	15	9
j	13	10	9
TOTALS	109	94	98
MEANS	10.9	9.4	9.8
CORRELATIONS	$r_{1,2} = 0.279$	$r_{1,3} = -0.021$	$r_{2,3} = -0.349$

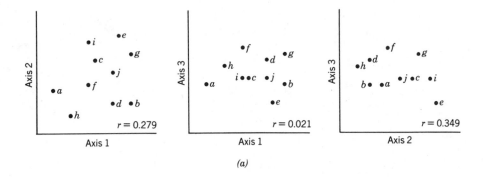

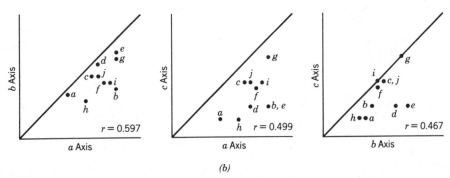

FIGURE 2.17 Scatter diagrams of axial lengths, in centimeters, of pebbles collected on a shingle beach. (*a*) Original measurements collected in random order. (*b*) Measurements sorted into *a*, *b*, and *c* axes, which causes all points to plot below diagonals of diagrams.

to proportions, apparently significant negative correlations will appear even though the original open data consisted entirely of independent variables. In the special case of a three-variable closed data set, the correlations between the closed variables are determined solely by the variances according to the following relationship:

$$r_{1,2} = \frac{s_3^2 - (s_1^2 + s_2^2)}{2s_1 s_2} \qquad (2.25)$$

Such intercorrelations are inherent in any geologic data that are plotted on triangular diagrams, such as sandstone-shale-limestone compositional triangles or ternary phase diagrams. These inverse relationships result solely from the fact that as the proportion of one constituent increases, the proportions of the other two constituents must decrease.

Figure 2.18 is a triangular diagram of the halite-anhydrite-shale sedimentary rock compositional system. The points plotted represent the estimated lithologic proportions in 5-ft intervals of the Hutchinson Salt member of the Permian

TABLE 2.8 Axial Lengths, in Centimeters, of Pebbles Collected on a Shingle Beach; Measurements are Sorted into Longest (*a*), Intermediate (*b*), and Shortest (*c*) Axes

Sample	a-Axis	b-Axis	c-Axis
a	8	7	3
b	16	8	5
c	12	10.	9
d	13	12	5
e	16	14	5
f	14	9	8
g	16	13	13
h	11	6	3
i	15	9	9
j	13	10	9
TOTALS	134	98	69
MEANS	13.4	9.8	6.9
CORRELATIONS	$r_{a,b} = 0.597$	$r_{a,c} = 0.499$	$r_{b,c} = 0.467$

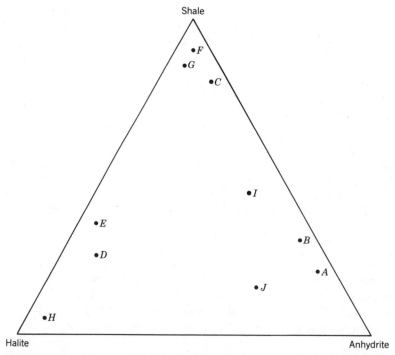

FIGURE 2.18 Triangular diagram of the halite-anhydrite-shale compositional system. Points represent average compositions of five-foot segments of the Wellington Formation (Permian) measured in a well in central Kansas.

Wellington Formation in a bore hole in central Kansas. The compositions were calculated from the responses of gamma ray, neutron, density, and sonic logging tools used to measure the petrophysical properties of the interval. The test well was drilled to investigate a potential disposal site for radioactive wastes.

Table 2.9 gives the composition of the ten intervals plotted on Figure 2.18, estimated to 5%. The table also gives the variances of the three mineralogical components, and the correlations calculated from these variances. Notice that the covariances are not necessary to calculate the correlations, as these are predetermined by the variances and the effect of closure.

Because constant-sum data are so prevalent in geology, many attempts have been made to devise ways of assessing the statistical significance of their intercorrelations. Koch and Link (1981, Volume II, Chapter 11) discuss a number of special statistical techniques appropriate for such data, and Chayes (1971) has written a book on the closure problem. Unfortunately, the proposed statistical procedures are not universally applicable (Kork, 1977, Aitchison, 1981), and at the present time there is no completely satisfactory way of evaluating the strengths of correlations between variables in closed data sets.

TABLE 2.9 Lithologic Composition, Estimated to 5%, of 5-ft Intervals in the Permian Wellington Formation of Central Kansas; Estimates Based on Petrophysical Measurements Made by Well-Logging Tools

Interval		Anhydrite	Shale	Halite
a		75	20	5
b		65	30	5
c		15	80	5
d		10	25	65
e		5	35	60
f		5	90	5
g		5	85	10
h		5	5	90
i		45	45	10
j		60	15	25
	TOTALS	290	430	280
	MEANS	29	43	28
	VARIANCES	832.22	962.22	1001.11
	STD. DEVIATIONS	28.85	31.02	31.64

$$r_{1,2} = \frac{1001.11 - (832.22 + 962.22)}{2 \cdot 28.85 \cdot 31.02} = \frac{-793.47}{1789.85} = -0.44$$

$$r_{1,3} = \frac{962.22 - (832.22 + 1001.11)}{2 \cdot 28.25 \cdot 31.64} = \frac{-871.11}{1787.66} = -0.48$$

$$r_{2,3} = \frac{832.22 - (962.22 + 1001.11)}{2 \cdot 31.02 \cdot 31.64} = \frac{-1131.11}{1962.95} = -0.58$$

Testing Normal Populations

Before proceeding, let us return for a moment to frequency distribution curves, specifically, to the normal distribution. If we could measure the lengths of a very large collection of *Composita,* rather than only the sample of six presented in Table 2.5, we would find that a frequency diagram would look something like Figure 2.19. The mean, say 14.2 mm, would have the greatest frequency, and progressively larger and smaller specimens would occur with decreasing frequency. Approximately two-thirds of the shell lengths would occur within a standard deviation, perhaps 4.7 mm, of the mean. Now consider the width measurements we made while examining this very large collection of *Composita.* The width frequency distribution is similar in form to the length distribution, but the mean and standard deviation are different. It might look, for example, like Figure 2.20, with a mean of 10.3 mm and a standard deviation of 3.6 mm.

Can we compare the two distributions with each other? They are measured in the same units, which makes the problem easier than if we wished to compare shell length to shell weight. We can draw the two distributions on the same millimeter scale, giving Figure 2.21.

It might be simpler to compare the two if they were centered about the same mean. We could center them about a common mean by subtracting enough from all observations of one population (or adding to the values of the other population) to move the means until they coincide. Instead, let us subtract the mean from each observation in each of the two populations. That is, the new measurements are $X_i' = X_i - \overline{X}$. This will move each of the distributions along the millimeter scale until they are centered about zero, which is the mean of both transformed distributions. This is shown in Figure 2.22.

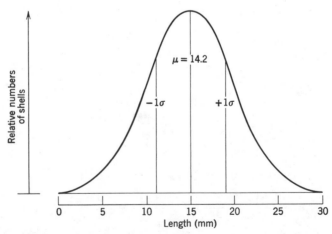

FIGURE 2.19 Hypothetical frequency distribution of the population of *Composita* lengths.

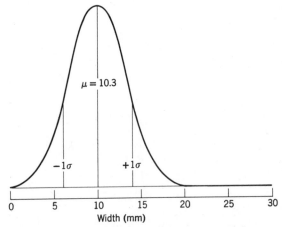

FIGURE 2.20 Hypothetical frequency distribution of the population of *Composita* widths.

Unfortunately, we are dependent upon the millimeter scale of the frequency distributions. This is no great problem with lengths and widths, but if we wish to compare the distributions to one describing weights of shells, we cannot do it. Is there an additional transformation that can be made which will make our normal distributions independent of measurement units? One extremely useful transformation is called *standardization,* and will result in new values for the individuals that not only have zero mean but are measured in units of standard deviations. This is done simply by subtracting the mean of the distribution from each obser-

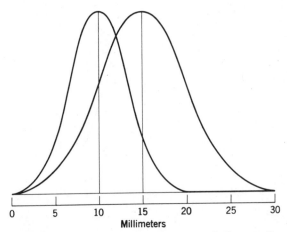

FIGURE 2.21 Frequency diagrams of *Composita* lengths and widths redrawn on the same millimeter scale.

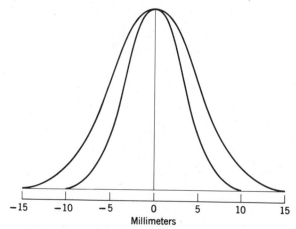

FIGURE 2.22 Frequency distributions of *Composita* lengths and widths, after both variables have been adjusted to have zero mean.

vation and dividing by the standard deviation of the distribution. The new variable Z has what is called the *standard normal form*.

$$Z_i = \frac{X_i - \overline{X}}{s} \tag{2.26}$$

Now, our frequency curves of different *Composita* measurements are identical, and have the form shown in Figure 2.23. The characteristics of the standard normal distribution are extremely well known, and tables of the areas under specified

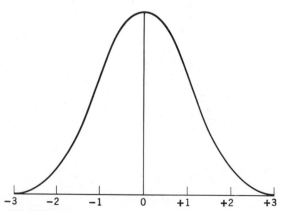

FIGURE 2.23 *Composita* frequency distributions after standardizing lengths and widths to have zero mean and unit standard deviation.

segments of the curve are available in almost all statistics books. Remember that the areas are directly expressible as probabilities. With the use of an abbreviated table such as Table 2.10, we can find the probability of encountering a sample, by random selection from a normal population, whose measurement falls within a specified range. We must know, however, the variance of the population.

Let us make the unrealistic assumption that we have examined the entire population of *Composita*, so we know that the mean and standard deviation of their lengths is 14.2 mm and 4.7 mm, respectively. What is the probability of finding,

TABLE 2.10 Cumulative Probabilities for the Standardized Normal Distribution

Standard Deviations from the Mean	Cumulative Probability	Standard Deviations from the Mean	Cumulative Probability
−3.0	0.0014	+0.0	0.5000
−2.9	0.0019	+0.1	0.5398
−2.8	0.0026	+0.2	0.5793
−2.7	0.0035	+0.3	0.6179
−2.6	0.0047	+0.4	0.6554
−2.5	0.0062	+0.5	0.6915
−2.4	0.0082	+0.6	0.7257
−2.3	0.0107	+0.7	0.7580
−2.2	0.0139	+0.8	0.7881
−2.1	0.0179	+0.9	0.8159
−2.0	0.0228	+1.0	0.8413
−1.9	0.0287	+1.1	0.8643
−1.8	0.0359	+1.2	0.8849
−1.7	0.0446	+1.3	0.9032
−1.6	0.0548	+1.4	0.9192
−1.5	0.0668	+1.5	0.9332
−1.4	0.0808	+1.6	0.9452
−1.3	0.0968	+1.7	0.9554
−1.2	0.1151	+1.8	0.9641
−1.1	0.1357	+1.9	0.9713
−1.0	0.1587	+2.0	0.9773
−0.9	0.1841	+2.1	0.9821
−0.8	0.2119	+2.2	0.9861
−0.7	0.2420	+2.3	0.9893
−0.6	0.2743	+2.4	0.9918
−0.5	0.3085	+2.5	0.9938
−0.4	0.3446	+2.6	0.9953
−0.3	0.3821	+2.7	0.9965
−0.2	0.4207	+2.8	0.9974
−0.1	0.4602	+2.9	0.9981
−0.0	0.5000	+3.0	0.9987

Source: Abridged from Table II, A. Hald, *Statistical Tables and Formulas,* John Wiley & Sons, Inc., New York, 1952.

by chance, a specimen shorter than 3 mm? To find the answer, we must convert 3 mm to units of standard deviation, and then examine Table 2.10.

$$Z = \frac{3.0 - 14.2}{4.7} \approx -2.4$$

The probability of finding a *Composita* smaller than -2.4 standard deviations is the cumulative probability to this point; from our table, we can see that it is 0.0082, which is very small indeed. Now, what is the probability of finding one longer than 20 mm? Again, converting to standard normal form:

$$Z = \frac{20.0 - 14.2}{4.7} \approx 1.2$$

Because the total area under the normal distribution curve is 1.00, the probability of obtaining a measurement of 1.2 standard deviations or greater than the mean is the same as 1.0 minus the cumulative probability of obtaining anything smaller. Stated in another way,

$$P(1.2 \text{ or larger}) = 1.0 - P(1.2 \text{ or smaller})$$

Table 2.10 will give us the cumulative probability up to 1.2, which is 0.8849. Therefore, the probability of finding a *Composita* longer than 20 mm is

$$1.0000 - 0.8849 = 0.1151$$

or slightly greater than one chance out of ten. Now, compute the probability of finding, at random, a *Composita* whose length falls in the size range from 15 to 20 mm:

for 15 mm

$$Z = \frac{15.0 - 14.2}{4.7} \approx 0.2$$

for 20 mm

$$Z = \frac{20.0 - 14.2}{4.7} \approx 1.2$$

$$P(1.2 \text{ or less}) = 0.8849$$
$$P(0.2 \text{ or less}) = \underline{0.5793}$$
$$P(\text{between 1.2 and 0.2}) = 0.3056$$

Approximately one out of three specimens will fall in this size range.

Central Limits Theorem

In this example, we have assumed that we are drawing samples from a normally distributed population. Unfortunately, we ordinarily do not know what shape the population distribution may have, and may even suspect that it is distinctly nonnormal. This does not mean that the normal distribution is of no use, however, because of a remarkable property called the *central limits theorem*. This states that if sets of random samples are taken from *any* population, and the means calculated for these samples, the sample means will tend to be normally distributed. The tendency toward normality becomes more pronounced for samples of larger size.

The central limits theorem may not seem reasonable at first glance; it is difficult to see why the means of samples should form a normal distribution if the samples were collected from a population quite different in form. However, a simulation experiment will show that the theorem does indeed hold true. Suppose we sample from a parent population having the distinctly nonnormal U-shape shown in Figure 2.24. Most of the individual observations in a sample will come from the two extremes of the distribution, which contain the bulk of the population. When these values are averaged together to form a mean, high values will tend to be counterbalanced by low values, resulting in a mean near the center of the distribution. Only in the very rare circumstance when all of our randomly selected observations happen to come from either the high end or the low end will we calculate a mean that differs much from the central value.

Note that the sample means are clustered near the central value of the hypothetical distribution in Figure 2.24. If this experiment were repeated a thousand or more times, we would find that the sample means would plot as the familiar, bell-shaped normal curve. Essentially the same results would be obtained if we began with almost any other form of original distribution, as shown in Figure 2.25, adapted from Lapin (1982).

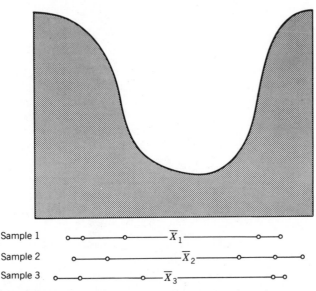

FIGURE 2.24 Three samples of five observations drawn randomly from U-shaped distribution. Means of samples are shown by $\overline{X}$.

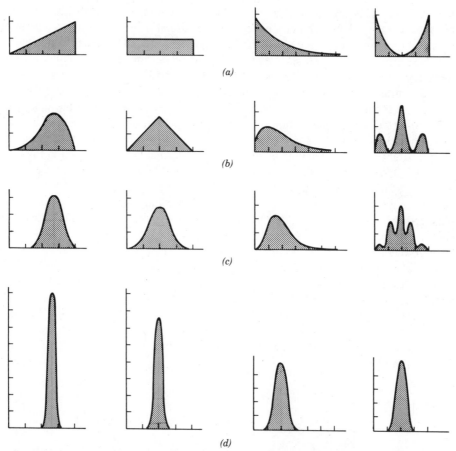

FIGURE 2.25 Distributions of $\overline{X}$ for large numbers of samples of size n, selected at random from nonnormal populations. Central limits theorem insures that distribution of $\overline{X}$ tends toward normal distribution as n increases. Parent populations from which samples are drawn are shown in (a). Distributions of $\overline{X}$ are shown for (b) samples of size $n = 2$, (c) samples of size $n = 4$, and (d) samples of size $n = 25$. After Lapin (1982).

Since the distribution of sample means tends to be normal, it may be described by only two statistics—its mean and variance. Both theoretical and empirical studies show that the mean of the sample means is equal to the population mean. That is,

$$\overline{X}_{\bar{x}} = \mu$$

The variance of the sample means is equal to the variance of the population divided by the size of the samples collected, or

$$s_{\bar{x}}^2 = \frac{\sigma^2}{n}$$

The standard deviation of the sample means is the square root of this quantity, and is called the *standard error of the estimate of the mean*, or simply the standard error. It describes the variability that can be expected in the means of samples by repeated random collection from the same population. The standard error is

$$s_e = \sqrt{\frac{\sigma^2}{n}}$$

or, equivalently,

$$s_e = \sigma \sqrt{\frac{1}{n}} \tag{2.27}$$

The central limits theorem allows us to formulate statistical tests based on the characteristics of the normal curve, and to apply these tests even in circumstances in which the population sampled is not normally distributed. Suppose our paleontologist who has been examining *Composita* is presented with a large slab covered with brachiopods. These fossils look like *Composita*, but are extremely large, the average length for ten specimens being 30.0 mm. Recall that we "know" that the mean and standard deviation of the population of *Composita* is 14.2 mm and 4.7 mm, respectively. Is it possible that the new sample of brachiopods is drawn from this population?

We can determine the difference between the mean of our new sample and the population mean. This difference can then be compared to the variation we expect to see in the means of samples selected at random from a specified population. This variation is given by the standard error, and is a function of both the variance of the population and the size of the sample.

The comparison between the difference in means and the standard error can be made in the following way:

$$Z = \frac{\bar{X} - \mu}{s_e} = \frac{\bar{X} - \mu}{\sigma \sqrt{1/n}} \tag{2.28}$$

Note that the test statistic is calculated in a manner exactly equivalent to that used to convert a variable to standardized form (see eq. 2.26). The test statistic, Z, is normally distributed with a mean of zero and a standard deviation of one, if the sample mean was indeed drawn from the hypothesized population. If Z is excessively large, we will tend to conclude that the sample was not taken from this population. A formal decision, however, requires that we establish a consistent procedure for evaluating the test statistic.

The initial step in statistical testing is the posing of an appropriate hypothesis or statement about the variable in question. Ordinarily, this is done in the form of a *null hypothesis*, symbolized H_0, which is a hypothesis of no difference. We may, as in the present example, speculate that a given sample of observations has been drawn from a parent population which has a certain specified mean. The null hypothesis is expressed in the form

$$H_0: \mu_1 = \mu_0 \tag{2.29}$$

which states that the mean μ_1 of the parent population from which the sample was drawn is equal to (or is not different from) the mean μ_0 of a population having a specified mean.

In our example, we are hypothesizing that the mean of the population from which the slab of brachiopods was taken is the same as the mean of the *Composita* population.

Having posed the null hypothesis, an *alternative* must be given. An appropriate alternative in this situation might be

$$H_1 : \mu_1 \neq \mu_0 \qquad (2.30)$$

stating that the mean of the population from which the sample was drawn does not equal the specified population mean. We now can devise procedures to test the hypothesis, with specified levels of probability of correctness. If the two parent populations are not the same, we must conclude that the slab of fossils was not drawn from the *Composita* population, but from the population of some other genus.

Once a hypothesis is expressed, we can make a decision to either accept or reject it on the basis of our statistical test. There are also two possible states of the hypothesis; it may be true or false. This combination produces four possible outcomes, of which two are correct and two incorrect. The possibilities can be graphically illustrated:

	Hypothesis is Correct	Hypothesis is Incorrect
Hypothesis is Accepted	Correct decision	Type II error β
Hypothesis is Rejected	Type I error α	Correct decision

Either acceptance of a true hypothesis or rejection of a false hypothesis will result in a correct decision. If a null hypothesis is rejected when it is in fact true, a type I error has been committed. Conversely, if an erroneous hypothesis is accepted, a type II error is committed. In terms of our example, the illustration above may be redrawn:

Hypothesis	Actuality	
	Slab is *Composita*	Slab is not *Composita*
μ of Slab $=$ μ of *Composita*	Correct decision	Type II error β
μ of Slab $\neq$ μ of *Composita*	Type I error α	Correct decision

Here, "μ of Slab" refers, of course, to the mean of the population from which the slab came.

In standard statistical procedures, the probability of committing a type I error is called the *level of significance* and is denoted by α; this probability must be set or specified before running the test. In order to minimize the possibility of committing a type II error, we write the null hypothesis with the intention that it will be rejected. If the null hypothesis is rejected, there is no chance for a type II error, and the probability of a type I error is known because it was specified. If, however, the test fails and does not reject the null hypothesis, the probability of a mistake in the form of a type II error remains. This probability, called β, generally is not known. Thus, if the hypothesis of equality is rejected, we can state that the two parent populations have different means. The probability that this statement is incorrect and that our decision is an error is equal to α. On the other hand, if H_0 is not rejected, the statement that the means of the two populations are equal is accompanied by an unknown probability, β, of a mistake.

The logic of statistical tests is based on the premise that the null hypothesis and the alternative are mutually exclusive and all inclusive. The null hypothesis is an explicit statement; therefore, the alternative must be general. If H_0 is rejected, we are stating that the specific relationship described by the null hypothesis does not exist. Rather, the true relationship is contained somewhere in the vast realm of possibilities encompassed by the alternative. We cannot determine what the true relationship is; we can only state what it is not. Some statisticians express the possible outcomes of statistical tests as "rejection of the null hypothesis" versus "failure to reject." Failure to reject, containing as it does an unassessed probability of error, is not equivalent to acceptance. Statistical tests, in one sense, cannot tell you what *is*, but only what *is not*.

Returning to the null hypothesis and alternative given in (2.29) and (2.30), suppose that it is decided that a probability level of a type I error of 5% would be appropriate. In other words, we are willing to risk rejecting the hypothesis when it is correct 5 times out of 100 trials.

We are assuming that we know the variance of the population against which we are checking. Our paleontologist has established that the variance of *Composita* lengths is 22.1 (recall that the standard deviation was 4.7). We now may set up a formal statistical test in the following manner.

1. The hypothesis and alternative:

$$H_0 : \mu_1 = \mu_0$$
$$H_1 : \mu_1 \neq \mu_0$$

2. The level of significance:

$$\alpha = 0.05$$

3. The test statistic:

$$Z = \frac{\overline{X} - \mu_0}{\sigma\sqrt{1/n}} \tag{2.31}$$

The test statistic, Z, has a frequency distribution that is a standardized normal distribution, provided the observations in the sample were selected randomly from a normal population whose variance is known. We have specified that we are willing to reject the hypothesis of equality of means when they actually are equal one time out of twenty; that is, we will accept a 5% risk of a type I error. On the standardized normal distribution curve, therefore, we wish to determine the extreme regions that contain 5% of the area of the curve. This part of the probability curve is called the *area of rejection* or the *critical region*. If the computed value of the test statistic falls into this area, we will reject the null hypothesis.

Because the alternative is simply one of nonequality, the hypothesis will be rejected if the test statistic is either too large or too small. That is, there are three possible true situations: $\mu_1 = \mu_0$, $\mu_1 > \mu_0$, or $\mu_1 < \mu_0$. We are not interested in distinguishing between the latter two possibilities. The critical region, therefore, occupies the extremities of the probability distribution and each subregion contains 2.5% of the total area of the curve.

The rationale of this particular test may be summarized as follows: We know the characteristics of the normal curve; these have been derived from theoretical considerations and their use has been empirically justified. The central limits theorem tells us that the means of samples will be normally distributed. Since we can find the standard error, we know what percentage of sample means will occupy various size ranges (for example, we know that two-thirds of the sample means will occur within one standard deviation of the population mean). If sample means are drawn from this population without bias (that is, by random selection of their constituent observations), the probabilities of obtaining a sample mean from within a specific range of the distribution curve is equal to the area within that portion of the curve. If a mean is drawn from a region of significantly low probability, we conclude that our sample was *not* obtained from the population specified by the null hypothesis, and we reject it. However, there is a finite possibility, equal to the area of rejection, that we *did* by chance obtain a sample that yielded a mean from this extreme region of the population.

Working through the *Composita* example, the outline takes the following form:

1. H_0: μ of slab $= 14.2$ mm
   H_1: μ of slab $\neq 14.2$ mm
2. α level $= 0.05$
3. $Z = \dfrac{30.0 - 14.2}{4.7\sqrt{1/6}} \approx 8.2$

We are prepared to reject the hypothesis of equality of means if the sample mean is either too large or too small. This leads to a *two-tailed* test, diagrammed in Figure 2.26. The critical region, which we have decided should contain 5% of the area of the normal distribution, is split into two equal parts, each containing 2.5% of the total area. If the computed value of Z falls into the left-hand region, the sample came from a population having a smaller mean than our known population. Conversely, if it falls into the right-hand region, the mean of the sample's parent population is larger than the mean of the known population. From Table 2.10, we find that approximately 2.5% of the area of the curve is to the left of a Z value of

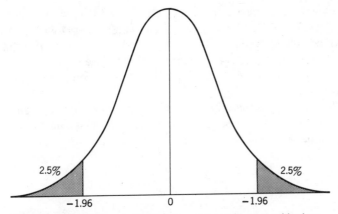

FIGURE 2.26 Normal distribution curve, with two critical regions (shown by shading) which contain a total of 5% of the area under the curve.

-1.9, and 97.5% ($100\% - 2.5\% = 97.5\%$) is to the left of $+1.9$. The computed test value of 8.2 exceeds 1.9, so we conclude that the means of the two populations are not equal, and the collection of fossils must represent some genus other than *Composita*.

It is important to note the assumptions that have been made in the application of this test. The Z test assumes:

1. The sample of brachiopods was selected randomly.
2. The population of lengths of *Composita* is known to be normally distributed.
3. The variance in lengths of *Composita* is known to be 22.1 mm.

If in a particular instance any of these test assumptions seem unwarranted, the results of the test are suspect. We must then look for an alternative procedure whose assumptions are more realistic.

Significance

Before continuing with additional statistical tests, a few comments on the choice of levels of significance may be helpful. Most statistics texts, particularly those concerned with agricultural statistics or industrial quality control, repeatedly use significance levels of one in twenty ($\alpha = 0.05$) or one in a hundred ($\alpha = 0.01$) in their examples and exercises. This practice may suggest that there is something particularly important about these specific levels, but this is not the case. Setting the level of significance is a responsibility of the researcher, who must decide what risk of rejecting a true hypothesis is appropriate.

In geology, we often deal with circumstances of great uncertainty, and it may be unrealistic to demand that a statistical test produce a decision that may be in error only one time in one hundred, or even one time in twenty. If extremely stringent levels of significance are set, we may find that our null hypothesis can never be rejected, and we always need greater and greater amounts of data, which we may not be able to produce. By setting more modest levels of significance, we may be able to come to conclusions more frequently, even though the possibility that these conclusions are erroneous may be high by comparison to standards in other areas.

Figure 2.27 illustrates the effect of setting different levels of significance for a hypothetical statistical test of petroleum prospects. We may imagine that a company has found some quantitative variable that can be measured prior to drilling to

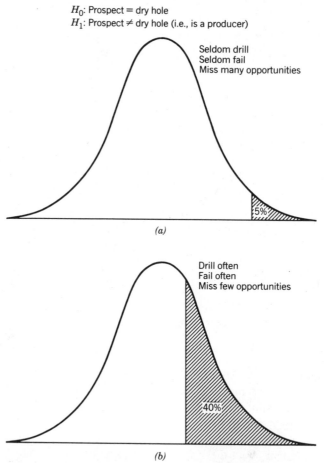

FIGURE 2.27 Distribution of test statistic with critical region specified for rejection of hypothesis that prospect is barren. (a) Critical region set at $\alpha = 0.05$. (b) Critical region set at $\alpha = 0.40$.

indicate whether or not a prospect may be productive. The company applies a statistical test to the variable to decide if it should drill a prospect or abandon it. The null hypothesis states that a prospect comes from the population of barren localities; the alternative is that it comes from the population of nonbarren, or producing, localities.

If a conventional level of significance such as $\alpha = 0.05$ is set as in Figure 2.27*a*, very few prospects will be found to differ from the barren null population. Those determined to be different will almost assuredly result in discoveries when drilled. The company will achieve a very high success ratio, but will drop many leases that may subsequently prove to be productive. In summary, the company will seldom drill, will seldom fail, and will leave many reservoirs unfound.

In contrast, a level of significance such as $\alpha = 0.40$ might be set, as in Figure 2.27*b*. Then, many prospects will be identified as drillable, but the failure rate will be much higher. With this decision criterion, the company will drill often, fail often, and will leave much less oil undiscovered.

The oil industry regards the consequences of failing to drill locations where oil exists (a type II error) as much more dire than drilling dry holes (a type I error). This is because the financial reward of a single large discovery may offset the cost of tens or even hundreds of dry holes. The industry's wildcat success ratio is about 10% in the United States. If these wells were being drilled on the basis of statistical tests, this would represent a level of significance of almost $\alpha = 0.90$!

This is perhaps an extreme example, but it does illustrate the point that the level of significance must be set at a value appropriate for the particular circumstances of the test. The α-level should be based on an assessment of the consequences of making a type I error. These consequences may be tangible and involve loss of money, time, or even lives, or they may be intangible and involve damage to professional reputation or personal pride. To preserve intellectual honesty, the investigator must decide at the outset the amount of risk he is willing to assume, and set the level of significance accordingly. Selecting a level of significance *after* a test has been run and the results are known is shameless gerrymandering, and the levels set may reflect the investigator's desire to accept or reject a hypothesis rather than a dispassionate evaluation of the risks involved.

The *t* Test

An assumption was made in the test given in eq. (2.31), which is seldom true in practical applications. We rarely know the parameters of a population. We have not examined the entire population of *Composita*, and it is obviously impossible to do so. Because μ and σ are not known, the best that can be done is to estimate them from samples. An amount of uncertainty is associated with such estimates, so decisions based upon them cannot be as precise.

The uncertainty introduced into estimates based on samples can be accounted for by using a probability distribution which has a wider "spread" than the normal

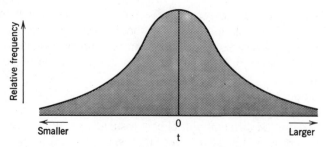

FIGURE 2.28   Distribution of Student's *t*, a sample-based probability distribution.

distribution. One distribution of this type is called the *t distribution*, which is similar to the normal distribution, but dependent upon the size of sample taken. A typical *t*-distribution curve is shown in Figure 2.28. The exact shape of the curve changes according to the number of observations in the sample being used to estimate the population. When the number of observations in the sample is infinite, the *t* distribution and the normal distribution are identical.

Degrees of Freedom

In tests based on samples, we must estimate a number of population parameters in order to calculate the test statistic. It seems intuitively unwise to both estimate the parameters and perform the test from the same set of data without somehow compensating for the double use of observations. This is done by considering a quantity called *degrees of freedom*, which may be defined as the number of observations in a sample, minus the number of parameters estimated from the sample. In other words, the degrees of freedom are the number of observations in excess of those necessary to estimate the parameters of the distribution. Degrees of freedom are symbolically indicated by the Greek letter ν (nu) and are always positive integers.

As an example, consider Figure 2.29, which represents the calculation of the mean and variance of a sample. The mean is estimated from the five independent observations and so has five degrees of freedom. The variance is estimated from the five squared differences, $(\overline{X} - X_i)^2$. However, note that if we determine four of these differences, we automatically know the fifth, since

$$\overline{X} - X_5 = 5\overline{X} - (X_1 + X_2 + X_3 + X_4)$$

Therefore, there are only four independent items of information from which to estimate the variance.

Unfortunately, the concept of degrees of freedom is seldom explained in beginning statistical texts, but rather is presented as an apparently arbitrary quantity such as $n - 1$. An excellent general discussion of degrees of freedom, in both a physical as well as statistical context, is given by Walker (1940). We will briefly discuss

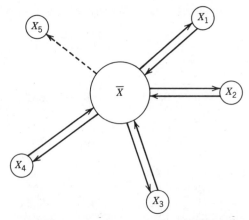

FIGURE 2.29 Diagrammatic representation of the calculation of mean and variance from five observations. Mean $\overline{X}$ is calculated from all observations. Variance is calculated from differences between observations and the mean. When four differences have been found, the fifth difference is known.

the reasons for the various degrees of freedom that are associated with different statistical tests as these arise.

Tables of the *t* distribution (and other sample-based distributions) are used in exactly the same manner as tables of the cumulative standard normal distribution, except that two entries are necessary to find a probability in the table. The two entries are the desired level of significance (α, the probability of a type I error) and the degrees of freedom (v). Table 2.11 is an abbreviated set of the *t* statistic; more extensive tables are contained in most of the references listed at the end of this chapter.

Tests, called *t* tests and based on the *t* probability distribution, are useful for establishing the likelihood that a given sample could be a member of a population with specified characteristics, or for testing hypotheses about the equivalency of two samples. These are the types of problems discussed in introductory statistics courses, and are fundamental in experimental sciences and quality-control fields.

For example, we may wish to test the hypothesis that a suite of Tensleep Sandstone samples, listed in Table 2.12, came from a parent population having a porosity of more than 18%. Assuming the samples were randomly collected from a normal population, the *t* statistic may be computed by

$$t = \frac{\overline{X} - \mu_0}{s_e}$$

$$= \frac{\overline{X} - \mu_0}{s\sqrt{1/n}} \tag{2.32}$$

TABLE 2.11   Critical Values of t for ν Degrees of Freedom and Selected Levels of Significance

Number of Degrees of Freedom, ν	Significance Level, α (%)					
	10	5	2.5	1	0.5	0.1
1	3.078	6.314	12.706	31.821	63.657	318.310
2	1.886	2.920	4.303	6.965	9.925	22.327
3	1.638	2.353	3.182	4.541	5.841	10.215
4	1.533	2.132	2.776	3.747	4.604	7.173
5	1.476	2.015	2.571	3.365	4.032	5.893
6	1.440	1.943	2.447	3.143	3.707	5.208
7	1.415	1.895	2.365	2.998	3.499	4.785
8	1.397	1.860	2.306	2.896	3.355	4.501
9	1.383	1.833	2.262	2.821	3.250	4.297
10	1.372	1.812	2.228	2.764	3.169	4.144
11	1.363	1.796	2.201	2.718	3.106	4.025
12	1.356	1.782	2.179	2.681	3.055	3.930
13	1.350	1.771	2.160	2.650	3.012	3.852
14	1.345	1.761	2.145	2.624	2.977	3.787
15	1.341	1.753	2.131	2.602	2.947	3.733
16	1.337	1.746	2.120	2.583	2.921	3.686
17	1.333	1.740	2.110	2.567	2.898	3.646
18	1.330	1.734	2.101	2.552	2.878	3.610
19	1.328	1.729	2.093	2.539	2.861	3.579
20	1.325	1.725	2.086	2.528	2.845	3.552
21	1.323	1.721	2.080	2.518	2.831	3.527
22	1.321	1.717	2.074	2.508	2.819	3.505
23	1.319	1.714	2.069	2.500	2.807	3.485
24	1.318	1.711	2.064	2.492	2.797	3.467
25	1.316	1.708	2.060	2.485	2.787	3.450
26	1.315	1.706	2.056	2.479	2.779	3.435
27	1.314	1.703	2.052	2.473	2.771	3.421
28	1.313	1.701	2.048	2.467	2.763	3.408
29	1.311	1.699	2.045	2.462	2.756	3.396
30	1.310	1.697	2.042	2.457	2.750	3.385
40	1.303	1.684	2.021	2.423	2.704	3.307
60	1.296	1.671	2.000	2.390	2.660	3.232
120	1.289	1.658	1.980	2.358	2.617	3.160
∞	1.282	1.645	1.960	2.326	2.576	3.090

Source: From Table 21, *The Penguin-Honeywell Book of Tables,* copyright F. W. Kellaway (ed.) and Honeywell Controls Ltd. (E.D.P. Division), 1968.

where

$$\overline{X} = \text{mean of the sample}$$
$$\mu_0 = \text{hypothetical mean of population (18\%)}$$
$$n = \text{number of observations}$$
$$s = \text{standard deviation of observations}$$
$$s_e = \text{standard error of mean}$$

Note that the test is essentially identical to eq. (2.31) except that we must estimate the standard error by $s_e = s\sqrt{1/n}$ rather than by $\sigma\sqrt{1/n}$ since we do not know the population variance. In formal statistical terms, we are testing the hypothesis

$$H_0 : \mu_1 \leqslant \mu_0$$

against the alternative

$$H_1 : \mu_1 > \mu_0$$

The first hypothesis states that the mean of the population from which the sample was drawn is equal to or less than a mean of 18%. The alternative hypothesis states that the parent population of the sample has a mean porosity greater than 18%.

Two items must be specified to obtain the critical value of t from Table 2.11. These are the desired level of significance and the degrees of freedom. In this particular test, one parameter is assumed (μ_0) and another is estimated (σ is estimated by s, the sample standard deviation). Therefore, in the sample of ten porosity measurements, there are nine degrees of freedom. We are interested in rejecting the null hypothesis only if the mean porosity significantly exceeds 18%; therefore, the critical region occurs only at high values of the test statistic, as shown in Figure

TABLE 2.12 Porosity Measurements on Ten Samples of Tensleep Sandstone, Pennsylvanian, from the Bighorn Basin, Wyoming

Sample Number	Porosity (%)
01	13
02	17
03	15
04	23
05	27
06	29
07	18
08	27
09	20
10	24

$$\text{TOTAL} = 213$$
$$\text{MEAN} = 21.3$$
$$s^2 = 30.46, \ s = 5.52$$
$$s_e = 0.57$$

2.30. Such a test is called *one-tailed,* because the critical region occupies only one extreme of the test distribution. If we wish to test this hypothesis with the probability of rejecting it when it is true only one time in twenty ($\alpha = 0.05$), the computed value of t must exceed 1.83 for a one-tailed test. The statistical test is given in the same manner as before:

1. $H_0: \mu_1 \leq 18\%$
   $H_1: \mu_1 > 18\%$
2. $\alpha = 0.05$
3. $t = \dfrac{21.3 - 18.0}{5.52\sqrt{1/10}} = 1.89$

The computed value of 1.89 exceeds the table value of t for nine degrees of freedom and the 5% ($\alpha = 0.05$) level of significance, and so lies in the critical region or region of rejection. On this basis we can reject the null hypothesis, leaving us with the alternative that the porosity of the population from which the Tensleep Sandstone samples were taken is indeed greater than 18%. If the computed value of t had been less than 1.83, we could only say that there is nothing in the sample to suggest that the population mean is greater than 18%. Note that we have not said that the mean is less than 18%, but only that there is no basis for suggesting that it is greater. As stated before, this indecisiveness is a consequence of the manner in which statistical tests are formulated. They can demonstrate, with specified probabilities, what things are not, but they cannot stipulate what they are.

From another area in Wyoming, ten additional measurements of porosity in the Tensleep Sandstone have been obtained. These are listed in Table 2.13. Are the means of the two sample collections the same? This is a somewhat different problem than the one previously considered. We now wish to compare statistics of two samples against one another, rather than against a proposed population parameter. The appropriate test is again a t test, but it is calculated in a more elaborate manner. The hypothesis we are testing is

$$H_0: \mu_1 = \mu_2$$

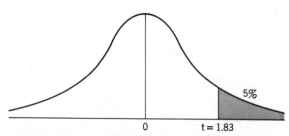

FIGURE 2.30   Student's t distribution for 9 degrees of freedom, with one critical region (shown by shading) that contains 5% of the area under the curve. Critical value of $t = 1.83$.

TABLE 2.13 Porosity Measurements on Ten Samples of Tensleep Sandstone, Pennsylvanian, from the Wind River Basin, Wyoming

Sample Number	Porosity (%)
11	15
12	10
13	15
14	23
15	18
16	26
17	24
18	18
19	19
20	21
	TOTAL = 189
	MEAN = 18.9
	s^2 = 23.21, s = 4.82

which states that the mean of the population from which the first sample was drawn is the same as the mean of the parent population of the second sample. This hypothesis is posed against the alternative

$$H_1 : \mu_1 \neq \mu_2$$

that the two population means are not equal. Again, we must specify a level of significance, say 10% ($\alpha = 0.10$). Our test statistic now has the form

$$t = \frac{\overline{X}_1 - \overline{X}_2}{s_e} \tag{2.33}$$

The standard error of the mean must be based on the characteristics of *both* samples, so we must generalize s_e to

$$s_e = s_p \sqrt{\frac{1}{n_1} + \frac{1}{n_2}}$$

Here, s_p is a *pooled* estimate of the standard deviation, found by combining the sample variances of the two data sets

$$s_p^2 = \frac{(n_1 - 1)s_1^2 + (n_2 - 1)s_2^2}{n_1 + n_2 - 2} \tag{2.34}$$

The subscripts refer, respectively, to the sample from the Bighorn Basin and the Wind River Basin. The process of pooling the two sample variances costs an additional degree of freedom, since two parameters (σ_1^2 and σ_2^2) must be estimated. The degrees of freedom for the *t* test of equivalency given in eq. (2.32) is therefore

$v = (n_1 + n_2 - 2)$. Is the difference between the two sample means significant at the 10% level?

$$s_p^2 = \frac{9(30.46) + 9(23.21)}{10 + 10 - 2} = \frac{483.03}{18}$$

$$= 26.84$$

$$s_p = 5.18$$

$$t = \frac{21.3 - 18.9}{5.18\sqrt{1/10 + 1/10}} = \frac{2.4}{2.32}$$

$$= 1.03$$

Because the table values of t for a two-tailed test with 18 degrees of freedom and 10% level of significance (5% in each tail) are -1.73 and $+1.73$, the computed value does not fall into either critical region, and the null hypothesis cannot be rejected. (Remember the critical region contains 10% of the area under the t-distribution curve.) We must conclude that there is no evidence to suggest that the two samples came from populations having different means.

Three assumptions are necessary to perform this test. One is that both samples were selected at random. The second is that the populations from which the samples were drawn are normally distributed. The third assumption is that the variances of the two populations are equal. The first assumption may be difficult to justify in geologic problems, and may be a serious source of error if the samples are strongly and systematically biased (as they might be if the porosity measurements were made only on samples from producing zones or from oil fields). A population may be tested for normality, but departures from normality are seldom a problem because of the central limits theorem, provided the sample collection is fairly large. The third assumption, equality of the variances of the two groups, is critical. In fact, almost all of the various statistical tests of equality require that the variances of the groups being compared must be equal. Fortunately, this assumption can easily be checked. Approximate tests are available if the variances of the two samples being compared prove to be significantly different. These are given in most introductory texts, including those listed in the Selected Readings.

Test of Correlation

Earlier, we introduced the correlation coefficient as a standardized measure of the linear relationship between two variables. However, we did not consider the question of the statistical significance of a given correlation coefficient. The sample correlation, r, is an estimate of the parameter ρ (rho), which expresses the relationship between two variables of a population. Provided both variables are normally distributed and the observations are chosen at random from the population, we can test the significance of r.

The most useful test is of the hypothesis and alternative

$$H_0: \rho = 0$$
$$H_1: \rho \neq 0$$

That is, we determine whether the observed sample correlation is significantly different from zero. The null hypothesis states that the two variables are independent, and that any nonzero value for r has arisen simply because of the vagaries of random sampling. A t test for the significance of r is given by

$$t = \frac{r\sqrt{n-2}}{\sqrt{1-r^2}} \tag{2.35}$$

which has $n - 2$ degrees of freedom.

As an example, we may test the significance of the correlations we have measured between axial lengths of pebbles from a shingle beach, using the data in Table 2.8. The first correlation, between the a and b axes, is $r_{a,b} = 0.597$, based upon ten pairs of measurements. The test statistic is therefore

$$t = \frac{0.597\sqrt{10-2}}{\sqrt{1-0.597^2}} = \frac{1.688}{0.802} = 2.10$$

The critical value for t with 8 degrees of freedom and a 10% level of significance is $t = 1.860$. Remember that the test is two-sided, as r may be significantly greater or smaller (negative) than zero, so our area of rejection must be split into upper and lower regions. Because the test statistic falls into the upper critical region, we must conclude that there is a real correlation between the lengths of the longest and the intermediate axes of the beach pebbles.

For the other two correlations in the data set of Table 2.8, we have $r_{a,c} = 0.499$ and $r_{b,c} = 0.467$. The corresponding test values are

$$t = \frac{0.499\sqrt{10-2}}{\sqrt{1-0.499^2}} = \frac{1.411}{0.866} = 1.629$$

$$t = \frac{0.467\sqrt{10-2}}{\sqrt{1-0.467^2}} = \frac{1.321}{0.884} = 1.494$$

The critical value remains the same, so we see that neither of these two correlations is significantly different than zero. In other words, the observed correlations could arise merely by chance in a random sample of ten pebbles, if the variables were completely independent of one another.

The F Test

Tests to determine equality of variances are based on a probability distribution called the *F distribution*. This is the theoretical distribution of values that would be expected by randomly sampling from a normal population and calculating, for all possible pairs of sample variances, the ratios

$$F = \frac{s_1^2}{s_2^2}$$

It seems reasonable that the sample variances will range more from trial to trial if the number of observations used in their calculation is small. Therefore, the shape of the F distribution would be expected to change with changes in sample size. This returns us to the idea of degrees of freedom, except in this situation the F distribution is dependent upon two values of v, one associated with each variance in the ratio. Also, it should be apparent that the distribution cannot be negative, because it is the ratio of two positive numbers. If the samples are large, the average of the ratios should be close to 1.0.

Because the F distribution describes the probabilities of obtaining specified ratios of sample variances drawn from the same population, it can be used to test the equality of variances which we obtain in statistical sampling.

We may hypothesize that two samples are drawn from populations having equal variances. After computing the F ratio, we then can ascertain the probability of obtaining, by chance, that specific value from two samples from one normal population. If it is unlikely that such a ratio could be obtained, we regard this as indicating that the samples came from different populations having different variances.

For any pair of variances, two ratios can be computed (s_1/s_2 and s_2/s_1); if we arbitrarily decide that the larger variance will always be placed in the numerator, the ratio will always be greater than 1.0 and the statistical tests can be simplified. Only one-tailed tests need be utilized, and the alternative hypothesis actually is a statement that the absolute difference between the two sample variances is greater than expected if the population variances are equal. This is shown in Figure 2.31, a typical F distribution curve in which the critical region or area of rejection has been shaded.

As an example of an elementary application of the F distribution, consider a comparison between the two sample sets of porosity measurements on the Tensleep Sandstone. We are interested in determining if the variation in porosity is the same in the two areas. For our purposes, we will be content with a level of significance of 5%. That is, we are willing to run the risk of concluding that the porosities are different when actually they are the same one time out of every twenty trials.

The variances of the two samples may be computed by (2.15). Then, the F ratio between the two may be calculated by

$$F = \frac{s_1^2}{s_2^2} \tag{2.36}$$

where s_1^2 is the larger variance and s_2^2 is the smaller. We now are testing the hypothesis

$$H_0: \sigma_1^2 = \sigma_2^2$$

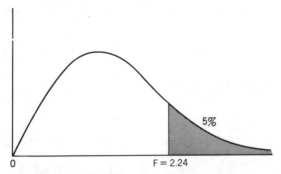

FIGURE 2.31   A typical *F* distribution with $v_1 = 10$ and $v_2 = 25$ degrees of freedom, with critical region (shown by shading) that contains 5% of the area under the curve. Critical value of *F* = 2.24.

against

$$H_1 : \sigma_1^2 \neq \sigma_2^2$$

The null hypothesis states that the parent populations of the two samples have equal variances; the alternative hypothesis states that they do not. Degrees of freedom associated with this test are $(n_1 - 1)$ for v_1 and $(n_2 - 1)$ for v_2. The critical value of *F* with $v_1 = 9$ and $v_2 = 9$ degrees of freedom and a level of significance of 5% ($\alpha = 0.05$) can be found from Table 2.14*a*; that value is 3.18.

The value of *F* calculated from (2.36) will fall into one of two areas. If the calculated value of *F* exceeds 3.18, the null hypothesis is rejected and we conclude that the variation in porosity is not the same in the two groups. If the calculated value is less than 3.18, we would have no evidence for concluding that the variances are different.

As an exercise, compute the variances for the two sets of Tensleep Sandstone porosity measures and determine if the variances are the same at the $\alpha = 0.05$ level of significance.

In most practical situations, we ordinarily have no knowledge of the parameters of the population except for estimates made from samples. In comparing two samples, it is appropriate to first determine if their variances are statistically equivalent. If they appear to be equal, and the samples have been selected without bias from a naturally occurring population, you probably are safe in proceeding to additional statistical tests.

As an example, consider the following problem: Snow and ice collected from permanently frozen parts of the Earth contain small quantities of micron-sized bits of dust called microparticles. Individual grains range in size from about 0.5 to 3.0 μ; these have been injected into the atmosphere by many agents, including volcanic eruptions, dust storms, and micrometeorite infall. The particles are so small they would remain suspended indefinitely, but they are scrubbed out by snow because they serve as nuclei for tiny ice crystals. The ice in turn is incorporated

TABLE 2.14a   Critical Values of F for ν_1 and ν_2 Degrees of Freedom and 5% ($\alpha = 0.05$) Level of Significance

Degrees of Freedom for Denominator, ν_2	Degrees of Freedom for Numerator, ν_1														
	1	2	3	4	5	6	7	8	9	10	12	15	20	24	∞
1	161.45	199.50	215.71	224.58	230.16	233.99	236.77	238.88	240.54	241.88	243.91	245.95	248.01	249.05	250.10
2	18.51	19.00	19.16	19.25	19.30	19.33	19.35	19.37	19.38	19.40	19.41	19.43	19.45	19.45	19.46
3	10.13	9.55	9.28	9.12	9.01	8.94	8.89	8.85	8.81	8.79	8.74	8.70	8.66	8.64	8.62
4	7.71	6.94	6.59	6.39	6.26	6.16	6.09	6.04	6.00	5.96	5.91	5.86	5.80	5.77	5.75
5	6.61	5.79	5.41	5.19	5.05	4.95	4.88	4.82	4.77	4.74	4.68	4.62	4.56	4.53	4.50
6	5.99	5.14	4.76	4.53	4.39	4.28	4.21	4.15	4.10	4.06	4.00	3.94	3.87	3.84	3.81
7	5.59	4.74	4.35	4.12	3.97	3.87	3.79	3.73	3.68	3.64	3.57	3.51	3.44	3.41	3.38
8	5.32	4.46	4.07	3.84	3.69	3.58	3.50	3.44	3.39	3.35	3.28	3.22	3.15	3.12	3.08
9	5.12	4.26	3.86	3.63	3.48	3.37	3.29	3.23	3.18	3.14	3.07	3.01	2.94	2.90	2.86
10	4.96	4.10	3.71	3.48	3.33	3.22	3.14	3.07	3.02	2.98	2.91	2.84	2.77	2.74	2.70
11	4.84	3.98	3.59	3.36	3.20	3.09	3.01	2.95	2.90	2.85	2.79	2.72	2.65	2.61	2.57
12	4.75	3.89	3.49	3.26	3.11	3.00	2.91	2.85	2.80	2.75	2.69	2.62	2.54	2.51	2.47
13	4.67	3.81	3.41	3.18	3.03	2.92	2.83	2.77	2.71	2.67	2.60	2.53	2.46	2.42	2.38
14	4.60	3.74	3.34	3.11	2.96	2.85	2.76	2.70	2.65	2.60	2.53	2.46	2.39	2.35	2.31
15	4.54	3.68	3.29	3.06	2.90	2.79	2.71	2.64	2.59	2.54	2.48	2.40	2.33	2.29	2.25
16	4.49	3.63	3.24	3.01	2.85	2.74	2.66	2.59	2.54	2.49	2.42	2.35	2.28	2.24	2.19
17	4.45	3.59	3.20	2.96	2.81	2.70	2.61	2.55	2.49	2.45	2.38	2.31	2.23	2.19	2.15
18	4.41	3.55	3.16	2.93	2.77	2.66	2.58	2.51	2.46	2.41	2.34	2.27	2.19	2.15	2.11
19	4.38	3.52	3.13	2.90	2.74	2.63	2.54	2.48	2.42	2.38	2.31	2.23	2.16	2.11	2.07
20	4.35	3.49	3.10	2.87	2.71	2.60	2.51	2.45	2.39	2.35	2.28	2.20	2.12	2.08	2.04
21	4.32	3.47	3.07	2.84	2.68	2.57	2.49	2.42	2.37	2.32	2.25	2.18	2.10	2.05	2.01
22	4.30	3.44	3.05	2.82	2.66	2.55	2.46	2.40	2.34	2.30	2.23	2.15	2.07	2.03	1.98
23	4.28	3.42	3.03	2.80	2.64	2.53	2.44	2.37	2.32	2.27	2.20	2.13	2.05	2.01	1.96
24	4.26	3.40	3.01	2.78	2.62	2.51	2.42	2.36	2.30	2.25	2.18	2.11	2.03	1.98	1.94
25	4.24	3.39	2.99	2.76	2.60	2.49	2.40	2.34	2.28	2.24	2.16	2.09	2.01	1.96	1.92
26	4.23	3.37	2.98	2.74	2.59	2.47	2.39	2.32	2.27	2.22	2.15	2.07	1.99	1.95	1.90
27	4.21	3.35	2.96	2.73	2.57	2.46	2.37	2.31	2.25	2.20	2.13	2.06	1.97	1.93	1.88
28	4.20	3.34	2.95	2.71	2.56	2.45	2.36	2.29	2.24	2.19	2.12	2.04	1.96	1.91	1.87
29	4.18	3.33	2.93	2.70	2.55	2.43	2.35	2.28	2.22	2.18	2.10	2.03	1.94	1.90	1.85
30	4.17	3.32	2.92	2.69	2.53	2.42	2.33	2.27	2.21	2.16	2.09	2.01	1.93	1.89	1.84
40	4.08	3.23	2.84	2.61	2.45	2.34	2.25	2.18	2.12	2.08	2.00	1.92	1.84	1.79	1.74
60	4.00	3.15	2.76	2.53	2.37	2.25	2.17	2.10	2.04	1.99	1.92	1.84	1.75	1.70	1.65
120	3.92	3.07	2.68	2.45	2.29	2.18	2.09	2.02	1.96	1.91	1.83	1.75	1.66	1.61	1.55
∞	3.84	3.00	2.60	2.37	2.21	2.10	2.01	1.94	1.88	1.83	1.75	1.67	1.57	1.52	1.46

Source: From Table 22, *The Penguin-Honeywell Book of Tables*, copyright F. W. Kellaway (ed.) and Honeywell Controls Ltd. (E.D.P. Division), 1968.

TABLE 2.14b Critical Values of F for ν_1 and ν_2 Degrees of Freedom and $2\frac{1}{2}\%$ ($\alpha = 0.025$) Level of Significance

Degrees of Freedom for Denominator, ν_2	Degrees of Freedom for Numerator, ν_1														
	1	2	3	4	5	6	7	8	9	10	12	15	20	24	∞
1	647.79	799.50	864.16	899.58	921.85	937.11	948.22	956.66	963.28	968.63	976.71	984.87	993.10	997.25	1001.4
2	38.51	39.00	39.17	39.25	39.30	39.33	39.36	39.37	39.39	39.40	39.41	39.43	39.45	39.46	39.46
3	17.44	16.04	15.44	15.10	14.88	14.73	14.62	14.54	14.47	14.42	14.34	14.25	14.17	14.12	14.08
4	12.22	10.65	9.98	9.60	9.36	9.20	9.07	8.98	8.90	8.84	8.75	8.66	8.56	8.51	8.46
5	10.01	8.43	7.76	7.39	7.15	6.98	6.85	6.76	6.68	6.62	6.52	6.43	6.33	6.28	6.23
6	8.81	7.26	6.60	6.23	5.99	5.82	5.70	5.60	5.52	5.46	5.37	5.27	5.17	5.12	5.07
7	8.07	6.54	5.89	5.52	5.29	5.12	4.99	4.90	4.82	4.76	4.67	4.57	4.47	4.41	4.36
8	7.57	6.06	5.42	5.05	4.82	4.65	4.53	4.43	4.36	4.30	4.20	4.10	4.00	3.95	3.89
9	7.21	5.71	5.08	4.72	4.48	4.32	4.20	4.10	4.03	3.96	3.87	3.77	3.67	3.61	3.56
10	6.94	5.46	4.83	4.47	4.24	4.07	3.95	3.85	3.78	3.72	3.62	3.52	3.42	3.37	3.31
11	6.72	5.26	4.63	4.28	4.04	3.88	3.76	3.66	3.59	3.53	3.43	3.33	3.23	3.17	3.12
12	6.55	5.10	4.47	4.12	3.89	3.73	3.61	3.51	3.44	3.37	3.28	3.18	3.07	3.02	2.96
13	6.41	4.97	4.35	4.00	3.77	3.60	3.48	3.39	3.31	3.25	3.15	3.05	2.95	2.89	2.84
14	6.30	4.86	4.24	3.89	3.66	3.50	3.38	3.29	3.21	3.15	3.05	2.95	2.84	2.79	2.73
15	6.20	4.77	4.15	3.80	3.58	3.41	3.29	3.20	3.12	3.06	2.96	2.86	2.76	2.70	2.64
16	6.12	4.69	4.08	3.73	3.50	3.34	3.22	3.12	3.05	2.99	2.89	2.79	2.68	2.63	2.57
17	6.04	4.62	4.01	3.66	3.44	3.28	3.16	3.06	2.98	2.92	2.82	2.72	2.62	2.56	2.50
18	5.98	4.56	3.95	3.61	3.38	3.22	3.10	3.01	2.93	2.87	2.77	2.67	2.56	2.50	2.44
19	5.92	4.51	3.90	3.56	3.33	3.17	3.05	2.96	2.88	2.82	2.72	2.62	2.51	2.45	2.39
20	5.87	4.46	3.86	3.51	3.29	3.13	3.01	2.91	2.84	2.77	2.68	2.57	2.46	2.41	2.35
21	5.83	4.42	3.82	3.48	3.25	3.09	2.97	2.87	2.80	2.73	2.64	2.53	2.42	2.37	2.31
22	5.79	4.38	3.78	3.44	3.22	3.05	2.93	2.84	2.76	2.70	2.60	2.50	2.39	2.33	2.27
23	5.75	4.35	3.75	3.41	3.18	3.02	2.90	2.81	2.73	2.67	2.57	2.47	2.36	2.30	2.24
24	5.72	4.32	3.72	3.38	3.15	2.99	2.87	2.78	2.70	2.64	2.54	2.44	2.33	2.27	2.21
25	5.69	4.29	3.69	3.35	3.13	2.97	2.85	2.75	2.68	2.61	2.51	2.41	2.30	2.24	2.18
26	5.66	4.27	3.67	3.33	3.10	2.94	2.82	2.73	2.65	2.59	2.49	2.39	2.28	2.22	2.16
27	5.63	4.24	3.65	3.31	3.08	2.92	2.80	2.71	2.63	2.57	2.47	2.36	2.25	2.19	2.13
28	5.61	4.22	3.63	3.29	3.06	2.90	2.78	2.69	2.61	2.55	2.45	2.34	2.23	2.17	2.11
29	5.59	4.20	3.61	3.27	3.04	2.88	2.76	2.67	2.59	2.53	2.43	2.32	2.21	2.15	2.09
30	5.57	4.18	3.59	3.25	3.03	2.87	2.75	2.65	2.57	2.51	2.41	2.31	2.20	2.14	2.07
40	5.42	4.05	3.46	3.13	2.90	2.74	2.62	2.53	2.45	2.39	2.29	2.18	2.07	2.01	1.94
60	5.29	3.93	3.34	3.01	2.79	2.63	2.51	2.41	2.33	2.27	2.17	2.06	1.94	1.88	1.82
120	5.15	3.80	3.23	2.89	2.67	2.52	2.39	2.30	2.22	2.16	2.05	1.94	1.82	1.76	1.69
∞	5.02	3.69	3.12	2.79	2.57	2.41	2.29	2.19	2.11	2.05	1.94	1.83	1.71	1.64	1.57

Source: From Table 22, *The Penguin-Honeywell Book of Tables*, copyright F. W. Kellaway (ed.) and Honeywell Controls Ltd. (E.D.P. Division), 1968.

TABLE 2.14c Critical Values of F for ν_1 and ν_2 Degrees of Freedom and 1% ($\alpha = 0.01$) Level of Significance

Degrees of Freedom for Denominator, ν_2	Degrees of Freedom for Numerator, ν_1														
	1	2	3	4	5	6	7	8	9	10	12	15	20	24	∞
1	4052.2	4999.5	5403.4	5624.6	5763.6	5859.0	5928.4	5981.1	6022.5	6055.8	6106.3	6157.3	6208.7	6234.6	6260.6
2	98.50	99.00	99.17	99.25	99.30	99.33	99.36	99.37	99.39	99.40	99.42	99.43	99.45	99.46	99.47
3	34.12	30.82	29.46	28.71	28.24	27.91	27.67	27.49	27.35	27.23	27.05	26.87	26.69	26.60	26.50
4	21.20	18.00	16.69	15.98	15.52	15.21	14.98	14.80	14.66	14.55	14.37	14.20	14.02	13.93	13.84
5	16.26	13.27	12.06	11.39	10.97	10.67	10.46	10.29	10.16	10.05	9.89	9.72	9.55	9.47	9.38
6	13.75	10.92	9.78	9.15	8.75	8.47	8.26	8.10	7.98	7.87	7.72	7.56	7.40	7.31	7.23
7	12.25	9.55	8.45	7.85	7.46	7.19	6.99	6.84	6.72	6.62	6.47	6.31	6.16	6.07	5.99
8	11.26	8.65	7.59	7.01	6.63	6.37	6.18	6.03	5.91	5.81	5.67	5.52	5.36	5.28	5.20
9	10.56	8.02	6.99	6.42	6.06	5.80	5.61	5.47	5.35	5.26	5.11	4.96	4.81	4.73	4.65
10	10.04	7.56	6.55	5.99	5.64	5.39	5.20	5.06	4.94	4.85	4.71	4.56	4.41	4.33	4.25
11	9.65	7.21	6.22	5.67	5.32	5.07	4.89	4.74	4.63	4.54	4.40	4.25	4.10	4.02	3.94
12	9.33	6.93	5.95	5.41	5.06	4.82	4.64	4.50	4.39	4.30	4.16	4.01	3.86	3.78	3.70
13	9.07	6.70	5.74	5.21	4.86	4.62	4.44	4.30	4.19	4.10	3.96	3.82	3.66	3.59	3.51
14	8.86	6.51	5.56	5.04	4.69	4.46	4.28	4.14	4.03	3.94	3.80	3.66	3.51	3.43	3.35
15	8.68	6.36	5.42	4.89	4.56	4.32	4.14	4.00	3.89	3.80	3.67	3.52	3.37	3.29	3.21
16	8.53	6.23	5.29	4.77	4.44	4.20	4.03	3.89	3.78	3.69	3.55	3.41	3.26	3.18	3.10
17	8.40	6.11	5.18	4.67	4.34	4.10	3.93	3.79	3.68	3.59	3.46	3.31	3.16	3.08	3.00
18	8.29	6.01	5.09	4.58	4.25	4.01	3.84	3.71	3.60	3.51	3.37	3.23	3.08	3.00	2.92
19	8.18	5.93	5.01	4.50	4.17	3.94	3.77	3.63	3.52	3.43	3.30	3.15	3.00	2.92	2.84
20	8.10	5.85	4.94	4.43	4.10	3.87	3.70	3.56	3.46	3.37	3.23	3.09	2.94	2.86	2.78
21	8.02	5.78	4.87	4.37	4.04	3.81	3.64	3.51	3.40	3.31	3.17	3.03	2.88	2.80	2.72
22	7.95	5.72	4.82	4.31	3.99	3.76	3.59	3.45	3.35	3.26	3.12	2.98	2.83	2.75	2.67
23	7.88	5.66	4.76	4.26	3.94	3.71	3.54	3.41	3.30	3.21	3.07	2.93	2.78	2.70	2.62
24	7.82	5.61	4.72	4.22	3.90	3.67	3.50	3.36	3.26	3.17	3.03	2.89	2.74	2.66	2.58
25	7.77	5.57	4.68	4.18	3.85	3.63	3.46	3.32	3.22	3.13	2.99	2.85	2.70	2.62	2.54
26	7.72	5.53	4.64	4.14	3.82	3.59	3.42	3.29	3.18	3.09	2.96	2.81	2.66	2.58	2.50
27	7.68	5.49	4.60	4.11	3.78	3.56	3.39	3.26	3.15	3.06	2.93	2.78	2.63	2.55	2.47
28	7.64	5.45	4.57	4.07	3.75	3.53	3.36	3.23	3.12	3.03	2.90	2.75	2.60	2.52	2.44
29	7.60	5.42	4.54	4.04	3.73	3.50	3.33	3.20	3.09	3.00	2.87	2.73	2.57	2.49	2.41
30	7.56	5.39	4.51	4.02	3.70	3.47	3.30	3.17	3.07	2.98	2.84	2.70	2.55	2.47	2.39
40	7.31	5.18	4.31	3.83	3.51	3.29	3.12	2.99	2.89	2.80	2.66	2.52	2.37	2.29	2.20
60	7.08	4.98	4.13	3.85	3.34	3.12	2.95	2.82	2.72	2.63	2.50	2.35	2.20	2.12	2.03
120	6.85	4.79	3.95	3.48	3.17	2.96	2.79	2.66	2.56	2.47	2.34	2.19	2.03	1.95	1.86
∞	6.63	4.61	3.78	3.32	3.02	2.80	2.64	2.51	2.41	2.32	2.18	2.04	1.88	1.79	1.70

Source: From Table 22, *The Penguin-Honeywell Book of Tables,* copyright F. W. Kellaway (ed.) and Honeywell Controls Ltd. (E.D.P. Division), 1968.

into the permanent snowfields of polar regions. It has been postulated that the concentration of microparticles in snow should be uniform over the Earth, because of mixing of the atmosphere and the manner by which microparticles are removed from the air. The theory, if true, has significance for the advisability of atmospheric testing of nuclear weapons, so two suites of snow samples have been collected carefully from the Greenland ice cap and from Antarctica. Under controlled conditions, the snow has been melted and the quantity of contained microparticles determined by an electron particle classifier. The volume concentration in parts per billion of microparticles in melted snow is given in Table 2.15. Do the two samples appear to be drawn from the same population, and do your conclusions tend to substantiate or refute the idea of atmospheric homogeneity?

Assuming the samples have been collected without bias and the distribution of microparticles is normal throughout the snowfields, the first step is to test the equality of variances in the two sample sets. This is done using (2.36). The hypothesis and alternative are

$$H_0 : \sigma_1^2 = \sigma_2^2$$
$$H_1 : \sigma_1^2 \neq \sigma_2^2$$

If the variances are not significantly different, the next step in the procedure is to test equality of means. The appropriate test is (2.33). For obvious reasons, the level of significance attached to this test cannot be higher than the significance attached to the test of equality of variances. The appropriate hypothesis and alternative are

$$H_0 : \mu_1 = \mu_2$$
$$H_1 : \mu_1 \neq \mu_2$$

because there is no reason to suppose that one region should have a larger mean than the other. If the variances and means cannot be distinguished (that is, the null hypotheses cannot be rejected), there is no statistical evidence to suggest that

TABLE 2.15 Concentration of Microparticles in Meltwater

Concentration (ppb)			
Antarctica, $n = 16$		Greenland, $n = 18$	
3.7	0.6	3.7	1.6
2.0	1.4	7.8	2.4
1.3	4.4	1.9	1.3
3.9	3.2	2.0	2.6
0.2	1.7	1.1	3.7
1.4	2.1	1.3	2.2
4.2	4.2	1.9	1.8
4.9	3.5	3.7	1.2
		3.4	0.8

microparticle concentrations in the two areas are derived from different populations. On the other hand, if either test rejects its null hypothesis, the question of atmospheric homogeneity is in serious doubt. [If (2.36) rejects the hypothesis of equality of variances, test (2.33) cannot be applied. Approximate tests, such as one described in Guenther (1973), are available for testing equality of means from samples of unequal variances, but they would serve little purpose in this problem.]

Analysis of Variance

Up to this point, we have only considered techniques for comparing two samples, yet many problems involve groups of observations. For example, suppose we have obtained five pieces of calcite-cemented sandstone. Each of the five rocks in our collection is lithologically somewhat different; one has conspicuously coarser sand grains, another appears to contain some clay, a third is slightly ferruginous, and so on. We wish to determine if the carbonate content is the same in each. We can consider this as a problem in the branch of statistics called *analysis of variance*.

In general, techniques in this field involve separating the total variance in a collection of measurements into various components or sources. The tests of equality operate by simultaneously considering both differences in means and in variances.

A possible experimental approach to the problem would be to break each rock into a number of fragments and determine the carbonate content of each fragment by measuring its weight loss after acid treatment. Each fragment is called a *replicate*. The purpose of breaking the original sample into replicates is to determine the variability within weight determinations on each sample. Obviously, if the variation within replicates of a single sample is large compared to the differences between samples, the differences will be difficult to detect.

Suppose we break the original rocks into six fragments and analyze each. The variation we see arises from several causes; variation of composition within the original rock sample, inadvertent differences in the manner the replicates were treated (perhaps the residue from one acid treatment was washed more vigorously than others), differences in weighing the samples (the replicates may retain different amounts of moisture, or the balance may vary in its response because of temperature changes during the day, etc.), and other subtle influences. These sources of variation all combine to produce what is known as *experimental error* or the variation not accounted for by differences between the samples.

To avoid the possibility of introducing a systematic error into the statistical analysis, the replicates must be treated in random order. This is known as *randomizing* the observations. The need for this step is apparent if there is some factor which consistently changes during the time of the experiment, perhaps continued drying of the acidized samples as they await their turn to be weighed. If we weigh all six replicates of sample 1, then all replicates of sample 2, and so on, a greater weight loss may be recorded for the last weighed samples simply because they dried over a longer period of time. One way to avoid this potential problem is to sequentially number each replicate and assign them to the analytical process according to a random number table. In fact, if the treatment proceeds in stages, it is best to randomly assign each specimen to each step of the treatment. Then, the various sources of error are mixed up, or *confounded,* over all of the replicates rather than being concentrated in a few.

Determining the equivalency of the five samples can be done by a technique called the *one-way analysis of variance.* The hypothesis and alternative are

$$H_0: \mu_1 = \mu_2 = \mu_3 = \mu_4 = \mu_5$$
$$H_1: \text{at least one mean is different}$$

Certain assumptions are necessary to perform a test to choose between the two hypotheses. The assumptions are (*a*) each set of replicates represents random samples from different populations, (*b*) each parent population is normally distributed, and (*c*) each parent population has the same variance. The data for our problem are given in Table 2.16. In the one-way analysis, total variance of the data set is broken into two parts—variance *within* each set of replicates and variance *among* the samples. Statisticians have developed a formalized procedure for analysis of variance which is contained within an ANOVA (*AN*alysis *O*f *VA*riance) table. This lists the sources of variation, a column of corrected sums of squares resulting from the various sources, degrees of freedom associated with each, a column called *mean squares* which are nothing more than the sample-based estimates of the variances, and the *F* value. The ANOVA table appropriate for our problem is outlined in the box below:

Source of Variation	Sum of Squares	Degrees of Freedom	Mean Squares	F Test
Among Samples	SS_A	$m - 1$	MS_A	MS_A/MS_W
Within Replications	SS_W	$N - m$	MS_W	
Total Variation	SS_T	$N - 1$		

The total variance of all observations (all replicates of all samples) is given by SS_T

TABLE 2.16 Carbonate Cement in Five Sandstone Samples with Replicates; Numbers in Brackets Signify Order of Analysis

Replicate	Carbon Cement (%)				
	Sample 1	Sample 2	Sample 3	Sample 4	Sample 5
1	19.2[11]	18.7[04]	12.5[28]	20.3[12]	19.9[21]
2	18.7[08]	14.3[19]	14.3[16]	22.5[30]	24.3[06]
3	21.3[09]	20.2[14]	8.7[20]	17.6[24]	17.6[18]
4	16.5[17]	17.6[07]	11.4[29]	18.4[03]	20.2[22]
5	17.3[26]	19.3[05]	9.5[27]	15.9[13]	18.4[12]
6	22.4[15]	16.1[25]	16.5[01]	19.0[02]	19.1[10]

and is

$$SS_T = \sum_{j=1}^m \sum_{i=1}^n X_{ij}^2 - \frac{(\sum_{j=1}^m \sum_{i=1}^n X_{ij})^2}{N} \tag{2.37}$$

where m = number of samples and n = number of replicates per sample. In this symbology, X_{ij} is the ith replicate of the jth sample. The double summation indicates that we first sum down each of the columns containing the n replicates, then sum the m column totals. The total number of observations, N, is equal to the number of replicates per sample times the number of samples, or $N = n \cdot m$. The final term in the equation is referred to as the *correction term* and appears in other equations as well. Similarly, the variance among the samples is found by

$$SS_A = \sum_{j=1}^m \left[\frac{(\sum_{i=1}^n X_{ij})^2}{n} \right] - \frac{(\sum_{j=1}^m \sum_{i=1}^n X_{ij})^2}{N} \tag{2.38}$$

which means that we sum the replicates within each sample ($\sum_{i=1}^n X_{ij}$); square each of the totals, divide by the number of replicates in each sample (n); sum the resulting figures for all samples; and, finally, subtract the correction term.

The second source of variance, that within samples, is

$$SS_W = \sum_{j=1}^m \sum_{i=1}^n X_{ij}^2 - \sum_{j=1}^m \frac{(\sum_{i=1}^n X_{ij})^2}{n} \tag{2.39}$$

Note that the first term of (2.39) is the first term of the equation for SS_T (2.37) and the last term is the first term in the equation of SS_A (2.38). Therefore, we can find SS_W by the operation

$$SS_W = SS_T - SS_A \tag{2.40}$$

The number of degrees of freedom for variance in the total data set is $N - 1$. Degrees of freedom for variance among the samples is $m - 1$, because we are estimating this variance from the means of the replicates within each sample. The difference between the two is the degrees of freedom attributable to variance within the samples.

Working through this problem will help clarify the analysis of variance procedure. First, the total variance of the data in Table 2.16 must be found. Using (2.37) for SS_T, we get

$$SS_T = 383.79$$

Next, we must determine the variance of the average values for the five samples. Following (2.38), we first sum all five columns, square the totals, divide each by six, and subtract the correction term. This gives us the sum of squares among the samples, or

$$SS_A = 237.42$$

Finally, we subtract SS_A from SS_T to give the sum of squares within the samples, or

$$SS_W = 146.37$$

Total degrees of freedom are $N - 1$, or 29. Because we are estimating the variance between the samples from five measurements (the five column means), degrees of freedom for SS_A is $m - 1$, or 4. The leftover degrees of freedom must be associated with the leftover sums of squares, or the "error" measure. The difference in degrees of freedom is $N - m$ or 25. Now the quantities SS_T, SS_A, and SS_W are corrected sums of squares and must be divided by the appropriate degrees of freedom to give variances (or mean squares, which is simply another name for the same thing).

$$\text{Total variance} = \frac{SS_T}{N - 1} = \frac{383.79}{29} = 13.23$$

$$\text{Variance among} = MS_A = \frac{SS_A}{m - 1} = \frac{237.42}{4} = 59.35$$

$$\text{Variance within} = MS_W = \frac{SS_W}{N - m} = \frac{146.37}{25} = 5.85$$

The rationale in analysis of variance may be clearer if we consider the extreme situation where replicates are all identical. Then, the means of the columns will be the same as the entries within the columns and the variance calculated by considering all observations will be the same as that based only on the column means. In other words, the error measure will vanish, there being *no* variance unaccounted for by the differences within the replicates. Such an unlikely result would indicate that the original samples are indeed different, each set of replicates having been drawn from separate populations having zero variances.

Our example is less extreme. Calculating the F test, we obtain the following critical value. Having chosen a critical region based on our selection of the desired level of significance and the degrees of freedom, we can now reject or accept the hypothesis.

$$F = \frac{MS_A}{MS_W} = 10.14$$

The one-way analysis of variance is appropriate if we wish to test the hypothesis that a number of populations, represented by samples, are identical. However, we must be able to randomly select replicates within the samples and make measurements in random order. This may be a severe restriction in certain situations and may lead to analyses in which too much information about the variance is lost. For example, it may be suspected that a certain step in the measurement technique is causing an increase in variance. Using a one-way model, we cannot assess the magnitude of the introduced variance because it is mixed with other sources of variance in the SS_W sum of squares. However, more elaborate statistical designs may allow us to isolate this cause of variation and estimate its magnitude.

Two-Way Analysis of Variance

Many tests, some of elaborate design, are described in texts of analysis of variance and design of experiments. Excellent descriptions of some designs most useful for geologists are contained in the books by Griffiths (1967) and Krumbein and Graybill (1965). Here we must limit our consideration to one additional example and the appropriate statistical design.

The St. Peter Sandstone of Ordovician age is a remarkably pure orthoquartzite which occurs in the upper Mississippi River Valley. Because the sand grains of this unit are so well rounded and sorted, the St. Peter is unusually homogeneous in character. For this reason, petroleum reservoirs in the St. Peter Sandstone should react during pumping in a manner that could be closely predicted by theoretical models of reservoir behavior, even though these are based on idealized conditions. Deviations from reality in model behavior might indicate erroneous assumptions in the model's structure.

A small oil field in southern Illinois appeared ideally suited for examining the coincidence of model behavior to actual production performance. Carefully documented development and production data were available for the field. However, before performing an extensive analysis of the reservoir's behavior, it seemed prudent to examine the fundamental assumption that St. Peter Sandstone is essentially homogeneous in its properties.

From the collection of drill cores taken during development of the field, ten were randomly selected for analysis. A 1-inch cube was cut at a random position on each core so that the proper vertical orientation of each sample was preserved. Using a liquid permeameter, two measurements of the rate of fluid flow through the samples were made; one vertically, or perpendicular to bedding, and one horizontally, parallel to bedding. Using the flow rates, sample permeabilities measured in millidarcies were calculated. The twenty permeability values are listed in Table 2.17. From these twenty values, we will attempt to determine two things; whether there are significant differences in permeability with location in the field (i.e., between wells), and if there are significant directional differences in permeability.

The problem may be considered as a *two-way analysis of variance*. We will be concerned with two main sources of variation; those arising because of differences

TABLE 2.17   Directional Permeability of Randomly Selected Cores of St. Peter Sandstone, Illinois

Directional Permeability (md)			
Vertical	Horizontal	Vertical	Horizontal
1037	1124	928	943
963	960	1108	1165
842	921	821	803
1121	1202	797	792
1043	1028	949	1004

between the cores, and those arising because of differences in the direction of flow in the permeability measurements. A third source of variation is the leftover, residual, or error variance, corresponding to variance within replicates in the one-way ANOVA. In this example, we will consider two hypotheses:

$$H_0 : \mu_{well1} = \mu_{well2} = \cdots = \mu_{well10}$$

$$H_0 : \mu_{vertical} = \mu_{horizontal}$$

The corresponding alternative hypotheses are that at least one well is different, and that vertical and horizontal permeabilities are not the same.

With one exception, this problem is much like the carbonate determination ANOVA which we just worked. However, rather than replicate measurements made in random fashion on the samples, we have made distinctly different types of measurements. These may be referred to as *treatments;* the term implies that the numbers generated with one treatment may be fundamentally different from those created by other treatments, even though the samples used are the same. Because the measurements are not completely randomized but are instead separated according to the treatment procedures, the data can be analyzed for differences between treatments as well as differences between samples. Thus, variance due to differences in treatments is no longer confounded with other sources of variation, but has been separated by the statistical design.

Four fundamental assumptions about the nature of the parent population are made in performing this test: (*a*) each combination of treatment and object is a random sample drawn from different populations, (*b*) each parent population is normal, (*c*) each parent population has the same variance, and (*d*) there is no interaction between different treatments and different samples. The last assumption is a statement that a particular combination of treatment and sample will not produce a greater variance than treatments and samples in other combinations. If we performed the ANOVA using replications (i.e., more than one permeability measurement on each direction/core combination), we could assess interaction, but in this simple design we assume that it does not occur. If interactions do exist, they cannot be detected by this experimental design, and their presence will invalidate the test results. Interactions between parameters erroneously assumed to be independent has led more than one researcher to grief. A good introduction to the effects of interactions is contained in Hicks (1973).

The appropriate ANOVA for two-way analysis without replications is given below. The SS_T is calculated as in (2.37); the SS_A is calculated according to (2.38). The SS_B is the sum of squares within treatments and is calculated by

$$SS_B = \sum_{i=1}^{n} \frac{(\sum_{j=1}^{m} X_{ij})^2}{m} - \frac{(\sum_{j=1}^{m} \sum_{i=1}^{n} X_{ij})^2}{N} \tag{2.41}$$

where m = number of samples and n = number of treatments applied to each sample. The error sum of squares, SS_e, is found by

$$SS_e = SS_T - (SS_A + SS_B) \tag{2.42}$$

From the new equations, you can see that SS_B is a measure of the variance of the treatments as determined from averages of the samples within each treatment. The error sum of squares is reduced by the amount assigned to this source of variation. Symbolic conventions are the same as in the one-way ANOVA, except that n is now the number of treatments rather than replications.

Source of Variation	Sum of Squares	Degrees of Freedom	Mean Squares	F Tests
Among Samples	SS_A	$m - 1$	MS_A	MS_A/MS_e[a]
Among Treatments	SS_B	$n - 1$	MS_B	MS_B/MS_e[b]
Error	SS_e	$(m - 1)(n - 1)$	MS_e	
Total Variation	SS_T	$N - 1$		

[a]Test of significance of differences between samples.
[b]Test of significance of differences between treatments.

After selecting the desired level of significance for the two hypotheses, we can use this statistical design to test the permeability data in Table 2.17. The questions to be answered are (a) is there a significant difference in permeabilities over the field, and (b) is there a significant difference in vertical and horizontal permeability? We can modify the one-way procedure to find the additional terms MS_B and the new error term MS_e, and then process the data to find both F values. Figure 2.32 may help visualize the proper modifications that must be made. From your results, what recommendations would you make concerning the advisability of using the field to test reservoir models?

Analyses of variance are among the most widely used statistical tests, especially in fields such as quality control, product development, and the testing of biological specimens. Consequently, computer programs are readily available which will perform ANOVAs of almost any degree of complexity.

The χ^2 Test

We must now introduce another probability distribution based on the properties of a normal population. If samples of size n are taken from a normal population having a mean μ and standard deviation σ, each observation within a sample can be standardized by (2.26) to the standard normal form (rewritten here to include population parameters rather than sample statistics):

$$Z = \frac{X - \mu}{\sigma} \tag{2.43}$$

If the standardized values of Z are squared and summed, they form a new statistic which we can denote ΣZ^2. That is,

$$\Sigma Z^2 = \sum_{i=1}^{n} \left(\frac{X_i - \mu}{\sigma} \right)^2 \tag{2.44}$$

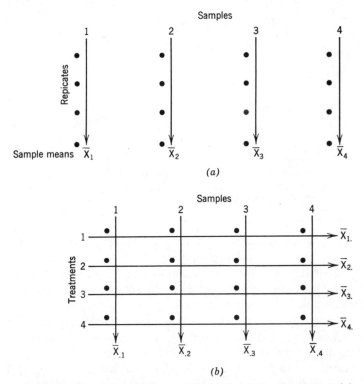

FIGURE 2.32 Pattern of summation in analysis of variance. (*a*) One-way analysis; summation proceeds down replicates to find sample means. (*b*) Two-way analysis; summation proceeds down treatments to find sample means and also across samples to find treatment means.

Because this is a sample-based statistic, it will vary from sample to sample. If we draw all possible samples of size n from a normal population and plot the values of ΣZ^2, they will form a χ^2 (chi-square) distribution. The distribution of χ^2 is dependent upon degrees of freedom that are related to the size of the samples involved in its creation. A typical χ^2 curve is shown in Figure 2.33, and tables of χ^2 for various degrees of freedom and levels of significance are given in Table 2.18. Note that the curve, like the F distribution, goes from zero to positive values approaching infinity. This approach to the derivation of χ^2 is developed further in Li (1964, Chapter 7).

The great utility of the χ^2 distribution is that it can be used for tests of nominal and ordinal data. Up to this point, we have considered only tests of data that are measured variables. We now will examine some methods to treat count data such as the number of echinoids per unit area on the sea floor, the number of plagioclase crystals encountered on traverses across a thin section, and the number of grains within certain size categories from a disaggregated sandstone.

A commonly encountered problem of elementary statistics is that of comparing

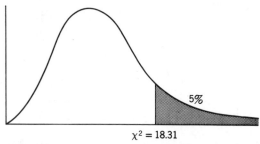

$$\chi^2 = 18.31$$

FIGURE 2.33 A typical χ^2 distribution for 10 degrees of freedom, with critical region (shown by shading) which contains 5% of area under the curve. Critical value of $\chi^2 = 18.31$.

a distribution of sample observations to some predefined standard distribution. As examples, a researcher may wish to apply statistical tests that assume his data are drawn from a population whose distribution has a specified shape, perhaps normal or lognormal. The frequency distribution of the sample may be compared to the hypothetical distribution to see if this assumption is warranted. In highway construction, allowable percentages of the various size classes of particles in aggregate may be specified; a geologist or engineer may wish to know if the distribution of particle sizes in a natural gravel deposit or in the output from a rock crusher meets these specifications. In both of these problems, it is necessary to measure the correspondence between the form of two distributions, one derived from a sample and the other assumed or specified. A probabilistic answer is needed to the question, "do the two distribution curves have the same shape?"

A similar problem arose in the course of a bottom-sampling project in Whitewater Bay, Florida, where 48 measurements of surface-water salinity were made. The measurements are listed in Table 2.19. Geographic considerations lead us to believe that the salinity variation in the sample area is random and should be normally distributed. If this hypothesis is true, it would imply a free mixing and interchange between open marine water and fresh water entering the bay. On the other hand, if some mechanism is operating which tends to keep fresh and saline waters separated in the bay, the distribution of salinity values would reflect this. It may in turn be possible to draw inferences about water circulation patterns and the expected distribution of bottom sediment types.

We can test how well the distribution of sample values conforms to a normal distribution by a test procedure called the *goodness-of-fit*. We hypothesize that the population of salinities from which our samples are drawn is normally distributed with an unknown mean μ and variance σ^2. The alternative to this hypothesis is, of course, that the parent population is not normally distributed. A test statistic may be devised by dividing a standard normal distribution into a number of segments. The probability that a random observation from a standard normal distribution will fall into one of the segments is equal to the area under the curve within that segment. From these probabilities, the number of observations that would be

TABLE 2.18 Critical Values of χ^2 for ν Degrees of Freedom and Selected Levels of Significance

Number of Degrees of Freedom, ν	Significance Level, α (%)				
	20	10	5	2.5	1
1	1.64	2.71	3.84	5.02	6.63
2	3.22	4.61	5.99	7.38	9.21
3	4.64	6.25	7.81	9.35	11.34
4	5.99	7.78	9.49	11.14	13.28
5	7.29	9.24	11.07	12.83	15.09
6	8.56	10.64	12.59	14.45	16.81
7	9.80	12.02	14.07	16.01	18.48
8	11.03	13.36	15.51	17.53	20.09
9	12.24	14.68	16.92	19.02	21.67
10	13.44	15.99	18.31	20.48	23.21
11	14.63	17.28	19.68	21.92	24.72
12	15.81	18.55	21.03	23.34	26.22
13	16.98	19.81	22.36	24.74	27.69
14	18.15	21.06	23.68	26.12	29.14
15	19.31	22.31	25.00	27.49	30.58
16	20.47	23.54	26.30	28.85	32.00
17	21.61	24.77	27.59	30.19	33.41
18	22.76	25.99	28.87	31.53	34.81
19	23.90	27.20	30.14	32.85	36.19
20	25.04	28.41	31.41	34.17	37.57
21	26.17	29.62	32.67	35.48	38.93
22	27.30	30.81	33.92	36.78	40.29
23	28.43	32.01	35.17	38.08	41.64
24	29.55	33.20	36.42	39.36	42.98
25	30.68	34.38	37.65	40.65	44.31
26	31.79	35.56	38.89	41.92	45.64
27	32.91	36.74	40.11	43.19	46.96
28	34.03	37.92	41.34	44.46	48.28
29	35.14	39.09	42.56	45.72	49.59
30	36.25	40.26	43.77	46.98	50.89
40	47.27	51.81	55.76	59.34	63.69
50	58.16	63.17	67.50	71.42	76.15
60	68.97	74.40	79.08	83.30	88.38
70	79.71	85.53	90.53	95.02	100.43
80	90.41	96.58	101.88	106.63	112.33
90	101.05	107.57	113.15	118.14	124.12
100	111.67	118.50	124.34	129.56	135.81

Source: Abridged from Table 24, *The Penguin-Honeywell Book of Tables,* copyright F. W. Kellaway (ed.) and Honeywell Controls Ltd. (E.D.P. Division), 1968.

TABLE 2.19 Measurements Recorded at 48 Stations in Whitewater Bay, Florida

Salinity (ppt)[a]									
46	53	58	60	60	49	59	48	46	78
37	58	46	46	47	48	42	50	63	48
62	49	47	36	40	39	61	43	53	42
59	60	52	34	40	36	67	44	40	
40	56	51	51	35	47	53	49	50	

[a]Parts per thousand.

expected in each segment can be calculated. The expected frequency of occurrence within each segment can be compared with the frequency of sample observations that fall within the segments. If the number in each segment deviates significantly from that expected, it seems unlikely that the sample was drawn from a normal population. By use of the χ^2 distribution, we can attach probabilistic meaning to the words "significant" and "unlikely" in this statement.

The test statistic is calculated by the equation

$$\chi^2 = \sum_{j=1}^{k} \frac{(O_j - E_j)^2}{E_j} \tag{2.45}$$

where O_j is the number of observations within the jth class, and E_j is the number of observations expected in that class. There are k classes or intervals.

In this problem, the test statistic is computed by dividing a standard normal curve into a number of segments, say 4, so that each contains the same area and hence the same probability of occurrence. The limits of these equal-sized areas are $-\infty$ to -0.67, -0.67 to 0.0, 0.0 to 0.67, and 0.67 to ∞. When our data are standardized, we expect approximately one-fourth of the standardized values to fall into each of the four categories. The number of samples falling into each of the categories is counted, the difference between this and the expected number is found, and the result is squared. The squared difference is divided by the expected number. The final values for each of the four categories are then summed. If the total exceeds a critical value, the null hypothesis is rejected and we may conclude that the distribution of salinity values is not normal.

The sampling distribution of χ^2 resembles Figure 2.33, although the exact shape depends upon the degrees of freedom. However, the degrees of freedom are not dependent upon the number of observations in the same sense as in previous tests. In the χ^2 problem, our "samples" actually are the four categories we are comparing against the four categories of the standard normal curve. The number of degrees of freedom are (number of categories $= k$) $- 3$, or in our example, one. We lose the two additional degrees of freedom because we must estimate μ by $\overline{X}$ and σ^2 by s^2 in order to standardize our observations. The critical value of $\chi^2 = 2.71$ is that for a 10% level of significance ($\alpha = 0.10$) and one degree of freedom, taken from Table 2.18.

Like the F distribution, χ^2 is not centered around zero, but is entirely positive. Because the deviations of expected from observed frequencies in each category are squared, negative numbers do not appear. Consequently, χ^2 tests are always one-tailed, and the region of rejection is on the right.

In the example problem, the normal distribution has been split into four categories of equal probability. If the salinity measurements are distributed normally, approximately twelve measurements should fall into each of the four categories when normalized. Samples consist of the number of observations (frequency of occurrence) actually in each group. Because we have four groups, we have only four samples, each with an expected value of twelve. The first step is to standardize the data by (2.26), repeated here:

$$Z_i = \frac{X_i - \overline{X}}{s} \tag{2.46}$$

The Whitewater Bay samples have a mean of $\overline{X} = 49.54$ and a standard deviation of $s = 9.27$, so the individual observations are standardized by

$$Z_i = \frac{X_i - 49.54}{9.27}$$

The standardized scores are listed in Table 2.20. They have been rearranged into

TABLE 2.20   Standardized Scores of Salinity Measurements from Whitewater Bay

Sample Number	Original Sample	Standardized Sample	Sample Number	Original Sample	Standardized Sample
1	46.00	−0.38	25	35.00	−1.57
2	37.00	−1.35	26	49.00	−0.06
3	62.00	1.34	27	48.00	−0.17
4	59.00	1.02	28	39.00	−1.14
5	40.00	−1.03	29	36.00	−1.46
6	53.00	0.37	30	47.00	−0.27
7	58.00	0.91	31	59.00	1.02
8	49.00	−0.06	32	42.00	−0.81
9	60.00	1.13	33	61.00	1.24
10	56.00	0.70	34	67.00	1.88
11	58.00	0.91	35	53.00	0.37
12	46.00	−0.38	36	48.00	−0.17
13	47.00	−0.27	37	50.00	0.05
14	52.00	0.27	38	43.00	−0.71
15	51.00	0.16	39	44.00	−0.60
16	60.00	1.13	40	49.00	−0.06
17	46.00	−0.38	41	46.00	−0.38
18	36.00	−1.46	42	63.00	1.45
19	34.00	−1.68	43	53.00	0.37
20	51.00	0.16	44	40.00	−1.03
21	60.00	1.13	45	50.00	0.05
22	47.00	−0.27	46	78.00	3.07
23	40.00	−1.03	47	48.00	−0.17
24	40.00	−1.03	48	42.00	−0.81

TABLE 2.21   Standardized Salinity Measurements Grouped into 25% Probability Categories for χ^2 Test of Normality

Category $-\infty$ to -0.67		Category -0.67 to 0.0	
-1.35	-1.14	-0.38	-0.17
-1.03	-1.46	-0.06	-0.27
-1.46	-0.81	-0.38	-0.17
-1.68	-0.71	-0.27	-0.60
-1.03	-1.03	-0.38	-0.06
-1.03	-0.81	-0.27	-0.38
-1.57		-0.06	-0.17
Number of entries = 13		Number of entries = 14	
Category 0.0 to $+0.67$		Category $+0.67$ to $+\infty$	
0.37	0.37	1.34	1.13
0.27	0.05	1.02	1.02
0.16	0.37	0.91	1.24
0.16	0.05	1.13	1.88
		0.70	1.45
		0.91	3.07
		1.13	
Number of entries = 8		Number of entries = 13	

the four categories in Table 2.21. If the samples are normally distributed, we expect approximately twelve observations per category.

Calculating the value of χ^2, we obtain

$$\chi^2 = \frac{(13 - 12)^2}{12} + \frac{(14 - 12)^2}{12} + \frac{(8 - 12)^2}{12} + \frac{(13 - 12)^2}{12}$$

$$= \frac{1}{12} + \frac{4}{12} + \frac{16}{12} + \frac{1}{12}$$

$$= \frac{22}{12} = 1.83$$

The calculated value of χ^2 is less than the critical value of 2.71 for one degree of freedom and a 10% level of significance. Therefore, there is no evidence to suggest that the measurements of surface-water salinities are not normally distributed.

Of course, we are not restricted to considering only normal distributions with the χ^2 statistic. We may test the goodness-of-fit of a data set against *any* specified curve, such as lognormal, exponential, or an arbitrary distribution. The test procedure remains the same, although the degrees of freedom must be adjusted to account for the number of parameters estimated. Cochran (1952) provides an extensive discussion of these tests.

The Lognormal Distribution and Other Transformations

Many geological variables very obviously do not follow a normal distribution. When, for example, the volumes of oil fields in a region are plotted, the result is a highly skewed distribution such as that shown in Figure 2.34. Most fields are small, but there are decreasing numbers of much larger fields, and a few rare giants that greatly exceed all others in volume. Geochemical variables such as the concentration of selenium in plant material assayed during a geochemical reconnaissance, or the concentrations of iodine detected in groundwater samples, also tend to follow similar skewed distributions. The grain-size distribution of sediments is pronouncedly skewed, and an entire system of size classification has been based on this fact (Krumbein and Pettijohn, 1938). Figure 2.35 is a histogram showing the concentration of copper in stream sediments from the Yukon. These exhibit a skewed distribution typical of many other geologic variables.

If the observations shown in Figures 2.34 and 2.35 are converted to logarithmic form (that is, we use $Y_i = \log X_i$ instead of X_i for each observation), we see that their distributions become nearly normal (Figures 2.36 and 2.37). Such variables are said to be *lognormal*. Because of its frequent occurrence in geology, the lognormal distribution is extremely important. However, if we confine our attention to the transformed variable Y_i rather than X_i itself, the properties of the lognormal distribution can be explained simply by reference to the normal distribution.

The mean and variance of a log transformed variable Y_i are found in the usual way:

$$\overline{Y} = \frac{\Sigma Y_i}{n}$$

and

$$s_Y^2 = \frac{\Sigma (Y_i - \overline{Y})^2}{n - 1} \tag{2.47}$$

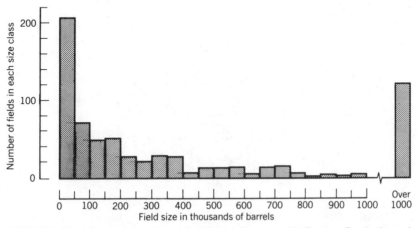

FIGURE 2.34 Histogram of sizes of oil fields discovered in Denver Basin through 1969.

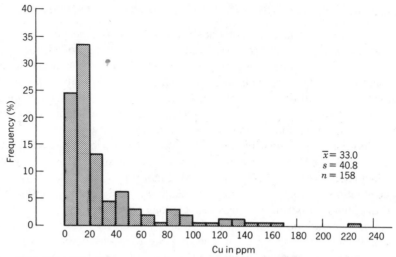

FIGURE 2.35 Histogram of copper content (in ppm) of stream sediments collected in Mt. Nansen area, Yukon. After Saager and Sinclair, 1974.

However, in terms of the original untransformed variable X_i, the mean $\overline{Y}$ corresponds to the nth root of the products of X_i,

$$\overline{Y} = \text{GM} = \sqrt[n]{\Pi X_i} \tag{2.48}$$

which is the *geometric mean*, GM. The symbol Π is analogous to Σ, except it means that all the elements in the indicated series are to be multiplied together, rather than added together. Π has limits exactly like those used with the summation

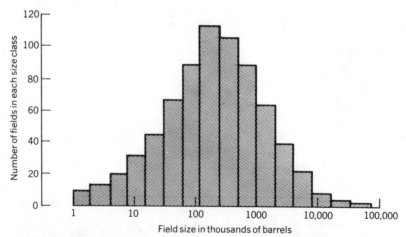

FIGURE 2.36 Histogram of sizes of oil fields discovered in Denver Basin through 1969, plotted on logarithmic scale. From Harbaugh, Doveton, and Davis (1977).

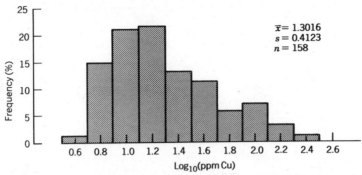

FIGURE 2.37 Histogram of copper content (in ppm) of stream sediments collected in Mt. Nansen area, Yukon, plotted on logarithmic scale. After Saager and Sinclair, 1974.

symbol, Σ. Sometimes the limits are omitted if they are clearly implied in the expression. As an example,

$$\Pi_{i=1}^{3} X_i, \qquad \text{where } X_1 = 2, \quad X_2 = 3, \quad X_3 = 4$$

is

$$\Pi X_i = 2 \times 3 \times 4 = 24$$

The variance of a logarithmic transformed variable is called the *geometric variance* and is equivalent to

$$s_Y^2 = s_g^2 = \sqrt[n-1]{\Pi\, 2\left(\frac{X_i}{GM}\right)} \tag{2.49}$$

In practice, of course, it is simplest to convert our observations into logarithms and then compute the mean and variance. If the geometric mean and variance are desired, they are found by taking the antilogarithms of $\overline{Y}$ and s_Y^2. As long as we work with the data in its transformed state, all of the statistical procedures that are appropriate for ordinary variables are applicable to log transformed variables.

In addition to converting a skewed variable into a more symmetric form, logarithmic transformation may also be useful in stabilizing the variance. Figure 2.38 is a plot of the volumes of petroleum reservoirs associated with salt domes in an area of the outer continental shelf of Louisiana. These reservoir volumes are plotted against the areas of closure of the structures in which the reservoirs are located. In general, there is a positive association between the two variables; that is, larger structures tend to contain larger reservoirs. However, the variation in reservoir size also increases as the sizes of structures increase, or in other words, the variance is proportional to the mean.

A logarithmic transformation may correct this condition, as can be seen in Figure 2.39. Here, both reservoir volume and size of structure are transformed into their logarithms, resulting in a log-log plot. The log variance in reservoir volumes remains almost constant for all values of the logarithms of structural size.

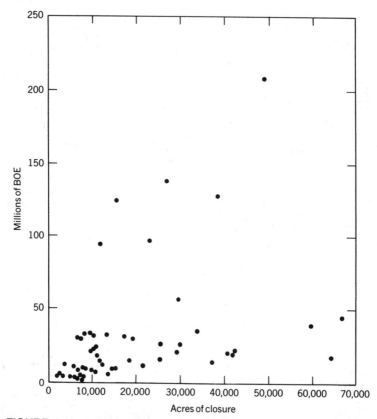

FIGURE 2.38 Volumes of petroleum reservoirs associated with salt domes in the Louisiana outer continental shelf, Gulf of Mexico, plotted against areas of closure as mapped from reflection seismic data.

The characteristics of the lognormal distribution are discussed at length in the monograph by Aitchison and Brown (1969) and in a geological context by Koch and Link (1981). The normal distribution often is explained as arising when repeated measurements are made of some fixed quantity, μ. Each individual measurement is perturbed by many random errors that add, sometimes in one direction and sometimes in the opposite, to the measurement. Usually these random errors cancel one another, and the final measurement is near the true value. But on rare occasions most of the random errors will have the same sign, and an extreme measurement results. The result is the familiar bell-shaped normal distribution of measurements.

A lognormal distribution may arise under the same circumstances if the random errors are multiplicative, rather than additive. Most of the random perturbations, when multiplied together, will produce an intermediate product near the geometric mean. Rarely, by random chance, all of the perturbations will be very small and their product will be near zero. Equally rarely, all of the perturbations will be large

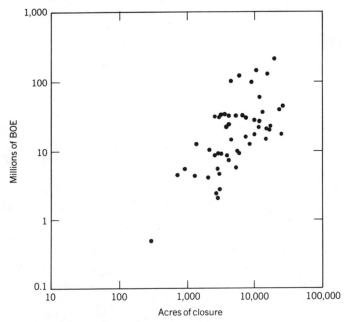

FIGURE 2.39 Data on reservoir volumes and areas of seismic closure from Figure 2.38, plotted as log-log graph.

and their product will be an extremely large value. The result of many random realizations will be a distribution that starts at zero, rises to a maximum, then slopes downward and extends to extremely large values.

Biological scientists often refer to the "law of proportionate effect," which states that the change in a variable at any step in a process is a random proportion of the previous value of the variable. Thus, for example, the probable change in size of colonies of microbes with time is proportional to the size of the colonies at the previous time. Large colonies will tend to expand (or shrink) a much greater amount than will small colonies. Perhaps oil pools accumulated in the same manner, so that during initial hydrocarbon migration, larger accumulations tended to increase at a proportionately greater rate than did small accumulations. Such processes will result in lognormal distributions.

Geologists may be more familiar with the "theory of breakage," which has been advanced to explain the lognormal distribution of particle size that is observed in both natural sediments and in the crushed material produced by mills and grinders. Suppose we begin with a collection of equal-sized particles and then break each one at random. In general, the result will be one smaller and one larger fragment from each original particle. If we then again break each of these at random, the small fragments will produce still smaller fragments, while each large fragment will tend to yield, again, a larger and a smaller piece. If the process is repeated again and again, the result will be an extremely large number of very small particles, and a few "select" grains whose dimensions approach those of the original particles. In other words, the lognormal size distribution we so commonly observe in sediments will be produced.

Other Transformations

Certain other transformations of X may make its distribution more nearly normal, make its variance more consistent, or yield some other statistically beneficial results. There is nothing at all sacred about our original scales of measurement, and we may feel free to change them if this seems a useful thing to do. However, we must keep in mind that our statistical analysis is now testing a different hypothesis, which pertains to the characteristics of the transformed variable and not necessarily to the original variable. Also, we should be careful not to befuddle ourselves with transformations so exotic that we lose sight of the original nature of the geologic properties we are attempting to understand.

If our data consist of integer counts such as the number of discovery wells in a township or the number of zircon grains in thin sections, the numbers may tend to follow a Poisson distribution. Rather than treating the data as discrete values, we can convert them to approximately normal form by the *square-root transformation*. That is, every value X_i is replaced by $Y_i = \sqrt{X_i}$. This transformation will make the variances more uniform, and will tend to shorten the long tail of the Poisson distribution. If the observed values X_i are less than about 10, the transformation $Y_i = \sqrt{X_i + 1/2}$ is better, especially if some of the counts are zero.

Robinson (1982) has suggested the use of *power transforms* such as $Y_i = X_i^2$, $Y_i = X_i^3$, and so on, for the enhancement of petrophysical logs measured in boreholes. Powering a variable will cause a much greater increase in larger values than in those that are smaller. If power transformation is followed by rescaling so that the range $(Y_{max} - Y_{min})$ is the same as the original range of the data $(X_{max} - X_{min})$, the effect on a borehole log will be to accentuate variations in parts of the log where readings are high, and to subdue and flatten out parts of the log where readings are low. A power transformation has the same effect on a data distribution, and may be used to correct a negative skew. However, it may also tend to inflate the variance and make it nonuniform.

Negatively skewed distributions sometimes can be made approximately normal by an arcsine transformation, $Y_i = \text{arcsine } X_i$. The original variables should be scaled within the range 0 to 1.00. Another arcsine transformation, described in Chapter 4, can be used to convert binomial distributions to approximately normal form.

Nonparametric Methods

All of the preceding statistical techniques are *parametric;* that is, they are based on the characteristics of test distributions whose parameters are known. These test distributions (the t, F, and χ^2) are all derived by sampling from a normal population.

We can invoke the central limits theorem to justify use of these tests in instances when the sampled population is not normal, provided our sample size is large and the population does not differ too much from normality. Sometimes, however, the underlying population may be distinctly nonnormal, or a small sample size cannot be increased. In such circumstances, we must turn to a category of test procedures called *nonparametric statistical tests* for assistance. These nonparametric tests use information of a lower rank, such as nominal or ordinal samples, rather than the metric data required by conventional tests. No assumptions about the form of the parent distribution are required, hence the name nonparametric. In general, nonparametric tests are less powerful than parametric equivalents in those circumstances when the sampled population does possess the characteristics assumed by parametric tests. However, if the sampled population does not have the specified characteristics, nonparametric procedures are more powerful.

Nonparametric tests have not been widely used in geology, and are not ordinarily presented in elementary statistics books. However, there are many excellent texts that describe nonparametric equivalents of the parametric procedures we have been discussing. These include Siegel (1956), Bradley (1968), and Conover (1980), among others.

Mann-Whitney Test

The Mann-Whitney test can be used as a nonparametric substitute for the *t* test of the equality of the means of two samples. Suppose we collect two samples of size n and m, and wish to test the hypothesis that both came from the same population. We combine the two sets of observations and sort them so they are arranged in order from smallest to largest. Each observation is assigned a rank; that is, the smallest is ranked 1, the next largest is ranked 2, and so on up to the largest observation, which is ranked $(n + m)$. If the two samples have been drawn randomly from the same population, we would expect that observations from one of the samples would be scattered more-or-less uniformly through the ranked sequence.

We will call X_i the ith observation in the first sample, and Y_i the ith observation in the second sample. The rank of an observation is denoted $R(X_i)$ or $R(Y_i)$. The Mann-Whitney test statistic has the form

$$T = \sum_{i=1}^{n} R(X_i) - \frac{n(n + 1)}{2} \tag{2.50}$$

The first term is simply the sum of the ranks of observations from the first sample. Critical values of T, abridged from Conover (1980), are given in Table 2.22.

As an example of the use of the Mann-Whitney test, we may examine the data given in Table 2.23. In the U.S. Gulf Coast region, oil and gas are produced from structural traps associated with salt domes. Prospects can be identified by mapping subsurface horizons which are detected by seismic methods. Potential hydrocarbon traps include tilted fault blocks on the flanks of salt piercements and closed anticlines over the crests of salt domes. Table 2.23 lists the areas of closure, in acres, of

TABLE 2.22 Critical Values for T for the Mann-Whitney Test; Values are for the Lower Critical Limit; Corresponding Limit for the Upper Critical Area is Given by $T_{1-\alpha} = nm - T_\alpha$

n	α	$m=2$	3	4	5	6	7	8	9	10	11	12	13	14	15	16	17	18	19	20
2	.01	0	0	0	0	0	0	0	0	0	0	0	1	1	1	1	1	1	2	2
	.05	0	0	0	1	1	1	2	2	2	2	3	3	4	4	4	4	5	5	5
	.10	0	1	1	2	2	2	3	3	4	4	5	5	5	6	6	7	7	8	8
3	.01	0	0	0	0	0	1	1	2	2	2	3	3	3	4	4	5	5	5	6
	.05	0	1	1	2	3	3	4	5	5	6	6	7	8	8	9	10	10	11	12
	.10	1	2	2	3	4	5	6	6	7	8	9	10	11	11	12	13	14	15	16
4	.01	0	0	0	1	2	2	3	4	4	5	6	6	7	9	8	9	10	10	11
	.05	0	1	2	3	4	5	6	7	8	9	10	11	12	13	15	16	17	18	19
	.10	1	2	4	5	6	7	8	10	11	12	13	14	16	17	18	19	21	22	23
5	.01	0	0	1	2	3	4	5	6	7	8	9	10	11	12	13	14	15	16	17
	.05	1	2	3	5	6	7	9	10	12	13	14	16	17	19	20	21	23	24	26
	.10	2	3	5	6	8	9	11	13	14	16	18	19	21	23	24	26	28	29	31
6	.01	0	0	2	3	4	5	7	8	9	10	12	13	14	16	17	19	20	21	23
	.05	1	3	4	6	8	9	11	13	15	17	18	20	22	24	26	27	29	31	33
	.10	2	4	6	8	10	12	14	16	18	20	22	24	26	28	30	32	35	37	39
7	.01	0	1	2	4	5	7	8	10	12	13	15	17	18	20	22	24	25	27	29
	.05	1	3	5	7	9	12	14	16	18	20	22	25	27	29	31	34	36	38	40
	.10	2	5	7	9	12	14	17	19	22	24	27	29	32	34	37	39	42	44	47
8	.01	0	1	3	5	7	8	10	12	14	16	18	21	23	25	27	29	31	33	35
	.05	2	4	6	9	11	14	16	19	21	24	27	29	32	34	37	40	42	45	48
	.10	3	6	8	11	14	17	20	23	25	28	31	34	37	40	43	46	49	52	55
9	.01	0	2	4	6	8	10	12	15	17	19	22	24	27	29	32	34	37	39	41
	.05	2	5	7	10	13	16	19	22	25	28	31	34	37	40	43	46	49	52	55
	.10	3	6	10	13	16	19	23	26	29	32	36	39	42	46	49	53	56	59	63
10	.01	0	2	4	7	9	12	14	17	20	23	25	28	31	34	37	39	42	45	48
	.05	2	5	8	12	15	18	21	25	28	32	35	38	42	45	49	52	56	59	63
	.10	4	7	11	14	18	22	25	29	33	37	40	44	48	52	55	59	63	67	71
11	.01	0	2	5	8	10	13	16	19	23	26	29	32	35	38	42	45	48	51	54
	.05	2	6	9	13	17	20	24	28	32	35	39	43	47	51	55	58	62	66	70
	.10	4	8	12	16	20	24	28	32	37	41	45	49	53	58	62	66	70	74	79
12	.01	0	3	6	9	12	15	18	22	25	29	32	36	39	43	47	50	54	57	61
	.05	3	6	10	14	18	22	27	31	35	39	43	48	52	56	61	65	69	73	78
	.10	5	9	13	18	22	27	31	36	40	45	50	54	59	64	68	73	78	82	87
13	.01	1	3	6	10	13	17	21	24	28	32	36	40	44	48	52	56	60	64	68
	.05	3	7	11	16	20	25	29	34	38	43	48	52	57	62	66	71	76	81	85
	.10	5	10	14	19	24	29	34	39	44	49	54	59	64	69	75	80	85	90	95
14	.01	1	3	7	11	14	18	23	27	31	35	39	44	48	52	57	61	66	70	74
	.05	4	8	12	17	22	27	32	37	42	47	52	57	62	67	72	78	83	88	93
	.10	5	11	16	21	26	32	37	42	48	53	59	64	70	75	81	86	92	98	103
15	.01	1	4	8	12	16	20	25	29	34	38	43	48	52	57	62	67	71	76	81
	.05	4	8	13	19	24	29	34	40	45	51	56	62	67	73	78	84	89	95	101
	.10	6	11	17	23	28	34	40	46	52	58	64	69	75	81	87	93	99	105	111
16	.01	1	4	8	13	17	22	27	32	37	42	47	52	57	62	67	72	77	83	88
	.05	4	9	15	20	26	31	37	43	49	55	61	66	72	78	84	90	96	102	108
	.10	6	12	18	24	30	37	43	49	55	62	68	75	81	87	94	100	107	113	120
17	.01	1	5	9	14	19	24	29	34	39	45	50	56	61	67	72	78	83	89	94
	.05	4	10	16	21	27	34	40	46	52	58	65	71	78	84	90	97	103	110	116
	.10	7	13	19	26	32	39	46	53	59	66	73	80	86	93	100	107	114	121	128

(Cont.)

Table 2.22 (*Continued*)

n	α	m = 2	3	4	5	6	7	8	9	10	11	12	13	14	15	16	17	18	19	20
18	.01	1	5	10	15	20	25	31	37	42	48	54	60	66	71	77	83	89	95	101
	.05	5	10	17	23	29	36	42	49	56	62	69	76	83	89	96	103	110	117	124
	.10	7	14	21	28	35	42	49	56	63	70	78	85	92	99	107	114	121	129	136
19	.01	2	5	10	16	21	27	33	39	45	51	57	64	70	76	83	89	95	102	108
	.05	5	11	18	24	31	38	45	52	59	66	73	81	88	95	102	110	117	124	131
	.10	8	15	22	29	37	44	52	59	67	74	82	90	98	105	113	121	129	136	144
20	.01	2	6	11	17	23	29	35	41	48	54	61	68	74	81	88	94	101	108	115
	.05	5	12	19	26	33	40	48	55	63	70	78	85	93	101	108	116	124	131	139
	.10	8	16	23	31	39	47	55	63	71	79	87	95	103	111	120	128	136	144	152

Abridged from Conover (1980), Table 8.

prospects selected from marine seismic maps of two regions of the Louisiana coast. We wish to know if there is a difference in areal extent of prospects between the two regions.

Table 2.23 also gives the ranks of the pooled observations of prospect areas in the two regions. The sum of the ranks of the first group is $\Sigma_{i=1}^{n} R(X_i) = 67$. The Mann-Whitney test statistic T is therefore

$$T = 67 - \frac{8(8 + 1)}{2} = 31$$

The critical values of T_α given in Table 2.22 are for the lower limit. That is, they are appropriate for testing the null hypothesis and alternative

$$H_0:E(X) \geq E(Y)$$
$$H_1:E(X) < E(Y)$$

(The symbol E means "expected value" and is the central measure of the distribution, expressed either as its mean or median.) The corresponding limits for the upper critical area can be found as

$$T_{1-\alpha} = nm - T_\alpha \tag{2.51}$$

These are appropriate for testing the null hypothesis and alternative

$$H_0:E(X) \leq E(Y)$$
$$H_1:E(X) > E(Y)$$

In our example, however, we wish to test a two-sided alternative, and to reject the hypothesis of equality of areas of prospects in the two regions if those of the first region are either significantly larger *or* significantly smaller than those of the second region. In more formal terms, our hypothesis and alternative are

$$H_0:E(X) = E(Y)$$
$$H_1:E(X) \neq E(Y)$$

TABLE 2.23 Areas of Oil and Gas Prospects, in Acres, on Salt Dome Structures in Two Regions off the Louisiana Coast; Areas Measured on Maps Made from Marine Seismic Surveys

Eastern Region Prospect	Size (Acres)	Rank
X_1	802	16
X_2	174	8
X_3	158	6
X_4	140	4
X_5	166	7
X_6	328	13
X_7	239	10
X_8	99	3
Western Region Prospect		
Y_1	312	12
Y_2	55	2
Y_3	220	9
Y_4	276	11
Y_5	154	5
Y_6	37	1
Y_7	478	14
Y_8	666	15

The limits of the two critical areas are found from Table 2.22 as $T_{\alpha/2}$ and $nm - T_{\alpha/2}$. If we specify a significance level of $\alpha = 10\%$, these limits are 16 and 56, since both n and $m = 8$. Our test statistic does not fall within either critical region, so this test does not suggest that sizes of prospects are different in the two regions.

The Mann-Whitney test appears, in slightly different forms, under several names in the statistical literature. These include the Wilcoxon test, Siegel-Tukey test, and Festinger's test. The Siegel-Tukey variant is especially interesting because it can be used to test the equivalency of variances in two samples, and thus is a nonparametric substitute for the simple F test.

It sometimes happens that two or more observations in a ranking test have the same values, leading to *tied ranks*. Then, the tied observations are all assigned the same rank, equal to the average of the ranks that would have been assigned if the observations were not exactly identical. For example, in Table 2.23 we might suppose that observation X_2 were 220 acres, rather than 174 acres. Then observations X_2 and Y_3 would be tied. Each would be assigned a rank of $(8 + 9)/2 = 8.5$. The test using tied ranks proceeds in the same manner as before.

Kruskal-Wallis Test

A nonparametric test of the equivalency of several samples has been devised by Kruskal and Wallis. In effect, it is a nonparametric alternative to the one-way analysis of variance. The procedure is very similar to that used in the Mann-Whitney test; the observations from the k samples are combined or pooled and then ranked from smallest to largest. For each sample, the sum of the ranks are found:

$$R_k = \sum_{i=1}^{n_k} R(X_{ik}) \tag{2.52}$$

where $R(X_{ik})$ represents the rank of the ith observation in the kth sample. The total number of observations is $N = \sum_{j=1}^{k} n_k$, where n_k is the number of observations in the kth sample.

The null hypothesis that we wish to test is that all of the k populations from which the samples are taken have identical distributions. The alternative is that at least one of the populations has a different central value. We must assume that all observations have been collected randomly, and that the samples are independent of one another.

From the sum of the ranks, we can compute the Kruskal-Wallis H statistic:

$$H = \frac{12}{N(N+1)} \sum_{j=1}^{k} \frac{[R_k - n_k(N+1)/2]^2}{n_k} \tag{2.53}$$

An algebraically equivalent form somewhat easier to use is:

$$H = \frac{12}{N(N+1)} \sum_{j=1}^{k} \frac{R_k^2}{n_k} - 3(N+1) \tag{2.54}$$

Critical values of H have been tabulated only for sets of three samples, each consisting of up to five observations [see, for example, Siegel's (1956) Table O]. Fortunately, H is approximately distributed as χ^2 with $k - 1$ degrees of freedom, so the test can readily be applied to larger problems.

Nonparametric Correlation

In earlier sections, the calculation of the correlation coefficient has been described at length, and a parametric test for the significance of a sample correlation coefficient was given in the section on the t test. However, there are many circumstances in which the conventional correlation coefficient (sometimes called the *Pearsonian product-moment correlation coefficient*) is not appropriate, yet some measure of relationship between variables is desirable.

The textural properties of sandstones are usually considered to reflect environmental conditions at the time of deposition. The high energy of a beach environment, for example, will cause winnowing and abrasion, resulting in well-sorted deposits of coarse, well-rounded grains. In low-energy environments, deposits are likely to consist of more poorly sorted material, with a finer grain size and more angular particles. Folk (1951) has defined *textural maturity* as the degree to which a sand is both well sorted and well rounded. These two properties are presumed to go hand-in-hand; sandstones that exhibit anomalous relationships (for example, a well-sorted angular sandstone) are said to show "textural inversion."

It would seem a simple matter to check the concept of textural maturity by collecting a suite of sandstone samples, measuring their textural properties, and then calculating a measure of the relationship between the properties. Unfortunately, roundness and degree of sorting are not usually measured on an interval or ratio scale; rather, ordinal scales are used. Sorting is usually expressed in ordinal classes as poor, moderate, or well-sorted, while roundness is classed as angular, suban-gular, subrounded, or rounded. The ordinary correlation coefficient cannot be used to measure the strength of the relationship between roundness and sorting when expressed in these terms.

An alternative measure is *Spearman's rank correlation coefficient*, which, as the name suggests, expresses the similarity between two sets of rankings. If we make two sets of ordinal measurements on a number of objects, we can designate one of the sets as X_i and the other as Y_i. We then rank each measurement, and call the two sets of ranks $R(X_i)$ and $R(Y_i)$. Spearman's coefficient measures the similarity between these two sets of ranks.

$$r' = 1 - \frac{6 \sum_{i=1}^{n} [R(X_i) - R(Y_i)]^2}{n(n^2 - 1)} \qquad (2.55)$$

The term inside the brackets of the numerator is simply the difference in ranks for each object.

Table 2.24 gives the textural properties of twelve reservoir sandstones. Each of the sandstones were compared to one another in order to rank them from most poorly sorted to best sorted, and from most angular to most rounded. The rankings

TABLE 2.24   Roundness and Sorting of Reservoir Sandstones

Formation	Age	Sorting[a]	Rank	Roundness[b]	Rank	$[R(X_i) - R(Y_i)]$	$[R(X_i) - R(Y_i)]^2$
Lakota Ss.	Cret.	P	4	SR	11	−7	49
Berea Ss.	Miss.	W	10	SA	9	1	1
Boise Ss.	Pliocene	P	2	A	1	1	1
Big Clifty Ss.	Miss.	M	8	SA	4	4	16
Clear Creek Ss.	Penn.	M	6	SA	6	0	0
Bromide Ss.	Ordov.	W	9	SR	12	−3	9
Noxie Ss.	Penn.	P	3	SA	8	−5	25
Green River Fm.	Eocene	M	7	SA	3	4	16
Reagan Ss.	Cambrian	W	11	SA	7	4	16
Peru Ss.	Devonian	W	12	SR	10	2	4
Bartlesville Ss.	Penn.	M	5	A	2	3	9
Mt. Simon Ss.	Cambrian	P	1	SA	5	−4	16
						TOTAL	162

[a]Categories of sorting are P = poor, M = moderate, W = well sorted.
[b]Categories of roundness are A = angular, SA = subangular, SR = subrounded.

are given in the table. As in the Mann-Whitney test, tied ranks are given the average of the ranks that would have been assigned if there were no ties.

Spearman's rank correlation between the two variables is

$$r' = 1 - \frac{6[162]}{12(12^2 - 1)}$$

$$= 1 - \frac{972}{1716}$$

$$= 0.43$$

The rank correlation r' is analogous to r in that it varies from $+1.0$ (perfect correspondence between the ranks) to -1.0 (perfect inverse relationship between the ranks). A rank correlation of $r' = 0.0$ indicates that the two sets of ranks are independent. A rank correlation of $r' = 0.43$ suggests there is a weak positive relationship between amount of grain roundness and degree of sorting.

The conventional t test of the significance of a correlation coefficient cannot be applied to r', because the t test assumes the sample is drawn from a bivariate normal population. Fortunately, a table of critical values is available that allows Spearman's rank correlation coefficients to be tested directly; part of this table is reproduced as Table 2.25. As in the t test, the null hypothesis is that the two variables are independent, or that $\rho' = 0$. The most common alternative is $\rho' \neq 0$, so the test is two-tailed, with either very large or very small (negative) correlations leading to rejection. Suppose we decide that a significance level of $\alpha = .05$ would be appropriate to test the correlation between roundness and sorting. Then, the upper critical value would correspond to $1 - \alpha/2 = 0.975$, or 0.5804 for $n = 12$. The lower critical value, corresponding to $\alpha/2 = 0.025$, is -0.5804. Our computed correlation, $r' = 0.43$, does not fall beyond either of these limits, so we cannot reject the hypothesis that roundness and sorting are independent of one another. If there is a relationship between these two properties, our sample of twelve observations is not adequate to detect it at the 5% level of significance.

Kolmogorov-Smirnov Tests

One extremely useful group of nonparametric procedures includes the *Kolmogorov-Smirnov* tests. Among their other applications, they can be used to test for goodness of fit, and thus are an alternative to the χ^2 methods discussed earlier. Although the χ^2 goodness-of-fit tests are also nonparametric in the sense that they can be applied to observations following any kind of distribution, the Kolmogorov-Smirnov tests are superior in certain circumstances. Their most obvious advantage is that it is not necessary to group the observations into arbitrary categories; for this reason they are more sensitive than the χ^2 test to deviations in the tails of distributions where frequencies are low.

Figure 2.40 illustrates how the Kolmogorov-Smirnov procedure works. We have selected a sample from some unknown population and wish to test its goodness of

TABLE 2.25 Critical Values of Spearman's Rank Correlation Coefficient, for Testing the Hypothesis that $\rho = 0$

n	α = .10	.05	.025	.01	.005	.001
4	.8000	.8000				
5	.7000	.8000	.9000	.9000		
6	.6000	.7714	.8286	.8857	.9429	
7	.5357	.6786	.7450	.8571	.8929	.9643
8	.5000	.6190	.7143	.8095	.8571	.9286
9	.4667	.5833	.6833	.7667	.8167	.9000
10	.4424	.5515	.6364	.7333	.7818	.8667
11	.4182	.5273	.6091	.7000	.7455	.8364
12	.3986	.4965	.5804	.6713	.7273	.8182
13	.3791	.4780	.5549	.6429	.6978	.7912
14	.3626	.4593	.5341	.6220	.6747	.7670
15	.3500	.4429	.5179	.6000	.6536	.7464
16	.3382	.4265	.5000	.5824	.6324	.7265
17	.3260	.4118	.4853	.5637	.6152	.7083
18	.3148	.3994	.4716	.5480	.5975	.6904
19	.3070	.3895	.4579	.5333	.5825	.6737
20	.2977	.3789	.4451	.5203	.5684	.6586
21	.2909	.3688	.4351	.5078	.5545	.6455
22	.2829	.3597	.4241	.4963	.5426	.6318
23	.2767	.3518	.4150	.4852	.5306	.6186
24	.2704	.3435	.4061	.4748	.5200	.6070
25	.2646	.3362	.3977	.4654	.5100	.5962
26	.2588	.3299	.3894	.4564	.5002	.5856
27	.2540	.3236	.3822	.4481	.4915	.5757
28	.2490	.3175	.3749	.4401	.4828	.5660
29	.2443	.3113	.3685	.4320	.4744	.5567
30	.2400	.3059	.3620	.4251	.4665	.5479

From Conover (1980), Table 10.

fit to a hypothetical model of the population. Both the sample and the hypothetical model are plotted together in cumulative form, each scaled so their cumulative sums are 1.0. We then look for the greatest difference between the two; this is the Kolmogorov-Smirnov statistic, K-S. Table 2.26 gives critical values for the K-S statistic, and can be used for either one-tailed or two-tailed hypotheses. The two-tailed null hypothesis states that classes of the distribution from which the sample is derived are *equal to* those of the hypothetical model for all values of X. The one-tailed null hypothesis states that all classes of the sample distribution are *equal to or less than* those of the hypothetical model (if we use the maximum positive difference as the test statistic), or that all classes of the sample distribution are *equal to or greater than* those of the hypothetical model (if we use the maximum negative difference as the test statistic). In most instances, we will use a two-tailed hypothesis and alternative.

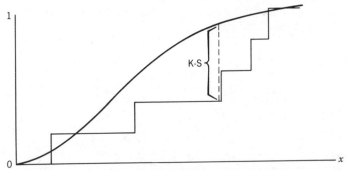

FIGURE 2.40 Kolmogorov-Smirnov procedure for testing good-ness-of-fit of sample distribution (light line) to hypothetical model of population (heavy line). Both are plotted in cumulative form from 0.0 to 1.0. Maximum difference is test statistic, K-S.

Ordinarily, the Kolmogorov-Smirnov test is used when the hypothetical model can be completely specified. That is, the parameters of the distribution are known (or assumed) from information other than that contained within the sample itself. A variation due to Lilliefors (1967), however, allows us to use the Kolmogorov-Smirnov procedure for testing the fit of a sample to a normal distribution with an unspecified mean and variance. This variation permits use of the procedure in a manner exactly analogous to the χ^2 procedure we used to test the data in Table 2.19.

In the Lilliefors procedure, we must first convert the observational data to standardized form, by the Z transformation.

$$Z_i = \frac{X_i - \overline{X}}{s}$$

The mean and variance of the sample are found in the usual manner. The standard normal distribution and the values of Z_i may then be plotted in cumulative form as graphs, as in Figure 2.41 of salinity measurements from Whitewater Bay. The maximum absolute difference between the two curves occurs at $Z = 0.37$, cor-responding to Sample 35, which has a salinity of 53 ppm. Critical values of the Kolmogorov-Smirnov test statistic are given in Table 2.26, although the table extends only to $n = 40$. Approximate values for larger n may be found by the formulas indicated on the table. For $n = 48$ and a level of significance of $\alpha = 0.10$, the critical value is K-S $= 0.17$. The computed test statistic is

$$\text{K-S} = |0.70 - 0.64| = 0.06$$

which does not fall into the critical region. Therefore, we cannot reject the null hypothesis that the samples were collected from a normally distributed population.

At this point we must end our discussion of elementary statistical procedures, not because we have considered them in all of their aspects, but simply to allow space for other topics. The material presented is an extreme condensation of what

TABLE 2.26 Critical Values of the Kolmogorov-Smirnov Test Statistic; Values are Given for Both One- and Two-Tailed Tests

One-Tailed Test α =	.10	.05	.025	.01	.005	α =	.10	.05	.025	.01	.005
Two-Tailed Test α =	.20	.10	.05	.02	.01	α =	.20	.10	.05	.02	.01
n = 1	.900	.950	.975	.990	.995	n = 21	.226	.259	.287	.321	.344
2	.684	.776	.842	.900	.929	22	.221	.253	.281	.314	.337
3	.565	.636	.708	.785	.829	23	.216	.247	.275	.307	.330
4	.493	.565	.624	.689	.734	24	.212	.242	.269	.301	.323
5	.447	.509	.563	.627	.669	25	.208	.238	.264	.295	.317
6	.410	.468	.519	.577	.617	26	.204	.233	.259	.290	.311
7	.381	.436	.483	.538	.576	27	.200	.229	.254	.284	.305
8	.358	.410	.454	.507	.542	28	.197	.225	.250	.279	.300
9	.339	.387	.430	.480	.513	29	.193	.221	.246	.275	.295
10	.323	.369	.409	.457	.489	30	.190	.218	.242	.270	.290
11	.308	.352	.391	.437	.468	31	.187	.214	.238	.266	.285
12	.296	.338	.375	.419	.449	32	.184	.211	.234	.262	.281
13	.285	.325	.361	.404	.432	33	.182	.208	.231	.258	.277
14	.275	.314	.349	.390	.418	34	.179	.205	.227	.254	.273
15	.266	.304	.338	.377	.404	35	.177	.202	.224	.251	.269
16	.258	.295	.327	.366	.392	36	.174	.199	.221	.247	.265
17	.250	.286	.318	.355	.381	37	.172	.196	.218	.244	.262
18	.244	.279	.309	.346	.371	38	.170	.194	.215	.241	.258
19	.237	.271	.301	.337	.361	39	.168	.191	.213	.238	.255
20	.232	.265	.294	.329	.352	40	.165	.189	.210	.235	.252
						Approximation for n > 40	$\dfrac{1.07}{\sqrt{n}}$	$\dfrac{1.22}{\sqrt{n}}$	$\dfrac{1.36}{\sqrt{n}}$	$\dfrac{1.52}{\sqrt{n}}$	$\dfrac{1.63}{\sqrt{n}}$

Adapted from Conover (1980), Table 14.

typically is a one-semester course in introductory statistics, plus selected other topics from more advanced courses. We cannot hope to have done justice to the subject in the short space allotted to it here. However, you should now have some familiarity with the basic concepts of statistical testing and an introduction to the jargon of the field.

This brief discussion of the testing of statistical hypotheses will serve as an introduction to more advanced tests and procedures presented in later chapters. This material by no means exhausts the potential of the discipline of statistics. As Wallis and Roberts (1956) state in their Preface, "Statistics is a lively and fascinating subject; but studying it is too often excruciatingly dull." By casting the topic in the context of the geological problems, hopefully this book may provide the motivation and special interest required to liven the subject for Earth scientists, and you will investigate the practical and rewarding field of statistics further.

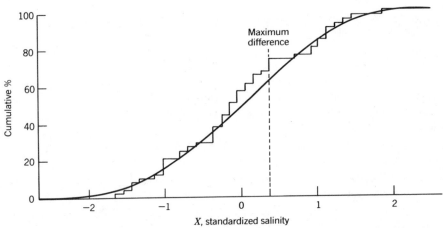

FIGURE 2.41 Cumulative graph of salinities from Whitewater Bay, standardized and scaled from 0.0 to 1.0. Maximum difference between cumulative sample distribution and cumulative normal distribution is K-S = 0.06.

SELECTED READINGS

Aitchison, J., 1981, A new approach to null correlations of proportions: *Jour. of Int'l. Assoc. Mathematical Geology*, **13**, p. 175–189.

Aitchison, J., and J. A. C. Brown, 1969, *The lognormal distribution: with special reference to its uses in economics:* Cambridge Univ. Press, Cambridge, 176 p.

Ash, R. B., 1970, *Basic probability theory:* John Wiley & Sons, Inc., New York, 337 p.

Contains an extensive discussion of the various philosophical bases of statistical analysis and probability.

Bradley, J. V., 1968, *Distribution-free statistical tests:* Prentice-Hall, Inc., Engelwood Cliffs, N.J., 388 p.

This book contains an especially complete set of tables for nonparametric tests.

Chayes, F., 1971, *Ratio correlation:* Univ. Chicago Press, Chicago, 99 p.

An extensive discussion of the statistical behavior of ratio and other closed forms of data. The examples are from geochemistry. Based on a lecture series, the book is available as an inexpensive paperback.

Cheeney, R. F., 1983, *Statistical methods in geology:* George Allen & Unwin Ltd., London, 169 p.

Four chapters of this book are devoted to nonparametric statistical tests, which the author advocates because they are easy to compute in the field.

Cochran, W. G., 1952, The χ^2 test of goodness-of-fit: *Annals of Mathematical Statistics*, **3**, p. 315–345.

Conover, W. J., 1980, *Practical nonparametric statistics*, 2nd ed.: John Wiley & Sons, Inc., New York, 493 p.

Most of the nonparametric tests mentioned in this chapter are discussed at length in this book. It contains a particularly thorough treatment of runs tests.

Duckworth, W. E., 1968, *Statistical techniques in technological research:* Methuen & Co., London, 303 p.

One of the best and most readable statistics books designed for the scientist who wishes to apply statistical techniques. Emphasis is on experimental designs, some of which are rather sophisticated. However, the excellent presentation manages to smooth out the going as much as possible.

Fisher, R. A., 1973, *Statistical methods and scientific inferences,* revised ed.: Hafner Press, New York, 175 p.

A very readable discussion of the evolution of statistical thought. Chapters 2 and 5 illustrate the basis, and some of the consequences, of different concepts of probability.

Folk, R. L., 1951, Stages of textural maturity in sedimentary rocks: *Jour. Sedimentary Petrology,* **21,** p. 127–130.

Freund, J. E., and F. J. Williams, 1966, *Dictionary/outline of basic statistics:* McGraw-Hill, Inc., New York, 195 p.

An essential requirement for every student or applicant of statistics. Available as an inexpensive paperback, this is a well-written dictionary of most statistical terms, compilation of common (and some uncommon) statistical formulas and tests, and set of statistical tables.

Griffiths, J. C., 1967, *Scientific method in the analysis of sediments:* McGraw-Hill, Inc., New York, 508 p.

Chapters 13 through 22 (approximately the second half of the book) actually are an introductory text on applied statistics. The book, unfortunately now out of print, is especially valuable for its consideration of sampling problems and emphasis on correct statistical design.

Guenther, W. C., 1973, *Concepts of statistical inference,* 2nd ed.: McGraw-Hill, Inc., New York, 512 p.

A concise introductory statistics text for those who want the maximum amount of information from a minimum of reading. Covers a wide spectrum of topics not usually found in an introductory text.

Hald, A., 1952, *Statistical tables and formulas:* John Wiley & Sons, Inc., New York, 97 p.

Hicks, C. R., 1973, *Fundamental concepts in the design of experiments,* 2nd ed.: Holt, Rinehart, and Winston, New York, 293 p.

An intermediate text on experimental design and analysis of variance. Chapter 6 contains a graphic explanation of interaction in two-way designs, with examples.

Kellaway, F. W., ed., 1968, *Penguin-Honeywell book of tables:* Penguin Books Ltd., Harmondsworth, England, 75 p.

Mathematical tables in this paperbound book were generated on a computer by the Electronic Data Processing Division of Honeywell Control Ltd. Output was recorded on magnetic tape, then used to drive a photographic typesetting machine. The tables are therefore "untouched by human hands" and free of the errors that usually creep into such material. Tables 2.11, 2.14, and 2.18 in this chapter are reproduced with permission from this book.

Koch, G. S., Jr., and R. F. Link, 1981, *Statistical analysis of geological data:* Dover, Inc., New York, 850 p.

A one-volume paperback reprint of the original two-volume edition, this contains a detailed discussion of statistical analysis of geologic data, especially mine assay values. Includes an extensive treatment of sample designs, the use of analysis of variance methods, and multiple regression. A number of specialized topics and examples are considered in detail. Of special interest is Chapter 11, dealing with the interpretation of constant-sum data.

Kork, J. O., 1977, Examination of the Chayes-Kruskal procedure for testing correlations between proportions: *Jour. of Int'l. Assoc. for Mathematical Geology*, **9**, p. 543–562.

Krumbein, W. C., and F. A. Graybill, 1965, *An introduction to statistical models in geology:* McGraw-Hill, Inc., New York, 475 p.

It is essential that any serious practitioner of geostatistics master this classic reference, now out-of-print. Concentrate first on Chapters 6 through 10.

Krumbein, W. C., and F. J. Pettijohn, 1938, *Manual of sedimentary petrography:* Appleton-Century-Crofts, Inc., New York, 549 p.

Lapin, L. L., 1982, *Statistics for modern business decisions,* 3rd ed.: Harcourt, Brace, Jovanovich, Inc., New York, 887 p.

Chapter 7 contains results of a simulation experiment to demonstrate the central limits theorem.

Li, J. C. R., 1964, *Statistical inference,* v. 1: Edwards Bros., Inc., Ann Arbor, Mich., 658 p.

The first volume of an encyclopedic coverage of elementary statistics. The approach is largely intuitive. The discussion of the χ^2 distribution in Chapter 7 is especially helpful.

Lilliefors, H. W., 1967, On the Kolmogorov-Smirnov test for normality with mean and variance unknown: *Jour. of the American Statistical Association,* **62**, p. 399–402.

McCray, A. W., 1975, *Petroleum evaluations and economic decisions:* Prentice-Hall, Inc., Englewood Cliffs, N.J., 448 p.

Chapter 3 is an extensive discussion of probability distributions that arise in petroleum exploration.

Parzen, E., 1960, *Modern probability theory and its application:* John Wiley & Sons, Inc., New York, 464 p.

Reyment, R. A., 1971, *Introduction to quantitative paleoecology:* Elsevier Publ. Co., Amsterdam, 226 p.

An extremely readable book on the application of elementary statistics to problems in ecology and paleoecology. Chapter 5 covers life tables, a topic we have not considered in this text. Many nonparametric methods are also described.

Robinson, J. E., 1982, *Computer applications in petroleum geology:* Hutchinson Ross Publ. Co., New York, 164 p.

An inexpensive paperback, with emphasis on mapping.

Saager, R., and A. J. Sinclair, 1974, Factor analysis of stream sediment geochemical data from the Mount Nansen area, Yukon Territory, Canada: *Mineralogica Deposita,* **9**, p. 243–252.

Siegel, S., 1956, *Nonparametric statistics for the behavioral sciences:* McGraw-Hill, Inc., New York, 312 p.

One of the first, and still most readable, of the texts on nonparametric statistics.

von Mises, R., 1981, *Probability, statistics, and truth:* Dover, Inc., New York, 244 p.

Although dated, this translation of a 1939 German classic is a basic introduction to the philosophy of statistics. The first three chapters are especially pertinent to the definition of probability.

Walker, H. M., 1940, Degrees of freedom: *Jour. of Educational Psychology,* **31,** p. 253–269.

An excellent tutorial on the concept of degrees of freedom.

Wallis, W. A., and H. V. Roberts, 1956, *Statistics, a new approach:* The Free Press of Glencoe, New York, 646 p.

An entertaining introduction to statistics, with many examples, stories, and anecdotes about the origin of statistical procedures and their applications. For those who prefer narratives to mathematics.

CHAPTER **3**

Matrix Algebra

This chapter is devoted to matrix algebra. Most of the methods we will discuss in subsequent chapters are based on matrix manipulations, especially as performed by computers. In this chapter, we will examine the mathematical operations that underlie such techniques as trend-surface analysis, principal components, and discriminant functions. These techniques are almost impossible to perform without the help of computers, because the calculations are complicated and must be performed repetitively. However, with matrix algebra we can express the basic principles involved in a manner that is succinct and easily understood. Once you master the rudiments of matrix algebra, you will be able to see the fundamental structure within the complex procedures we will examine later.

Most geologists probably have not taken a course in matrix algebra. This is unfortunate, because the subject is not difficult and probably is more useful than most mathematics courses. College courses in matrix algebra usually are liberally sprinkled with numerous theorems and their proofs. Such an approach is certainly beyond the scope of this short chapter, so we will confine ourselves to those topics pertinent to techniques which we will utilize later. Rather than giving derivations and proofs, the material will be presented by examples.

The Matrix

A *matrix* is a rectangular array of numbers, exactly the same as a table of data. In matrix algebra, the array is considered to be a single unit rather than a collection of individual entries, and is operated upon as a unit. This results in a great simplification of the statement of complicated procedures and relationships. Individual numbers within a matrix are identified by subscripts, in which the first subscript identifies the row and the second identifies the column. These individuals may be variances and covariances, sums of observations, terms in a series of simultaneous equations, in fact, any set of numbers.

As one example, in Chapter 2 you were asked to compute the variances and covariances of trace-element data given in Table 2.3. Your answers can be arranged in the form of the matrix below.

$$
\begin{bmatrix}
s_{Cr}^2 & cov_{Cr-Ni} & cov_{Cr-V} \\
cov_{Ni-Cr} & s_{Ni}^2 & cov_{Ni-V} \\
cov_{V-Cr} & cov_{V-Ni} & s_V^2
\end{bmatrix}
=
\begin{bmatrix}
570 & 537.5 & 663.75 \\
537.5 & 562.5 & 718.75 \\
663.75 & 718.75 & 1007.5
\end{bmatrix}
$$

A collection of numbers (say, values of a variable x) arranged into matrix form is symbolically represented by $[X]$, **X**, (X), or $\|X\|$. We will use square brackets in this book, and designate individual entries, or *elements*, of the matrix by subscripted lower case letters such as x_{ij}. In other books you may encounter conventions where these are represented by $[x_{ij}]$, $[x]_{ij}$, or by Greek letter equivalents such as χ_{ij}. The symbol x_{ij} is the element in the ith row and the jth column. For example, if $[X]$ is the 3×3 matrix

$$
\begin{bmatrix}
1 & 4 & 7 \\
2 & 5 & 8 \\
3 & 6 & 9
\end{bmatrix}
$$

x_{33} is 9, x_{13} is 7, x_{21} is 2, and so on. If the number of rows equals the number of columns, the matrix is *square*. Entries whose subscripts are equal (i.e., $i = j$) are called *diagonal elements*. In the trace-element matrix, the diagonal elements are the variances of the elements. Off-diagonal elements are covariances. The diagonal elements in the matrix above are 1, 5, and 9. Square matrices are the most commonly encountered form, but nonsquare forms are possible. However, severe restrictions are placed upon the manipulation of nonsquare matrices. Two forms of nonsquare matrices which have special importance are the $1 \times m$ and $m \times 1$, which are called row and column *vectors*.

Two square matrices have special importance also. The first of these is the *symmetric matrix*, in which all observations $x_{ij} = x_{ji}$, as for example

$$
\begin{bmatrix}
1 & 2 & 3 \\
2 & 4 & 5 \\
3 & 5 & 6
\end{bmatrix}
$$

The variance–covariance matrix of trace elements given earlier is another example of a matrix that is symmetrical about the diagonal.

The *identity matrix*,

$$
\begin{bmatrix}
1 & 0 & 0 \\
0 & 1 & 0 \\
0 & 0 & 1
\end{bmatrix}
$$

is a symmetric matrix in which all diagonal elements are 1 and all off-diagonal elements are 0.

Elementary Matrix Operations

Addition and subtraction of matrices obey the laws of algebra of ordinary numbers, with one important additional characteristic. The two matrices being added or subtracted must have the same number of rows and columns.

To perform the operation $[C] = [A] + [B]$, every element of $[A]$ is added to its corresponding element in $[B]$. If the matrices are not of equal size, there will be leftover elements, and the operation cannot be completed. Subtraction, such as $[C] = [A] - [B]$, proceeds in exactly the same manner, with every element of $[B]$ subtracted from its corresponding element in $[A]$.

As an illustration, Table 3.1 contains 1964 bentonite production figures for three mining districts in Wyoming. Three major grades of clay are produced; that used for drilling mud, as foundry clay, and a miscellaneous category that includes cattle feed binder, drug and cosmetic use, and pottery clay. These same data can be expressed as a 3×3 matrix, $[A]$:

$$[A] = \begin{bmatrix} 105 & 63 & 5 \\ 218 & 80 & 2 \\ 220 & 76 & 1 \end{bmatrix}$$

Production figures for the following year may be expressed in the same manner, giving the matrix, $[B]$:

$$[B] = \begin{bmatrix} 84 & 102 & 4 \\ 240 & 121 & 1 \\ 302 & 28 & 0 \end{bmatrix}$$

Total production for the two years in the three districts is the sum, $[C]$, of the two matrices $[A]$ and $[B]$:

$$\underset{[A]}{\begin{bmatrix} 105 & 63 & 5 \\ 218 & 80 & 2 \\ 220 & 76 & 1 \end{bmatrix}} + \underset{[B]}{\begin{bmatrix} 84 & 102 & 4 \\ 240 & 121 & 1 \\ 302 & 28 & 0 \end{bmatrix}} = \underset{[C]}{\begin{bmatrix} 189 & 165 & 9 \\ 458 & 201 & 3 \\ 522 & 104 & 1 \end{bmatrix}}$$

TABLE 3.1 Bentonite Production in Wyoming, 1964

District	Clay (100,000 tons)		
	Drilling Mud	Foundry Clay	Miscellaneous
Eastern	105	63	5
Montana Border	218	80	2
Central	220	76	1

Similarly, the change in production can be found by subtracting:

$$
\begin{array}{ccc}
[B] & - & [A] & = & [C]
\end{array}
$$

$$
\begin{bmatrix} 84 & 102 & 4 \\ 240 & 121 & 1 \\ 302 & 28 & 0 \end{bmatrix} - \begin{bmatrix} 105 & 63 & 5 \\ 218 & 80 & 2 \\ 220 & 76 & 1 \end{bmatrix} = \begin{bmatrix} -21 & 39 & -1 \\ 22 & 41 & -1 \\ 82 & -48 & -1 \end{bmatrix}
$$

Note that [A] was subtracted from [B] simply to show increases in production as positive values.

As in ordinary algebra [A] + [B] = [B] + [A], and ([A] + [B]) + [C] = [A] + ([B] + [C]), provided all are $n \times m$ matrices. The order of subtraction is, of course, mandatory.

A matrix may be multiplied by a constant by multiplying every element in the matrix by the value. For example.

$$
3 \cdot \begin{bmatrix} 1 & 4 \\ 2 & 5 \\ 3 & 6 \end{bmatrix} = \begin{bmatrix} 3 & 12 \\ 6 & 15 \\ 9 & 18 \end{bmatrix}
$$

Similarly, a matrix may be divided by a constant.

As a simple example of these techniques, consider Table 3.2, which contains measurements of the a, b, and c axes of chert pebbles collected in a glacial till. The measurements were recorded in inches and we wish to convert them to millimeters. If the data are expressed in the form of the matrix [A], we may multiply [A] by a constant 25.4 to obtain a matrix containing the measurements in millimeters.

$$
\begin{array}{ccc}
25.4 \cdot & [A] & = & [C]
\end{array}
$$

$$
25.4 \cdot \begin{bmatrix} 3.4 & 2.2 & 1.8 \\ 4.6 & 4.3 & 4.2 \\ 5.4 & 4.7 & 4.7 \\ 3.9 & 2.8 & 2.3 \\ 5.1 & 4.9 & 3.8 \end{bmatrix} = \begin{bmatrix} 86.36 & 55.88 & 45.72 \\ 116.84 & 109.22 & 106.68 \\ 137.16 & 119.38 & 119.38 \\ 99.06 & 71.12 & 58.42 \\ 129.54 & 124.46 & 96.52 \end{bmatrix}
$$

TABLE 3.2 Measurements of Axes of Pebbles Collected from Glacial Till

Sample	Axis (in.)		
	a	b	c
1	3.4	2.2	1.8
2	4.6	4.3	4.2
3	5.4	4.7	4.7
4	3.9	2.8	2.3
5	5.1	4.9	3.8

Matrix Multiplication

Recall the coin-flipping problem from Chapter 2, where we considered the probability of obtaining a succession of heads if the probability of heads on one flip was 1/2. The probability that we would get three heads in a row was $1/2 \cdot 1/2 \cdot 1/2$, or $1/2^3$. We can develop an equivalent set of probabilities for lithologies encountered in a stratigraphic section. Suppose we have measured a rock section and identified the units as sand, shale, or limestone. At every foot, the rock type can be categorized and the type immediately above noted. We would eventually build a matrix of frequencies similar to that below. This is called a transition frequency matrix and tells us, for example, that sand is followed by shale 18 times, but by limestone only 2 times. Similarly, limestone follows shale 41 times, succeeds itself 51 times, but follows sandstone only 2 times.

		To		
		Sand	Shale	Limestone
	Sand	59	18	2
From	Shale	14	86	41
	Limestone	4	34	51

We can convert these frequencies to probabilities by dividing each element in a row by the total of the row. This will give the transition probability matrix shown below, from which the probability of proceeding from one state to another can be assessed. This subject will be considered in detail in a later chapter, where its use in time series analysis will be examined. Now, however, we are interested in the matrix of probabilities, which is analogous to the single probability associated with the flip of a coin.

		To		
		Sand	Shale	Limestone
	Sand	0.74	0.23	0.03
From	Shale	0.10	0.61	0.29
	Limestone	0.05	0.38	0.57

Just as we can find the probability of producing a string of heads in a coin-flipping experiment by powering the probability associated with a single flip, we can determine the probability of attaining specified states at successive intervals by powering the transition probability matrix. That is, the probability matrix $[P']$ after n steps through the succession is equal to $[P]^n$. The nth power of a matrix is simply the matrix times itself n times. To perform this operation, however, we must know the special procedures of matrix multiplication.

The simplest form of multiplication involves two square matrices, $[A]$ and $[B]$, of equal size, producing the product matrix, $[C]$. An easy method of performing this operation is to arrange the matrices in the following manner:

$$[B]$$
$$[A] \quad [\text{Ans.}]$$

To obtain the value of an element c_{ij}, multiply each element of row i of $[A]$, starting at the left, by each element of column j of $[B]$, starting at the top. All the products are summed to obtain the c_{ij} element of the answer. The steps in multiplication are demonstrated below on the two matrices,

$$\begin{bmatrix} 1 & 4 & 7 \\ 2 & 5 & 8 \\ 3 & 6 & 9 \end{bmatrix} \times \begin{bmatrix} 1 & 2 & 3 \\ 3 & 4 & 5 \\ 5 & 6 & 7 \end{bmatrix}$$

First, multiply a_{11} by $b_{11} = 1$,

$$\begin{bmatrix} \textcircled{1} & 4 & 7 \\ 2 & 5 & 8 \\ 3 & 6 & 9 \end{bmatrix} \begin{bmatrix} \textcircled{1} & 2 & 3 \\ 3 & 4 & 5 \\ 5 & 6 & 7 \end{bmatrix}$$

Then, a_{12} by $b_{21} = 12$,

$$\begin{bmatrix} 1 & \textcircled{4} & 7 \\ 2 & 5 & 8 \\ 3 & 6 & 9 \end{bmatrix} \begin{bmatrix} 1 & 2 & 3 \\ \textcircled{3} & 4 & 5 \\ 5 & 6 & 7 \end{bmatrix}$$

Finally, a_{13} by $b_{31} = 35$,

$$\begin{bmatrix} 1 & 4 & \textcircled{7} \\ 2 & 5 & 8 \\ 3 & 6 & 9 \end{bmatrix} \begin{bmatrix} 1 & 2 & 3 \\ 3 & 4 & 5 \\ \textcircled{5} & 6 & 7 \end{bmatrix}$$

The entry c_{11} is the sum of these three values, $1 + 12 + 35 = 48$. These steps can be summarized in the diagram below. Note that each entry c_{ij} in the answer matrix results from multiplying and summing the products of elements in the ith row of matrix $[A]$ by elements in the jth column in matrix $[B]$.

To find element c_{11} To find element c_{32}

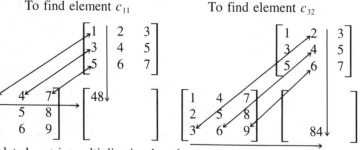

The completed matrix multiplication has the appearance

$$\begin{bmatrix} 1 & 2 & 3 \\ 3 & 4 & 5 \\ 5 & 6 & 7 \end{bmatrix}$$

$$\begin{bmatrix} 1 & 4 & 7 \\ 2 & 5 & 8 \\ 3 & 6 & 9 \end{bmatrix} \begin{bmatrix} 48 & 60 & 72 \\ 57 & 72 & 87 \\ 66 & 84 & 102 \end{bmatrix}$$

If the order of multiplication is reversed to $[B] \cdot [A] = [C]$, a different answer will, in general, be obtained:

$$\begin{bmatrix} 1 & 4 & 7 \\ 2 & 5 & 8 \\ 3 & 6 & 9 \end{bmatrix}$$

$$\begin{bmatrix} 1 & 2 & 3 \\ 3 & 4 & 5 \\ 5 & 6 & 7 \end{bmatrix} \begin{bmatrix} 14 & 32 & 50 \\ 26 & 62 & 98 \\ 38 & 92 & 146 \end{bmatrix}$$

In the operation $[A] \cdot [B] = [C]$, the matrix $[B]$ is said to be *premultiplied* by $[A]$. Similarly, the matrix $[A]$ can be said to be *postmultiplied* by $[B]$. This is simply a verbal way of specifying the order of multiplication.

If two square matrices are multiplied, the product is a square matrix of the same size. However, if an $m \times n$ matrix is multiplied by an $n \times r$ matrix, the result is an $m \times r$ matrix. That is, the product matrix has the same number of rows as the left-hand matrix and the same number of columns as the right-hand matrix. For example, multiplying a 5×3 matrix by a 3×2 matrix results in a 5×2 matrix:

$$\begin{bmatrix} 1 & 2 & 4 \\ 2 & 1 & 2 \\ 3 & 1 & 1 \\ 2 & 3 & 1 \\ 1 & 2 & 0 \end{bmatrix} \cdot \begin{bmatrix} 3 & 4 \\ 2 & 1 \\ 0 & 1 \end{bmatrix} = \begin{bmatrix} 7 & 10 \\ 8 & 11 \\ 11 & 14 \\ 12 & 12 \\ 7 & 6 \end{bmatrix}$$

However, the 3×2 matrix cannot be multiplied by the 5×3 matrix, because the number of columns in the left matrix (two) does not equal the number of rows (five) in the right matrix.

Multiplication of an $m \times n$ matrix by an $n \times m$ matrix results in a square matrix product. Here, the order of multiplication determines the size of the product:

$$\begin{bmatrix} 1 & 2 & 3 \\ 4 & 5 & 6 \end{bmatrix} \cdot \begin{bmatrix} 9 & 6 \\ 8 & 5 \\ 7 & 4 \end{bmatrix} = \begin{bmatrix} 46 & 28 \\ 118 & 73 \end{bmatrix}$$

But reversing the order of multiplication gives

$$\begin{bmatrix} 9 & 6 \\ 8 & 5 \\ 7 & 4 \end{bmatrix} \cdot \begin{bmatrix} 1 & 2 & 3 \\ 4 & 5 & 6 \end{bmatrix} = \begin{bmatrix} 33 & 48 & 63 \\ 28 & 41 & 54 \\ 23 & 34 & 45 \end{bmatrix}$$

The equation for the general case of matrix multiplication is

$$c_{ij} = \sum_{k=1}^{n} a_{ik} b_{kj} \tag{3.1}$$

In a series of multiplications, the sequence in which the multiplications are accomplished is not mandatory, if the arrangement is not changed. That is,

$$[A] \cdot [B] \cdot [C] = ([A] \cdot [B]) \cdot [C] = [A] \cdot ([B] \cdot [C])$$

A matrix can be raised to a power, because powering is simply a series of multiplications. So,

$$[A]^2 = [A] \cdot [A]$$

and

$$[A]^3 = [A]^2 \cdot [A]$$
$$= [A] \cdot [A] \cdot [A]$$

Matrices also can be raised to a fractional power, most commonly to the one-half power. This is equivalent to finding the square root of a matrix. That is, $[A]^{1/2}$ is a matrix $[X]$ whose square is $[A]$.

$$[A]^{1/2} = [X]$$
$$[X]^2 = [A]$$

Finding fractional powers of matrices can be troublesome. Fortunately, we will only need to find the fractional powers of *diagonal matrices*. These are square matrices whose elements are all zero except those on the major diagonal, and they have special properties that make it easy to raise them to a fractional power. If we raise the diagonal matrix $[A]$ to the one-half power, the result is a diagonal matrix whose nonzero elements are equal to the square roots of the equivalent elements in $[A]$. For example, if $[A]$ is 3×3,

$$\begin{bmatrix} a_{11} & 0 & 0 \\ 0 & a_{22} & 0 \\ 0 & 0 & a_{33} \end{bmatrix}^{1/2} = \begin{bmatrix} \sqrt{a_{11}} & 0 & 0 \\ 0 & \sqrt{a_{22}} & 0 \\ 0 & 0 & \sqrt{a_{33}} \end{bmatrix}$$

The *identity matrix* is a special diagonal matrix with a very useful property. If a matrix is multiplied by an identity matrix, the resulting product is exactly the same as the initial matrix:

$$\begin{bmatrix} 1 & 4 & 7 \\ 2 & 5 & 8 \\ 3 & 6 & 9 \end{bmatrix} \cdot \begin{bmatrix} 1 & 0 & 0 \\ 0 & 1 & 0 \\ 0 & 0 & 1 \end{bmatrix} = \begin{bmatrix} 1 & 4 & 7 \\ 2 & 5 & 8 \\ 3 & 6 & 9 \end{bmatrix}$$

Thus, the identity matrix corresponds to the 1 of ordinary multiplication. This property is especially important in operations in the following sections.

Convolution

Convolution is a vector operation that is widely used in time series analysis, and especially the processing of seismic signals. In a physical sense, convolution describes the manner in which two energy wavelets combine. Consider two vectors, $[A] = [a_0 \quad a_1 \quad a_2 \quad \ldots]$ and $[B] = [b_0 \quad b_1 \quad b_2 \quad \ldots]$. Their convolution is indicated by the operator *, as in

$$[C] = [A] * [B]$$

and leads to the vector $[C] = [c_0 \quad c_1 \quad c_2 \quad . . .]$. The terms of $[C]$ are given by

$$c_i = \sum_{j=0}^{i} a_j b_{i-j} \tag{3.2}$$

For example, suppose we wish to convolve the two vectors $[A]$ and $[B]$, when both contain two terms. That is, $[A] = [a_0 \quad a_1]$ and $[B] = [b_0 \quad b_1]$. The first term of the convolution is

$$c_0 = a_0 b_{0-0}$$
$$= a_0 b_0$$

The second term is

$$c_1 = a_0 b_{1-0} + a_1 b_{1-1}$$
$$= a_0 b_1 + a_1 b_0$$

The third term is

$$c_2 = a_0 b_{2-0} + a_1 b_{2-1} + a_2 b_{2-2}$$
$$= a_0 b_2 + a_1 b_1 + a_2 b_0$$

But neither a_2 nor b_2 exist; therefore, the third term is simply

$$c_2 = a_1 b_1$$

So, the convolution of $[A]$ and $[B]$ in this instance is

$$[A] * [B] = [C] = [a_0 b_0 \quad a_0 b_1 + a_1 b_0 \quad a_1 b_1]$$

Note that although both $[A]$ and $[B]$ contain two elements, the convolution $[C]$ contains three. In general, if the first vector being convolved contains n elements and the second vector contains m, the convolution will contain $n + m - 1$ elements.

Robinson and Treitel (1980) present an interesting graphic explanation of convolution. Using their method, we will convolve two vectors, each containing three elements. We first construct a table or matrix in which one vector defines the rows and the other vector the columns. The entries in the matrix are the row-column products.

$$
\begin{array}{c}
 & \begin{array}{ccc} b_0 & b_1 & b_2 \end{array} \\
\begin{array}{c} a_0 \\ a_1 \\ a_2 \end{array} &
\left[\begin{array}{ccc}
a_0 b_0 & a_0 b_1 & a_0 b_2 \\
a_1 b_0 & a_1 b_1 & a_1 b_2 \\
a_2 b_0 & a_2 b_1 & a_2 b_2
\end{array} \right]
\end{array}
$$

The matrix is now divided into diagonal strips, from lower left to upper right, such that each element in a strip is one column to the right and one row above the preceding element. That is,

$$\begin{bmatrix} a_0b_0 & a_0b_1 & a_0b_2 \\ a_1b_0 & a_1b_1 & a_1b_2 \\ a_2b_0 & a_2b_1 & a_2b_2 \end{bmatrix}$$

The terms in $[C]$ consist of the sums of the successive strips, beginning in the upper left corner. For this example,

$$[C] = [a_0b_0 \quad a_1b_0 + a_0b_1 \quad a_2b_0 + a_1b_1 + a_0b_2 \quad a_2b_1 + a_1b_2 \quad a_2b_2]$$

These are exactly the terms that will be found using the equation. Note that, as expected, the number of terms in $[C]$ is $3 + 3 - 1 = 5$.

We will now work a numerical example, demonstrating at the same time that one vector may be convolved with another of different size. Our two vectors are

$$[A] = [1 \quad 3 \quad 5 \quad 7 \quad 2]$$
$$[B] = [6 \quad 2 \quad 4]$$

The convolution $[C]$ will have seven terms. Using the graphic method, these can be found as

$$\begin{array}{cc} & \begin{array}{ccc} 6 & 2 & 4 \end{array} \\ \begin{array}{c} 1 \\ 3 \\ 5 \\ 7 \\ 2 \end{array} & \begin{bmatrix} 6 & 2 & 4 \\ 18 & 6 & 12 \\ 30 & 10 & 20 \\ 42 & 14 & 28 \\ 12 & 4 & 8 \end{bmatrix} \end{array}$$

So,

$$c_0 = 6$$
$$c_1 = 18 + 2 \qquad = 20$$
$$c_2 = 30 + 6 + 4 = 40$$
$$c_3 = 42 + 10 + 12 = 64$$
$$c_4 = 12 + 14 + 20 = 46$$
$$c_5 = 4 + 28 \qquad = 32$$
$$c_6 = 8$$

or

$$[C] = [6 \quad 20 \quad 40 \quad 64 \quad 46 \quad 32 \quad 8]$$

Convolution is not widely used in most statistical operations, except in time series analysis. We will encounter the procedure again when we discuss filtering in Chapter 4.

Matrix Inversion and Simultaneous Equations

Division of one matrix by another, in the sense of ordinary algebraic division, cannot be performed. However, by utilizing the rules of matrix multiplication, an operation can be performed that is equivalent to solving the equation

$$[A] \cdot [X] = [B]$$

for the unknown matrix $[X]$. This is one of the most important techniques in matrix algebra, and it is essential for the solution of simultaneous equations such as those of trend-surface analysis and discriminant functions. The techniques of *matrix inversion* will be encountered again and again in the next chapters of this book.

The equation given above is solved by finding the inverse of matrix $[A]$. The *inverse matrix* (or reciprocal matrix) $[A]^{-1}$ is one that satisfies the relationship $[A] \cdot [A]^{-1} = [I]$. Because multiplication by $[I]$ has no effect on a matrix, both sides of an equation can be multiplied by $[A]^{-1}$, effectively removing the quantity $[A]$ from the left side. At the same time, $[B]$ is converted into a quantity that is the value of the unknown matrix $[X]$. The matrix $[A]$ must be a square matrix.

Beginning with

$$[A] \cdot [X] = [B]$$

multiply both sides by the inverse of $[A]$, or $[A]^{-1}$:

$$[A]^{-1} \cdot [A] \cdot [X] = [A]^{-1} \cdot [B]$$

Since

$$[A]^{-1} \cdot [A] = [I]$$

and

$$[I] \cdot [X] = [X]$$

the equation reduces to

$$[X] = [A]^{-1} \cdot [B] \tag{3.3}$$

Thus, the problem of division by a matrix reduces to one of finding a matrix that satisfies the reciprocal relationship. In some situations, an inverse cannot be found, because division by zero is encountered during the inversion process. A matrix with no inverse is called a *singular matrix*, and presents problems beyond the scope of this chapter.

The inversion procedure may be illustrated by solving the following pair of simultaneous equations in matrix form. The unknown coefficients are $x_1 = 2$ and $x_2 = 3$. We will attempt to recover them by a process of matrix inversion and multiplication:

$$4x_1 + 10x_2 = 38$$
$$10x_1 + 30x_2 = 110$$

This is a set of equations of the general type

$$[A] \cdot [X] = [B]$$

where $[A]$ is a matrix of coefficients, $[X]$ is a column vector of unknowns, and $[B]$ is a column vector of right-hand sides of the equations. In the specific set of equations given above, we have

$$\begin{bmatrix} 4 & 10 \\ 10 & 30 \end{bmatrix} \cdot \begin{bmatrix} x_1 \\ x_2 \end{bmatrix} = \begin{bmatrix} 38 \\ 110 \end{bmatrix}$$

To solve the equation, the matrix $[A]$ will be inverted and $[B]$ will be multiplied by $[A]^{-1}$ to give the solution for $[X]$.

It may not be apparent why the set of simultaneous equations can be set into the matrix form shown. You can satisfy yourself on this point, however, by multiplying the two terms $[A] \cdot [X]$ to obtain the left-hand side of the simultaneous equation set:

$$\begin{bmatrix} 4 & 10 \\ 10 & 30 \end{bmatrix} \cdot \begin{bmatrix} x_1 \\ x_2 \end{bmatrix} = \begin{bmatrix} 4x_1 + 10x_2 \\ 10x_1 + 30x_2 \end{bmatrix}$$

Working through this multiplication, you will see that all of the terms are associated with the proper coefficients. By the rules of matrix multiplication,

$$\begin{bmatrix} 4 & 10 \\ 10 & 30 \end{bmatrix} \begin{bmatrix} x_1 \\ x_2 \end{bmatrix} = \begin{bmatrix} 4x_1 + 10x_2 \end{bmatrix}$$

Then, multiplying the bottom row,

$$\begin{bmatrix} 4 & 10 \\ 10 & 30 \end{bmatrix} \begin{bmatrix} x_1 \\ x_2 \end{bmatrix} = \begin{bmatrix} 10x_1 + 30x_2 \end{bmatrix}$$

We will solve the simultaneous equation set by first inverting the term $[A]$. Place the $[A]$ matrix beside an identity matrix $[I]$ and perform all operations simultaneously on both matrices. The purpose of each operation is to convert the diagonal elements of $[A]$ to ones and the off-diagonal elements to zeros. This is done by dividing rows of the matrix by constants and subtracting (or adding) rows of the matrix from other rows.

1. $\begin{bmatrix} 4 & 10 \\ 10 & 30 \end{bmatrix}$ $\begin{bmatrix} 1 & 0 \\ 0 & 1 \end{bmatrix}$ The matrix $[A]$ is placed beside an identity matrix $[I]$

2. $\begin{bmatrix} 1 & 2.5 \\ 10 & 30 \end{bmatrix}$ $\begin{bmatrix} 0.25 & 0 \\ 0 & 1 \end{bmatrix}$ Row one is divided by 4, the first element in the row, to produce 1 at $[a_{11}]$

3. $\begin{bmatrix} 1 & 2.5 \\ 0 & 5 \end{bmatrix}$ $\begin{bmatrix} 0.25 & 0 \\ -2.5 & 1 \end{bmatrix}$ 10 times row one is subtracted from row two, to reduce $[a_{21}]$ to zero

4. $\begin{bmatrix} 1 & 2.5 \\ 0 & 1 \end{bmatrix}$ $\begin{bmatrix} 0.25 & 0 \\ -0.5 & 0.2 \end{bmatrix}$ Row two is divided by 5, to give 1 at $[a_{22}]$

$4x_1 + 10x_2 = 38$

X_1

5. $\begin{bmatrix} 1 & 0 \\ 0 & 1 \end{bmatrix}$ $\begin{bmatrix} 1.5 & -0.5 \\ -0.5 & 0.2 \end{bmatrix}$ 2.5 times row two is subtracted from row one, to reduce the final off-diagonal element to zero

The matrix is now inverted. Work may be checked by multiplying the original matrix [A] by the inverted matrix [A]$^{-1}$, which should yield the identity matrix.

$$\begin{bmatrix} 1.5 & -0.5 \\ -0.5 & 0.2 \end{bmatrix} \cdot \begin{bmatrix} 4 & 10 \\ 10 & 30 \end{bmatrix} = \begin{bmatrix} 1 & 0 \\ 0 & 1 \end{bmatrix}$$

Because [A]$^{-1}$ × [A] = [I], the following identities hold:

$$[A]^{-1} \times [A] \times [X] = [A]^{-1} \times [B]$$
$$[I] \times [X] = [A]^{-1} \times [B]$$
$$[X] = [A]^{-1} \times [B]$$

By multiplying the inverted matrix [A]$^{-1}$ by the matrix [B], the unknown matrix, [X], is solved.

$$[A]^{-1} \quad \times \quad [B] \quad = \quad [X]$$
$$\begin{bmatrix} 1.5 & -0.5 \\ -0.5 & 0.2 \end{bmatrix} \cdot \begin{bmatrix} 38 \\ 110 \end{bmatrix} = \begin{bmatrix} 2 \\ 3 \end{bmatrix}$$

The column vector contains the unknown coefficients we have found to be equal to $x_1 = 2$ and $x_2 = 3$. You will recall that it was stated that these were the coefficients originally in the equation set, so we have recovered the proper values.

As an additional example of the solution of simultaneous equations by matrix inversion, we can set the equations below into matrix form and solve for x_1 and x_2 by inversion:

$$2x_1 + x_2 = 4$$
$$3x_1 + 4x_2 = 1$$

The steps in the inversion process can briefly be written out:

1. $\begin{bmatrix} 2 & 1 \\ 3 & 4 \end{bmatrix} \cdot \begin{bmatrix} x_1 \\ x_2 \end{bmatrix} = \begin{bmatrix} 4 \\ 1 \end{bmatrix}$

2. $\begin{bmatrix} 2 & 1 \\ 3 & 4 \end{bmatrix}^{-1} = \begin{bmatrix} 4/5 & -1/5 \\ -3/5 & 2/5 \end{bmatrix}$

3. $\begin{bmatrix} 4/5 & -1/5 \\ -3/5 & 2/5 \end{bmatrix} \cdot \begin{bmatrix} 4 \\ 1 \end{bmatrix} = \begin{bmatrix} 3 \\ -2 \end{bmatrix}$

Therefore, the unknown coefficients are $x_1 = 3$ and $x_2 = -2$.

It may be noted that the procedure just described is almost exactly the same as the classical algebraic method of solving two simultaneous equations. In fact, the solution of simultaneous equations is probably the most important application of matrix inversion. The advantage of matrix manipulation over the "try it and see" approach of ordinary algebra is that it is more amenable to computer programming. Almost all of the techniques described in subsequent chapters of this book involve the solution of sets of simultaneous equations. These can be expressed conveniently in the form of matrix equations and solved in the manner just described.

Matrix inversion can, of course, be applied to square matrices of any size, and not just the 2×2 examples we have investigated so far. Demonstrate this to yourself by inverting the 3×3 matrix below.

$$\begin{bmatrix} 1 & 2 & 3 \\ 2 & 6 & 5 \\ 3 & 5 & 6 \end{bmatrix}$$

If we need the inverse of a diagonal matrix, the problem is much simpler. The inverse is simply a diagonal matrix whose nonzero elements are the reciprocals of the corresponding elements of the matrix to be inverted. Considering the 3×3 matrix $[A]$,

$$\begin{bmatrix} a_{11} & 0 & 0 \\ 0 & a_{22} & 0 \\ 0 & 0 & a_{33} \end{bmatrix}^{-1} = \begin{bmatrix} \dfrac{1}{a_{11}} & 0 & 0 \\ 0 & \dfrac{1}{a_{22}} & 0 \\ 0 & 0 & \dfrac{1}{a_{33}} \end{bmatrix}$$

Certain combinations of otherwise complicated operations become very simple when the matrices involved are diagonal matrices. For example, consider the multiplication

$$[A]^{-1} \cdot [A]^{1/2} = [A]^{-1/2}$$

If $[A]$ is 3×3, the product is

$$\begin{bmatrix} a_{11} & 0 & 0 \\ 0 & a_{22} & 0 \\ 0 & 0 & a_{33} \end{bmatrix}^{-1/2} = \begin{bmatrix} \dfrac{1}{\sqrt{a_{11}}} & 0 & 0 \\ 0 & \dfrac{1}{\sqrt{a_{22}}} & 0 \\ 0 & 0 & \dfrac{1}{\sqrt{a_{33}}} \end{bmatrix}$$

In some applications, the inverse may not be required, but only the solutions to a set of simultaneous equations. In the hand-worked example, we wanted the values of the matrix $[X]$ in the equation

$$\begin{bmatrix} 4 & 10 \\ 10 & 30 \end{bmatrix} \cdot \begin{bmatrix} x_1 \\ x_2 \end{bmatrix} = \begin{bmatrix} 38 \\ 110 \end{bmatrix}$$

To find this, we inverted $[A]$, and then multiplied $[A]^{-1}$ by $[B]$ to give $[X]$. We could have instead found $[X]$ directly by operating on $[B]$ as $[A]$ was transformed into an identity matrix. To do this, we utilize what is called an *augmented matrix*, having one more column than row. The column vector [B] then occupies the $(n + 1)$ column of the matrix, and the remaining $(n \times n)$ part is inverted. Repeating the same problem:

1. $\begin{bmatrix} 4.0 & 10.0 & | & 38.0 \\ 10.0 & 30.0 & | & 110.0 \end{bmatrix}$ Matrices [A] and [B] are combined in an $n \times (n + 1)$ matrix

2. $\begin{bmatrix} 1.0 & 2.5 & | & 9.5 \\ 1.0 & 3.0 & | & 11.0 \end{bmatrix}$ Row one is divided by 4
Row two is divided by 10

3. $\begin{bmatrix} 1.0 & 2.5 & | & 9.5 \\ 0.0 & 0.5 & | & 1.5 \end{bmatrix}$ Row one is subtracted from row two

4. $\begin{bmatrix} 1.0 & 0.0 & | & 2.0 \\ 0.0 & 0.5 & | & 1.5 \end{bmatrix}$ 5 times row two is subtracted from row one

5. $\begin{bmatrix} 1.0 & 0.0 & | & 2.0 \\ 0.0 & 1.0 & | & 3.0 \end{bmatrix}$ Row two is divided by 0.5

So, the $(n + 1)$ column of the matrix contains the solution to the simultaneous equation set, and our original matrix has been replaced by an identity matrix.

Few mathematical procedures have received the attention given to matrix inversion. Dozens of methods have been devised to solve series of simultaneous equations, and hundreds of programmed versions exist. Some are especially tailored to deal with special types of matrices, such as those containing many zero elements (such matrices are called "sparse") or possessing certain types of symmetry. Most computing centers will have many matrix manipulation routines in their program libraries. The possible advantages of these library routines should be investigated before embarking on a project that requires extensive manipulation of large arrays.

Transposition

If we exchange rows and columns in a matrix, the operation is called *transposition*. For example, the matrix

$$\begin{bmatrix} 1 & 2 & 3 \\ 4 & 5 & 6 \end{bmatrix}$$

has a transpose

$$\begin{bmatrix} 1 & 4 \\ 2 & 5 \\ 3 & 6 \end{bmatrix}$$

Note that the first row has become the first column of the transpose, and the second row has become the second column. If we designate a matrix as [A], its transpose is denoted by a prime, as [A]′ or by a superscript T, as in $[A]^T$. An element a_{ij} becomes element a_{ji} in the transpose. In a set of problems we will consider in Chapter 6, a row vector [A] becomes a column vector [A]′ when transposed, and vice versa. The row and column vectors

$$[1 \quad 2 \quad 3 \quad 4] \qquad \begin{bmatrix} 1 \\ 2 \\ 3 \\ 4 \end{bmatrix}$$

are the transpose of each other.

We will later encounter numerous situations where a matrix is multiplied by its transpose. If a matrix $[X]$ has n rows and m columns, its transpose $[X]'$ will have m rows and n columns. Premultiplying $[X]$ by its transpose will result in an $m \times m$ square symmetric matrix called the *minor product matrix* of $[X]$. Postmultiplying $[X]$ by its transpose yields an $n \times n$ square symmetric matrix called the *major product matrix* of $[X]$.

Determinants

Before discussing our final topic, which is eigenvalues and eigenvectors and how they are obtained, we must examine an additional property of a matrix called the *determinant*. A determinant is a single number extracted from a square matrix by a series of operations, and is symbolically represented by det A, $|A|$, or by

$$\begin{vmatrix} a_{11} & a_{12} \\ a_{21} & a_{22} \end{vmatrix}$$

It is defined as the sum of $n!$ terms of the form

$$(-1)^k a_{1i_1} a_{2i_2} \cdots a_{ni_n} \tag{3.4}$$

where n is the number of rows (or columns) in the matrix, the subscripts i_1, i_2, ..., i_n are equal to 1, 2, ..., n, taken in any order, and k is the number of exchanges of two elements necessary to place the i subscripts in the order 1, 2, ..., n. Each term contains one element from each row and each column. The process of obtaining a determinant from a matrix is called *evaluating the determinant*.

We begin the process of evaluating the determinant by selecting one element from each row of the matrix to form a term or combination of elements. The elements in a term are selected in order from row 1, 2, 3, ..., n, but each combination can contain only one element from each column. For example, we might select the combination $a_{12}a_{21}a_{33}$ from a 3×3 matrix. Note that the method of selection places the elements in proper order according to their first, or row, subscript. The term contains one and only one element from each row and each column. We must find all possible combinations of elements that can be formed in this way. If a matrix is $n \times n$, there will be $n!$ combinations which contain one element from each row and column, and whose first subscripts are in the order 1, 2, 3, ..., n.

Now the order of multiplication of a series of numbers makes no difference in the product, that is, $a_{11}a_{22}a_{33} = a_{22}a_{11}a_{33} = a_{33}a_{22}a_{11}$, and so on, so we can rearrange our combinations without changing the result. We wish to rearrange each combination until the *second*, or column, subscript of each element is in proper numerical order. The rearranging may be performed by swapping any *two* adjacent elements. As the operation is performed, we must keep track of the number of exchanges or transpositions necessary to get the second subscript in the correct order. If an *even* number of transpositions are required (i.e., 0, 2, 4, 6, etc.), the product is given a positive sign. If an *odd* number of transpositions are necessary (1, 3, 5, 7, etc.), the product is negative.

In a 2×2 matrix

$$\begin{bmatrix} a_{11} & a_{12} \\ a_{21} & a_{22} \end{bmatrix}$$

we can find two combinations of elements that contain one and only one element from each row and each column. These are

$$a_{11}a_{22}$$

and

$$a_{12}a_{21}$$

The second subscripts in $a_{11}a_{22}$ are in correct numerical order, and no rearranging is necessary. The number of transpositions is zero, so the sign of the product is positive. However, $a_{12}a_{21}$ must be rearranged to $a_{21}a_{12}$ before the second subscripts are in numerical order. This requires one transposition, so the product is negative. The determinant of a 2×2 matrix is therefore

$$\begin{vmatrix} a_{11} & a_{12} \\ a_{21} & a_{22} \end{vmatrix} = +a_{11}a_{22} - a_{12}a_{21}$$

For a real example, we will consider the matrix

$$\begin{bmatrix} 2 & 1 \\ 4 & 3 \end{bmatrix}$$

The determinant is

$$\begin{vmatrix} 2 & 1 \\ 4 & 3 \end{vmatrix} = +(2 \cdot 3) - (1 \cdot 4) = 2$$

Next, let us consider a more complex example, a 3×3 determinant:

$$\begin{vmatrix} a_{11} & a_{12} & a_{13} \\ a_{21} & a_{22} & a_{23} \\ a_{31} & a_{32} & a_{33} \end{vmatrix}$$

There are 3!, or $3 \cdot 2 \cdot 1 = 6$ combinations of the elements in a 3×3 matrix that contain one element from each row and column, and whose first subscripts are in the order of 1, 2, 3. Start with the top row and pick an entry from each row. Be sure to choose in the order first row, second row, third row, . . . , nth row, with no more than one entry from each column. All possible combinations that satisfy these conditions in a 3×3 matrix are

$$a_{11}a_{22}a_{33} \quad a_{11}a_{23}a_{32}$$
$$a_{12}a_{23}a_{31} \quad a_{12}a_{21}a_{33}$$
$$a_{13}a_{21}a_{32} \quad a_{13}a_{22}a_{31}$$

To determine the signs of each of these terms, we must see how many transpositions are necessary to get the *second* subscripts in the order 1, 2, 3. For $a_{11}a_{22}a_{33}$, no transpositions are necessary, so $k = 0$ and the term is positive. Transpositions for the others and the resulting signs are given below.

$a_{11}a_{23}a_{32} = a_{11}a_{32}a_{23}$	so $k = 1$	sign = −
$a_{12}a_{23}a_{31} = a_{12}a_{31}a_{23} = a_{31}a_{12}a_{23}$	so $k = 2$	sign = +
$a_{12}a_{21}a_{33} = a_{21}a_{12}a_{33}$	so $k = 1$	sign = −
$a_{13}a_{21}a_{32} = a_{21}a_{13}a_{32} = a_{21}a_{32}a_{13}$	so $k = 2$	sign = +
$a_{13}a_{22}a_{31} = a_{13}a_{31}a_{22} = a_{31}a_{13}a_{22} = a_{31}a_{22}a_{13}$	so $k = 3$	sign = −

Thus, there are three negative and three positive terms in the determinant. Summing according to the signs just found yields a single number, which is $+\ a_{11}a_{22}a_{33} - a_{11}a_{23}a_{32} + a_{12}a_{23}a_{31} - a_{12}a_{21}a_{33} + a_{13}a_{21}a_{32} - a_{13}a_{22}a_{31}$.

We can now try a matrix of real values:

$$\begin{vmatrix} 4 & 3 & 2 \\ 2 & 4 & 1 \\ 1 & 0 & 3 \end{vmatrix}$$

The six terms possible are

$$(4 \cdot 4 \cdot 3) = 48$$
$$(4 \cdot 1 \cdot 0) = 0$$
$$(3 \cdot 1 \cdot 1) = 3$$
$$(3 \cdot 2 \cdot 3) = 18$$
$$(2 \cdot 2 \cdot 0) = 0$$
$$(2 \cdot 4 \cdot 1) = 8$$

The first, third, and fifth of these require an even number of transpositions for proper arrangement of the second subscript and so are positive. The others require an odd number of transpositions and are therefore negative. Summing, we have

$$\det A = 48 - 0 + 3 - 18 + 0 - 8 = 25$$

This method of evaluating a determinant is described by Pettofrezzo (1978). A more conventional approach (Gere and Weaver, 1982) uses what is called the *method of cofactors*, but the two can be shown to be equivalent.

We now have at our command a system for reducing a square matrix into its determinant, but no clear grasp of what is a determinant "really is." Determinants arise in many ways, but they appear most conspicuously during the solution of sets of simultaneous equations. You may not have noticed them there, however, because they have been hidden in the inversion process we have been using.

Consider the two sets of equations:

$$a_{11}x_1 + a_{12}x_2 = b_1 \qquad (3.5a)$$
$$a_{21}x_1 + a_{22}x_2 = b_2$$

Expressed in matrix form, this becomes

$$\begin{bmatrix} a_{11} & a_{12} \\ a_{21} & a_{22} \end{bmatrix} \cdot \begin{bmatrix} x_1 \\ x_2 \end{bmatrix} = \begin{bmatrix} b_1 \\ b_2 \end{bmatrix} \qquad (3.5b)$$

and we have discussed how the vector of unknown x's can be solved by matrix inversion. However, with algebraic rearrangement, the unknowns also can be found by the equations

$$x_1 = \frac{b_1 a_{22} - a_{12} b_2}{a_{11} a_{22} - a_{12} a_{21}} \qquad (3.6)$$

and

$$x_2 = \frac{a_{11} b_2 - b_1 a_{21}}{a_{11} a_{22} - a_{12} a_{21}} \qquad (3.7)$$

You will note that the denominators are the same for both unknowns. They also are the determinants of the matrix $[A]$. That is,

$$|A| = \begin{vmatrix} a_{11} & a_{12} \\ a_{21} & a_{22} \end{vmatrix} = a_{11} a_{22} - a_{12} a_{21} \qquad (3.8)$$

Furthermore, the numerators can be expressed as determinants. For the equation of x_1, the numerator is the determinant of the matrix

$$|B \cdot A_{i2}| = \begin{vmatrix} b_1 & a_{12} \\ b_2 & a_{22} \end{vmatrix} = b_1 a_{22} - b_2 a_{12} \qquad (3.9)$$

and for x_2, it is the determinant of

$$|A_{1i} \cdot B| = \begin{vmatrix} a_{11} & b_1 \\ a_{21} & b_2 \end{vmatrix} = a_{11} b_2 - a_{21} b_1 \qquad (3.10)$$

The circumstance can be generalized to any set of simultaneous equations, and provides one common method for their solution. This procedure for solving equations is expressed as *Cramer's rule*. The rule states that the solution for any unknown x_i in a set of simultaneous equations is equal to the ratio of the two determinants. The denominator is the determinant of the coefficients (in our example, the a's). The numerator is the same determinant except that the ith column is replaced by the vector of right-hand terms (the vector of b's). Let us check the rule with an example used before:

$$\begin{bmatrix} 4 & 10 \\ 10 & 30 \end{bmatrix} \cdot \begin{bmatrix} x_1 \\ x_2 \end{bmatrix} = \begin{bmatrix} 38 \\ 110 \end{bmatrix}$$

The denominators of the ratios for both unknown coefficients are the same:

$$\begin{vmatrix} 4 & 10 \\ 10 & 30 \end{vmatrix} = (4 \cdot 30) - (10 \cdot 10) = 20$$

The numerator of x_1 is the determinant

$$\begin{vmatrix} 38 & 10 \\ 110 & 30 \end{vmatrix} = (38 \cdot 30) - (110 \cdot 10) = 40$$

so $x_1 = 40/20 = 2$. For x_2, the numerator is the determinant

$$\begin{vmatrix} 4 & 38 \\ 10 & 110 \end{vmatrix} = (4 \cdot 110) - (10 \cdot 38) = 60$$

so $x_2 = 60/20 = 3$. These are the same unknowns we recovered by matrix inversion.

Eigenvalues and Eigenvectors

The topic we will consider next usually is regarded as the most difficult topic in matrix algebra, the determination of eigenvalues and eigenvectors. The difficulty is not in their calculation, which is cumbersome but no more so than many other mathematical procedures. Rather, difficulties arise in developing a "feel" for the meaning of these quantities, especially in an intuitive sense. Unfortunately, many textbooks provide no help in this regard, placing their discussions in strictly mathematical terms, which may be difficult for nonmathematicians to interpret.

A lucid discussion and geometric interpretation of eigenvectors and eigenvalues was prepared by Peter Gould for the benefit of geography students at Pennsylvania State University. The following discussion leans heavily on his prepared notes and a subsequent article (Gould, 1967). We will consider a real matrix of coordinates of points in space, and interpret the eigenvalues and associated functions as geometric properties of the arrangement of these points. This approach limits us, of course, to small matrices, but the insights gained can be extrapolated to larger systems even though hand computation becomes impractical. In this regard, it may be noted that we are entering a realm where the computational powers of even the largest computers often are inadequate to solve real problems.

Having worked through determinants, we can use them to develop eigenvalues and eigenvectors. Consider a hypothetical set of simultaneous equations expressed in the following matrix form:

$$[A] \cdot [X] = \lambda[X] \tag{3.11}$$

This equation states that the matrix of coefficients (the a_{ij}'s) times the vector of unknowns (the x_i's) is equal to some constant (λ) times the unknown vector itself. The problem is the same as in the solution of the simultaneous equation set

$$[A] \cdot [X] = [B] \tag{3.12}$$

except now

$$[B] = \lambda[X]$$

Our concern is to find values of λ that satisfy this relationship. Equation (3.11) can be rewritten into the form

$$([A] - \lambda[I]) \cdot [X] = [0] \tag{3.13}$$

where $\lambda[I]$ is nothing more than an identity matrix (of the same size as $[A]$) times the quality λ. That is,

$$\lambda[I] = \begin{bmatrix} \lambda & 0 & 0 \\ 0 & \lambda & 0 \\ 0 & 0 & \lambda \end{bmatrix} \tag{3.14}$$

for a 3×3 matrix. Written in conventional form, the equivalent of the three simultaneous equations is

$$\begin{aligned} (a_{11} - \lambda)x_1 + a_{12}x_2 + a_{13}x_3 &= 0 \\ a_{21}x_1 + (a_{22} - \lambda)x_2 + a_{23}x_3 &= 0 \\ a_{31}x_1 + a_{32}x_2 + (a_{33} - \lambda)x_3 &= 0 \end{aligned} \tag{3.15}$$

Let us assume that there are solutions to these equations other than the trivial case where all the unknown x's $= 0$. Look back at Cramer's rule for the solution of simultaneous equations, in which the unknowns are expressed as the ratio of two determinants. Because the numerator in our present example would contain a column of zeros, the determinant of the numerator also will be zero. That is, the solution for the $[X]$ vector is

$$[X] = \frac{[0]}{|A|}$$

rewriting, this becomes

$$|A| \cdot [X] = [0] \tag{3.16}$$

If the vector $[X]$ is not zero, it follows that the determinant of the matrix $[A]$ must be zero, or

$$|A - \lambda I| = \begin{vmatrix} a_{11} - \lambda & a_{12} & a_{13} \\ a_{21} & a_{22} - \lambda & a_{23} \\ a_{31} & a_{32} & a_{33} - \lambda \end{vmatrix} = 0 \tag{3.17}$$

We usually know the coefficients a_{ij} of the matrix, so this equation can be used to determine the values of λ that satisfy all of these various conditions. This is done by expanding the determinant to yield a polynomial equation. Looking first at a 2×2 determinant,

$$\begin{vmatrix} a_{11} - \lambda & a_{12} \\ a_{21} & a_{22} - \lambda \end{vmatrix} = 0$$

Expanding gives

$$(a_{11} - \lambda)(a_{22} - \lambda) - a_{21}a_{12} = 0$$

Multiplying out the first term,

$$(a_{11} - \lambda)(a_{22} - \lambda) = (a_{11}a_{22}) - (a_{11}\lambda) - (a_{22}\lambda) + \lambda^2$$

so we have

$$(a_{11}a_{22}) - (a_{21}a_{12}) - (a_{11}\lambda) - (a_{22}\lambda) + \lambda^2 = 0$$

Because we know the various values of the elements a_{ij}, we can collect all of these terms together in the form of an equation such as

$$\lambda^2 + \alpha_1\lambda + \alpha_2 = 0 \tag{3.18}$$

where the α's represent the sum of the numerical values of the appropriate a_{ij}'s. You should recognize that this is a quadratic equation of the general form

$$ax^2 + bx + c = 0$$

which can be solved for the unknown terms by factoring. The general solution to a quadratic equation is

$$x = \frac{-b \pm \sqrt{b^2 - 4ac}}{2a} \tag{3.19}$$

If this seems unfamiliar, review the sections in an elementary algebra book that deal with factoring and quadratic equations. Now, we can try the procedures just outlined to find the eigenvalues of the 2×2 matrix:

$$[A] = \begin{bmatrix} 17 & -6 \\ 45 & -16 \end{bmatrix}$$

First, we must set the matrix in the form

$$[A] - \lambda[I] = \begin{bmatrix} 17 - \lambda & -6 \\ 45 & -16 - \lambda \end{bmatrix}$$

Equating the determinant to zero,

$$\begin{vmatrix} 17 - \lambda & -6 \\ 45 & -16 - \lambda \end{vmatrix} = 0$$

We can expand the determinant

$$\begin{vmatrix} 17 - \lambda & -6 \\ 45 & -16 - \lambda \end{vmatrix} = (17 - \lambda)(-16 - \lambda) - (-6)(45) = 0$$

Multiplying out gives

$$-272 - 17\lambda + 16\lambda + \lambda^2 + 270 = 0$$

which can be collected to give

$$\lambda^2 - \lambda - 2 = 0$$

This can be factored into

$$(\lambda - 2)(\lambda + 1) = 0$$

so, the two eigenvalues associated with the matrix $[A]$ are

$$\lambda_1 = +2, \qquad \lambda_2 = -1$$

This example was deliberately chosen for ease in factoring. We can try a somewhat more difficult example by using the set of simultaneous equations we have solved earlier. This is the 2×2 matrix

$$[A] = \begin{bmatrix} 4 & 10 \\ 10 & 30 \end{bmatrix}$$

Repeating the sequence of steps yields the determinant

$$\begin{vmatrix} 4 - \lambda & 10 \\ 10 & 30 - \lambda \end{vmatrix} = 0$$

which is then expanded into

$$\begin{vmatrix} 4 - \lambda & 10 \\ 10 & 30 - \lambda \end{vmatrix} = (4 - \lambda)(30 - \lambda) - 100 = 0$$

or

$$\lambda^2 - 34\lambda + 20 = 0$$

There are no obvious factors in the quadratic equation, so we must apply the rule for a general solution:

$$x = \frac{-b \pm \sqrt{b^2 - 4ac}}{2a}$$

$$= \lambda = \frac{-(-34) \pm \sqrt{-34^2 - 4 \cdot 1 \cdot 20}}{2 \cdot 1}$$

$$= \frac{34 \pm \sqrt{1076}}{2}$$

$$\lambda_1 = 33.4, \qquad \lambda_2 = 0.6$$

We can check our work by substituting the eigenvalues back into the determinant to see if it is equal to zero, within the error introduced by round-off:

$$\begin{vmatrix} 4 - 33.4 & 10 \\ 10 & 30 - 33.4 \end{vmatrix} = (-29.4)(-3.4) - (10)(10) = -0.04$$

and

$$\begin{vmatrix} 4 - 0.6 & 10 \\ 10 & 30 - 0.6 \end{vmatrix} = (3.4)(29.4) - (10)(10) = -0.04$$

So, the eigenvalues we have found are correct within two decimal places.

Before we leave the computation of eigenvalues of 2×2 matrices, we should consider one additional complication that may arise. Suppose we want the eigenvalues of the matrix

$$[A] = \begin{bmatrix} 2 & 4 \\ -6 & 3 \end{bmatrix}$$

Expressed as a determinant equal to zero, we have

$$\begin{vmatrix} 2 - \lambda & 4 \\ -6 & 3 - \lambda \end{vmatrix} = 0$$

which expands to

$$\begin{vmatrix} 2 - \lambda & 4 \\ -6 & 3 - \lambda \end{vmatrix} = (2 - \lambda)(3 - \lambda) + 24 = 0$$

or

$$\lambda^2 - 5\lambda + 30 = 0$$

Roots of this equation are

$$\lambda_1, \lambda_2 = \frac{5 \pm \sqrt{25 - 120}}{2}$$

But this leads to equations involving the square roots of negative numbers:

$$\lambda_1 = \frac{5 + \sqrt{-95}}{2} = 2.5 + 4.9i$$

$$\lambda_2 = \frac{5 - \sqrt{-95}}{2} = 2.5 - 4.9i$$

These are complex numbers, containing both real parts and imaginary parts which include the imaginary number $i = \sqrt{-1}$. Fortunately, a symmetric matrix always yields real eigenvalues, and most of our computations involving eigenvalues and eigenvectors will utilize covariance, correlation, or similarity matrices which are always symmetrical.

Next, we will consider the eigenvalues of the third-order matrix

$$\begin{bmatrix} 20 & -4 & 8 \\ -40 & 8 & -20 \\ -60 & 12 & -26 \end{bmatrix}$$

The determinant of the matrix is then set to zero, giving

$$\begin{vmatrix} 20 - \lambda & -4 & 8 \\ -40 & 8 - \lambda & -20 \\ -60 & 12 & -26 - \lambda \end{vmatrix} = 0$$

Expanding out the determinant and combining terms yields

$$-\lambda^3 + 2\lambda^2 + 8\lambda = 0$$

This is a cubic equation having three roots that must be found. In this instance, the polynomial can be factored into

$$(\lambda - 4)(\lambda - 0)(\lambda + 2) = 0$$

and the roots are directly obtainable:

$$\lambda_1 = +4, \qquad \lambda_2 = 0, \qquad \lambda_3 = -2$$

Although the techniques we have been using are extendible to any size matrix, finding the roots of large polynomial equations can be an arduous task. Usually, eigenvalues are not found by solution of a polynomial equation, but rather by matrix manipulation methods that involve refinement of a successive series of approximations to the eigenvalues. These methods are practical only because of the great computational speed of digital computers. Utilizing this speed, a researcher can compress literally a lifetime of trial solutions and refinements into a few minutes.

Now that we have an idea of the manipulative operations that produce eigenvalues, we may try to get some insight into their nature. A matrix can be regarded as a representation of the coordinates of points in n-dimensional space. By restricting our consideration to 2×2 matrices, we can confine this space to a flat plane, such as the surface of this page. Thus, matrix operations can be represented geometrically. The matrix

$$[A] = \begin{bmatrix} 4 & 8 \\ 8 & 4 \end{bmatrix}$$

can be regarded as two points with the coordinates 4,8 and 8,4. These define two vectors from the origin of the coordinate system, as shown in Figure 3.1. We could use the columns as coordinates rather than rows with no change in our analysis. However, for consistency we will work with rows.

We can think of the two points as lying on the boundary of an ellipse whose center is the origin of the coordinate system. The ellipse is, in effect, an envelope which just encloses the points. The eigenvalues represent the magnitudes, or lengths,

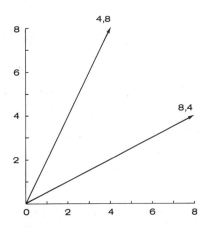

FIGURE 3.1 Two vectors defined by entries in a 2×2 matrix.

of the major and minor axes. In this example, the eigenvalues of our matrix are

$$\lambda_1 = 12, \qquad \lambda_2 = -4$$

Gould refers to the relative lengths of the axes as a measure of the "stretchability" of the enclosing ellipse. The relationships are graphically shown in Figure 3.2. The first eigenvalue represents the major axis whose length from center to edge is 12 units. The second eigenvalue represents the minor axis, in a negative part of the coordinate system, whose length from center to edge is -4 units.

If the two points were closer together, the ratio of the two axes would change. For example, if the coordinates of the points form the matrix

$$[A] = \begin{bmatrix} 6 & 8 \\ 8 & 6 \end{bmatrix}$$

the eigenvalues are

$$\lambda_1 = 14, \qquad \lambda_2 = -2$$

These relations are shown graphically in Figure 3.3. The major axis is highly elongated compared to the minor axis. If the two points coincide so the rows of the matrix are identical, the second eigenvalue will become zero and the enclosing ellipse will collapse to a straight line. This can be illustrated with the matrix

$$[A] = \begin{bmatrix} 4 & 8 \\ 4 & 8 \end{bmatrix}$$

which can be reduced to the equation

$$\lambda^2 - 12\lambda + 0 = 0$$

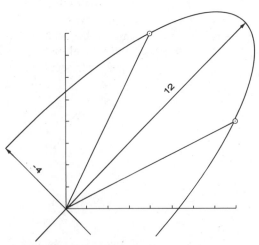

FIGURE 3.2 Ellipse defined by entries in a 2 × 2 matrix. Eigenvectors of matrix correspond to principal axes of ellipse.

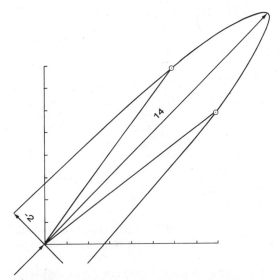

FIGURE 3.3 Elongated ellipse representing a matrix whose entries define points closer together than in Figure 3.2. Eigenvectors correspond to principal axes.

giving the eigenvalues:

$$\lambda_1 = 12, \qquad \lambda_2 = 0$$

This is graphically shown in Figure 3.4. The two points coincide, the major axis passes through these points, and the perpendicular minor axis has zero length. The opposite extreme might occur if our matrix consists of two vectors perpendicular to one another and equal in magnitude.

$$[A] = \begin{bmatrix} -4 & 8 \\ 8 & 4 \end{bmatrix}$$

The polynomial equation is

$$\lambda^2 + 0\lambda - 80 = 0$$

whose roots are

$$\lambda_1 = +8.95, \qquad \lambda_2 = -8.95$$

This relationship can be graphically illustrated as in Figure 3.5. The vectors to the coordinates are radii of the inscribing ellipse, which has become a circle. The eigenvalues also are equal in length to these vectors, and also form radii.

One final note on eigenvalues: As you can verify by checking the previous examples, the sum of the eigenvalues is always equal to the sum of the diagonal elements, or the *trace*, of the original matrix. This property is useful for checking the proper operation of an eigenvalue routine and also will be useful in principal components problems.

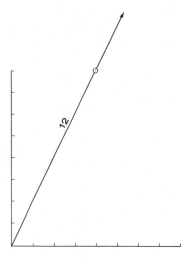

FIGURE 3.4 Matrix containing identical rows produces ellipse collapsed to a line.

You will recall that (3.13) defined eigenvalues as quantities arising in a set of simultaneous equations. Now that we have found these quantities, we can substitute back into the simultaneous equations and proceed to solve for the unknown values of x. For a 2×2 matrix, we can rewrite (3.13) in the form

$$\begin{bmatrix} a_{11} - \lambda_1 & a_{12} \\ a_{21} & a_{22} - \lambda_1 \end{bmatrix} \cdot \begin{bmatrix} x_1 \\ x_2 \end{bmatrix} = \begin{bmatrix} 0 \\ 0 \end{bmatrix} \tag{3.20}$$

for the first eigenvalue and into an equivalent equation for the second eigenvalue. After determining the numerical values of the diagonal elements, we can use the techniques of matrix inversion to produce the unknown vector of coefficients. The vector is called the *eigenvector* (or *characteristic*, *proper*, *latent*, or *principal*

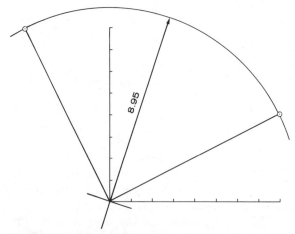

FIGURE 3.5 Matrix whose entries define vectors at right angles to one another yields an ellipse of constant radius or a circle.

vector) associated with its respective eigenvalue. There will be as many of these as there are eigenvalues, or as many as there are rows and columns in the matrix.

In summary, to extract eigenvalues and eigenvectors from an $n \times n$ matrix, we must find the determinant of the matrix, the n roots of its characteristic polynomial equation, and solve n sets of n simultaneous equations! Fortunately, our 2×2 examples are reasonably small, so we can start on the matrix:

$$[A] = \begin{bmatrix} 17 & -6 \\ 45 & -16 \end{bmatrix}$$

You will recall we found the eigenvalues of this matrix to be

$$\lambda_1 = +2, \qquad \lambda_2 = -1$$

Substituting the first eigenvalue into the matrix gives

$$\begin{bmatrix} 17-2 & -6 \\ 45 & -16-2 \end{bmatrix} = \begin{bmatrix} 15 & -6 \\ 45 & -18 \end{bmatrix}$$

This is analogous to the simultaneous equation set

$$15x_1 - 6x_2 = 0$$
$$45x_1 - 18x_2 = 0$$

or, in matrix notation,

$$\begin{bmatrix} 15 & -6 \\ 45 & -18 \end{bmatrix} \cdot \begin{bmatrix} x_1 \\ x_2 \end{bmatrix} = \begin{bmatrix} 0 \\ 0 \end{bmatrix}$$

In either form, you can see that the second equation is simply three times the first. We need, therefore, only to solve one of the equations, and the other also is satisfied. A brief inspection will show that the solutions are

$$x_1 = 2, \qquad x_2 = 5$$

This is the eigenvector associated with the first eigenvalue. Actually, there are an infinite number of solutions, because the equations are also satisfied by

$$\begin{bmatrix} x_1 \\ x_2 \end{bmatrix} = \beta \begin{bmatrix} 2 \\ 5 \end{bmatrix}$$

where β is any constant. However, we are concerned here only with the special situation where $\beta = 1$. As we will see later, we will be interested primarily in the ratios between elements of the vector, and these are not changed by multiplication by a constant.

The solution for the second eigenvector is

$$\begin{bmatrix} 17-(-1) & -6 \\ 45 & -16-(-1) \end{bmatrix} = \begin{bmatrix} 18 & -6 \\ 45 & -15 \end{bmatrix}$$

Both of these also reduce to a single equation

$$3x_1 - x_2 = 0$$

which is satisfied by the vector

$$\begin{bmatrix} x_1 \\ x_2 \end{bmatrix} = \beta \begin{bmatrix} 1 \\ 3 \end{bmatrix}$$

The methods just developed are directly extendible to $n \times n$ matrices, although the procedure becomes increasingly cumbersome with larger matrices. Before considering computer approaches to the eigenvector problem, we will work with some of the matrices used for geometric analysis of eigenvalues and see if the same approach can yield insight into the nature of eigenvectors.

First, consider the matrix

$$[A] = \begin{bmatrix} 4 & 8 \\ 8 & 4 \end{bmatrix}$$

with eigenvalues

$$\lambda_1 = 12, \qquad \lambda_2 = -4$$

Substituting into the original matrix gives

$$\begin{bmatrix} 4 - 12 & 8 \\ 8 & 4 - 12 \end{bmatrix} = \begin{bmatrix} -8 & 8 \\ 8 & -8 \end{bmatrix}$$

whose solution is the eigenvector

$$\begin{bmatrix} x_1 \\ x_2 \end{bmatrix} = \begin{bmatrix} 1 \\ 1 \end{bmatrix}$$

If we refer back to Figure 3.2, we can interpret the eigenvector as the *slope* of the major axis of our ellipse. If we regard the elements of the eigenvector as coordinates, the first eigenvector defines an axis which bisects the angle between the two points and the center of the ellipse, and whose length is equal to the first eigenvalue. With the second eigenvalue, the matrix becomes

$$\begin{bmatrix} 4 - (-4) & 8 \\ 8 & 4 - (-4) \end{bmatrix} = \begin{bmatrix} 8 & 8 \\ 8 & 8 \end{bmatrix}$$

whose solution gives the second eigenvector:

$$\begin{bmatrix} x_1 \\ x_2 \end{bmatrix} = \begin{bmatrix} -1 \\ 1 \end{bmatrix}$$

In Figure 3.2, this will plot as the vector direction $-1/1 = 135°$, perpendicular to the major axis of the ellipse. Its magnitude also is negative defining the second or minor axis. Now we will consider the second example, the matrix of the points whose coordinates are closer together:

$$[A] = \begin{bmatrix} 6 & 8 \\ 8 & 6 \end{bmatrix}$$

For the first eigenvector,

$$\begin{bmatrix} (6-14) & 8 \\ 8 & (6-14) \end{bmatrix} = \begin{bmatrix} -8 & 8 \\ 8 & -8 \end{bmatrix}$$

$$\begin{bmatrix} x_1 \\ x_2 \end{bmatrix} = \begin{bmatrix} 1 \\ 1 \end{bmatrix}$$

so the slope of the first eigenvector also is 45°, bisecting the angle between the two points and the center of the ellipse. The magnitude of this major axis is equal to 14, the first eigenvalue. As we would expect, the vector parallels that found for the first set of coordinates, but is of greater magnitude. A similar analysis shows that the eigenvector associated with the second eigenvalue is

$$\begin{bmatrix} x_1 \\ x_2 \end{bmatrix} = \begin{bmatrix} -1 \\ 1 \end{bmatrix}$$

These techniques are directly extendible to larger matrices, even though the operations become tedious. As an example, we will consider the matrix

$$[A] = \begin{bmatrix} 5 & 2 & 6 \\ 2 & 4 & 3 \\ 6 & 3 & 2 \end{bmatrix}$$

A perspective diagram of the three vectors defined by the rows of this matrix is shown in Figure 3.6. Even before we begin our analysis, we know something of the properties of the eigenvectors and eigenvalues. Because the matrix is symmetrical, the three eigenvalues will be real. The major eigenvector will pass somewhere within the solid angle 0, X_1, X_2, X_3 defined by the three original vectors. The sum of the eigenvalues will be equal to 11, the trace of the matrix. For the matrix, the following set of eigenvalues and eigenvectors can be found:

$$\lambda_1 = 11.3, \qquad \lambda_2 = -2.9, \qquad \lambda_3 = 2.6$$

	Vector 1	Vector 2	Vector 3
$X_1 =$	0.69	-0.56	-0.45
$X_2 =$	0.43	-0.19	0.88
$X_3 =$	0.58	0.81	-0.11

These are shown as vectors of the proper magnitude and orientation in Figure 3.7. The orientation is specified by the eigenvector and the length by the eigenvalue. Note that the conditions we specified are fulfilled. Extrapolation of this analysis to larger matrices is mathematically straightforward, but we leave the realm where graphic analysis can be utilized. However, you should be able to appreciate the existence of points that occur in more than three dimensions, and the reality of a set of more than three lines that are mutually perpendicular.

At this point, we may point out a unique property of eigenvectors of symmetric matrices. They always are at right angles to each other, or *orthogonal*. This is not true

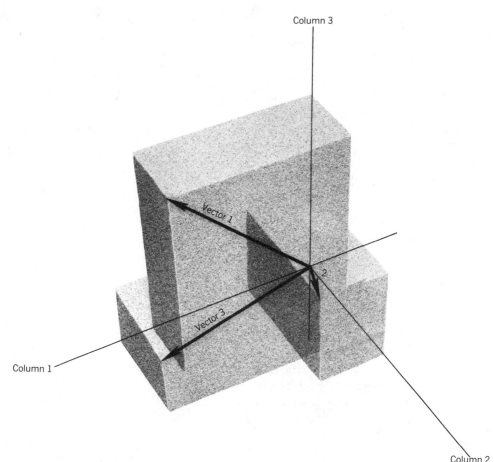

FIGURE 3.6   Vectors in three-dimensional space, defined by entries in a 3 × 3 matrix. X_1, X_2, and X_3 axes represent first, second, and third columns of matrix. Vectors are defined by elements in a specific row.

of eigenvectors of matrices in general, but only of symmetric matrices. This property will be a great help in factor analysis.

If you have persisted through this chapter and have faithfully worked examples (and thought of the possibilities and complexities involved in larger problems), you are ready to go on to the application of modern computational methods to geologic problems. We have attempted to present, in as painless a manner as possible, the rudiments of beginning matrix algebra. We stated at the conclusion of Chapter 2 that statistics was too large a subject to be covered in one chapter, or even one book. Matrix algebra also is an impossibly large subject to encompass in these few pages. However, you should now have some insight into matrix methods that will enable you to understand the computational basis of techniques we will cover in the remainder of this book.

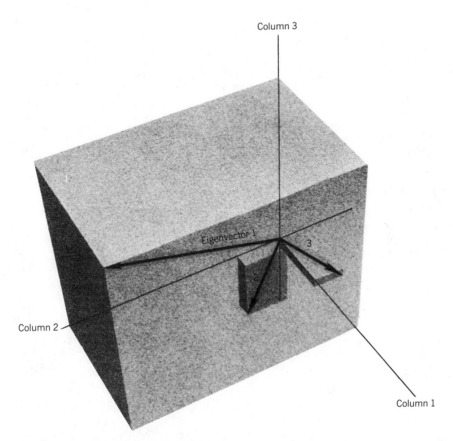

Column 3

Eigenvector 1

3

2

Column 2

Column 1

FIGURE 3.7 Plot of eigenvectors of a 3 × 3 matrix, in the same spatial arrangement as in Figure 3.6. Note that the dominant eigenvector extends through the center of the space defined by the three original vectors.

SELECTED READINGS

Davis, P. J., 1973, *The mathematics of matrices*, 2nd ed.: John Wiley & Sons, Inc., New York, 348 p.

One of the most readable texts on matrices, with a minimum of mathematical jargon and a maximum of examples and applications.

Gere, J. M., and W. Weaver, Jr., 1982, *Matrix algebra for engineers*, 2nd ed.: Brooks-Cole Publ. Co., Monterey, Ca., 175 p.

This inexpensive paperback is one of the best-written matrix algebra texts available.

Gould, P., 1967, On the geographic interpretation of eigenvalues: an initial exploration: *Trans. Inst. British Geographers*, no. 42, p. 53–86.

An intuitive look at eigenvalues and vectors by geometric analogy. Part of this chapter is derived from this excellent exposition for students.

Maron, M. J., 1982, *Numerical analysis—a practical approach:* Macmillan Publ. Co., Inc., New York, 471 p.

Gives procedures and algorithms for matrix operations, especially different methods for inversion, solution of simultaneous equations, and extraction of eigenvalues.

Pettofrezzo, A. J., 1978, *Matrices and transformations:* Dover, Inc., New York, 133 p.

This paperback reprint of a classic text covers the traditional material for a one-semester matrix algebra course. It is liberally sprinkled with worked examples and problems.

Reiner, I., 1971, *Introduction to matrix theory and linear algebra:* Holt, Rinehart, and Winston, Inc., New York, 154 p.

An introductory paperback text with a how-to-do-it emphasis.

Robinson, E. A., and S. Treitel, 1980, *Geophysical signal analysis:* Prentice-Hall, Inc., Englewood Cliffs, N.J., 466 p.

Chapter 3 contains a short but clear discussion of convolution.

CHAPTER 4

Analysis of Sequences of Data

Geologic Measurements in Sequences

In this chapter, we will consider ways of examining data that are characterized by their position along a single line. That is, they form a sequence, and the position at which a data point occurs within the sequence is important. Data sets of this type are common in geology, and include measured successions of lithologies, geochemical or mineralogical assays along traverses or drill-holes, electric logs of oil wells, and chart recordings from instruments. Also in this general category are measurements separated by the flow of time, such as a sequence of water quality determinations at a river station, or the production history of a flowing gas well. Techniques for examining data having a single positional characteristic traditionally are considered part of the field of *time series analysis,* although we will take the broader view that time and space relationships can be considered interchangeably.

Before proceeding to some geological examples and appropriate methods of examination, we must consider the nature of different types of sequences apt to be encountered by geologists. At one extreme, we may have a record which is quite precise, both in the variable which is measured and in the scale along which successive observations are located. Examples might include an electrical resistivity log from a borehole, or the production history of a commercial well. In the former, the variable is a measured attribute expressed in ohms (Ω) and the scale is measured in feet. In the latter example, the variable again is a measured attribute, barrels (bbl) of oil, and the scale is measured in days, months, or years. There are two important characteristics in either record. First, the variable being measured is expressed in units of an interval or ratio scale; 1000 bbl of oil is twice as large a quantity as 500 bbl, and a measurement of 10 Ω is ten times the resistance of 1 Ω. Second, the scales along which the data points are located also are expressed in units having magnitude. A depth of 3000 ft in a well is ten times a depth of 300 ft, and the decade between the years 1940 and 1950 has the same duration as the interval between 1950 and 1960. These may seem obvious or even trivial points to emphasize, but as we shall see, not all geologic sequences have such well-behaved characteristics.

At the opposite extreme, we can consider a stratigraphic sequence consisting of the lithologic states encountered in a sedimentary succession. Such a sequence might be a cyclothem of limestone–shale–limestone–shale–sandstone–coal–shale–limestone, from bottom to top. We are interested in the significance of the succession, but we cannot put a meaningful scale on the sequence itself. Obviously, the succession of lithologies represents changes that occurred through time, but we have no way of estimating the time scale involved. We could use thickness, but this may change dramatically from location to location even though the sequence is not altered. If thickness is considered, it may obscure our examination of the succession, which is the subject of our interest. Thus, the fact that limestone is the third state in the section and coal is the sixth has no significance that can be expressed numerically (that is, position 6 is not "twice" position 3). Likewise, the lithologic states of the units cannot be expressed on a numerical scale. We might code the sequences just given as 1-2-1-2-3-4-2-1, where limestone is equated to 1, shale is 2, sandstone is 3, and coal is 4, but such a convention is purely arbitrary and expresses no meaningful relations between the states. It is obvious that this sequence poses different problems to the analyst than do the first examples.

There also are intermediate possibilities. For example, we may be interested in some measurable attribute contained in successive stages of a sequence. Perhaps we have measured the boron content of each lithologic unit in the cyclothem just discussed. We can utilize a distance scale of feet between samples and consider this a problem related to depth or distance. Alternatively, we can consider the relationship between the boron measurements and the sequence of states.

A closely related problem is the analysis of a sequence characterized by the presence or absence of some variable or variables at points along a line. We might be interested, for example, in the repeated recurrence of certain environment-dependent microfossils in the chips recovered during the drilling of a well. Another class of problems may be typified by the succession of mineral grains encountered on traverses across a thin section. In this case, we can use millimeters as a convenient spatial scale, but we have no way of evaluating whether olivine rates a higher number than plagioclase.

Data having the characteristic of being arranged along a continuum, either of time or space, often are referred to as forming a *series, sequence, string,* or *chain.* The nature of the data and the chain determine the questions that we can consider. Obviously, we cannot extract information about time intervals from stratigraphic succession data, because the time scale accompanying the succession is not known. We often substitute spatial scales for a time scale in stratigraphic problems, but our conclusions are no better than our fundamental assumptions about the length of time required to deposit the interval we have measured.

Table 4.1 is a classification of the various data-analysis techniques discussed in this chapter. We can consider three types of sequences. In the first, the distance between observations varies and must be specified for every point. Next, the points are assumed to be equally and regularly spaced; the numerical value of the spacing

TABLE 4.1　Techniques Discussed in this Chapter Classified by the Nature of the Variable and Its Spacing along a Line

Nature of Variable	Observations Irregularly Spaced	Observations Equally and Regularly Spaced	Spacing Not Considered
Variable Measured on Interval or Ratio Scales	Interpolation Polynomial regression Splines	Orthogonal polynomial regression Moving averages Filtering and smoothing Zonation Autocorrelation and cross-correlation Semivariograms Spectral analysis	Autocorrelation and cross-correlation
Variable Measured on Nominal or Ordinal Scales	Series of events K-S tests	Autoassociation and cross-association Substitutability analysis Markov chains Runs tests	Autoassociation and cross-association Substitutability analysis Markov chains Runs tests

does not enter into the analyses except as a constant. Finally, the spacing may not be considered at all, and only the sequence of the observations is important.

The techniques also may be classified on the type of observations they require. Some necessitate interval or ratio observations; the variate must be measured on a scale and expressed in real numbers. Other methods accept nominal or ordinal data, and observations need only to be categorized in some fashion. In the methods discussed in this chapter, the classes are not ranked; that is, state A is not "greater" or "larger" in some sense than states B or C. Nominal data may be represented by integers, alphabetic characters, or symbols.

In the remainder of this chapter, we are going to examine the mathematical techniques required to analyze data in sequences. The methods described here do not exhaust the possibilities by any means. Rather, these are a collection of operations that have proved valuable in quantitative problem-solving in the Earth sciences, or that seem especially promising. Other methods may be more appropriate or powerful in specific situations or for certain data sets. However, a familiarity with the techniques discussed here will provide an introduction to a diverse field of analytical tools. Unfortunately, many of these methods were developed in scientific specialities alien to most geologists, and the description of an application in radar engineering, stock market analysis, speech therapy, or cell biology may be difficult to relate to a geologic problem. Some of the methods involve nonparametric statistics, and these are not widely considered in introductory statistics courses. Because of the general unfamiliarity of most Earth scientists with developments in the numerical analysis of data sequences, we have thought it best to present a potpourri of techniques and approaches. As you can see from Table 4.1, these cover a variety of sequences of different types, and are designed to answer different kinds of questions. None of the techniques can be considered exhaustively in this short space, but from the examples and applications presented, one or another may suggest themselves to the geologist with a problem to solve. The list of Selected Readings can then provide a discussion of a specific subject in more detail.

These methods provide answers to the following broad categories of questions: Are the observations random, or do they contain evidence of a trend or pattern? If a trend exists, what is its form? Can cycles or repetitions be detected and measured? Can predictions or estimations be made from the data? Can variables be related or their effectiveness measured? Although such questions may not be explicitly posed in each of the following discussions, you should examine the nature of the methods and think about their applicability and the type of problems they may help solve. The sample problems are only suggestions from the many that could be used.

Geologists are concerned not only with the analysis of data in sequences, but also with the comparison of two or more sequences. An obvious example is stratigraphic correlation, either of measured sections or electric well logs. A geologist's motive for numerical correlation may be a simple desire for speed, as in the production of geologic cross-sections from digitized logs stored in data banks. Alternatively, he may be faced with a correlation problem where the recognition of equivalency is beyond his ability. Subtle degrees of similarity, too slight for unaided detection, may provide the clues that will allow him to make a decision where none is otherwise possible. Numerical methods allow the geologist to consider many variables simultaneously, a powerful extension of his pattern-recognition facilities. Finally, because of the absolute invariance in operation of a computer program, mathematical correlation provides a challenge to the human interpreter. If a geologist's correlation disagrees with that established by the machine, the geologist must determine the reason for the discrepancy. The forced scrutiny may reveal complexities or biases not apparent during the initial examination. This is not to say that the geologist should bend his interpretation to conform with that of the computer. Quite the contrary, because presently available programs for automatic correlation are crude and unsophisticated compared to the mental processes of a human interpreter. However, as we continue to examine the process of correlation, increasingly elegant and useful algorithms and programs will be developed, releasing the geologist for more significant aspects of research.

Most techniques for comparing two or more sequences can be grouped into two broad categories. In the first of these, the data sequences are assumed to match at one position only, and we wish to determine the degree of similarity between the two sequences. An example is the comparison of an X-ray diffraction chart with a set of standards in an attempt to identify an unknown mineral. The chart and standards can be compared only in one position, where intensities at certain angles are compared to intensities of the standards at the same angles. Nothing is gained, for example, by comparing X-ray intensity at 20° 2θ with the intensity at 30° 2θ on another chart. Although the correspondence may be high, it is meaningless.

The fact that data such as these are in the form of sequences is irrelevant, because each data point is considered to be a separate and distinct variable. The intensity of diffracted radiation at 20° 2θ is one variable, and the intensity at 30° 2θ is another. We will consider methods for the comparisons of such sequences in greater detail in Chapter 6, when we discuss multivariate measures of similarity and problems of classification and discrimination. In this class of problems, an observation's location in a sequence merely serves to identify it as a specific variable, and its location has no other significance.

In contrast, some of the techniques we will discuss in this chapter regard data sequences as samples from a continuous string of possible observations. There is no *a priori* reason why one position of comparison should be better than any other. These methods, such as cross-correlation and cross-association, are closest in their operation to the mental process of geologic correlation. Unfortunately, they are limited in their application because they cannot assume distortions in the scale of one section as compared to another. In many types of data sequences, some of which are discussed in this chapter, distortion of the scale of the sequences cannot occur and so this is not a problem. Sedimentation rates, however, are not constants, and stratigraphic records are difficult sequences to correlate by any presently available algorithms.

As we emphasized in Chapter 1, the computer is a powerful tool for the analysis of complex problems. However, it is mindless and will accept unreasonable data and return nonsense answers without a qualm. A bundle of programs for analyzing sequences of data can readily be obtained from many sources. If you utilize these as a "black box" without understanding their operation and limitations, you may be badly led astray. It is our hope in this chapter that the discussions and examples will indicate the areas of appropriate application for each method, and that the programs will be sufficiently straightforward so that their operation is clear. However, in the final analysis, the researcher must be his own guide. When confronted with a problem involving data along a sequence, you may ask yourself the following questions to aid in planning your research: (*a*) What question(s) do I want to answer? (*b*) What is the nature of my observations? (*c*) What is the nature of the sequence in which the observations occur? You may quickly discover that the answer to the first question requires that the second and third be answered in specific ways. Therefore, you avoid unnecessary work if these points are carefully thought out before your investigation begins. Otherwise, the manner in which you gather your data may predetermine the techniques that can be used for interpretation, and may seriously limit the scope of your investigation.

Interpolation Procedures

Many of the following techniques require data that are equally spaced; the observations must be taken at regular intervals on a traverse or line. Of course, this often is not possible when dealing with natural phenomena over which you have little control. Many stratigraphic measurements, for example, are recorded bed-by-bed rather than foot-by-foot. This also may be true of analytical data from drill-holes, or from samples collected on traverses across regions which are incompletely exposed. We must, therefore, estimate the variable under consideration at regularly spaced points from its values at irregular intervals. Estimation of regularly spaced points will also be considered in Chapter 5, when we discuss contouring of map data. Contouring programs operate by creating a regular grid of control points estimated from irregularly spaced observations. The appearance and fidelity of the finished map is governed to a large extent by the fineness of the grid system and the algorithm used to estimate values at the grid intersections. We are now considering a one-dimensional analogy of this same problem.

The data in Table 4.2 consist of analyses of the magnesium concentration in stream samples collected along a river. Because of the problems of accessibility, the samples were collected at irregular intervals up the winding stream channel. Sample localities were carefully noted on aerial photographs, and later the distances between samples were measured.

Although there are many methods whereby regularly spaced data might be estimated from these data, we will consider only two in detail. The first and most obvious technique consists of *simple linear interpolation* between data points to estimate intermediate points. This approach is illustrated in Figure 4.1. Assume Y_1 and Y_2 are observed values at points X_1 and X_2; we wish to estimate the value

TABLE 4.2 Measurements of Magnesium Concentration in Stream Water at Twenty Locations; Distances Given Are from Stream Mouth to Sample Locations

Distance (m)	Magnesium Concentration (ppm)	Distance (m)	Magnesium Concentration (ppm)
0.0	6.44	11,098	2.86
1,820	8.61	11,922	1.22
2,542	5.24	12,530	1.09
2,889	5.73	14,065	2.36
3,460	3.81	14,937	2.24
4,586	4.05	16,244	2.05
6,020	2.95	17,632	2.23
6,841	2.57	19,002	0.42
7,232	3.37	20,860	0.87
10,903	3.84	22,471	1.26

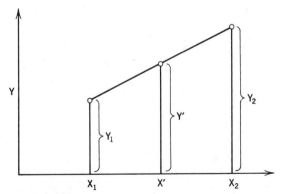

FIGURE 4.1 Linear interpolation between data points along a sequence.

of Y' at point X'. If we assume that a straight linear relation exists between sample points, intermediate values can be calculated from the geometric relationship

$$Y' = \frac{(Y_2 - Y_1)(X' - X_1)}{X_2 - X_1} + Y_1 \qquad (4.1)$$

Expressed in other words, the difference between values of two adjacent points is assumed to be a function of the distance separating them. The value of a point halfway between two observations is exactly intermediate between the values of the two enclosing points. The closer a point is to an observation, the closer its value is to that of the observation.

Although linear interpolation is simple, it possesses certain drawbacks in many applications. If the number of equally spaced points is approximately the same as the number of original points, and the original points are somewhat uniformly spaced, the technique will give satisfactory results. However, if there are many more original points than interpolated points, most of the original data will be ignored because only two surrounding points determine an interpolated value. If the original data possess a large random component which causes values to fluctuate widely, interpolated points may also fluctuate unacceptably. Both of these objections may be met by techniques that consider more than two of the original values, perhaps by fitting a linear function that extends over several adjacent values. Wilkes (1966) devotes an entire chapter to various interpolation procedures.

If the original data are sparse and several values must be estimated between each pair of observations, linear interpolation will perform adequately provided the idea of uniformity of slope between points is reasonable. In any problem where points are interpolated between observations, however, you must always remember that you cannot create data by estimation using any method. The validity of your result is controlled by the density of the original values and no amount of interpolation will allow refinement of the analysis beyond the limitations of the data. For example, we could estimate the magnesium content of the river at 500-m intervals,

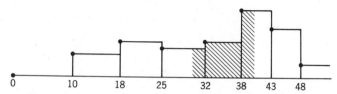

FIGURE 4.2 Data sequence considered as a step function or "rectangular curve."

or even at every 5 m, but it is obvious that these new values would provide no additional information on the distribution of the metal in the stream.

We will next consider a method that produces equally spaced estimates of a variable and considers all observations between successive points of estimation. The technique is called *rectangular integration*. If we regard the original data as a rectangular curve or step function in which the interval from one observation to the succeeding observation has a constant value, a data set might have the form shown in Figure 4.2. If we wish to create an equally spaced approximation to this distribution, we can generate another step function of rectangles of equal length whose areas equal the total areas of the original rectangles. This is shown graphically in Figure 4.3, with the resulting sequence of equally spaced values derived from the data in the preceding figure. The shaded area under the curve is the same in both illustrations. This procedure has the advantage of considering all data within an interval in estimating a point. Also, because the area under the estimated curve is equal to the area under the original curve, observations used in the estimation of a point are weighted proportionally to the length of interval they represent.

Calculation of an estimate by rectangular integration is easy in theory but presents a somewhat difficult programming challenge. Starting at one estimated point, the distance to the next observation must be calculated, multiplied by the magnitude of the observation to give the rectangular area, and the process repeated through all successive observations up to the next estimated point. That point is determined by summing the areas just found and dividing by the equally spaced interval to give the estimated value. The initial estimated point in a sequence is taken as the same as the first preceding data point.

An obvious difference in the two interpolation procedures is apparent when original data are sparse and more than one point must be estimated between two observations. Using linear interpolation, values will be created which lie on a

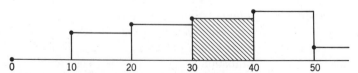

FIGURE 4.3 Equally spaced sequence created from unevenly spaced data by rectangular integration. Shaded region has same area as shaded region in Figure 4.2.

straight line between two surrounding data points. In contrast, rectangular integration will create estimates that are equal to the first observation.

In the study of a metamorphic halo around an intrusive, a diamond drill core was taken perpendicular to the intrusive wall. The entire core was split and all garnet crystals exposed on the split surface were removed, individually crushed, and analyzed for iron content by a rapid spectrochemical method. Both the spacing between successive crystals and their iron content fluctuate through a wide range. Data from this core are listed in Table 4.3. A generalized picture of compositional changes is desired, but the data seem too erratic for direct interpretation. As a preparatory step to further analysis, the data may be approximated by equally spaced estimates. The desired spacing interval is 0.5 m. Here we are presented with a situation that is different from the river data; observations are more abundant than estimates and we wish to preserve as much of the original information as possible. Rectangular integration seems more appropriate in this instance than linear interpolation. As an exercise, these data may be interpolated by both techniques and the results compared.

In geology, equal spacing procedures have been most widely used to pretreat

TABLE 4.3 Iron Content of Garnets Removed from Diamond Drill Core

Depth (cm)	Fe content (%)	Depth (cm)	Fe content (%)
0	14.21	283	16.67
3	19.35	297	18.56
10	17.22	322	18.87
14	15.87	335	20.81
23	13.62	351	24.52
30	16.31	370	25.03
36	14.13	408	25.11
48	13.95	416	23.28
59	15.00	419	22.56
66	14.23	425	19.00
68	16.81	429	20.53
81	15.93	443	19.08
94	16.02	447	22.83
96	17.85	465	21.06
102	17.02	474	24.96
115	15.87	493	19.12
121	19.84	502	22.24
130	16.94	522	26.88
163	16.72	550	21.15
168	19.20	558	28.92
205	20.41	571	27.96
239	16.88	586	25.03
251	18.74	596	26.27

stratigraphic data (measured sections, drilling-time logs, and similar records) prior to filtering or time-trend analysis. Time-trend procedures, which essentially are smoothing operations, require equally spaced data for their operation. Time series methods, such as autocorrelation and spectral analysis, also require equally spaced data. Time series techniques are inherently more powerful than other analytical methods for examining sequential data, and their use has become widespread. However, they require long strings of data, which has restricted their application to geophysics, well log analysis, and the study of stratigraphic sequences and diamond drill cores through ore deposits. Some work also has been done on mineral successions along traverses across thin sections. These applications will be considered in greater detail later in this chapter.

Markov Chains

In many geologic investigations, data sequences may be created that consist of ordered successions of mutually exclusive states. An example is a point-count traverse across a thin section, where the states are the minerals noted at succeeding points. Measured stratigraphic sections also have the form of series of lithologies, as may drill-holes through zoned ore bodies where the rocks encountered are classified into different types of ore and gangue. Observations along a traverse may be taken at equally spaced intervals, as in point counting, or they may be taken wherever a change in state occurs, as is commonly done in the measurement of stratigraphic sections. In the first instance, we would expect runs of the same state; that is, several successive observations could conceivably fall in the same category. This obviously cannot happen if observations are taken only where states change.

The data contained in these sequences are essentially the same as those used for autoassociation and cross-association. However, our interest now is in the nature of transitions from one state to another, rather than in the relative positions of states in the sequence. We will consider techniques that sacrifice all information about the position of observations within the succession, but that provide in return information on the tendency of one state to follow another. The data in Table 4.4 represent a stratigraphic section, shown in Figure 4.4, in which the sedimentary rock has been classified at successive points spaced 1 foot apart. The lithologies include four mutually exclusive states—sandstone, limestone, shale, and coal, arbitrarily designated A, B, C, and D, respectively. A 4×4 matrix can be constructed, showing the number of times a given rock type is succeeded, or overlain, by another. A matrix of this type is called a *transition frequency matrix* and is shown below. The measured stratigraphic section contains 63 observations, so there are $(n - 1) = 62$ transitions. The matrix is read "from rows to columns," meaning, for example, that a transition from state A to state C is counted as an entry in element $a_{1,3}$ of the matrix. That is, if we read *from* the row labelled A *to* the column labelled C, we see that we move from state A into state C five times in the sequence. Similarly, there are five transitions from state C to state A in the

TABLE 4.4 Stratigraphic Succession Shown in Figure 4.4 Coded into Four Mutually Exclusive States of Sandstone (*A*), Limestone (*B*), Shale (*C*), and Coal (*D*); Observations Taken at 1-Ft Intervals

Top

C	C	B	C	A	A
C	C	B	C	A	A
C	C	B	C	A	A
A	A	B	C	C	A
A	A	B	A	C	A
A	C	C	A	D	A
A	C	C	A	C	C
A	D	C	A	C	Bottom
A	D	B	A	D	
C	C	B	C	D	
C	C	C	C	C	

sequence; this number appears as the matrix element defined by row *C* and column *A*. The transition frequency matrix is a concise way of expressing the incidence of one state following another.

		To				Row Totals
		A	*B*	*C*	*D*	
	A	⎡18	0	5	0⎤	23
From	*B*	0	5	2	0	7
	C	5	2	18	3	28
	D	⎣0	0	3	2⎦	5
Column Totals		23	7	28	5	63 Grand Total

Note that the row totals and the column totals will be the same, provided the section begins and ends with the same state; otherwise two rows and columns will differ by one. Also note that, unlike most matrices we have calculated before, the transition frequency matrix is asymmetric and in general $a_{i,j} \neq a_{j,i}$ if the sequence begins and ends with a different state.

The tendency for one state to succeed another can be emphasized in the matrix by converting the frequencies to decimal fractions or percentages. If each element in the *i*th row is divided by the total of the *i*th row, the resulting fractions express the relative number of times state *i* is succeeded by the other states. In a probabilistic sense, these are estimates of the conditional probability $P(j|i)$, the probability that state *j* will be the next state to occur, *given* that the present state is *i*. (We here introduce the unconventional, but equivalent notation $P(i \rightarrow j)$, which can be read

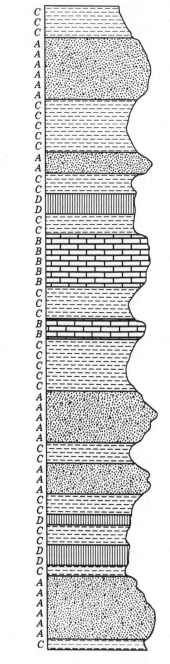

FIGURE 4.4 Measured stratigraphic section in which lithologies have been classed into four mutually exclusive states of sandstone (*A*), limestone (*B*), shale (*C*), and coal (*D*).

as the probability that state i will be followed by state j. This alternative notation will be useful later.)

		To			Row Totals
	A	B	C	D	
A	0.78	0	0.22	0	1.00
B	0	0.71	0.29	0	1.00
From C	0.18	0.07	0.64	0.11	1.00
D	0	0	0.60	0.40	1.00

Here, for example, we see that if we are in state C at one point, the probability is 64% that the lithology 1 foot up will also be state C. The probability is 18% that the lithology will be state A, 7% that it will be state B, and 11% that it will be state D. Since the four states are mutually exclusive and exhaustive, the lithology must be one of the four and so their sum, given as the row total, is 100%.

If we divide the row totals of the transition frequency matrix by the total number of transitions, we obtain the relative proportions of the four lithologies that are present in the section. This is called the *marginal*, or *fixed*, *probability vector*.

$$\begin{array}{c} A \\ B \\ C \\ D \end{array} \begin{bmatrix} 0.37 \\ 0.11 \\ 0.44 \\ 0.08 \end{bmatrix}$$

You will recall from Chapter 2 that the joint probability of two events A and B is

$$P(A,B) = P(B|A)P(A)$$

rearranging,

$$P(B|A) = \frac{P(A,B)}{P(A)}$$

So, the probability that state B will follow, or overlie, state A is the probability that both state A and B occur, divided by the probability that state A occurs. If the occurrence of states A and B are independent, or unconditional,

$$P(A,B) = P(A)P(B)$$

and

$$P(B|A) = \frac{P(A)P(B)}{P(A)} = P(B)$$

That is, the probability that state B will follow state A is simply the probability that state B occurs in the section, which is given by the appropriate element in the fixed probability vector. If the occurrence of all the states in the section are independent, the same relationship holds for all possible transitions; so, for example,

$$P(B|A) = P(B|B) = P(B|C) = P(B|D) = P(B)$$

This allows us to predict what the transition probability matrix should look like if the occurrence of a lithologic state at one point in the stratigraphic interval were completely independent of the lithology at the immediately underlying point. The expected transition probability matrix would consist of rows that were all identical to the fixed probability vector. For our stratigraphic example, this would appear as

$$\text{From} \begin{array}{c} \\ A \\ B \\ C \\ D \end{array} \overset{\overset{\text{To}}{\begin{array}{cccc} A & B & C & D \end{array}}}{\begin{bmatrix} 0.37 & 0.11 & 0.44 & 0.08 \\ 0.37 & 0.11 & 0.44 & 0.08 \\ 0.37 & 0.11 & 0.44 & 0.08 \\ 0.37 & 0.11 & 0.44 & 0.08 \end{bmatrix}}$$

We can compare this expected transition probability matrix to the transition probability matrix we actually observe to test the hypothesis that all lithologic states are independent of the immediately preceding states. This is done using a χ^2 test, first converting the probabilities to expected numbers of occurrences by multiplying each row by the corresponding total number of occurrences:

Expected Transition
Probabilities   Totals   Expected Frequencies

$$\begin{bmatrix} 0.37 & 0.11 & 0.44 & 0.08 \\ 0.37 & 0.11 & 0.44 & 0.08 \\ 0.37 & 0.11 & 0.44 & 0.08 \\ 0.37 & 0.11 & 0.44 & 0.08 \end{bmatrix} \begin{array}{c} \times\ 23\ = \\ \times\ \ 7\ = \\ \times\ 28\ = \\ \times\ \ 5\ = \end{array} \begin{bmatrix} 8.5 & 2.5 & 10.1 & 1.8 \\ 2.6 & 0.8 & 3.1 & 0.6 \\ 10.4 & 3.1 & 12.3 & 2.2 \\ 1.9 & 0.6 & 2.2 & 0.4 \end{bmatrix}$$

The χ^2 test is similar in form to the test described in Chapter 2. Each element in the transition frequency matrix constitutes a category, with both an observed and an expected number of transitions. These are compared by

$$\chi^2 = \sum \frac{(O - E)^2}{E} \tag{4.2}$$

where O is the observed number of transitions from one state to another, and E is the number of transitions expected if the successive states are independent. The test has $(m - 1)^2$ degrees of freedom, where m is the number of states (a degree of freedom is lost from each row because the probabilities in the rows sum to 1.00). As with other types of χ^2 tests, each category must have an expected frequency of at least five transitions. This is not the case in this example, but we can still make a conservative test of independence by calculating the test statistic using the four categories whose expected frequency is greater than five. The remaining categories can be combined until their expected frequencies exceed five.

The categories include the transitions $A \rightarrow A$, $A \rightarrow C$, $C \rightarrow A$, and $C \rightarrow C$. Combined categories can be formed of all elements in the B row, all elements in the D row, and the combination of transitions $A \rightarrow B$, $A \rightarrow D$, $C \rightarrow B$, and $C \rightarrow D$. The resulting χ^2 statistic is

$$\chi^2 = \frac{(18 - 8.5)^2}{8.5} + \frac{(5 - 10.4)^2}{10.4} + \frac{(5 - 10.1)^2}{10.1}$$

$$+ \frac{(18 - 12.3)^2}{12.3} + \frac{(7 - 7.0)^2}{7.0} + \frac{(5 - 5.0)^2}{5.0} + \frac{(5 - 9.8)^2}{9.8}$$

$$= 20.99$$

The critical value of χ^2 for nine degrees of freedom and a 5% level of significance is 16.92; the test value comfortably exceeds this, so we may conclude that the hypothesis of independence of successive states is not correct. There is a statistically significant tendency for certain states to be preferentially followed by certain other states.

A sequence in which the state at one point is partially dependent, in a probabilistic sense, on the preceding state is called a *Markov chain*. Named after the Russian statistician A. A. Markov, it is intermediate between deterministic sequences and completely random sequences. Our stratigraphic section exhibits *first-order* Markov properties; that is, the statistical dependency exists between points and their immediate predecessors. Higher order Markov properties can exist as well. For example, a second-order Markov sequence exhibits a significant conditional relationship between points that are two steps apart.

From the transition probability matrix we can estimate what the lithology will be 2 feet (that is, two observations) above a given point. Suppose we start in limestone (state B). The following probabilities estimate the lithology to be encountered at the next point upward:

State A (sandstone)   0%
State B (limestone)   71%
State C (shale)   29%
State D (coal)   0%

Suppose the next point actually falls in a shale; we can then determine the probable lithology of the following point:

State A (sandstone)   18%
State B (limestone)   7%
State C (shale)   64%
State D (coal)   11%

So, the probability that the lithologic sequence will be limestone $\rightarrow$ shale $\rightarrow$ limestone is

$$P(B \rightarrow C) \times P(C \rightarrow B) = 29\% \times 7\% = 2\%$$

However, there is another way to reach the limestone state in two steps. The sequence limestone $\rightarrow$ limestone $\rightarrow$ limestone is also possible. The probability attached to this sequence is

$$P(B \rightarrow B) \times P(B \rightarrow B) = 71\% \times 71\% = 50\%$$

Since the other transitions limestone → sandstone and limestone → coal have zero probability, these two sequences are the only possible ones which lead from limestone and back again in two steps. The probability that the lithology two steps above a limestone will also be a limestone, regardless of the intervening lithology, is the sum of all possibilities. That is,

$$P(B \rightarrow A \rightarrow B) = 0\%$$
$$P(B \rightarrow B \rightarrow B) = 50\%$$
$$P(B \rightarrow C \rightarrow B) = 2\%$$
$$P(B \rightarrow D \rightarrow B) = \underline{0\%}$$
$$52\%$$

The same reasoning can be applied to determine the probability of any lithology two steps hence, from any starting lithology. However, all of the various sequences do not have to be worked out individually, because the process of multiplying and summing is exactly that used for matrix multiplication. If the transition probability matrix is multiplied by itself (that is, the matrix is squared), the result is the second-order transition probability matrix describing the second-order Markov properties of the succession.

$$
\begin{bmatrix}
0.78 & 0 & 0.22 & 0 \\
0 & 0.71 & 0.29 & 0 \\
0.18 & 0.07 & 0.64 & 0.11 \\
0 & 0 & 0.60 & 0.40
\end{bmatrix}^2
=
\begin{bmatrix}
0.64 & 0.02 & 0.31 & 0.02 \\
0.05 & 0.52 & 0.39 & 0.03 \\
0.26 & 0.09 & 0.54 & 0.11 \\
0.11 & 0.04 & 0.62 & 0.23
\end{bmatrix}
$$

Note that the rows of the squared matrix also sum to 100%.

The existence of a significant second-order property can be checked in exactly the same manner as we checked for independence between successive states, by using a χ^2 test. If you repeat the test performed earlier, but using the second-order transition probability matrix, you should find that the sequence has no significant second-order properties.

We can estimate the probable state to be encountered at any step in the future simply by powering the transition probability matrix the appropriate number of times. If the matrix is raised to a sufficiently high power, it reaches a stable state in which the rows all become equal to the fixed probability vector, or in other words, becomes an independent transition probability matrix and will not change with additional powering.

You will note in the example that the highest transition probabilities are from one state to itself, particularly from sandstone to sandstone, from limestone to limestone, and from shale to shale. It is obvious that these transition probabilities are related to the thicknesses of the stratigraphic units being sampled and the distance between the sample points. For example, the frequencies along the main diagonal of the transition frequency matrix would be doubled while off-diagonal frequencies remained unchanged if observations were made every half-foot. This would greatly enhance the Markovian property, but in a specious manner. Selecting the appropriate distance between sampling points can be a vexing problem; if observations are too closely spaced, the transition matrix reflects mainly the thickness of the more massive stratigraphic units. If the spacing is too great, thin units may be entirely missed.

Embedded Markov Chains

The difficulty of selecting an appropriate sampling interval can be avoided if observations are taken only when there is a change in state. A stratigraphic section, for example, would be recorded as a succession of beds, each one of a different lithology than the immediately preceding bed. Table 4.5 contains the record of successive rock types penetrated by a well drilled in the Midland Valley of Scotland. The well was drilled through 1600 feet of Coal Measures of Pennsylvanian age, consisting of interbedded shales, siltstones, sandstones, and coal beds or root zones. These sediments are interpreted as having been deposited in a delta plain environment subject to repeated flooding, so we would expect that certain lithologies would occur in preferred relations to others. The data are taken from one of a large number of wells studied by Doveton (1971).

The four-state transition frequency matrix for the section in the Scottish well is given below. One obvious difference between this matrix and the one we have considered previously is that all the diagonal terms must be zero, since a state cannot succeed itself. The transition probability matrix, computed by dividing each element of the transition frequency matrix by the appropriate row total, shares this same characteristic. Sequences in which transitions from a state to itself are not permitted are called *embedded Markov chains,* and their analysis presents special problems that have not always been appreciated by geologists studying stratigraphic records.

		To					Row Totals
		A	B	C	D	E	
	A	0	11	36	21	52	120
	B	28	0	4	4	0	36
From	C	34	2	0	45	13	94
	D	29	1	45	0	3	78
	E	28	23	9	8	0	68
Column Totals		119	37	94	78	68	396 Grand Total

The lithologic states have been coded as (A) unfossiliferous shale and mudstone, (B) shales containing nonmarine bivalves, (C) siltstone, (D) sandstone, and (E) coals and root zones. The corresponding transition probability matrix is

		To					Row Totals
		A	B	C	D	E	
	A	0	0.09	0.30	0.18	0.43	1.00
	B	0.78	0	0.11	0.11	0.00	1.00
From	C	0.36	0.02	0	0.48	0.14	1.00
	D	0.37	0.01	0.58	0	0.04	1.00
	E	0.41	0.34	0.13	0.12	0	1.00

TABLE 4.5 Successive Lithologic States Encountered in a Drill Hole Through the Coal Measures in the Midland Valley of Scotland; Mutually Exclusive States are Barren Shale (*A*), Shale with Fossils of Nonmarine Bivalves (*B*), Siltstone (*C*), Sandstone (*D*), and Coal or Root Zone (*E*); Read Down the Columns

Top

B	D	C	B	E	B	B	A	C	E	A	C	D	A	E	E	A	A	A	D	
E	C	D	E	A	A	E	C	E	C	B	D	A	C	A	A	C	E	D	C	
A	A	C	A	C	C	A	D	A	A	E	C	E	A	D	D	E	A	E	D	
E	E	A	D	D	A	C	C	D	B	A	D	C	B	A	C	D	E	C	C	
A	D	B	C	C	C	A	D	A	C	B	C	D	C	B	D	A	C	D	A	
D	C	A	A	D	A	C	A	D	E	A	A	C	D	E	A	B	D	C	D	
A	A	E	E	C	B	B	B	C	C	B	C	D	A	A	E	D	A	A	A	
C	D	D	C	D	A	E	E	A	A	E	A	E	E	C	A	B	B	E	B	
D	C	C	D	C	B	C	A	E	D	A	C	C	A	D	C	A	E	A	A	
C	A	A	C	A	E	A	C	A	B	B	E	D	C	E	D	E	A	E	B	
D	E	E	A	B	A	D	D	C	E	A	A	C	D	A	C	A	B	A	E	
C	C	C	B	E	C	C	C	D	A	B	C	E	C	D	A	C	E	C	A	
A	D	A	A	A	D	A	A	A	D	E	D	A	E	C	D	A	A	D	D	
B	C	D	E	B	C	C	B	E	C	C	C	C	A	A	A	E	E	A	B	
E	B	E	A	A	D	D	A	A	D	A	D	A	C	B	E	C	A	E	A	
A	E	A	D	B	C	C	B	E	E	C	C	E	B	E	A	D	C	C	E	
D	A	D	E	A	D	E	E	A	A	D	A	A	E	A	D	C	D	D	A	
C	D	A	A	B	C	A	A	C	D	A	B	C	A	D	A	D	E	B		Bottom
D	C	C	D	E	A	C	D	D	A	E	E	A	C	C	D	C	A	E		
C	D	A	C	A	C	D	A	C	C	A	A	E	A	D	C	D	D	A		

After Doveton (1971).

The marginal probability vector is

$$
\begin{array}{c}
A \\
B \\
C \\
D \\
E
\end{array}
\begin{bmatrix}
0.30 \\
0.09 \\
0.24 \\
0.20 \\
0.17
\end{bmatrix}
$$

A χ^2 test, identical to eq. (4.2), can be used to check for the Markov property in an embedded sequence. This is done by comparing the observed transition frequency matrix to the matrix expected if successive states are independent. However, the fixed probability vector cannot be used to estimate the columns of the expected transition probability matrix. This would result in the expectation of transitions from a state to itself, which are forbidden. Rather, we must use a somewhat roundabout procedure to estimate the frequencies of transitions between independent states, subject to the constraint that states cannot succeed themselves.

We begin by imagining that our sequence is actually a censored sample taken from an ordinary succession in which transitions from a state to itself can occur. The transition frequency matrix of this succession would look like the one we observe except that the diagonal elements would contain values other than zero. If we were to compute a transition probability matrix from this frequency matrix and then raise it to an appropriately high power, it would estimate the transition probability matrix of a sequence in which successive states were independent. If the diagonal elements were then discarded and the off-diagonal probabilities recalculated, the result would be the expected transition probability matrix for an embedded sequence whose states are independent.

How do we estimate the frequencies of transitions from each state to itself, when this information is not available? We do this by trial-and-error, searching for those values that, when inserted on the diagonal of the transition frequency matrix, do not change when the matrix is powered. The off-diagonal elements, however, will change until a stable configuration is reached, corresponding to the independent events model.

In practice it is not necessary to calculate the off-diagonal probabilities at all. We begin by assigning some arbitrarily large number, say 1000, to the diagonal positions of the observed transition frequency matrix. The fixed probability vector is found, by summing each row and dividing by the grand total, and then is used as an estimate of the transition probabilities along the diagonal. These probabilities are powered by squaring and multiplied by the grand total to obtain new estimates of the diagonal frequencies. These new estimates are inserted into the original transition frequency matrix and the process repeated. We can work through the first cycle of the procedure.

Step 1. Initial estimate of transition frequency matrix, with 1000 inserted in each diagonal position.

		To					Row Totals	
		A	*B*	*C*	*D*	*E*		
	A	1000	11	36	21	52	1120	
	B	28	1000	4	4	0	1036	
From	*C*	34	2	1000	45	13	1094	
	D	29	1	45	1000	3	1078	
	E	28	23	9	8	1000	1068	
							5396	Grand Total

Step 2. Estimate of transition probabilities of diagonal elements, found by dividing row totals by grand total.

		A	*B*	To *C*	*D*	*E*	Row Totals
	A	0.208					0.208
	B		0.192				0.192
From	*C*			0.203			0.203
	D				0.200		0.200
	E					0.198	0.198

Step 3.   Second estimate of transition frequency matrix using new diagonal elements calculated by multiplying probabilities on the diagonal by the grand total of 5396. Off-diagonal terms are the original observed frequencies. New row totals and grand total are then found.

		To					Row Totals
		A	*B*	*C*	*D*	*E*	
	A	233	11	36	21	52	353
	B	28	199	4	4	0	235
From	*C*	34	2	222	45	13	316
	D	29	1	45	215	3	294
	E	28	23	9	8	212	280
							1478 Grand Total

The process is repeated again and again, until the estimated transition frequencies along the diagonal do not change from time to time. This generally requires about 10 to 20 iterations, depending upon how closely the initial guesses were to the final, stable estimates. In this example, the estimates do not change after 12 iterations. The final form of the transition frequency matrix with estimated diagonal frequencies is given below.

		To					Row Totals
		A	*B*	*C*	*D*	*E*	
	A	66	11	36	21	52	186
	B	28	3	4	4	0	39
From	*C*	34	2	29	45	13	123
	D	29	1	45	17	3	95
	E	28	23	9	8	13	81
Column Totals		185	40	123	95	81	524 Grand Total

This matrix could be converted into an expected transition probability matrix of the hypothetical Markov sequence by dividing each element by the corresponding row total. However, such a matrix is of little interest because it pertains to the hypothetical sequence rather than the observed embedded sequence. The marginal totals are another matter, because they are required to compute the marginal probability vector:

$$
\begin{array}{c}
A \\
B \\
C \\
D \\
E
\end{array}
\begin{bmatrix}
0.355 \\
0.074 \\
0.235 \\
0.181 \\
0.155
\end{bmatrix}
$$

We may now calculate the expected probabilities and expected frequencies of a hypothetical sequence of independent states from the marginal probability vector. We are testing the hypothesis of independence between successive states by noting that, for example, if state A is independent of state B, then $P(A|B) = P(A)P(B)$. As $P(A)$ and $P(B)$ are given by the appropriate elements of the marginal probability vector, the estimated conditional probability that state A will follow state B is $P(A|B) = (0.355)(0.074) = 0.026$. The expected probabilities for all transitions are given below.

$$
\text{From} \quad
\begin{array}{c}
 \\ A \\ B \\ C \\ D \\ E
\end{array}
\begin{array}{ccccc}
\multicolumn{5}{c}{\text{To}} \\
A & B & C & D & E \\
\left[\begin{array}{ccccc}
0.125 & 0.026 & 0.083 & 0.064 & 0.055 \\
0.026 & 0.006 & 0.017 & 0.013 & 0.012 \\
0.083 & 0.017 & 0.055 & 0.043 & 0.036 \\
0.064 & 0.013 & 0.043 & 0.033 & 0.028 \\
0.055 & 0.012 & 0.036 & 0.028 & 0.024
\end{array}\right]
\end{array}
$$

The expected frequencies are found by multiplying this matrix by the grand total, 524.

$$
\text{From} \quad
\begin{array}{c}
 \\ A \\ B \\ C \\ D \\ E
\end{array}
\begin{array}{ccccc}
\multicolumn{5}{c}{\text{To}} \\
A & B & C & D & E \\
\left[\begin{array}{ccccc}
65.5 & 13.6 & 43.5 & 33.5 & 28.8 \\
13.6 & 3.1 & 8.9 & 6.8 & 6.3 \\
43.5 & 8.9 & 28.8 & 22.5 & 18.9 \\
33.5 & 6.8 & 22.5 & 17.3 & 14.7 \\
28.8 & 6.3 & 18.9 & 14.7 & 12.6
\end{array}\right]
\end{array}
$$

Note that the matrix is symmetrical and the diagonal elements remain unchanged, within the limits of rounding error. The off-diagonal elements are the expected frequencies of transitions within the embedded sequence, assuming independence between successive states. If the diagonal elements are stripped from the matrix, it may be compared directly to the observed transition frequency matrix because the row and column totals of the two are the same, again within rounding limits.

The comparison by χ^2 methods yields a test statistic of $\chi^2 = 172$. The test has $\nu = (m - 1)^2 - m$ degrees of freedom, where m is the number of states, or in this example, $\nu = 11$. The critical value of χ^2 for 11 degrees of freedom and an $\alpha = .05$ level of significance is 19.68, which is far exceeded by the test statistic. Therefore, we must conclude that successive lithologies encountered in the Scottish well are not independent, but rather exhibit a strong first-order Markovian property.

If tests determine that a sequence exhibits partial dependence between successive states, the structure of this dependence may be investigated further. Simple graphs of the most significant transitions may reveal repetitive patterns in the succession; these also may be detected by autoassociation techniques. Modified χ^2 procedures are available to test the significance of individual transition pairs. Some authors have found that the eigenvalues extracted from the transition probability matrix are useful indicators of cyclicity. (It should be noted, however, that extracting the eigenvectors from an asymmetric matrix such as the transition probability matrix may not be an easy task.) These topics will not be pursued further in this book; the interested reader should refer to the text by Kemeny and Snell (1960) and the book on quantitative sedimentology by Schwarzacher (1975). Chi-square tests appropriate for embedded sequences are discussed by Goodman (1968). In a geological context, the articles by Doveton (1971) and Doveton and Skipper (1974) as well as the comment by Türk (1979) are recommended.

Series of Events

An interesting type of time series we will now consider is called a *series of events*. Geological examples of this type of data sequence include the historical record of earthquake occurrences in California, the record of volcanic eruptions in the Mediterranean area, and the incidence of landslides in the Tetons. The characteristics of these series are (*a*) the events are distinguishable by where they occur in time; (*b*) the events are essentially instantaneous; and (*c*) the events are so infrequent that no two occur in the same time interval. A series of events is therefore nothing more than a sequence of the intervals between occurrences. Our data may consist of the duration between successive events, or the cumulative length of time over which the events occur. One form may be directly transformed into the other.

Series-of-events models may be appropriate for certain types of spatially distributed data. We might, for example, be interested in the occurrence of a rare mineral encountered sporadically on a traverse across a thin section or in the appearance of bentonite beds in a vertical succession of sedimentary rocks. Justification for applying series-of-events models to spatial data may be tenuous, however, and depends on the assumption that the spatial sequence has been created at a constant rate. This assumption probably is reasonable in the first example, but the second requires that we assume that the sedimentation rate remained constant through the series.

The historic record of eruptions of the volcano Aso in Kyushu, Japan, has been kept since 1229 (Kuno, 1962), and is given in Table 4.6. Aso is a complex strato-volcano, but all historic eruptions have been explosive, ejecting ash of andesitic composition. Although the ancient monastic records contain an indication of the relative violence and duration of some eruptions, for all practical purposes we must regard the record as one of indistinguishable, instantaneous explosive events. Analysis of volcanic histories may shed some light on the nature of eruptive mechanisms and can even lead to physical models of the structure of volcanoes (Wickman, 1966). Of course, we would also hope that such studies might lead to predictive tools to forecast future eruptions.

Studies of series of events may have several objectives. Usually, an investigator is interested in the *mean rate of occurrence*, or number of events per interval of time. In addition, it may be necessary to examine the series in more detail, in order to estimate any pattern that may exist in the events. This additional information can be used to determine the precision of the estimate of the rate of occurrence, to assess the appropriateness of the sampling scheme, to detect a trend, and to detect other systematic features of the series.

Because series of events are very simple, in the sense that they consist of nominal occurrences (presence-absence), simple analytical techniques may prove

TABLE 4.6 Years of Eruptions of the Volcano Aso for the Period 1229–1962

1229	1376	1583	1780	1927
1239	1377	1584	1804	1928
1240	1387	1587	1806	1929
1265	1388	1598	1814	1931
1269	1434	1611	1815	1932
1270	1438	1612	1826	1933
1272	1473	1613	1827	1934
1273	1485	1620	1828	1935
1274	1505	1631	1829	1938
1281	1506	1637	1830	1949
1286	1522	1649	1854	1950
1305	1533	1668	1872	1951
1324	1542	1675	1874	1953
1331	1558	1683	1884	1954
1335	1562	1691	1894	1955
1340	1563	1708	1897	1956
1346	1564	1709	1906	1957
1369	1576	1765	1916	1958
1375	1582	1772	1920	1962

to be the most effective. Cox and Lewis (1966) described a variety of graphical tools that are useful in examining series of events. These are illustrated using the data on the eruptions of Aso from Table 4.6.

A cumulative plot of the total number of events (N_t) to have occurred at or before time t, against time t, is given in Figure 4.5. This plot is especially good for showing changes in the average rate of occurrence. The slope of a straight line connecting any two points on the cumulative plot is the average number of events per unit of time for the interval between the two points.

Figure 4.6 is a histogram of the number of events occurring in successive equal intervals of time. This histogram directly indicates local periods of fluctuation from the average rate of occurrence. The pattern shown by the histogram is sensitive to the length of the chosen intervals, so more than one histogram may be useful in examining a series.

The *empirical survivor function* is obtained by plotting the percent "survivors," or Y = proportion of time intervals longer than X, against X = length of time interval. The function estimates the probability that an event has not occurred before time X. In Figure 4.7, the points represent the percentage of intervals between eruptions which are longer than the specified number of years. If events occur randomly in time, the survivor function will be exponential in form.

This same function can be plotted in logarithmic form, as log Y against X. The *log empirical survivor function* is especially good for showing departures from randomness, which appear as deviations from the straight-line form of the plot (Fig. 4.8).

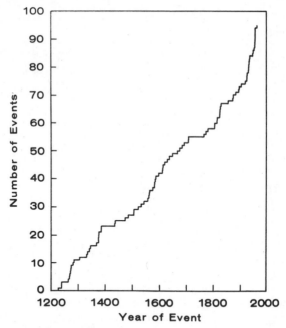

FIGURE 4.5 Cumulative number of eruptions of the volcano Aso, plotted against years of eruptions.

A scatter diagram of the *serial correlation*, or first-order autocorrelation, of successive intervals between events is shown in Figure 4.9. The degree of correspondence between the length of an interval and the length of the immediately preceding interval is shown by plotting $X_i = t_{i+1} - t_i$ against $Y_i = t_i - t_{i-1}$, where t_i is the time of occurrence of the ith event. This plot reveals any tendency for intervals to be followed by intervals of similar length. A scatter diagram with

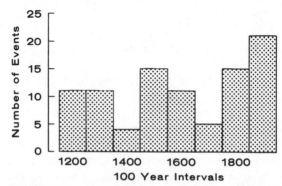

FIGURE 4.6 Histogram of number of eruptions of the volcano Aso occurring in successive 100-year intervals.

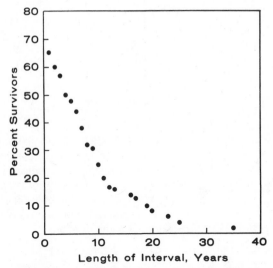

FIGURE 4.7 Empirical survivor function for the volcano Aso. The vertical axis gives the percent of intervals between eruptions that are more than a specified duration, versus the duration in years along the horizontal axis.

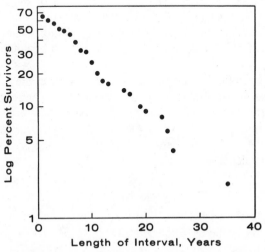

FIGURE 4.8 Log empirical survivor function of the volcano Aso. The vertical axis of Figure 4.7 is expressed in logarithmic form.

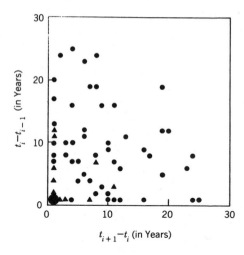

FIGURE 4.9 Serial correlation of durations between successive eruptions of the volcano Aso. Vertical axis is duration of quiet before ith eruption, and horizontal axis duration after ith eruption. Triangles indicate more than one point plotted at the same location.

large dispersion and relatively high concentrations of points near the axes is typical of random series of events.

In most series of events studies, we hope that we can describe the basic features of the series in a way that will suggest a physical mechanism for the lengths of the intervals between occurrences. First we must consider the possibility of a trend in the data. We may check for a trend in two ways. A series may be subdivided into a number of segments of equal length, provided each segment contains several observations. The number of events within each segment are taken to be observations located at the midpoints of the segments. A regression then can be run with these numbers as the dependent variable Y_i and the locations of the midpoints of the segments as values of X_i. The slope coefficient of the regression can be tested by the ANOVA given in Table 4.12 to determine if it is significantly different from zero. The process is illustrated in Figure 4.10. Unfortunately, this test is not particularly efficient because degrees of freedom are lost by subdividing the series into segments.

There are tests specifically designed to detect a trend in the rate of occurrence of events by comparing the midpoint of the sequence to its centroid. If the sequence is relatively uniform, the two will be very similar, but if there is a trend the centroid will be displaced in the direction of increasing rate of occurrence. If t_i is the time or distance from the start of the series to the ith event and N is the total number of events, we can calculate the centroid S by

$$S = \frac{\sum_{i=1}^{N} t_i}{N} \qquad (4.3)$$

This statistic can in turn be used in eq. (4.4),

$$Z = \frac{S - 1/2T}{T/\sqrt{12N}} \qquad (4.4)$$

where T is the total length of the series, Z is the standardized normal variate, and the significance of the test result can be determined by normal tables such as Table 2.10.

The test is very sensitive to changes in the rate of occurrence of events. Specifically, if the events are considered to be the result of a process

$$Y_t = e^{a + \beta t} \tag{4.5}$$

the null hypothesis states that $\beta = 0$. You will recognize that the model is exponential; if β has any value other than zero, the rate of occurrence of Y_t will change with t. It is this possibility that we are testing.

If no trends are detected in the rate of occurrence, we may conclude that the series of events is stationary. We can next check to see if successive occurrences are independent. This can be done by computing the autocorrelation of the lengths between events. That is, we regard the intervals between events as a variable X_i located at equally spaced points. If the intervals are not independent, there will be a tendency for large values of X (long intervals between events) to be succeeded by large values; similarly, there will be a tendency for small values of X (short intervals) to be followed by other small values. We can compute autocorrelation coefficients for successive lags and test these for significance. Usually only the first few lags will be of interest. If the autocorrelation coefficients are not significantly different from zero, as tested by methods that will be developed later in this chapter, we can conclude that the events are occurring independently in time or space.

If we have established that the series is neither autocorrelated nor contains a trend, we may wish to test the possibility that the events are distributed according to a *Poisson distribution*. You will recall from Chapter 2 that the Poisson is a discrete probability distribution that can be regarded as the limiting case of the binomial when n, the number of trials, becomes very large, and p, the probability

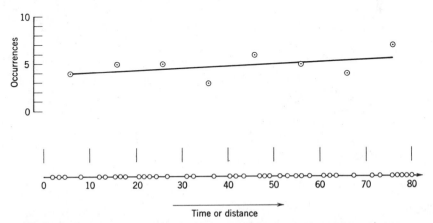

FIGURE 4.10 A series of events that occur as "instantaneous" happenings along a continuum of time or distance. Time or distance scale is divided into segments 10 units long, and number of events per segment is plotted. Upper line is regression of number of events per segment on the midpoints of segments.

of success on any one trial, becomes very small. We can imagine that our time series is subdivided into n intervals of equal duration. If events occur randomly, the number of intervals that contain exactly zero, one, two, . . . , x events will follow the binomial distribution. As we make the lengths of the intervals progressively shorter, n becomes progressively larger and the probabilities of occurrence decline. The binomial distribution becomes difficult to compute, but the Poisson can be readily used because it does not require either n or p directly. Instead, the product $np = \lambda$ is all that is needed, which is given by the *rate of occurrence* of events.

The Poisson probability model assumes that (*a*) the events occur independently, (*b*) the probability that an event occurs does not change with time, (*c*) the probability that an event will occur in an interval is proportional to the length of the interval, and (*d*) the probability of more than one event occurring at the same time is vanishingly small.

The equation for the Poisson distribution in this instance is

$$p(X) = e^{-\lambda}\lambda^X/X! \tag{4.6}$$

Note that the rate of occurrence, λ, is the only parameter of the distribution. Typical Poisson frequency distributions are shown in Figure 4.11. The distribution is applicable to such problems as the rate that telephone calls come to a switchboard or the length of time between failures in a computer system. It seems reasonable that it also may apply to the series of geological events described at the beginning of this section. If we can determine that our series follows a Poisson distribution, we can use the characteristics of the distribution to make probabilistic forecasts of the series.

The Kolmogorov-Smirnov test provides a simple way to test the goodness-of-fit of a series of events to that expected from a Poisson distribution. First, the series must be converted to a cumulative form

$$Y_i = \frac{t_i}{T}$$

where t_i is the time from the start of the series to the ith event, and T is the total length of the series. Three estimates can then be calculated

1. $\mathrm{KS}^+ = \sqrt{n} \max \left\{ \dfrac{i}{n} - Y_i \right\}$

2. $\mathrm{KS}^- = \sqrt{n} \max \left\{ Y_i - \dfrac{i-1}{n} \right\}$ $\qquad$ (4.7)

3. $\mathrm{KS} = \max |\mathrm{KS}^+, \mathrm{KS}^-|$

The first test is simply the maximum positive difference between the observed series and that expected from a Poisson, the second is the maximum negative difference, and the third is the larger of the absolute values of the two. The test statistic KS can be compared to two-tailed critical values given in Table 2.26. If

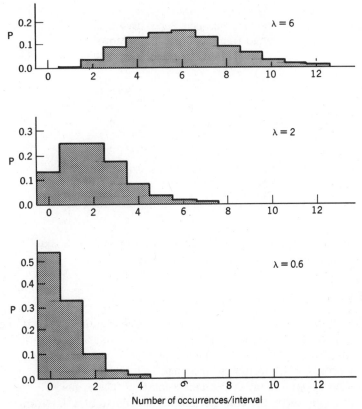

FIGURE 4.11 Poisson probability distributions with different rates of occurrence, λ, expressed as numbers of occurrences per interval.

the statistic exceeds the critical value, the maximum deviation is larger than that expected in a sample collected at random from a Poisson distribution.

The Mowry Shale is a black, siliceous shale of Early Cretaceous age, occurring in Colorado, Wyoming, and Montana. The interval in characterized by numerous bentonite beds, which are mined at several locations in Wyoming and Montana for drilling mud and foundry clay. Bentonite is composed almost entirely of montmorillonite, developed as an alteration product of rhyolitic or andesitic volcanic ash. Table 4.7 gives the thickness between successive bentonite beds measured in an outcrop of Mowry Shale in Fremont County, Wyoming. These beds represent ash falls from explosive eruptions of volcanoes in western Idaho. If it is assumed that the enclosing black shale was deposited at an approximately uniform rate, it may be possible to analyze this sequence of thicknesses as a series of events analogous to the historical series formed by the eruptions of Aso.

Test these data for a trend in the rate of occurrence. If none is observed, test for autocorrelation of successive intervals between events. Comment on (*a*) the

TABLE 4.7 Thickness, in Feet, of Intervals Between Successive Bentonite Beds in Cretaceous Mowry Shale in Fremont County, Wyoming

(Top)	47	6
29	7	3
11	8	17
6	23	4
5	10	5
10	15	4
5	2	26
17	35	4
14	4	(Bottom)

possible effects of unequal rates of sedimentation of the black shale, and (*b*) the possibility that more than one volcano was active.

Runs Tests

The simplest type of sequence is a succession of observations arranged in order of occurrence, where the observations are two mutually exclusive categories or states. Consider a rock collector cracking open concretions in a search for fossils. The breaking of a concretion constitutes a trial, and each trial has two mutually exclusive outcomes: The concretion either contains a fossil or it does not. The sequence of successes and failures by the collector during the course of a day forms a special type of time series. We can experimentally create a similar succession by flipping pennies and noting the occurrence of heads or tails. The sequence generated might resemble this set of twenty trials:

HTHHTHTTTHTHTHHTTHHH

We intuitively expect, of course, that about ten heads will appear, and we can determine the probability of obtaining this (or any other) number of heads. Here we obtained eleven heads; assuming the coin is unbiased, the probability of obtaining this number in twenty trials is 0.16 or about one in six. We would expect similar trials to contain nine, ten, or eleven heads slightly more than one third of the time. Results of this experiment follow the binomial distribution, discussed in Chapter 2.

However, one aspect that we have not considered is the order in which the heads appear. We probably would regard a sequence such as

HHHHHHHHHHHHTTTTTTTTT

as being very strange, although the probability of obtaining this many heads in twenty trials is the same as in the preceding example. At the other extreme, the regular alternation of heads and tails

HTHTHTHTHTHTHTHTHTHH

would also appear very unusual to us, although the probability of the number of heads is unchanged. What arouses our suspicions is not the proportion of heads but the order in which they appear. We assume that heads and tails will occur at random; in the two preceding examples, it seems very unlikely that they have.

We can test these sequences for randomness of occurrence by examining the number of runs. *Runs* are defined as uninterrupted sequences of the same state. The first set of trials contains 13 runs, the second only two, and the third contains 19. Runs in the sequence shown are underlined:

$$\underline{\text{H}} \; \underline{\text{T}} \; \underline{\text{HH}} \; \underline{\text{T}} \; \underline{\text{H}} \; \underline{\text{TTT}} \; \underline{\text{H}} \; \underline{\text{T}} \; \underline{\text{H}} \; \underline{\text{T}} \; \underline{\text{HH}} \; \underline{\text{TT}} \; \underline{\text{HHH}}$$
$$\text{1} \; \text{2} \;\; \text{3} \;\;\; \text{4} \; \text{5} \;\;\; \text{6} \;\;\;\; \text{7} \; \text{8} \; \text{9} \; \text{10} \;\; \text{11} \;\; \text{12} \;\;\; \text{13}$$

We can calculate the probability that a given sequence of runs was created by the random occurrence of two states (heads and tails, in this example). This is done by enumerating all possible ways of arranging n_1 items of state 1 and n_2 items of state 2. The total number of runs in a sequence is denoted U; tables are available which give critical values of U for specified n_1, n_2 and level of significance, α. However, if n_1 and n_2 each exceed ten, the distribution of U can be closely approximated by a normal distribution, and we can use tables of the standard normal variate Z for our statistical tests. The expected mean number of runs in a randomly generated sequence of n_1 items of state 1 and n_2 items of state 2 is

$$\overline{U} = \frac{2n_1 n_2}{n_1 + n_2} + 1 \tag{4.8}$$

The expected variance in the mean number of runs is

$$\sigma_{\overline{U}}^2 = \frac{2n_1 n_2 (2n_1 n_2 - n_1 - n_2)}{(n_1 + n_2)^2 (n_1 + n_2 - 1)} \tag{4.9}$$

By these equations, we can determine the mean number of runs and the standard error of the mean number of runs in all possible arrangements of n_1 and n_2 items. Having calculated these, we can create a Z test by (4.10), where U is the observed number of runs:

$$Z = \frac{U - \overline{U}}{\sigma_{\overline{U}}} \tag{4.10}$$

You will recognize that this is simply (2.28) rewritten to include the runs statistics. We can formulate a variety of statistical hypotheses which can be tested with this statistic. For example, we may wish to see if a sequence contains more than the expected number of runs from a random arrangement; the null hypothesis and alternative are

$$H_0 : U \leq \overline{U}$$
$$H_1 : U > \overline{U}$$

and too many runs leads to rejection. The test is one-tailed. Conversely, we may wish to determine if the sequence contains an improbably low number of runs.

The appropriate alternatives are

$$H_0 : U \geqslant \overline{U}$$
$$H_1 : U < \overline{U}$$

and too few runs will cause rejection of the null hypothesis. Again, the test is one-tailed. We may wish to reject either form of nonrandomness. A two-tailed test is appropriate, with hypotheses

$$H_0 : U = \overline{U}$$
$$H_1 : U \neq \overline{U}$$

We can work through the test procedure for the first series of coin flips and determine the likelihood of achieving this sequence by a random process. The null hypothesis states that there is no difference between the observed number of runs and the mean number of runs from random sequences of the same size. We will use a two-tailed test, and reject if there are too many or too few runs in the sequence. Therefore, the proper alternative is

$$H_1 : U \neq \overline{U}$$

Using a 5% ($\alpha = 0.05$) level of significance, our critical regions are bounded by -1.96 and $+1.96$. We first calculate the expected mean and standard deviation of runs for random sequences having n_1 heads ($n_1 = 11$) and n_2 tails ($n_2 = 9$):

$$\overline{U} = \frac{2 \cdot 11 \cdot 9}{11 + 9} + 1 = 10.9$$

$$\sigma^2_{\overline{U}} = \frac{(2 \cdot 11 \cdot 9)(2 \cdot 11 \cdot 9 - 11 - 9)}{(9 + 11)^2 (9 + 11 - 1)} = 4.6$$

The test statistic is

$$Z = \frac{U - \overline{U}}{\sigma_{\overline{U}}} \approx \frac{13 - 10.9}{2.1} = 1.0$$

The number of runs in the sequence is one standard deviation from the mean of all runs possible in such a sequence, and does not fall within the critical region. Therefore, the number of runs does not suggest that the sequence is nonrandom. The other sequences, in contrast, yield very different test results. Because n_1 and n_2 are the same for all three sequences, $\overline{U}$ and $\sigma_{\overline{U}}$ also are the same. For the second sequence, the test statistic is

$$Z = \frac{2 - 10.9}{2.1} = -4.2$$

and for the third,

$$Z = \frac{19 - 10.9}{2.1} = 3.9$$

Both of these values lie within the critical region, and we would reject the hypothesis that they contain the number of runs expected in random sequences.

Geologic applications of this test may not be obvious, because we ordinarily must consider more than two states in a succession. Stratigraphic sections or traverses across thin sections, for example, usually include at least three states and these cannot be ranked in a meaningful way. We will consider ways that certain sequences can be reduced to dichotomous states, but first we will examine a geologic application of the runs test to a traverse through a two-state system.

Simple pegmatites originate by crystallization of the last, volatile-laden substances squeezed off from solidifying granitic magma. Their textures result from simultaneous crystallization of quartz and feldspar at the eutectic point. If the solidifying pegmatite is undisturbed, we might suppose that quartz and feldspar begin to appear at random locations within the cooling body. This situation may persist, with grains crystallizing at random until the entire mass is solid. However, the presence of one crystal, perhaps feldspar, might stimulate the local crystallization of additional crystals of feldspar, eventually producing a patchwork texture. Alternately, growth of a crystal of one state might locally deplete the magma of that constituent, retarding crystallization and resulting in a highly alternating mosaic of quartz and feldspar. A large slab of polished pegmatite used as a window ledge in the washroom of a geology building provides a way for students to investigate these alternative possibilities. The polished surface allows easy discrimination of adjacent grains, so a line drawn on the ledge produces a sequence through the quartz and feldspar grains in the pegmatite. The line on the polished slab may be regarded as a random sample of possible successions through the pegmatite body from which the slab was quarried. The quartz–feldspar sequence along the line is listed in Table 4.8. Our problem is to determine if the alternations between quartz and feldspar form a random pattern; if there is a systematic tendency for one state to succeed itself; or alternatively, a tendency for one state to immediately succeed the other. Perform a runs test on this data and evaluate the three possibilities.

We will now consider a related statistical procedure for examining what are called *runs up and down*. We are concerned, not with two distinct states, but whether an observation exceeds or is smaller than the preceding observation. Figure 4.12 shows a typical sequence that can be analyzed by the method of runs up and down.

The segment *ABC* is a run up, because each observation is larger than the preceding one; similarly, the segment *GHI* is a run down. Segment *CDEF* is a run down even though the difference between *D* and *E* is zero. This is because the interval *DE* lies between segments *CD* and *EF*, both of which run downward;

TABLE 4.8 Sequence of 100 Feldspar (F) and Quartz (Q) Grains Encountered along Traverse through Pegmatite

(Start)	FQQFQQFFQFQFFFFFFFFQQFQFFFQFFFFQFFFQ
	QFQFQQQFFFFFQFFFFFQQQQFFQQQFFFFFFFQ
	FQFFFFFQFQFQFFQFFFFFQFFFQQFQFQFFQ (End)

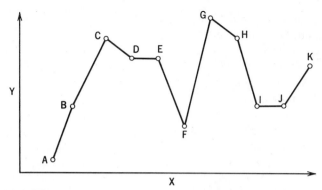

FIGURE 4.12 Sequence of data points to be analyzed by the method of runs up and down.

therefore, the entire segment *CDEF* can be considered as a single downward run. The interval *IJ* can be considered either as part of the run down *GHIJ* or the run up *IJK*, as the total number of runs remains the same in either case. In this example, we are assuming that the successive points have integer values. If the observations are expressions of magnitude, they ordinarily will contain fractional parts, and ties (two successive points with identical values) are unlikely.

By considering only differences in magnitude between successive points, we have reduced the data sequence to a string having only two states (or three, if ties occur). We can rewrite the sequence in Figure 4.12 to the following form:

$$+ + + - 0 - + - - 0 +$$

Regarding the first zero as " $-$ " gives a total of five runs, three of " $+$ " and two of " $-$ " (it makes no difference in the number of runs if we call the second zero " $+$ " or " $-$ "). We can now apply test procedures outlined for the case of sequences of two dissimilar items (4.8)–(4.10). We must have a large sample to utilize the normal approximation method presented here, but in most geologic problems, adequate numbers of samples will be available.

In the study of a silicified shale unit in the Rocky Mountains, it was noted that the rock contained unusual numbers of well-preserved radiolarian tests. Their presence in the silicified shale suggested a causal relationship, so a sequence of samples was collected at approximately equal intervals in an exposure through the unit. Thin sections were made of the samples, and the number of radiolarian tests in a 10 × 10 mm area of the slides was counted. Data for 50 samples are given in Table 4.9. Does the abundance of radiolarians vary at random through the section? A computer program can be written that will perform the necessary calculations, but the programming effort probably exceeds the difficulty of computing the test statistic by hand.

In this procedure, observations are dichotomized by comparing their magnitude to the preceding observations. Actually, runs tests may be applied to data dichotomized by any arbitrary scheme, provided the hypothesis tested reflects the di-

TABLE 4.9 Numbers of Radiolarian Tests per Square Centimeter in
Thin Sections of Siliceous Shale

(Bottom of section)	1	2	10	
	2	2	12	
	3	1	14	
	2	0	22	
	3	2	17	
	5	3	19	
	7	2	14	
	9	0	4	
	9	3	2	
	11	3	1	
	10	4	0	
	12	9	0	
	7	10	8	
	4	10	14	
	3	8	16	
	2	9	27	(Top of section)
	3	12		

chotomizing method. For example, a common test procedure is to dichotomize a series by subtracting each observation from the median of all observations, and testing the signs for randomness of runs about the median.

A large area of Precambrian anorthosite in the Laramie Range of Wyoming contains several bodies of magnetite. One of these is mined and the rock crushed for use as high-density drilling mud. An igneous petrology class has visited the quarry and collected a suite of equally spaced samples along a traverse across the magnetite body. The samples vary, some being predominantly anorthosite, others mostly olivine, some almost entirely magnetite, and others mixtures of the three minerals. Class members disagree as to whether a systematic variation in mineralogy occurs across the magnetite body. To test the possibility, the specific gravities of the samples are determined and are given in Table 4.10. Is the variation in specific gravity along the traverse that which would be expected if the composition varied randomly about a central value?

We also can test the randomness of runs about the mean, and we will use this as a test of residuals from trends later in this chapter. Runs tests are another example of the nonparametric procedures introduced in Chapter 2.

There are a number of variants on the runs tests described here. Information about these tests may be found in texts on nonparametric statistics, such as Bradley (1968, Chapters 11 and 12). Conover (1980, pp. 122–142), and Siegel (1956, pp. 52–59). Examples of the geologic application of runs tests are included in Miller and Kahn (1962, Chapter 14). Some consider the length of the longest run as an indicator of nonrandomness, and others use the number of *turning points*, which are points in the sequence where the signs of successive observations change. In certain instances, these tests may be more appropriate than the procedures described

TABLE 4.10 Specific Gravities of Samples Collected along a
Traverse across Magnetite Body in Laramie Range, Wyoming

(West margin)	3.57	4.58	4.22	
	3.63	5.02	3.52	
	2.86	4.68	2.91	
	2.94	4.37	3.87	
	3.42	4.88	3.52	
	2.85	4.52	3.77	
	3.67	4.80	3.84	
	3.78	4.55	3.92	
	3.86	4.61	4.09	
	4.02	4.93	3.86	
	4.56	4.60	4.13	
	4.62	4.51	3.92	
	4.31	3.98	3.54	(East margin)

here. The runs up and down test generally is regarded as the most powerful of the runs tests, because it utilizes changes in magnitude of every point with respect to adjacent points. Other dichotomizing schemes reflect only changes with respect to a single value such as the median or mean.

Runs tests are appropriate when the cause of nonrandomness is the object of investigation. They test for a form of nonrandomness expressed by the presence of too few or too many runs and do not identify overall trends. It should be emphasized that randomness itself cannot be proven, as the condition of random occurrence is implied in the null hypothesis. Rather, we can demonstrate, at specified levels of significance, that the null hypothesis is incorrect and the sequence is therefore not random, or we can fail to reject the null hypothesis, implying that we have failed to find an indication of nonrandomness. We will next consider procedures for detecting trends, or systematic changes in average value, and will find that runs tests may be used to good advantage in conjunction with these procedures.

Least-Squares Methods and Regression Analysis

In many types of problems, we are concerned not only with changes along a sequence, but we are also interested in where these changes occur. To examine these problems, we must have a collection of measurements of a variable, and we must also know the locations of the measurement points. Both the variable and the scale along the sequence must be expressed in units having magnitude; it is not sufficient to know simply the order of succession of points. We are interested in the general tendency of the data in most of the examples we will consider now. This tendency will be used to interpolate between data points, extrapolate beyond the data sequence, infer the presence of trends, or estimate characteristics that may be of interest to the geologist. If certain assumptions can justifiably be made about the distribution of the populations from which the samples are collected, statistical tests called *regression analyses* can be performed.

It must be emphasized that we are now using the expression "sequence" in the broadest possible sense. Regression methods are useful for much more than the analysis of observations arranged in order in time or space; they can be used to analyze *any* bivariate data set when it is useful to consider one of the variables as a function of the other. It is as though one variable forms a scale along which observations of the other variable are located, and we want to examine the nature of changes in this variable as we move up or down the scale.

The data in Table 4.11 are the moisture contents of samples from a core through Recent marine muds accumulating in a small inlet on the Gulf Coast in eastern Louisiana. The measurements are made by comparing the weight of a sample immediately after it is removed from the core barrel with its weight after forced drying. Moisture content is expressed as grams of water per 100 g of dried sediment. If we plot the measurements against depth, as in Figure 4.13, we can see that moisture content drops rapidly in the upper layers of sediment, but decreases slowly, if at all, in sediment near the bottom of the core. We will now consider various ways of representing the relationship implicit in these observations.

The value of 47.75 indicated on Figure 4.13 represents the mean moisture content of the samples. This value is the single number about which there is the smallest variance. That is, the sum of the squared deviations of moisture content is the smallest about this point. You will recall from Chapter 2 that, if certain sampling precautions are observed, this value also is an unbiased and efficient estimator of the population mean, and is the "best guess" or predictor of additional samples drawn from the same population. However, the mean clearly is not adequate to represent data in Figure 4.13. This is because the samples were taken sequentially, and hence are not independent. Rather than a point estimate, we need a line that will express the relation between moisture content and depth throughout the range of both variables. It seems intuitively reasonable to construct the line so that deviations from it are minimized in some way. One choice might be to minimize

TABLE 4.11 Moisture Content of Core Samples of Recent Mud in Louisiana Estuary

Depth (ft)	Moisture (grams water/100 g dried solids)
0	124
5	78
10	54
15	35
20	30
25	21
30	22
35	18

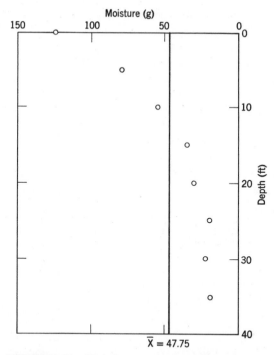

FIGURE 4.13   Plot of moisture content (grams water/ 100 grams dry weight) versus depth below sediment—water interface. Data collected from a core through Recent estuary mud in Louisiana bay. Note that orientation of plot corresponds to correct geologic orientation and not to standard mathematical form.

the sum of squared deviations from the line, reasoning by analogy to the mean. (The mean is the value about which the variance, and hence the sum of squared deviations, is the smallest.) We then could construct a unique line about which the variance is a minimum. If the value of this line is subtracted at the appropriate points from the corresponding observations, the collection of resulting numbers has a mean of zero and smaller variance than the set of deviations that could be calculated from any other straight line through the data.

There are, however, several alternative ways that deviations from the fitted line can be defined and measured. For example, we might consider deviations in moisture content, deviations in depth, or some combination of the two. In Figure 4.14, line A represents the deviation of moisture content from a fitted line, and C is deviation of depth from the line. The deviation B is measured perpendicular to the fitted line. It is possible to construct lines in which any of these criteria of deviation are used; however, we should examine the implications of each in the light of our specific problem. If we minimize deviations of moisture content, we are stating

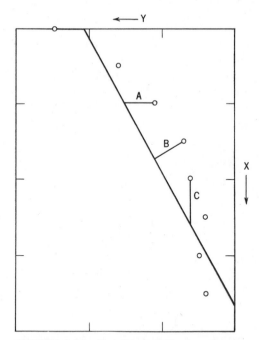

FIGURE 4.14 Possible criteria for minimization of deviations from fitted lines: A, minimization of deviations in moisture content; B, minimization of joint deviations; C, minimization of deviations in depth.

that the resulting line is the best estimator of moisture at specified depths. In contrast, if we minimize deviations of depth, we are estimating depth from moisture content. The third alternative is an expression of the joint relationship of the two variables. In the particular class of problems we are examining in this chapter, time or distance is known and a second variable is distributed along this continuum. Therefore, the first alternative is appropriate. In other words, moisture content, Y, is a random variable, and depth, X, is fixed. Therefore, the problem is to predict Y from X. Other approaches will be considered in later sections of this book.

Having agreed on the characteristics of the desired trend line, we now must define some terms. The variable being examined is the *dependent* or *regressed* variable, designated Y_i. Deviations of Y_i from the fitted line will be minimized. The other variable is the *independent* or *regressor* variable and is denoted X_i. The fitted line will cross the Y-axis at a point b_0 (the intercept) and will have a slope b_1. The equation of the line is

$$\hat{Y}_i = b_0 + b_1 X_i \tag{4.11}$$

$\hat{Y}_i$ is the estimated value of Y_i at specified values of X_i. The deviation we are considering, therefore, is $\hat{Y}_i - Y_i$ and our problem becomes one of finding a method such that

$$\sum_{i=1}^{n} (\hat{Y}_i - Y_i)^2 = \text{minimum} \tag{4.12}$$

Derivation of the necessary technique requires differential calculus, so we will not consider a proof. Rather, we will present what are called the *normal equations* that give values of b_0 and b_1 defining a line having the desired characteristics:

$$\sum_{i=1}^{n} Y_i = b_0 n + b_1 \sum_{i=1}^{n} X_i \tag{4.13}$$

$$\sum_{i=1}^{n} X_i Y_i = b_0 \sum_{i=1}^{n} X_i + b_1 \sum_{i=1}^{n} X_i^2 \tag{4.14}$$

Rewriting, we obtain

$$b_1 = \frac{\sum_{i=1}^{n} X_i Y_i - \left(\sum_{i=1}^{n} X_i \sum_{i=1}^{n} Y_i\right)\Big/ n}{\sum_{i=1}^{n} X_i^2 - \left(\sum_{i=1}^{n} X_i\right)^2 \Big/ n} = \frac{SP_{xy}}{SS_x} \tag{4.15}$$

and

$$b_0 = \frac{\sum_{i=1}^{n} Y_i}{n} - b_1 \frac{\sum_{i=1}^{n} X_i}{n} = \bar{Y} - b_1 \bar{X} \tag{4.16}$$

We can use these to obtain the coefficients of the line, but you should recognize that (4.13) and (4.14) are a pair of simultaneous equations that can be solved by methods developed in Chapter 3. The two equations can be rewritten in matrix form:

$$\begin{bmatrix} n & \Sigma X \\ \Sigma X & \Sigma X^2 \end{bmatrix} \cdot \begin{bmatrix} b_0 \\ b_1 \end{bmatrix} = \begin{bmatrix} \Sigma Y \\ \Sigma XY \end{bmatrix} \tag{4.17}$$

Although there is hardly any advantage to the matrix expression in this simple case, matrix methods will be necessary to fit more complex lines. Therefore, we will solve the problem of moisture content versus depth by matrix algebra, and we will use this method throughout the chapter. The necessary quantities are $n = 8$, $\Sigma X = 140$, $\Sigma Y = 382$, $\Sigma XY = 3870$, and $\Sigma X^2 = 3500$. In matrix form,

$$\begin{bmatrix} 8 & 140 \\ 140 & 3500 \end{bmatrix} \cdot \begin{bmatrix} b_0 \\ b_1 \end{bmatrix} = \begin{bmatrix} 382 \\ 3870 \end{bmatrix}$$

Solving gives $b_0 = 94.67$ and $b_1 = -2.68$. We can use this line to estimate values of moisture content at various depths. The estimated values at the sample points ($\hat{Y}_i$) provide a measure of how well the least-squares line conforms to the raw data. If the line passed exactly through each sample point, $\hat{Y}_i$ and Y_i would correspond and the sum of the squared deviations from the line would be zero. Of course, in this example they do not. Both $\hat{Y}_i$ and Y_i are plotted in Figure 4.15.

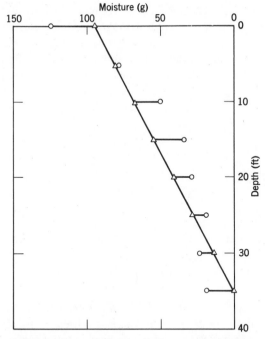

FIGURE 4.15 Observed values of moisture content and estimated values predicted by a straight line fitted by least squares.

We can define three terms that express variation of the dependent variable. The first of these is measured by the *total sum of squares* (SS_T) of Y:

$$SS_T = \sum_{i=1}^{n} Y_i^2 - \frac{\left(\sum_{i=1}^{n} Y_i\right)^2}{n} = \sum_{i=1}^{n} (Y_i - \bar{Y})^2 \qquad (4.18)$$

This quantity, divided by $(n - 1)$, gives the variance of Y:

$$s^2 = \frac{SS_T}{n - 1} = \frac{n \sum_{i=1}^{n} Y_i^2 - \left(\sum_{i=1}^{n} Y_i\right)^2}{n(n - 1)} \qquad (4.19)$$

The second measure of variation is the *sum of squares due to regression* (SS_R):

$$SS_R = \sum_{i=1}^{n} \hat{Y}_i^2 - \frac{\left(\sum_{i=1}^{n} \hat{Y}_i\right)^2}{n} = \sum_{i=1}^{n} (\hat{Y}_i - \bar{Y})^2 \qquad (4.20)$$

As the expression on the right implies, the estimated values of $\hat{Y}_i$ have the same mean as the original values. The sum of squares of these estimates $\hat{Y}_i$ provides a measure of the variation of the regression line around the mean. If $\hat{Y}_i$ and Y_i correspond for all observations, the sums of squares calculated by (4.18) and (4.20) will be the same. Otherwise, the sum of squares due to regression will be smaller, and there will be "leftover" variation we can call the *sum of squares due to deviations* (SS_D):

$$SS_D = SS_T - SS_R \qquad (4.21)$$

This is a measure of the failure of the least-squares line to fit the data points. It can also be calculated by

$$SS_D = \sum_{i=1}^{n} (\hat{Y}_i - Y_i)^2 \qquad (4.22)$$

which is algebraically equivalent to (4.21).

The *goodness-of-fit* of the line to the points can be defined by

$$R^2 = \frac{SS_R}{SS_T} \qquad (4.23)$$

If the line is a good estimator of the data, this ratio will be near unity; we will later discuss tests of "how good" this value is. Often R^2 is expressed as a percentage. The same terminology is commonly used in trend-surface analysis, which we will see is a direct extension of this method. A most useful relation is that the square root of goodness-of-fit is the *multiple correlation coefficient*, R:

$$R = \sqrt{R^2} = \sqrt{SS_R/SS_T} \qquad (4.24)$$

This definition is algebraically equivalent to the definition of the correlation coefficient given in Chapter 2:

$$r = \frac{SS_{xy}}{\sqrt{SS_x \cdot SS_y}} \qquad (4.25)$$

In calculating the equation of the least-squares line between moisture content and depth, we have found all of the entries necessary to determine the various sums of squares, goodness-of-fit, and correlation. Compute the quantities SS_T, SS_R, SS_D, R^2, and R for the data in Table 4.11.

It is obvious from the appearance of the fitted line that a straight line does not closely approximate the data, even though the correlation is high. Poor fit may arise from several causes, including high variance in the dependent variable (excessive scatter in the data), or from the fitting of an inappropriate model. In this example, we would tend to suspect the latter, because the points seem to approximate a smooth curve rather than a straight line. We will investigate fitting curves after considering statistical tests that can be performed, provided the data meet certain requirements.

If Y_i is a random variable observed at specified intervals of X_i, we can assume

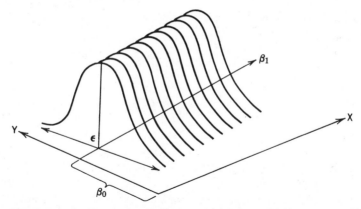

FIGURE 4.16 Components of the regression model $Y_i = \beta_0 + \beta_1 X + \epsilon_i$. Error is assumed to be normally distributed about the regression line.

that our data follow the theoretical population model

$$Y_i = \beta_0 + \beta_1 X_i + \epsilon_i \tag{4.26}$$

where i represents successive observations. The quantity ϵ is a normally distributed random variable with mean zero and unknown variance σ^2 independent of the value of Y_i. That is, an observed value of Y_i is assumed equal to the sum of a constant related to the mean (if both Y_i and X_i are converted to deviations around their means, β_0 vanishes), plus a linear function of X_i plus a random error or deviation ϵ. This relationship is shown in Figure 4.16. At every point, a normal frequency distribution of possible values of Y_i is assumed to exist around the line of regression. By the method of least squares, we can estimate the population regression parameters [the β's in model (4.26)] by sample regression coefficients [the b's in the computational equation (4.17)]. If the assumptions we have just made are true, the least-squares method will give b's that are the most likely estimate of the regression parameters, and the line we compute will be closer to the true regression line than any other. If the linear equation we fit to the data is a true model of the regression, the variance of ϵ is equal to the variance around the regression line. On the other hand, if the model is not correct, the variance around the regression will be greater than the variance of ϵ.

We can use the sums of squares to compute variances that in turn can be used to test the two alternatives. In particular, SS_D can be used to estimate variance about the regression. We can estimate σ^2 adequately only if we have replicate measurements of Y_i at each point X_i, because this is the only way we can obtain a measure of the variance in Y independently of variance in X. However, SS_R will give an estimate of σ^2 when our model is correct and will estimate σ^2 plus the amount of bias if our model is incorrect. Using SS_R, an analysis of variance can be constructed that will lead to rejection if either the observations are too variable

TABLE 4.12 ANOVA for Simple Linear Regression

Source of Variation	Sum of Squares	Degrees of Freedom	Mean Squares	F Test
Linear Regression	SS_R	1	MS_R	MS_R/MS_D
Deviation	SS_D	$n - 2$	MS_D	
Total Variation	SS_T	$n - 1$		

for reliable judgment or the postulated model is incorrect. Table 4.12 shows the form of the analysis of variance.

As in Chapter 2, mean squares are variance estimates made by dividing the appropriate sums of squares by their degrees of freedom. The MS_R has one degree of freedom because it is based on two "observations," the coefficients b_0 and b_1. The total variance has $(n - 1)$ degrees of freedom. Therefore, MS_D must have degrees of freedom equal to the difference between the two, or $(n - 1) - 1 = n - 2$. We can complete the ANOVA for the problem as we have done in Table 4.13. We are testing the hypothesis and alternative

$$H_0:\beta_1 = 0$$

$$H_1:\beta_1 \neq 0$$

The regression line is constrained to pass through the means of X and Y. If the slope coefficient β_1 is not significantly different from zero, this is equivalent to saying that the scatter in values of Y about the regression line is no less than their scatter about $\overline{Y}$. We will use a 5% level of significance ($\alpha = 0.05$). The test statistic follows an F distribution with degrees of freedom $v_1 = 1$ and $v_2 = 6$, so the critical region will consist of values exceeding $F = 5.99$. The computed test value falls well in the critical region, so we must reject the hypothesis that the variance about the regression line is no different than the variance in the observations. However, even though a significant linear trend exists in the data, a plot of the data suggests that we should be able to do better.

Fifty feet from the first core, a second core was taken through the mud sequence in the estuary. The water content of samples from the core can provide us with a

TABLE 4.13 Completed ANOVA for Significance of Regression of Water Content on Depth

Source of Variation	Sum of Squares	Degrees of Freedom	Mean Squares	F Test
Regression	7,546.88	1	7,546.88	23.07[a]
Deviation	1,962.62	6	327.10	
Total	9,509.50	7		

[a]Significant at the $\alpha = 5\%$ level of significance.

set of replicates of Y_i, allowing direct estimation of σ^2. We then can determine if the poor correlation between water content and depth is due to high variance or lack of fit to our model equation. Data from the second core are given in Table 4.14. Plot these points on a graph and compare their distribution to those in Table 4.11.

Measurements from Table 4.11 can be combined with those of Table 4.14 and a regression computed using all of the observations. We will calculate SS_T, SS_R, and SS_D in exactly the same manner as before except, of course, we now have twice as many observations. Because we have replicate measurements, we also can calculate a *sum of squares due to lack of fit* (SS_{LF}) and a *sum of squares due to pure error* (SS_{PE}). These divide the sum of squares of deviation into two parts. In the case of pairs of replicates we can find SS_{PE} by

$$SS_{PE} = 1/2 \sum_{i=1}^{n} (Y_{i1} - Y_{i2})^2 \qquad (4.27)$$

which has one degree of freedom for every replicated point. The remaining sum of squares, SS_{LF}, is found by subtraction, as are its degrees of freedom:

$$SS_{LF} = SS_D - SS_{PE} \qquad (4.28)$$

It is not necessary that we measure replicates of every point, but the analysis is more powerful if all are replicated. It is possible to use more than two replicates of Y_i for each value of X_i, but the calculation of SS_{PE} becomes somewhat more complex. These and other modifications are described in books on regression such as those by Li (1964, Chapter 30) and by Draper and Smith (1981, pp. 33–42, 80–85) and will not be considered further here.

Our modified analysis of variance table has the form shown in Table 4.15. Using the combined raw data from the two cores, complete the ANOVA and calculate SS_{PE} and SS_{LF}. The mean square of SS_{PE} is an estimate of $\sigma_{Y \cdot X}^2$, the variance around the regression line. It is found by

$$MS_{PE} = \frac{SS_{PE}}{k} \qquad (4.29)$$

where k is the number of data points that are replicated. In our case, we have taken replicates of all points, so k is equal to $n/2$, as half of the observations of Y_i are

TABLE 4.14 Moisture Content of Second Core

Depth (ft)	Moisture (grams water/100 g dried solids)
0	137
5	84
10	50
15	32
20	28
25	24
30	23
35	20

TABLE 4.15   ANOVA for Simple Linear Regression with Replicates; Number of Observations of Y_i is n, Number of Points Replicated is k

Source of Variation	Sum of Squares	Degrees of Freedom	Mean Squares	F Test
Linear Regression	SS_R	1	MS_R	MS_R/MS_D[a]
Deviation	SS_D	$n - 2$	MS_D	
Lack of Fit	SS_{LF}	$(n - 2) - k$	MS_{LF}	MS_{LF}/MS_{PE}[b]
Pure Error	SS_{PE}	k	MS_{PE}	
Total Variation	SS_T	$n - 1$		

[a]Tests for goodness-of-fit.
[b]Tests for appropriateness of model.

replicates. We stated that SS_D was a measure of the variance around the regression plus any bias that might result from an inappropriate model, so the mean square of SS_{LF} is an estimate of the bias alone. We can test the appropriateness of the model equation by

$$F = \frac{MS_{LF}}{MS_{PE}} \tag{4.30}$$

If the test value falls into the critical region, we may conclude that the model appears to be inadequate. If the test does not lead to rejection of the model, we can combine both variance estimates $MS_{LF} + MS_{PE} = MS_D$ and test for goodness-of-fit as we have done previously. Calculate this F ratio by completing ANOVA Table 4.15 and determine if the simple linear model is appropriate. The four possible outcomes for the two tests, one of the appropriateness of the model, and the other of goodness-of-fit, are graphically shown in Figure 4.17.

Curvilinear Regression

After calculating the F test for the appropriateness of your model, you may conclude that a straight line is inadequate. What you do next depends upon the objectives of the study and your knowledge (or beliefs) of the relationship between X and Y. On one hand, you may be able to make definitive statements about the relation between the two variables. For example, if a bomb is dropped from an airplane, its theoretical path, ignoring wind resistance, is a curve defined by the forward velocity of the plane and the downward acceleration of gravity (Fig. 4.18); a parabola can justifiably be fitted to observations of the bomb's location. On the other hand, we may know nothing of the physical relation between the two variables X and Y (indeed, none may exist), but simply want a concise expression of one in terms of the other. Usually our problems lie between these two extremes; we suspect a causal relationship, but do not know its form. In the latter two cases, we can fit *approximating equations* to the data in the hope that these will shed

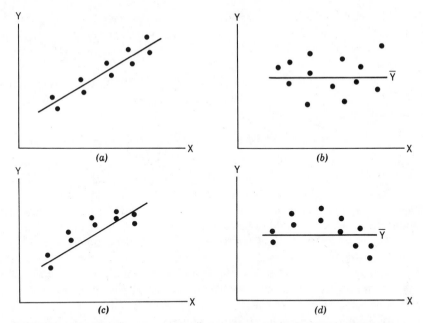

FIGURE 4.17   Possible straight-line regression situations. (*a*) Significant linear regression, no lack of fit. (*b*) Linear regression not significant, no lack of fit. (*c*) Significant linear regression, significant lack of fit. (*d*) Linear regression not significant, significant lack of fit. After Draper and Smith (1981).

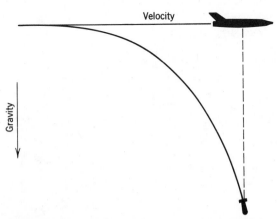

FIGURE 4.18   Theoretical path of a projectile dropped from an airplane.

some light on the underlying relations, or at least describe the form of the X–Y relationship in a concise way. These equations are chosen because they are capable of approximating many classes of functions and commonly are used when the correct form of a function is not known.

There are several approximating equations available, but the one most commonly used is a *polynomial expansion*. This is a summation of integer powers of the independent variable:

$$Y_i = b_0 + b_1 X_i + b_2 X_i^2 + b_3 X_i^3 + \cdots + b_m X_i^m \qquad (4.31)$$

An equation in which all terms are added together is called a *linear function*, because the relationships between the dependent variable and any one of the independent variables graph as straight lines if the other variables are held constant. Expansion of our original equation by adding successive powers allows our original straight line to bend. One additional term (X^2) allows the line to reverse its slope; a second additional term (X^3) allows two changes in the direction of the slope, and so on. Increased flexibility means that the line can conform more closely to the data. In fact, when the number of additional powers reaches ($n - 1$), the line will go exactly through every data point. There is little purpose in computing such a line, because it is no more efficient than the original data. Hopefully, the essential aspects of the data array can be preserved with only a few terms in the polynomial equation. Figure 4.19 shows typical polynomial curves incorporating various powers of X. The maximum power used in the polynomial defines the *degree* of the equation. That is, $Y_i = b_0 + b_1 X_i + b_2 X_i^2 + b_3 X_i^3$ is a third-degree polynomial. So is $Y_i = b X_i^3$, because this is the same as stating that b_0, b_1, and b_2 are equal to zero. The polynomial equations are fitted to the observations by least-square methods, and the process is called *curve fitting*. If certain statistical assumptions are valid, the goodness-of-fit and appropriateness of the curve can be tested by extensions of the regression techniques we have just considered. The statistical procedures are grouped under *curvilinear regression analysis*.

To fit a second-degree (or quadratic) curve to data, we must expand the normal equations to include additional terms. The two normal equations (4.13) and (4.14) become a set of three simultaneous equations:

$$\Sigma Y = b_0 n + b_1 \Sigma X + b_2 \Sigma X^2$$
$$\Sigma XY = b_0 \Sigma X + b_1 \Sigma X^2 + b_2 \Sigma X^3 \qquad (4.32)$$
$$\Sigma X^2 Y = b_0 \Sigma X^2 + b_1 \Sigma X^3 + b_2 \Sigma X^4$$

All summations are understood to extend over the observations from 1 to n. Rewriting these into matrix form gives

$$\begin{bmatrix} n & \Sigma X & \Sigma X^2 \\ \Sigma X & \Sigma X^2 & \Sigma X^3 \\ \Sigma X^2 & \Sigma X^3 & \Sigma X^4 \end{bmatrix} \cdot \begin{bmatrix} b_0 \\ b_1 \\ b_2 \end{bmatrix} = \begin{bmatrix} \Sigma Y \\ \Sigma XY \\ \Sigma X^2 Y \end{bmatrix} \qquad (4.33)$$

which can be readily solved by the matrix algebra procedure given in Chapter 3. Note that high powers of the independent variable are required in the equations.

The largest power in the matrix is twice the degree of the equation being fitted. This can be a major source of errors in computer programs that fit polynomials, because elements in the lower right part of the coefficient matrix may be many orders of magnitude greater than those in the upper left corner of the matrix. This may lead to round-off errors and loss of significance in critical digits, resulting in unstable or unreliable solutions to the simultaneous equations. An extensive discussion of these problems is contained in the book by Westlake (1968).

The structure of the coefficient matrix can be seen if we utilize a dummy variable, X^0, which we will define as being equal to 1 for every observation X_i. We can label the rows and columns in the matrix equation in the following manner:

$$
\begin{array}{c}
\begin{array}{cccccc} X^0 & X^1 & X^2 & X^3 & \cdots & X^m \end{array} \\
\begin{array}{c} X^0 \\ X^1 \\ X^2 \\ X^3 \\ \vdots \\ X^m \end{array}
\begin{bmatrix} & & & & & \\ & & & & & \\ & & & & & \\ & & & & & \\ & & & & & \\ & & & & & \end{bmatrix}
\end{array}
\cdot
\begin{bmatrix} b \\ \\ \\ \\ \\ \end{bmatrix}
=
\begin{bmatrix} Y \\ \\ \\ \\ \\ \end{bmatrix}
\qquad (4.34)
$$

Entries within the body of the coefficient matrix and within the matrices of b coefficients and right-hand parts are sums of the cross-products of the row and column labels. By our definition of X^0, the a_{11} entry becomes $\sum_{i=1}^{n} 1 \cdot 1 = n$, and the other entries in the top row are equal to 1 times the column label. Element a_{43} in the matrix is, for example, $\sum X^3 \cdot X^2 = \sum X^5$. Remember that multiplication of exponents consists of adding the powers. That is, $X^a \cdot X^b = X^{a+b}$.

To demonstrate the computations used in curvilinear regression, we can analyze the combined data from Tables 4.11 and 4.14. We will work through the problem using a quadratic fit to demonstrate an additional procedure that tests to see if the increase in the degree of the polynomial has significantly improved the fit of the regression. The second-degree polynomial curve fit to the data is shown in Figure 4.20. The regression equation is

$$
Y_i = \beta_0 + \beta_1 X_i + \beta_2 X_i^2
$$
$$
= 122.9 - 7.9X_i + 0.1X_i^2
$$

Statistics necessary to perform an analysis of variance include

$$
\begin{array}{ll}
SS_T = 21{,}363.0 & SS_R = 20{,}673.2 \\
SS_D = 689.8 & SS_{PE} = 126.0 \\
R^2 = 0.97 & R = 0.98
\end{array}
$$

You will note that SS_T and SS_{PE} are the same as in the linear fit to this data, because they do not contain the estimated values $\hat{Y}$. As we would expect, the more flexible quadratic equation fits the moisture content data much more closely than does a straight line. The sum of squares due to deviations from the regression has been reduced from 5,177.8 to 689.8. It seems obvious that this is a significant reduction,

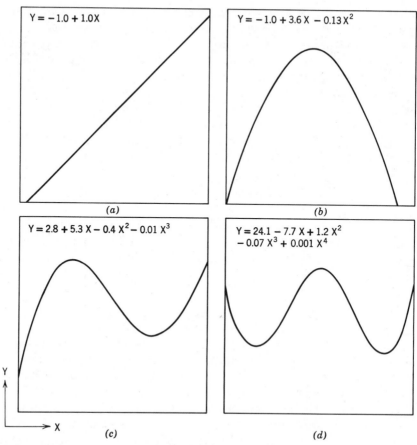

FIGURE 4.19 Polynomial regression curves of increasing powers of X. (a) Linear or first-degree curve. (b) Quadratic or second-degree curve. (c) Cubic or third-degree curve. (d) Quartic or fourth-degree curve.

but this distinction is not always so apparent. The analysis of variance table can be expanded again to test this possibility. This new ANOVA is given in Table 4.16.

As you can see from the table, a sum of squares is created by subtracting the regression sum of squares for a linear fit (SS_{R1}) from the equivalent sum of squares for a quadratic fit (SS_{R2}). This new sum of squares is a measure of the increase in fit resulting from the additional regression term. In test (b) of the ANOVA table, this quantity, which we have designated SS_{2-1}, is used to estimate the additional regression variance. Its significance is tested in exactly the same manner as the significance of the regression itself. If the resulting F value falls in the critical region, the added term is making a significant contribution to the regression and should be retained. If the test is not significant, the additional power is not contributing to the regression. Note that test (a) may be significant when test (b) is

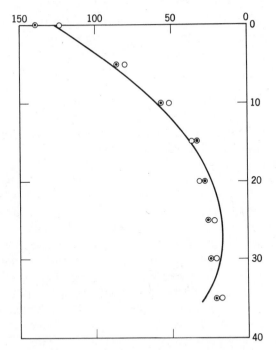

FIGURE 4.20   Second-degree polynomial regression fit to moisture data from Tables 4.11 and 4.14.

not. This is because test (a) actually is examining the significance of the linear and quadratic terms combined; the linear fit may be highly significant but the quadratic contribution very low. Test (a) will then be significant because of the strength of the linear term alone. In different situations, we may find that either, or both, or neither terms may be significant.

You should also note that the correlation always will increase with the addition

TABLE 4.16   ANOVA for Significance of Added Terms in Curvilinear Regression

Source of Variation	Sum of Squares	Degrees of Freedom	Mean Square	F Test
Linear Regression	SS_{R1}	1	MS_{R1}	
Quadratic Regression	SS_{R2}	2	MS_{R2}	MS_{R2}/MS_{D2}[a]
Addition by Quadratic	SS_{2-1}	1	MS_{2-1}	MS_{2-1}/MS_{D2}[b]
Quadratic Deviation	SS_{D2}	$n-3$	MS_{D2}	
Total Variation	SS_T	$n-1$		

[a]Tests for significance of the quadratic fit.
[b]Tests for significance of increase of quadratic over linear fit.

of terms. When the number of terms becomes equal to $(n - 1)$, the correlation will equal 1.00, regardless of how wildly the data points are scattered. However, the tests outlined above may show that increases in correlation are not statistically meaningful. The F ratio for significance of fit may decrease because the mean squares due to deviation may increase. This variance estimator is dependent in part on the number of observations used in its creation, or its degrees of freedom. These are being reduced constantly as we add coefficients to the regression equation. Remember from Chapter 2 that we lose one degree of freedom for each parameter we estimate, and the b's of the polynomial equation are estimates of β's, the population regression parameters.

This procedure for testing the significance of added terms can be extended to successively higher powers in the polynomial regression, provided the statistical assumptions are fulfilled. These tests can also be combined with the tests for lack of fit and pure error if replicates are available. Table 4.17 is the completed ANOVA for a quadratic regression on the combined moisture content data.

In some problems, we are more concerned with the value of the intercept or the slope of the regression line than we are with estimates of points or deviations from these. As an example of a problem of this type, we will consider a stratigraphic succession from a long core through Lower Paleozoic rocks in eastern Oklahoma. A geologist has studied a thick clastic unit cut by the drill in an attempt to interpret environmental conditions prevailing at the time of deposition. The unit consists of interbedded siltstones and sandstones believed to be offshore marine deposits. The geologist has postulated that the basin gradually filled, and as the shoreline advanced toward the location of the well, successively deposited sand layers should be increasingly thicker. The unit contains thousands of beds, and it would be extremely time-consuming to measure each of them. Instead, the thickness of a single sandstone bed was measured at successive 10-ft intervals. These measurements are listed in Table 4.18. The geologist wishes to determine if the slope of individual bed thickness versus thickness of accumulated sediment is a real phenomenon.

TABLE 4.17 Completed ANOVA for Significance of the Quadratic Regression of Moisture Content of Sediment on Depth of Burial

Source of Variation	Sum of Squares	Degrees of Freedom	Mean Square	F Test
Linear Regression	16,185.19	1	16,185.19	
Quadratic Regression	20,673.24	2	10,336.62	$MS_{R2}/MS_{D2} = 227.83$[a]
Addition from Quadratic	4,488.05	1	4,488.05	$MS_{2-1}/MS_{D2} = 98.92$[b]
Quadratic Deviation	689.76	13	45.37	
Total Variation	21,363.00	15		

[a]The quadratic regression is highly significant.
[b]The quadratic regression is a highly significant improvement over the linear regression alone.

TABLE 4.18 Thickness of Sandstone Beds Measured at Intervals through a Lower Paleozoic Clastic Unit in Oklahoma

Interval X (ft)	Thickness Y (inch)	Interval X (ft)	Thickness Y (inch)
10.0	9.2	260.0	8.5
20.0	7.1	270.0	8.9
30.0	5.9	280.0	10.7
40.0	3.7	290.0	14.4
50.0	6.2	300.0	15.2
60.0	4.1	310.0	12.1
70.0	3.9	320.0	15.3
80.0	5.0	330.0	9.0
90.0	4.4	340.0	11.2
100.0	6.8	350.0	8.9
110.0	5.9	360.0	9.0
120.0	6.1	370.0	6.5
130.0	7.7	380.0	11.0
140.0	7.0	390.0	13.9
150.0	5.5	400.0	9.1
160.0	9.8	410.0	11.2
170.0	6.9	420.0	17.3
180.0	5.2	430.0	15.8
190.0	6.8	440.0	11.1
200.0	8.5	450.0	11.8
210.0	7.1	460.0	18.9
220.0	10.4	470.0	9.6
230.0	6.7	480.0	17.9
240.0	8.6	490.0	12.8
250.0	6.4	500.0	15.0

Note that cumulative thickness, X, is measured at fixed locations, while individual bed thickness is a random variable; therefore, a regression model is appropriate. In other words, the geologist must test the hypothesis that the slope coefficient, b_1, of a linear regression is significantly greater than zero.

After we have calculated the regression equation $Y_i = b_0 + b_1 X_i$ by the procedure just outlined, we can estimate the variance about the regression by MS_D. This can in turn be used to calculate the t test statistic:

$$t = \frac{b_1}{\sqrt{MS_D/SS_X}} \tag{4.35}$$

The mean square due to deviation (MS_D) is equal to SS_D divided by $(n - 2)$ degrees of freedom, as shown in Table 4.12. The corrected sums of squares of X, SS_X, is found by

$$SS_X = \sum_{i=1}^{n} X_i^2 - \frac{(\sum_{i=1}^{n} X_i)^2}{n} \tag{4.36}$$

This is a test of one of the hypotheses

1. $H_0 : \beta_1 = 0$
2. $H_0 : \beta_1 \leqslant 0$
3. $H_0 : \beta_1 \geqslant 0$

against their respective alternatives:

1. $H_1 : \beta_1 \neq 0$
2. $H_1 : \beta_1 > 0$
3. $H_1 : \beta_1 < 0$

The first null hypothesis requires a two-tailed test, because either significant positive or negative slopes lead to rejection. The other tests are one-tailed. Our geologist is only interested in establishing if bed thickness is increasing up the section; that is, if the slope of the regression between bed thickness and cumulative thickness is positive. Therefore, the second hypothesis and its alternative are the appropriate choices. The test will be one-tailed, with the critical region on the right.

We can quickly calculate the necessary entries for the test. Some of the numerical values required for this test are listed below.

$$\text{Regression equation:} \quad Y = 4.25 + 0.020X$$

$$SS_T = 730.94 \qquad SS_R = 425.18$$

$$SS_D = 305.76 \qquad SS_X = 1{,}041{,}250.00$$

$$R^2 = 0.58 \qquad R = 0.76$$

There are 50 observations in the data set, so there are 48 degrees of freedom associated with SS_D. At a significance level of 5% ($\alpha = 0.05$), the critical value of t with $\nu = 46$ degrees of freedom is 1.68. The test statistic is

$$t = \frac{0.02}{\sqrt{6.37/1{,}041{,}250.00}} = 8.09$$

This value lies within the critical region, so we must reject the hypothesis that the slope of the regression is zero or negative. A small but definite increase in bed thickness occurs through the sequence.

The test just presented is a special case of

$$t = \frac{b_1 - \beta_1}{\sqrt{MS_D/SS_X}} \tag{4.37}$$

which tests the hypothesis that the regression slope is equal to some predefined value β_1. In test (4.35), β_1 is zero. This test, or a version of it, is very important in time series analysis. Time series procedures are based on the assumption that there is no trend in the data, or that the slope of the regression on time (or distance) is zero. If a trend is present, it must be removed, or the analysis of the time series is invalid. A series having no significant linear trend is said to be *stationary*. If a persistent trend or "drift" is present in the data, it is called *evolutionary* or *nonstationary*.

One of the assumptions of linear regression is that the variance is constant about the regression line. This assumption can be tested by examining the residuals from the fitted line. If the variance is constant, the residuals will form a more-or-less uniform band around the regression line. If there is a progressive change in the width of the band of deviations, the variance may not be constant. These two conditions are known by the somewhat terrifying names of *homoscedasticity* for constant variance, and *heteroscedasticity* for changing variance. A "quick and dirty" way to check the consistency of variance about a regression is to perform a linear regression on the absolute values of the deviations. A change in variance along the sequence will appear as a significant slope.

Another assumption made in regression analysis is that deviations or residuals from the regression are free from autocorrelation. *Autocorrelation* in this context means that residuals tend to occur as "clumps" of adjacent deviations on the same side of the regression line. The presence of large sequences of autocorrelated residuals may indicate that the regression model is inappropriate. It also happens that autocorrelated deviations may suggest phenomena of geologic interest. This subject will be pursued more thoroughly when we discuss trend-surface analysis, where autocorrelated positive residuals may be an indicator of economic potential in the form of oil or other mineral accumulations. Testing for autocorrelation is easy; we need only apply the runs test to the signs of the deviations from the regression, or we can use one of the procedures discussed in the section on autocorrelation.

Both consistency of variance and autocorrelation of residuals are of interest in an analysis of a mining prospect in northern Quebec. A long trench has been cut by bulldozer across a gold prospect. Samples have been taken at intervals along the trench and the gold value of each determined by assay. A trend in the values seems apparent, and deviations from the regression seem more extreme near one end of the trench. This is suggestive of the "bonanza" characteristic of many gold deposits. These often have extremely low values through most of the mineralized zone, but occasional rich veins occur. In addition, large positive deviations seem to occur in clusters or runs, which also suggest that the trench transects mineralized veins. From the data given in Table 4.19, determine the trend of gold values, and examine the consistency of variance along the traverse. Use a runs test of the signs of the deviations to determine if they are distributed randomly about the regression. From the results of your analysis, does it seem possible that mineralized veins have been cut by the trench? Does extrapolation of gold values beyond the limits of the trench seem wise in view of the behavior of deviations from the regression?

Orthogonal Polynomial Regression

Fitting a high-degree polynomial curve to data by least squares involves solving a large set of simultaneous equations—an onerous task prior to the development of computers. As a consequence, early workers avoided the general regression method whenever possible and used a computationally simpler approach called *orthogonal polynomial regression*. To apply this method, the data must be collected at equal increments in X. In addition to simplifying the calculations, the coefficients of an

TABLE 4.19 Gold Assay Values of Samples Collected in a Prospect Trench in Northern Quebec

Distance (ft)	Values (dwt/ton)[a]	Distance (ft)	Values (dwt/ton)[a]
(North end of trench)			
3.0	0.9	66.0	9.0
9.2	1.2	67.0	12.0
13.0	0.5	68.1	10.4
18.9	1.7	71.1	5.2
22.3	1.4	73.0	1.4
23.1	1.3	74.1	1.2
25.5	1.0	76.0	1.1
28.6	1.1	76.1	1.0
30.1	12.0	80.4	6.5
30.9	9.1	82.2	11.9
33.0	4.9	84.0	15.6
36.4	1.9	86.6	6.9
39.8	1.1	87.6	1.1
42.9	1.9	90.5	1.1
46.0	1.4	92.5	15.9
50.1	1.7	93.9	9.9
53.9	2.2	94.4	3.8
55.8	0.9	96.3	1.6
60.0	1.3	98.7	2.7
64.9	1.3	100.1	0.8
		(South end of trench)	

[a]Pennyweights (dwt) per ton; 20 dwt = 1 oz.

orthogonal polynomial are independent. This means that adding a new term to the fitted equation will not change the values of terms already computed.

Orthogonal polynomials date back to the work of Tchebycheff in the nineteenth century, although modern procedures were introduced by Fisher in 1925. Extensive discussions of orthogonal polynomials are given by Fisher (1970), Draper and Smith (1981), and Morrison (1983), among others. Although computers have made the use of orthogonal polynomials less appealing, they can be very helpful in the analysis of data collected by instruments that sample at regular intervals. We will consider some of these applications when we discuss filtering of time series.

An ordinary polynomial regression equation, such as

$$Y_i = \beta_0 + \beta_1 X_i + \beta_2 X_i^2 + \cdots + \beta_m X_i^m + \epsilon_i \qquad (4.38)$$

can also be expressed as

$$Y_i = \alpha_0 + \alpha_1 \xi_{1i} + \alpha_2 \xi_{2i} + \cdots + \alpha_m \xi_{mi} + \epsilon_i \qquad (4.39)$$

where the α are the least-squares coefficients and the ξ_m are terms of the orthogonal polynomial. The numerical values of ξ_m can be determined from the sequence in which the observations of X occur and from the degree of the fitted equation.

Usually, however, the orthogonal polynomial terms are simply found using tables such as Table 4.20. The polynomial terms are always integers and one term is required for each observation in the sequence being fitted. This suggests a major problem—if a sequence consists of many observations, a very large table of polynomial terms must be available. (A table for up to 75 observations is given by Fisher and Yates, 1963.)

The α coefficients are found by the following equations:

$$\alpha_0 = \frac{\Sigma Y_i}{n} = \bar{Y} \qquad\qquad (4.40)$$

$$\alpha_m = \frac{\Sigma Y_i \xi_m}{\Sigma \xi_m^2} \qquad\qquad (4.41)$$

The measurements of moisture in a core sample given in Table 4.11 are ideal for analysis by orthogonal polynomials, because the observations are equally spaced 5 ft apart down the core. The measurements, together with orthogonal polynomial terms ξ_1 for a linear fit to eight observations (taken from Table 4.20), are given below. Also shown are the products of Y and ξ_1.

$Y_i =$	124	78	54	35	30	21	22	18
$\xi_1 =$	-7	-5	-3	-1	1	3	5	7
$Y_i\xi_1 =$	-868	-390	-162	-35	30	63	110	126

The α_0 coefficient of the fitted line is simply the mean of values of Y_i,

$$\alpha_0 = \frac{382}{8} = 47.75$$

The α_1 coefficient is found by multiplying each observation Y_i by the corresponding polynomial term, summing, then dividing by the sum of the squares of the terms.

$$\alpha_1 = \Sigma Y_i \xi_1 / \Sigma \xi_1^2$$
$$= -1126/168$$
$$= -6.70$$

The linear regression of moisture content on depth is therefore

$$\hat{Y}_i = 47.75 - 6.70\xi_{1i}$$

Using ordinary regression, we had found the relationship

$$\hat{Y}_i = 94.67 - 2.68X_i$$

The two differ because the orthogonal polynomial is given in terms of ξ_{1i} rather than X_i. If we evaluate the two equations, however, both will yield the same estimate of $\hat{Y}_i$. For example, suppose we calculate the predicted moisture content in the core at a depth of 30 ft, which is the seventh measurement in the sequence. The orthogonal polynomial term corresponding to the seventh measurement is $+5$, so

TABLE 4.20 Orthogonal Polynomial Terms for Degrees 1 Through 4 and for 3 to 12 Observations

First-Degree (Linear) Terms

n	3	4	5	6	7	8	9	10	11	12
	−1	−3	−2	−5	−3	−7	−4	−9	−5	−11
	0	−1	−1	−3	−2	−5	−3	−7	−4	−9
	1	1	0	−1	−1	−3	−2	−5	−3	−7
		3	1	1	0	−1	−1	−3	−2	−5
			2	3	1	1	0	−1	−1	−3
				5	2	3	1	1	0	−1
					3	5	2	3	1	1
						7	3	5	2	3
							4	7	3	5
								9	4	7
									5	9
										11

Second-Degree (Quadratic) Terms

n	3	4	5	6	7	8	9	10	11	12
	1	1	2	5	5	7	28	6	15	55
	−2	−1	−1	−1	0	1	7	2	6	25
	1	−1	−2	−4	−3	−3	−8	−1	−1	1
		1	−1	−4	−4	−5	−17	−3	−6	−17
			2	−1	−3	−5	−20	−4	−9	−29
				5	0	−3	−17	−4	−10	−35
					5	1	−8	−3	−9	−35
						7	7	−1	−6	−29
							28	2	−1	−17
								6	6	1
									15	25
										55

Third-Degree (Cubic) Terms

n	3	4	5	6	7	8	9	10	11	12
		−1	−1	−5	−1	−7	−14	−42	−30	−33
		3	2	7	1	5	7	14	6	3
		−3	0	4	1	7	13	35	22	21
		1	−2	−4	0	3	9	31	23	25
			1	−7	−1	−3	0	12	14	19
			5	−4	−1	−7	9	−12	0	7
					1	−5	−13	−31	−14	−7
						7	−7	−35	−23	−19
							14	−14	−22	−25
								42	−6	−21
									30	−3
										33

TABLE 4.20 (*Continued*)

Fourth-Degree (Quartic) Terms

n	3	4	5	6	7	8	9	10	11	12
			1	1	3	7	14	18	6	33
			−4	−3	−7	−13	−21	−22	−6	−27
			6	2	1	−3	−11	−17	−6	−33
			−4	2	6	9	9	3	−1	−13
			1	−3	1	9	18	18	4	12
				1	−7	−3	9	18	6	28
					3	13	−11	3	4	28
						7	−21	−17	−1	12
							14	−22	−6	−13
								18	−6	−33
									6	−27
										33

the two alternative equations are

$$\hat{Y} = 94.67 - 2.68(30)$$
$$= 14.27$$

and

$$\hat{Y} = 47.75 - 6.70(5)$$
$$= 14.25$$

The two are equivalent within rounding error.

Another advantage of orthogonal polynomials becomes apparent if we wish to expand the fitted equation to a higher degree. All that is necessary is to repeat the computational steps, but substituting the polynomial terms appropriate for the new degree. The coefficients already found will remain unchanged. For example, to expand the regression of moisture content on depth to a quadratic relationship, we select the second-degree terms for eight observations from Table 4.20.

$$Y_i = 124 \quad 78 \quad 54 \quad 35 \quad 30 \quad 21 \quad 22 \quad 18$$
$$\xi_2 = 7 \quad 1 \quad -3 \quad -5 \quad -5 \quad -3 \quad 1 \quad 7$$
$$Y_i\xi_2 = 868 \quad 78 \quad -162 \quad -175 \quad -150 \quad -63 \quad 22 \quad 126$$

The sum of the products is $\Sigma Y_i \xi_2 = 544$, so the quadratic coefficient is

$$\alpha_2 = \frac{544}{168} = 3.24$$

The second-degree regression of moisture on depth is therefore

$$\hat{Y}_i = 47.75 - 6.70\xi_{1i} + 3.24\xi_{2i}$$

Successively higher degree polynomials can be fitted to the data in the same manner, up to a seventh-degree curve that would pass exactly through each value of Y_i. The orthogonal coefficients α can be converted into ordinary regression coefficients β by substitution in the equations that are used to determine the orthogonal polynomial terms. Details are given in Draper and Smith (1981) and Ostle and Mensing (1975), who also provide equations for direct determination of the various sums of squares necessary for analyses of variance and tests of significance for polynomial regressions of higher degrees.

Reduced Major Axis

The regression procedures we have been discussing involve fitting a line to a collection of bivariate observations so that the squared deviations of one of the variables from the line is a minimum. If the deviations in the Y direction are minimized, one set of linear regression coefficients are obtained, and if the deviations are minimized in the X direction, another set of coefficients will result. When the two lines are plotted, as in Figure 4.21, they will cross at $\overline{X}, \overline{Y}$. The

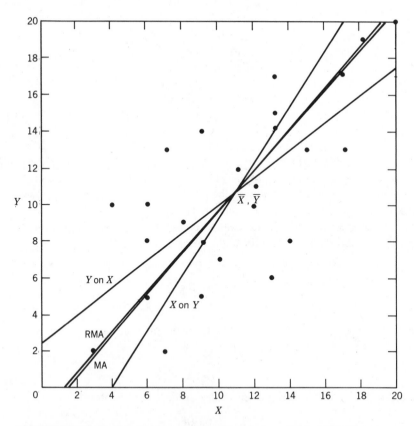

FIGURE 4.21   Scatter diagram of bivariate data from Table 6.19. Also shown are regression line of Y on X, regression line of X on Y, reduced major axis line (RMA), and the major axis (MA).

cosine of the angle between them is directly related to the correlation between X and Y.

Sometimes, either the physical circumstances leading to the observations dictate which variable should be considered dependent, or the purpose of the analysis clearly indicates which should be regressed onto the other. However, it may not be possible to rationally decide which variable should be X and which should be Y. This occurs, for example, in biometry, where it may be useful to know the relationship between two sets of measurements, such as the lengths and widths of shells, but it is not obvious which set of measurements should be expressed as a function of the other. Similar circumstances arise in petrophysics where a common problem is relating two sets of measurements made by different logging tools, such as sonic transit times and neutron density measurements. Both measurements are subject to errors, and neither can be regarded as a function of the other, yet it is extremely useful to be able to cross-plot the two variables and express their mutual relationship in some manner.

An appealing solution would be to fit a line that minimizes the deviations of the observations from the line in both the X and Y directions simultaneously. Such a line would split the difference between the regression lines of X on Y and Y on X. It would conform more closely to the visual impression of the trend in the observations, and it would attribute the scatter of the data points to both variables, rather than assigning all of the deviations from the fitted line to a single variable.

There are two ways in which such a line could be defined. One method involves minimizing the squared deviations from the line in both the X and Y directions simultaneously. By the Pythagorian theorem, this is equivalent to minimizing the squared perpendicular deviations from the fitted line (Fig. 4.22). Such a line is

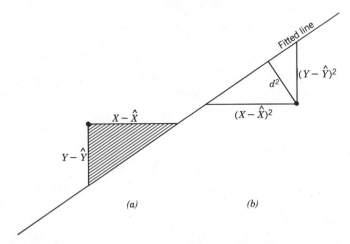

FIGURE 4.22 Criteria for fitting RMA and major axis lines. (a) Reduced major axis minimizes the product of deviations $(X - \hat{X})$ and $(Y - \hat{Y})$ from the fitted line, equivalent to minimizing the areas of triangles (shaded). (b) Major axis minimizes the sum of the squared deviations $(X - \hat{X})^2$ and $(Y - \hat{Y})^2$, in effect minimizing the squared perpendicular deviations d^2.

called the *major axis,* and can be found as the principal eigenvector of the variance-covariance matrix of X and Y. We will discuss computation of the major axis at length in Chapter 6, under the heading of principal components analysis.

The second procedure minimizes the product of the deviations in the X and Y directions. This in effect minimizes the sum of the areas of the triangles formed by the observations and the fitted line (Fig. 4.22) resulting in what is called the *reduced major axis,* often referred to simply as the RMA line. Most articles on the reduced major axis have been published in the two journals *Biometrics* and *Biometrika,* reflecting the popularity of the technique among scientists concerned with growth in organisms. Although the properties of the reduced major axis have received scant attention from statisticians, they have been investigated by Kermack and Haldane (1950) and by Kruskal (1953). A summary for geologists is given by Till (1974) and a more extensive treatment is contained in Miller and Kahn (1962).

The reduced major axis is defined by an ordinary linear equation having two coefficients, one representing the intercept and the other the slope:

$$Y = b_0 + b_1 X$$

The slope is defined as the ratio of the standard deviations of the two variables, X and Y, or

$$b_1 = s_Y / s_X \tag{4.42}$$

Since n is the same for both standard deviations, b_1 may be found by the equivalent equation

$$b_1 = \sqrt{\frac{SS_Y}{SS_X}} \tag{4.43}$$

The intercept of the reduced major axis is given by

$$b_0 = \overline{Y} - b_1 \overline{X} \tag{4.44}$$

The computation of the reduced major axis can be demonstrated using the observations shown on Figure 4.21 and listed in Table 6.19, which also will be used

TABLE 4.21 Sums of Squares and Other Quantities for Data Listed in Table 6.19

$n_X = 25$	$n_Y = 25$
$\Sigma X = 272$	$\Sigma Y = 267$
$\overline{X} = 10.88$	$\overline{Y} = 10.68$
$s_X^2 = 20.3$	$s_Y^2 = 24.1$
$s_X = 4.51$	$s_Y = 4.91$
$SS_X = 487.2$	$SS_Y = 578.4$

$$\text{cov}_{XY} = 15.6$$
$$SP_{XY} = 374.4$$
$$r = 0.71$$

later to illustrate the calculation of principal components (or, as they would be referred to in the present context, the major axes). Sums, sums of squares and cross-products, means, variances, and covariances are given in Table 4.21. From these, we may first calculate the ordinary regressions of Y on X and X on Y. For the regression of Y on X:

$$b_1 = \frac{SP_{XY}}{SS_X} = \frac{374.4}{487.2} = 0.77$$

$$b_0 = \bar{Y} - b_1\bar{X} = 10.68 - 0.77(10.88) = 2.43$$

So, the regression equation is

$$Y = 2.43 + 0.77X$$

For the regression of X on Y:

$$b_1 = \frac{SP_{XY}}{SS_Y} = \frac{374.4}{578.4} = 0.65$$

$$b_0 = \bar{X} - b_1\bar{Y} = 10.88 - 0.65(10.68) = 3.97$$

Yielding the regression equation

$$X = 3.97 + 0.65Y$$

For the reduced major axis,

$$b_1 = \sqrt{\frac{SS_Y}{SS_X}} = \sqrt{\frac{578.4}{487.2}} = 1.09$$

$$b_0 = \bar{Y} - b_1\bar{X} = 10.68 - 1.09(10.88) = -1.18$$

The equation of the RMA line is therefore

$$Y = -1.18 + 1.09X$$

For comparative purposes, the first eigenvector of the variance-covariance matrix of X and Y is

$$I = \begin{bmatrix} 0.66 \\ 0.75 \end{bmatrix}$$

which means that the eigenvector has a slope of 0.75 units in Y for 0.66 units in X, equivalent to a b_1 coefficient of 1.14. The intercept is

$$b_0 = \bar{Y} - b_1\bar{X} = 10.68 - 1.14(10.88) = -1.72$$

The equation of the major axis can be written as

$$Y = -1.72 + 1.14X$$

In addition to the two regression lines, the major axis and reduced major axis are also shown on Figure 4.21. Note that the reduced major axis and the major axis are very similar to each other. The reduced major axis bisects the angle between the line of regression of Y on X and the line of regression of X on Y; the major axis responds to the somewhat greater variance of Y by swinging to a slightly steeper angle.

The standard errors of both reduced major axis coefficients can easily be computed, and from them approximate tests of significance can be formulated. However, there are no equivalents to the more soundly based analyses of variance that can be performed in conventional regression. The standard error of the RMA slope is

$$se_{b_1} = b_1 \sqrt{\left(\frac{1 - r^2}{n}\right)} \tag{4.45}$$

The equivalency of the slopes, b_1 and b_2, of two reduced major axis lines may be tested by

$$Z = \frac{b_1 - b_2}{\sqrt{se_{b_1}^2 - se_{b_2}^2}} \tag{4.46}$$

which will be recognized as a variant of one of the elementary tests discussed in Chapter 2. The test statistic Z is approximately normally distributed and its significance can be determined from a table of the standardized normal distribution.

The standard error of the intercept is

$$se_{b_0} = s_Y \sqrt{\frac{1 - r^2}{n} \left(1 + \frac{\overline{X}^2}{s_X^2}\right)} \tag{4.47}$$

Equation (4.47) may be used to construct an approximate confidence interval around the computed value of b_0. Similarly, the standard error of the slope coefficient may be used to determine an approximate confidence interval around b_1. Other, essentially ad hoc, tests of the coefficients of the reduced major axis have been used. Because of the lack of theoretical underpinnings, however, the reduced major axis should be used primarily for descriptive purposes and not for tests of statistical significance.

Splines

Some data may be thought of conveniently as strings of coordinate pairs. That is, the observations consist of measurements on two properties, and can be envisioned as defining a sequence of points in a two-dimensional space. For purposes of presentation or analysis, it may be desirable to connect these points with a smooth, continuous line. We can do this by the use of *spline functions*.

Splines are one of a large class of piecewise functions that can be used to represent curves in two or three dimensions. The mathematical spline gains its name from a physical counterpart, the flexible drafting spline made from a narrow strip of wood or plastic that can be bent to conform to an irregular shape. A drafting spline is held by lead weights called "ducks," which fix the position of the spline at their points of attachment. Between the ducks, the spline flexes into a smooth, continuous form. A mathematical spline is similarly constrained at defined points, but between the points it flexes in a manner that results in a smoothly varying line.

Splines are not analytical functions, nor are they statistical models such as the polynomial regressions described earlier. Rather, they are purely arbitrary and devoid of any theoretical basis except that which defines the characteristics of the lines themselves. They are, however, extremely useful for interpolation and are important in software for generating computer displays. Interactive computer graphics systems are becoming more widely used for geological and geophysical modelling. Spline fitting plays an important role in these systems.

Splines are piecewise polynomials that are constrained to have continuous derivatives at the joints between the pieces or segments. The most common spline consists of cubic polynomials, which are functions of the form

$$Y = \beta_1 + \beta_2 X + \beta_3 X^2 + \beta_4 X^3$$

The curve defined by a cubic polynomial can pass exactly through four points, but in order to fit a longer sequence it is necessary to use a succession of polynomial segments. To insure that there are no abrupt changes in slope or curvature between successive segments, the polynomial function is not fitted to four points, but only to two. This allows us to use additional constraints which will insure that the resulting spline has continuous first derivatives between segments (the slope of the line will be the same on either side of a joint) and continuous second derivatives (the rate of change in the slope of the line will not change across a joint). In general, a spline of degree m will have continuous derivatives across the points up to order $m - 1$.

Developing the spline equations requires knowledge of differential calculus, a skill not presumed for this book. Therefore, we will simply present the necessary equations in computational form and work through their application; those interested in their derivation are referred to the excellent introductory text on computer graphics by Rogers and Adams (1976), and to the monograph on geologic applications of surface fitting techniques by Tipper (1979).

The mathematical notation used with spline functions is somewhat confusing, but can be clarified with the help of Figure 4.23, which shows a set of four observations connected by a piecewise spline function. The observations are plotted as points, and are symbolically indicated as P_i with the understanding that P is actually a vector of Cartesian coordinates. That is, $P_i = [X_i, Y_i]$. The intervals between successive points are referred to as *spans;* the chord or straight-line distance between two points is indicated as t_i with i assuming the value of the second point. A cubic spline function covers a single span or two points; the illustration, therefore, shows three successive splines, one from point P_1 to P_2, from P_2 to P_3, and from point P_3 to P_4.

In general form, the spline equation may be written as

$$\hat{P}_t = \beta_1 + \beta_2 t + \beta_3 t^2 + \beta_4 t^3 \tag{4.48}$$

which states that the coordinates of the spline at some distance t along a span are equal to a cubic polynomial function of t. To determine the coefficients we must know the coordinates of the points, which define the ends of the splines, and the

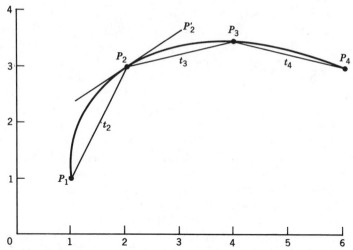

FIGURE 4.23 Four points connected by cubic spline function. Original observations are labeled P_i. Chord distances between points are t_i. Tangent to spline at interior point P_2 is indicated as P_2'.

slopes of tangent lines at the points. In addition, we must specify end conditions that determine the behavior of the line in the first and last spans. The point coordinates, of course, are given. The slopes, expressed as tangent vectors, must be determined. Various end conditions can be selected, depending upon the desired shape of the line at its terminus. We will consider only what is called *relaxed* or *natural* end conditions that do not require specifying tangent vectors at the end points.

To find the tangent vectors at the interior points (P_2 and P_3 of Fig. 4.23), we must solve a set of simultaneous equations of the form

$$[M][P'] = [B] \qquad (4.49)$$

where the unknown vector of P' coefficients are the desired tangents. The left-hand matrix that must be inverted is *tridiagonal;* that is, all elements are zero except for the diagonal elements and those immediately adjacent to the diagonal. For relaxed end conditions, $[M]$ has the form of an n by n matrix:

$$[M] = \begin{bmatrix} 1.0 & 0.5 & 0 & 0 & 0 & \cdots & 0 \\ t_3 & 2(t_2 + t_3) & t_2 & 0 & 0 & \cdots & 0 \\ 0 & t_4 & 2(t_3 + t_4) & t_3 & 0 & \cdots & 0 \\ 0 & 0 & t_5 & 2(t_4 + t_5) & t_4 & \cdots & 0 \\ 0 & 0 & 0 & t_6 & 2(t_5 + t_6) & \cdots & 0 \\ \cdot & \cdot & \cdot & \cdot & \cdot & \cdots & \cdot \\ \cdot & \cdot & \cdot & \cdot & \cdot & \cdots & \cdot \\ \cdot & \cdot & \cdot & \cdot & \cdot & \cdots & \cdot \\ 0 & 0 & 0 & 0 & \cdots & 2 & 4 \end{bmatrix}$$

$$(4.50)$$

The right-hand vector $[B]$ has the form

$$[B] = \begin{bmatrix} \dfrac{3}{2t_2}(P_2 - P_1) \\[2ex] \dfrac{3}{t_2 t_3}[t_2^2(P_3 - P_2) + t_3^2(P_2 - P_1)] \\[2ex] \dfrac{3}{t_3 t_4}[t_3^2(P_4 - P_3) + t_4^2(P_3 - P_2)] \\[2ex] \dfrac{3}{t_4 t_5}[t_4^2(P_5 - P_4) + t_5^2(P_4 - P_3)] \\[2ex] \dfrac{3}{t_5 t_6}[t_5^2(P_6 - P_5) + t_6^2(P_5 - P_4)] \\[2ex] \vdots \\[1ex] \dfrac{6}{t_n}(P_n - P_{n-1}) \end{bmatrix} \qquad (4.51)$$

The matrix equation is solved by inverting $[M]$ and post-multiplying the inverse by $[B]$. Note that since the point coordinates P_i are double-valued (that is, each consists of a value for X and a value for Y), the right-hand matrix $[B]$ is $n \times 2$, where n is the number of points to be included in the set of splines. Equation (4.51) shows the form of the terms in $[B]$. The first column of $[B]$ is found by inserting the appropriate chord distances t_k and the X coordinates of the observations. The second column is formed in the same manner, but using the Y coordinates.

Similarly, the solution matrix $[P']$ is also $n \times 2$. Each row of $[P']$ represents the slope of a tangent to the spline at the observation point, given in terms of X and Y.

To find the four β coefficients that define the kth spline (that is, the span between point P_k and point P_{k+1}), we set

$$\begin{aligned} \beta_1 &= P_k \\[1ex] \beta_2 &= P_k' \\[1ex] \beta_3 &= \frac{3(P_{k+1} - P_k)}{t_{k+1}^2} - \frac{2P_k'}{t_{k+1}} - \frac{P_{k+1}'}{t_{k+1}} \\[1ex] \beta_4 &= \frac{2(P_k - P_{k+1})}{t_{k+1}^3} + \frac{P_k'}{t_{k+1}^2} + \frac{P_{k+1}'}{t_{k+1}^2} \end{aligned} \qquad (4.52)$$

Finally, when the four coefficients are found for the kth span, points along the curve within this interval can be determined. The length of the chord between points k and $k + 1$ can be divided into a convenient number of parts and these successive distances inserted as t in eq. (4.48). This will provide a set of equally spaced coordinates that can be connected to form the spline curve. The process is repeated for each segment of the piecewise spline, using the appropriate slopes at the interior points, chord lengths, and point coordinates to find a new set of β coefficients for each span.

To demonstrate the fitting of cubic splines, we will use four points shown in Figure 4.24 whose coordinates are

$$[P] = \begin{bmatrix} 1 & 1 \\ 1 & 3 \\ 4 & 3 \\ 3 & 1 \end{bmatrix}$$

The chord lengths are $t_2 = 2.0$, $t_3 = 3.0$, and $t_4 = 2.236$. These distances are all that are required to form the matrix $[M]$, defined in eq. (4.50):

$$[M] = \begin{bmatrix} 1.0 & 0.5 & 0 & 0 \\ 3.0 & 10.0 & 2.0 & 0 \\ 0 & 2.236 & 10.472 & 3.0 \\ 0 & 0 & 2.0 & 4.0 \end{bmatrix}$$

The inverse of $[M]$ is

$$[M]^{-1} = \begin{bmatrix} 1.1875 & -0.0625 & 0.0139 & -0.0104 \\ -0.3749 & 0.1250 & -0.0279 & 0.0209 \\ 0.0934 & -0.0311 & 0.1184 & -0.0888 \\ -0.0467 & 0.0156 & -0.0592 & 0.2944 \end{bmatrix}$$

We must also determine the right-hand vector $[B]$. The necessary information to find the elements of $[B]$ consists of the chord lengths and the coordinates of the points. Since each point consists of two coordinates, the vector $[B]$ has two columns, the first for X and the second for Y:

$$[B] =$$

$$\begin{bmatrix} \dfrac{3}{2 \cdot 2}(1-1) & \dfrac{3}{2 \cdot 2}(3-1) \\ \dfrac{3}{2 \cdot 3}[2^2(4-1)+3^2(1-1)] & \dfrac{3}{2 \cdot 3}[2^2(3-3)+3^2(3-1)] \\ \dfrac{3}{3 \cdot 2.236}[3^2(3-4)+2.236^2(4-1)] & \dfrac{3}{3 \cdot 2.236}[3^2(1-3)+2.236^2(3-3)] \\ \dfrac{6}{2.236}(3-4) & \dfrac{6}{2.236}(1-3) \end{bmatrix}$$

$$= \begin{bmatrix} 0 & 1.5 \\ 6 & 9 \\ 2.683 & -8.050 \\ -2.683 & -5.367 \end{bmatrix}$$

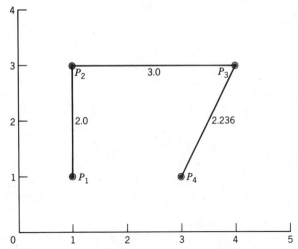

FIGURE 4.24 Four points to be fitted by cubic spline function. Chord lengths of spans between points are indicated.

Multiplying $[B]$ by $[M]^{-1}$ yields $[P']$:

$$[P'] = \begin{bmatrix} -.3097 & 1.2026 \\ .6187 & .6750 \\ .3723 & -.6160 \\ -.8552 & -1.0328 \end{bmatrix}$$

We now have all of the terms necessary to calculate the spline coefficients for each of the spans in our example. For the first span, we may substitute the appropriate values of t, P, and P' into eq. (4.52) to obtain:

For the X coordinate,

$$\beta_1 = 1$$

$$\beta_2 = -.3097$$

$$\beta_3 = \frac{3(1-1)}{2^2} - \frac{2(-.3097)}{2} - \frac{.6187}{2} = .0004$$

$$\beta_4 = \frac{2(1-1)}{2^3} + \frac{(-.3097)}{2^2} + \frac{.6187}{2^2} = .0773$$

For the Y coordinate,

$$\beta_1 = 1$$

$$\beta_2 = 1.2026$$

$$\beta_3 = \frac{3(3-1)}{2^2} - \frac{2(1.2026)}{2} - \frac{.6750}{2} = -.0401$$

$$\beta_4 = \frac{2(1-3)}{2^3} + \frac{1.2026}{2^2} + \frac{.6750}{2^2} = -.0306$$

or

$$[B] = \begin{bmatrix} 1 & 1 \\ -.3097 & 1.2026 \\ .0004 & -.0401 \\ .0773 & -.0306 \end{bmatrix}$$

In a similar manner, we can determine the spline coefficients for spans 2 and 3. These are

$$\begin{bmatrix} 1 & 3 \\ .6187 & .6750 \\ .4634 & -.2447 \\ -.1121 & .0066 \end{bmatrix} \begin{bmatrix} 4 & 3 \\ .3723 & -.6160 \\ -.5506 & -.1872 \\ .0823 & .0280 \end{bmatrix}$$

Finally, we can use the spline coefficients to determine the coordinates of intermediate points on the splines between each observation. If we calculate a large number of such points and connect them with straight lines, the visual result will be an apparently smooth, continuous curve. This is the way in which a computer graphics system calculates and draws smoothly curving lines. For purposes of demonstration, we will content ourselves with only three intermediate points on each spline.

To find these intermediate points, we first divide each chord into four parts; the distances at $t_k/4$, $2t_k/4$, and $3t_k/4$ will define the values of t to be inserted into the spline equation. For the first spline, these distances are 0.5, 1.0, and 1.5.

Inserting into eq. (4.48), first for X and then for Y,

$$\hat{P}_{.5X} = 1 - .3097(.5) + .0004(.5^2) + .0773(.5^3) = .8549$$

$$\hat{P}_{.5Y} = 1 + 1.202(.5) - .0401(.5^2) - .0306(.5^3) = 1.5874$$

In a similar manner we may compute the coordinates of the first spline at distances $t = 1.0$ and $t = 1.5$. These are

$$\text{for } t_{1.0}, \quad [.7679 \quad 2.1319]$$

$$\text{for } t_{1.5}, \quad [.7969 \quad 2.6104]$$

The process is now repeated for the second and third splines, yielding the sets of coordinates

$$\begin{array}{c} [1.6774 \quad 3.3714] \\ \text{for spline 2,} \quad [2.5924 \quad 3.4841] \\ [3.4612 \quad 3.3548] \end{array}$$

$$\begin{array}{c} [4.050 \quad 2.6020] \\ \text{for spline 3,} \quad [3.8431 \quad 2.1163] \\ [3.4642 \quad 1.5722] \end{array}$$

These are shown plotted in Figure 4.25. Also shown is the smooth spline generated

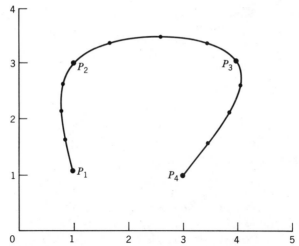

FIGURE 4.25 Smooth spline function consisting of 30 segments between each point of Figure 4.24. The three intermediate points on each spline calculated in text are shown as small dots.

by calculating 30 intermediate points between each joint. Although the process of determining the spline coefficients is involved, once they have been found it is a relatively trivial matter to calculate as many points as desired along the curve.

Segmenting Sequences

Zonation

Zonation is the dividing of a sequence into relatively uniform segments, each of which is distinctive from adjacent segments. Paleontologists, for example, may want to zone a stratigraphic sequence on the basis of consistent abundances of microfossils. Well logs may be subdivided into relatively uniform intervals that represent zones of constant lithology, corresponding to stratigraphic units. Airborne radiometric traverses may be subdivided into zones that can be interpreted as belts of uniform rock composition or mineralization.

There are basically two contrasting approaches to zonation. The simplest procedures can be referred to as "local boundary hunting," searching for abrupt changes in average values, or equivalently, for the steepest gradients in the sequence. Webster (1973) developed a "split-moving window" for defining boundaries between soil zones along a transect. A sequence is examined by iteratively moving a short interval along the sequence. The moving interval is called a window and is split into two parts, a segment from point $(i + h)$ on the sequence to point i, and another segment from point i to point $(i - h)$.

A measure called the *generalized distance* is calculated for the difference between the segments within the two halves of the window. The generalized difference is the ratio formed by dividing the squared difference between the average values of the two segments by the pooled variance of the sequences in the segments. We can denote the mean of the segment from x_i to x_{i+h} as $\overline{X}_1$ and its variance as s_1^2; the mean of the segment from x_i to x_{i-h} is $\overline{X}_2$ and the variance is s_2^2. Then, the generalized difference is

$$D^2 = \frac{(\overline{X}_1 - \overline{X}_2)^2}{s_1^2 + s_2^2} \tag{4.53}$$

Note that the pooled variance is simply the sum of the variances of the two segments, because both segments contain the same number of observations. Also note that the first and last h of the points in the sequence cannot be separated into different zones.

A plot of h versus D^2 will result in transformation of the original traverse into a new sequence in which zone boundaries appear as sharp spikes. Figure 4.26a shows an original 6-km-long transect across the upper Thames Valley in England (Webster, 1973). Soil samples were collected at 20-m intervals and analyzed for 27 soil properties. These multiple measurements were compressed by principal components analysis, which will be discussed in detail in Chapter 6. Here, we may simply note that the transect shown represents the most efficient possible linear

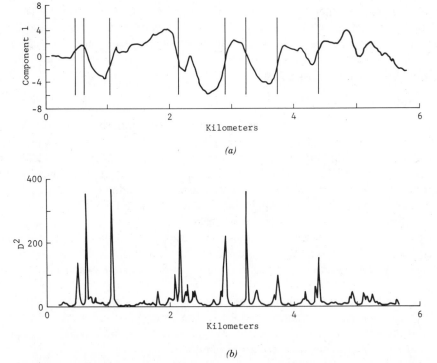

FIGURE 4.26 Transect showing variation in soil properties along a 6-km line in the upper Thames Valley, England. (*a*) Variation in first principal component of 27 soil properties. (*b*) Values of D^2 along traverse. Maxima define boundaries shown on (*a*). After Webster (1973).

combination of the original variables. Figure 4.26b is a plot of D^2 against distance along the transect, computed using a split-moving window that spans 18 points.

Webster noted that the performance of the procedure depends upon the variability of the original sequence and the length of the moving window. A long window will average across small zones and may miss short intervals. However, it will subdue the erratic variation of a noisy original record. A short window is more sensitive and will identify small zones, but may create an irregular, uninterpretable plot of D^2. Webster (1980) has published a FORTRAN program that finds zone boundaries by this method.

One objection to local boundary hunting procedures is that they may find an inordinate number of boundaries, particularly within a highly variable part of the sequence. Global zonation is a different approach, using procedures that break the sequence into a specified number of segments which are as internally homogeneous as possible and as distinct as possible from adjacent segments.

One of the first, and still most practical, of these procedures was devised by Gill (1970), who used an iterative analysis of variance approach. First, the sequence is divided into two segments, a very short initial part, and the remainder of the sequence. The sum of squares within the segments, SS_w, is computed as

$$SS_w = \left. \sum_{j=1}^{m} \sum_{i=1}^{n_j} (x_{ij} - \overline{X}_{\cdot j})^2 \middle/ \sum_{j=1}^{m} n_j - m \right. \tag{4.54}$$

where x_{ij} is the ith point within segment j, $\overline{X}_{\cdot j}$ is the mean of the jth segment, n_j is the number of points in the jth segment, and m is the number of segments. The sum of squares between segments, SS_b, is a measure of the variance of the segment means about $\overline{X}_{\cdot\cdot}$, the grand mean of the total sequence, or

$$SS_b = \left. \sum_{j=1}^{m} (\overline{X}_{\cdot j} - \overline{X}_{\cdot\cdot})^2 \middle/ m - 1 \right. \tag{4.55}$$

The partition between the two segments is moved along the sequence to successive positions and the two quantities SS_w and SS_b recomputed at each position. For every possible position of the boundary, the ratio

$$R = \frac{SS_b - SS_w}{SS_b} \tag{4.56}$$

is calculated. The position corresponding to the maximum value of R is chosen as the location of the first zonal boundary.

Next, the two zones are themselves partitioned by repeating the process to insert an additional boundary which again maximizes the quantity R. The zonation is repeatedly run until the entire sequence is divided into the specified number of zones, or until the quantity R no longer increases with the addition of new boundaries.

Gill's procedure has been used as a way of automatically zoning digitized well logs. A more recent, but philosophically similar, procedure has been published by Hawkins and Merriam (1973, 1974). They use global optimization, based on methods of dynamic programming. Their algorithm is iterative, as is Gill's, but is also recursive and takes advantage of Bellman's principle of optimality to insure that the final set of zone boundaries is the best possible of all sets of partitions that might have been chosen. With a nonrecursive procedure, it is always possible that the position selected as the best boundary between two zones is no longer the best when another boundary is inserted into one of the zones.

Hawkins and Merriam calculate a quantity that is the sum of the within-zone variances, equivalent to Gill's SS_w. If this quantity is computed for all possible partitions of the log into two segments, the result is a table of SS_w for $(n - 1)$ possible locations for the first boundary. For each possible first partition, a new value of SS_w is then computed for all possible positions of a second boundary, which would divide the log into three zones. By selecting the smallest value of SS_w for the second partition, the associated location for the first boundary is optimal, and will remain the best location no matter how many additional boundaries are inserted.

The process now iterates for the third cycle, and for every combination of the optimum first boundary with all possible second boundaries, all possible third boundary positions are found and their values of SS_w calculated. Selection of the smallest SS_w value then determines the optimum position for the second boundary. The process repeats again and again, until the specified number of boundaries is found.

Because of the recursive nature of the algorithm, the final set of zones is guaranteed to have the smallest internal variance of any possible set of the same number of zones covering the total interval. Unfortunately, the computational cost of achieving this optimality is very high, and the method is not practical for very long log sequences.

Seriation

The terms *ordination* and *seriation* mean the putting of observations in some logical order, on the basis of their relative similarities. If the observations are characterized by multiple variables, this essentially means projecting them by some means onto a single line where their position is a logical expression of their place in the data set. This can be done by any of several techniques, such as principal components analysis and factor analysis, discussed in the final chapter. Seriation has the additional connotation of chronological order; the term is widely used by archaeologists. Unfortunately, there is no guarantee that a sequence of observations arranged in order of similarity will also be arranged in a chronologically meaningful way.

The concepts of ordination and seriation have not been widely used in geology, except in applications of numerical taxonomy to paleontology (Sneath and Sokal, 1973). However, there is one area where the seriation concept seems useful, and that is in the geologic correlation of two stratigraphic sequences.

Two petrophysical well logs can be matched together on the basis of similarity in log response by "slotting," a dynamic programming procedure that shuffles the two sections together like a deck of cards (Gordon and Reyment, 1979). Every point in one sequence is paired with the most similar point in the other sequence, subject to the constraint that stratigraphic order must be preserved in the two sections. In this way, true seriation is achieved, because the final arrangement is meaningful both lithologically and chronologically. Additional constraints can be introduced that will force specified points on the two sequences to match, or that will force a specified segment of one sequence to match with a point on the other. Thus, if marker beds are identified within two well logs being compared, these beds can be forced to correspond. The remainders of the logs are correlated on the basis of greatest similarity, subject to the constraints that correlation lines cannot cross, and that the marker beds *must* correlate.

The algorithm, published by Gordon and Reyment (1979), is similar to the zoning algorithm of Hawkins and Merriam (1974). First, every point in the first well sequence is compared to every point in the second well sequence. If there are n observations on the first log and m on the second, this results in an $n \times m$ array of comparisons. A number of comparative measures could be used, but Gordon and Reyment (1979) use the simple dissimilarity measure

$$D_{j,k} = \sum_{l=1}^{p} \omega_l(u_{l,j} - v_{l,k}) \qquad (4.57)$$

where $u_{l,j}$ is the response for log variable l at depth j in the first well, and $v_{l,k}$ is the response for the same log variable at depth k in the second hole. Weights (ω_l) can be assigned to the different log variables if desired.

The dynamic programming algorithm now seeks to trace a single path through this array, from the upper left corner to the lower right, such that the sum of the dissimilarities is a minimum. Stratigraphic order is preserved by requiring the search to move only down, or to the right. A small example is shown in Figure 4.27, slotting three intervals from one log with four intervals from another. Note that the outermost rows and columns of the matrix are repeated; this allows the algorithm to trace down one log or the other at the beginning and ending of the process. Some example paths through the matrix are shown in Figure 4.28, with their corresponding stratigraphic correlations.

To find the optimum path, a recursive procedure is used. Beginning at (Start) on the upper left, the first interval can be either u_1 or v_1; the dissimilarity is the same, $2 + 2$, along either path. If u_1 is chosen, the next interval may be either

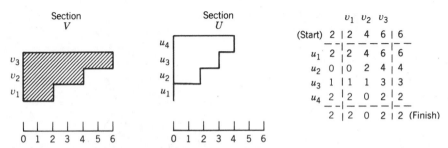

FIGURE 4.27 Artificial stratigraphic sections to be slotted together. Section V contains three intervals and section U contains four intervals. Characteristic measured on each interval ranges from 1 to 6. Matrix contains simple dissimilarity measures between all possible pairs of intervals in sections U and V.

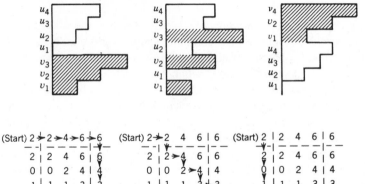

(Start) 2→2→4→6→6   (Start) 2→2 4 6 | 6   (Start) 2 | 2 4 6 | 6
  --|-----|↓-   --|↓----|--   ↓-|-----|--
  2 | 2 4 6 | 6   2 | 2→4 6 | 6   2 | 2 4 6 | 6
  0 | 0 2 4 | 4   0 | 0 2→4 | 4   0 | 0 2 4 | 4
  1 | 1 1 3 | 3   1 | 1 1 3→3   1 | 1 1 3 | 3
  2 | 2 0 2 | 2   2 | 2 0 2 | 2   2 | 2 0 2 | 2
  --|-----|↓-   --|-----|↓-   ↓-|-----|--
  2 | 2 0 2 | 2 (Finish)   2 | 2 0 2 | 2 (Finish)   2→2→0→2→2 (Finish)

     Σ = 37   Σ = 26   Σ = 16

FIGURE 4.28 Results obtained by arbitrarily slotting sequences U and V together. Arrows in matrix show path of succession from bottom to top. Total dissimilarity is indicated by Σ.

u_2 with a total dissimilarity of $2 + 2 + 0 = 4$, or v_1, with a total dissimilarity of $2 + 2 + 2 = 6$. Alternatively, if v_1 is chosen as the first step, the second step could be either to v_2 or u_1. The path to v_2 results in a total dissimilarity of $2 + 2 + 4 = 8$; the path to u_1 gives a dissimilarity of $2 + 2 + 2 = 6$. Of these alternatives, the path (Start) $\rightarrow u_1 \rightarrow u_2$ has the minimum dissimilarity, so the proper first step is to u_1.

Having set the initial point as u_1, with two possible second steps to either u_2 or v_1, the possible third steps must be examined. Again there are four possibilities: from u_2 to u_3 with a total dissimilarity of $2 + 2 + 0 + 1 = 5$; from u_2 to v_1 ($2 + 2 + 0 + 0 = 4$); from v_1 to v_2 ($2 + 2 + 2 + 4 = 10$); and from v_1 to u_2 ($2 + 2 + 2 + 0 = 6$). The smallest total occurs along the path (Start) $\rightarrow u_1 \rightarrow u_2 \rightarrow v_1$, so the optimal second step is to the point u_2. The procedure then iterates again, examining the outcomes of the four possible paths from u_2.

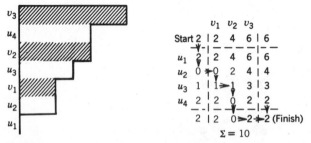

                    v_1 v_2 v_3
          Start 2 | 2 4 6 | 6
            --|↓----|--
          u_1 2 | 2 4 6 | 6
          u_2 0 | 0→0 2 4 | 4
          u_3 1 | 1 1→1 3 | 3
          u_4 2 | 2 0 2 | 2
            --|--↓---|↓-
                2 | 2 0→2→2 (Finish)
                   Σ = 10

FIGURE 4.29 Optimal slotting of sequences U and V. Arrows in matrix indicate path of succession in slotted sequence. Total dissimilarity is only 10, the lowest possible for any succession.

The minimum-value path defines the optimum step after u_2. The process repeats until the lower left (Finish) point is reached. In the example, the optimum path is

$$(\text{Start}) \rightarrow u_1 \rightarrow u_2 \rightarrow v_1 \rightarrow u_3 \rightarrow v_2 \rightarrow u_4 \rightarrow v_3 \rightarrow (\text{Finish})$$

with a total dissimilarity of 10. The matrix and the resulting slotted stratigraphic section is shown in Figure 4.29.

Because of its flexibility, slotting seems potentially to be a very powerful tool for correlation. Unfortunately, it is demanding of computer resources, even with the efficiency of dynamic programming, and the cost of slotting together long sequences probably is prohibitive for all except research purposes.

Autocorrelation

Figure 4.30 shows a gamma ray log of part of the Pennsylvanian section as measured in an oil well in western Kansas. The interval penetrated consists of alternating limestones and shales. Because of the radiation emitted by potassium-40 in clay minerals, the shales are marked by relatively high log responses while the limestones are characterized by low radioactivity. This particular section has been noted for the existence of cyclothems, which are more-or-less regular repetitions of lithologies. A brief inspection of the log will show that the limestones do appear to be separated by shales that have about the same thickness.

Repetitions, as well as other properties, of a sequence can be found by computing a measure of the self-similarity of the sequence. That is, the sequence can be compared to itself at successive positions, and the degree of similarity between the corresponding intervals computed. If every point in the sequence is compared successively to every other point, all positions of good correspondence will be detected, and also the degree of dissimilarity at other positions will be determined.

To perform this operation, the time series must have certain characteristics. It must consist of a sequence of observations of a variable Y, measured at successive instants in time or at points in space. Each observation must be separated from the preceding observation by an interval of time or distance that is constant for the series. We indicate the position of an observation within the series by a subscript, such as Y_t. It is therefore not necessary to explicitly consider the time or distance variable X, because it is implied by the subscript and can be found if needed by $X = \Delta t$, where Δ is the spacing between points. The entire time series contains n points and has a total length of $T = \Delta(n - 1)$.

The separation between any two points Y_t and $Y_{t+\tau}$ is referred to as a *lag* of length τ, where τ is the number of intervals between the points. It is the displacement between the time series and itself at a previous time or location. An analogy may be drawn between a time series and a chain. Each link in the chain corresponds to an observation in the series. If we lay two identical chains side by side and compare each link, we are making a cross-comparison at lag 0. If we move one of the chains so the first link in one is matched to the second link in the other, then all of the other links are also offset by one. This position of comparison is lag 1. The chains can be offset by one more link, and the cross-comparison will then be for lag 2, and so on.

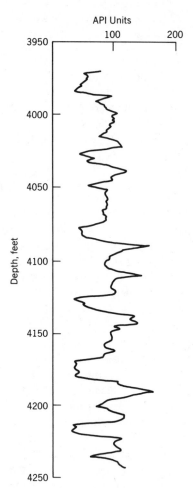

API Units

FIGURE 4.30   Gamma ray log trace through part of the Pennsylvanian sequence in an oil well in western Kansas.

The *autocovariance* for lag τ is the covariance between all observations Y_t and observations $Y_{t+\tau}$. That is, the covariance is calculated between a series and itself displaced by a lag of length τ. The definitional equation of the autocovariance is

$$\text{COV}_\tau = \frac{1}{n-\tau} \sum_{t=1+\tau}^{n} Y_t Y_{t-\tau} - \overline{Y}_t \overline{Y}_{t-\tau} \tag{4.58}$$

The autocovariance at lag 0 is simply the variance of the time series. If the series is very long and the lag τ is short, the mean of the series and that of the lagged series are essentially identical and eq. (4.58) can be simplified. However, if τ is an appreciable fraction of the length of the time series, the differences between the means become important. A computational equivalent of eq. (4.58) is

$$\text{COV}_\tau = \frac{[n - \tau(\sum_{t=1+\tau}^{n} Y_t Y_{t-\tau}) - \sum_{t=1+\tau}^{n} Y_t \sum_{t=1+\tau}^{n} Y_{t-\tau}]}{(n-\tau)(n-\tau-1)} \tag{4.59}$$

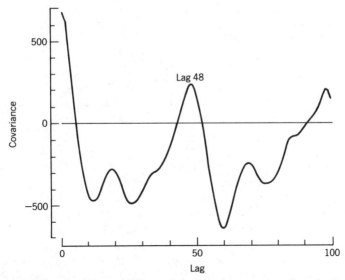

FIGURE 4.31 Autocovariance function of gamma ray log shown in Figure 4.30. Lag 48 corresponds to spacing in depth of 48 feet.

Conventionally, the autocovariance is calculated for lags from 0 to about $n/4$.- The resulting values can be displayed as an *autocovariogram* or *autocovariance function*, which is a plot of autocovariance against lag. Figure 4.31 shows the autocovariance function for the well log of Figure 4.30. The plot begins at a maximum value of 672 for lag $\tau = 0$, then drops and rises again at around lag 48, which corresponds to a spacing of about 48 ft since the gamma ray log was digitized at 1-ft intervals. This is the approximate vertical distance between successive limestone beds in the sequence shown in Figure 4.30.

The units of autocovariance are the squares of the measurements of the time series; in our example, API units^2. This means that the autocovariance is sensitive to changes in the scale of the time series, which makes it difficult to compare two autocovariograms. However, if the time series is standardized by subtracting the mean from each observation and dividing by the standard deviation, the series will be in units of standard deviation and the autocovariance will be in standardized form. As noted in Chapter 2, the covariance of a standardized variable is the correlation, and the same relationship is true for autocovariances.

Rather than first standardizing our time series, we may compute the autocorrelation directly by dividing the autocovariance by the variance of the time series. That is,

$$r_\tau = \frac{\text{cov}_\tau}{\text{var } Y} = \frac{\sum_{t=1+\tau}^{n} Y_t Y_{t-\tau} - \overline{Y}_t \overline{Y}_{t-\tau}}{\sum_{t=1}^{n} (Y_t - \overline{Y})^2} \tag{4.60}$$

The autocorrelation of a time series of finite length n and lag τ, which is some appreciable fraction of n, may be found by using eq. (4.59) as the numerator and the estimate $\sqrt{\text{var } Y_t \text{ var } Y_{t+\tau}}$ for the denominator. Because the summations in the numerator and denominator extend over the same limits, the resulting equation can be simplified by cancelling out the number of observations used, $n - \tau$. As in eq. (4.59) the limits of the summations extend from $t = 1 + \tau$ to n.

$$r_\tau = \frac{\Sigma\, Y_t Y_{t-\tau} - \Sigma\, Y_t\, \Sigma\, Y_{t-\tau}}{\sqrt{[\Sigma\, Y_t^2 - (\Sigma\, Y_t)^2][\Sigma\, Y_{t-\tau}^2 - (\Sigma\, Y_{t-\tau})^2]}} \qquad (4.61)$$

Figure 4.32 is the autocorrelogram of the gamma ray log shown in Figure 4.30. Note that it is identical in form to the autocovariance function except that it has been scaled between the limits $+1.0$ and -1.0.

Unfortunately, conflicts exist in time series terminology. Some authors refer to the autocorrelation as a parameter of a population and use the term *serial correlation* as the equivalent statistic calculated from a sample. Others use serial correlation as meaning the correlation between two time series. Still others call this cross-correlation. We will use the terms autocorrelation and cross-correlation and will make no distinction in terminology between statistics and parameters.

It should be noted that the correlogram is actually two-sided, with a negative part that is the mirror image of the positive part. We can see how this arises if we return to our analogy of two chains. If one of the chains is designated A and the other B, and we successively move chain A ahead of chain B, we can regard the resulting autocorrelations as forming the positive part of the correlogram. But if we move chain B ahead of A and compute the autocorrelations, the lags will be negative because the relative movement of the two sequences is reversed. As chains A and B are identical, it makes no difference which direction the two are shifted because

$$r_{-\tau} = r_\tau$$

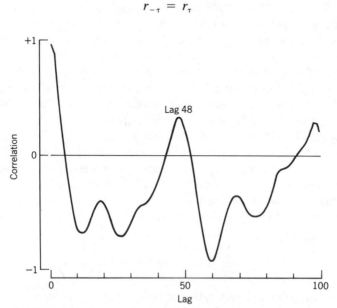

FIGURE 4.32 Autocorrelogram of gamma ray log shown in Figure 4.30.

In practice we do not consider the negative part of the autocorrelogram, but it does play a role in the development of the Fourier transform.

Correlograms help to reveal the characteristics of times series. This is done by comparing the correlogram of the time series to the correlograms of idealized models of the series, and determining how well they match. The simplest model that can be proposed for a time series is that successive observations are independent and are normally distributed. This means that there is no relationship between observations at one time t and observations at any other time $t + \tau$. This is the behavior that we would expect if the time series had been generated by a random process. The expected autocorrelation for this model is

$$\rho_\tau = 0$$

for all lags τ greater than zero. The theoretical correlogram will plot as a flat line through $r = 0$.

Other models assume some dependence between successive observations. For example, suppose a process generates random, normally distributed observations which are averaged together as they are created.

$$Y_t = \sum_{t-w}^{t} Z_t / w$$

This is a moving average model and will have a correlogram of the form

$$\rho_\tau = 1 - \frac{\tau}{w} \tag{4.62}$$

There are a wide variety of models of increasing complexity that can be proposed. A good introduction to these is given in Yule and Kendall (1969), and in a hydrological context by Yevjevich (1972). Figure 4.33 illustrates how a complex time series can be built up by the combination of simpler elements. In Figure 4.33a, a regular sine wave and its correlogram are shown. The correlogram drops from $+1.0$ to 0 and then to -1.0 as the series moves out of phase with itself, or when peaks are matched with troughs. The correlation increases again until it reaches $+1.0$ when the signal is shifted exactly one wavelength. Figure 4.33b shows a signal created by the first model discussed, a sequence of random numbers. The correlogram drops immediately from $+1.0$ at 0 lag, then fluctuates slightly about 0. Both of these series are *stationary;* that is, no significant trend exists in the observations. A nonstationary signal is shown in Figure 4.33c, where the observations increase steadily in value along the sequence. The correlogram shows steadily decreasing correlation. Figure 4.33d shows a combination of Figures 4.33a and 4.33b, a sine wave with superimposed noise. Perfect autocorrelation does not occur except at 0 lag, but the periodic component of the time series is revealed by the peak in the correlogram following the second zero crossing. Figure 4.33e is a combination of Figures 4.33a–4.33c, a sine wave combined with a linear trend with superimposed noise. Note that the trend further reduces our ability to discern the periodic component in the signal.

As noted earlier, the expected or mean autocorrelation of a sequence of random numbers is zero. The expected variance in the autocorrelation of a random sequence

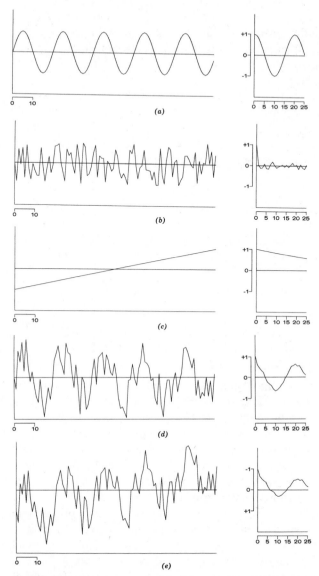

FIGURE 4.33   Some idealized time series and their autocorrelation functions. (a) Sine wave with wavelength of 20
units. (b) Sequence of random numbers or "noise." (c) Sequence of linearly increasing numbers or "trend." (d) Sine
wave plus random noise (sequence a plus sequence b). (e)
Sine wave plus random noise plus linear trend (sequence
a plus sequence b plus sequence c).

at any lag τ is

$$\sigma_\tau^2 = \frac{1}{n - \tau + 3} \tag{4.63}$$

These two parameters define the population of a random time series of a specified length n (Yule and Kendall, 1969, p. 639–641). You will recall from Chapter 2 that we can determine the probability of drawing a specific observation from a normal population having a known mean and variance by

$$Z = \frac{X - \mu}{\sigma} \tag{4.64}$$

Taking the square root of eq. (4.63) gives the expected standard deviation of autocorrelations, which can be substituted with the expected mean into eq. (4.64):

$$Z_\tau = \frac{r_\tau - 0}{1/\sqrt{n - \tau + 3}} = r_\tau \sqrt{n - \tau + 3} \tag{4.65}$$

This can be used as a conservative test of the hypothesis that an autocorrelation r_τ is zero, provided the length of the sequence, n, is large and the lag, τ, is small. "Large" and "small" are relative terms and difficult to define exactly. As a rule of thumb, n should exceed 50 and τ should not exceed $n/4$. (Some authors advocate more conservative limits, down to $n/10$ or less.) These restrictions are based on the fact that as the lag increases, r_τ is based on fewer and fewer observations. This not only causes an increase in the variance of r_τ but also results in an increasing violation of the assumption that the autocorrelation is a sample from an infinitely long time series. For these reasons, little importance should be attached to high autocorrelations at long lag intervals, unless the time series itself is many times larger.

The data in Table 4.22 represent an unusual phenomenon in our science; the values are observations of a true time series from the geologic past. Only in very special circumstances are datable occurrences recorded in the rocks, allowing us to establish a definitive time scale for a geologic sequence. The Eocene lake deposits of the Rocky Mountains consist of thinly laminated dolomitic oil shales hundreds of feet thick. It has been well established that the laminations are varves, or layered deposits caused by seasonal climatic changes in the lake basins. By measuring the thickness of these laminations, we record annual changes in the rate of deposition through the lake's history. Table 4.22 contains the thickness in millimeters of a varved section deposited near the western shore of one of these major lakes. We may attempt to answer several questions with these data. For example, was there a trend in the rate of deposition of dolomite through time, perhaps caused by a gradual climatic change? Is there evidence of cyclicity in the thickness of the laminae, possibly related to astronomical phenomena? Because the cyclicity we are seeking may have periods of many years (sunspot cycles, for example, last 11 years), 101 observations have been made and are given in the table. Process the varve data and determine if significant trends or periodicities in varve thickness

TABLE 4.22 Thickness of Successive Varves of a Section Through the Green River Oil Shale

	Thickness (mm)			
(Top of section)	6.0	8.6	10.8	4.2
	7.2	9.0	9.5	4.5
	7.1	12.0	8.1	5.9
	7.1	13.7	7.2	7.3
	7.2	14.0	7.1	7.3
	7.4	13.6	6.8	6.7
	8.0	12.1	7.0	6.0
	8.6	12.9	7.1	5.8
	10.0	12.8	5.6	5.7
	11.4	11.1	3.8	6.5
	12.0	9.0	3.4	8.2
	11.0	7.5	4.2	10.2
	9.6	7.5	4.8	12.3
	8.7	8.4	4.5	13.2
	7.6	8.4	3.6	13.2
	7.2	7.9	3.0	12.4
	7.2	7.0	2.8	9.7
	7.8	6.7	4.1	9.2
	8.1	6.8	6.8	9.3
	7.8	7.3	8.1	8.3
	7.1	7.3	7.8	6.0
	7.2	7.2	6.4	5.7
	7.1	8.1	4.6	6.1
	7.0	9.8	3.7	6.3
	7.0	11.0	4.0	6.3 (Bottom of section)
	7.7			

exist. Remember, the data must be stationary, so any significant linear trend must be removed prior to autocorrelation analysis.

A current major research area in applied geology is earthquake prediction. Studies are heavily funded by government, and initial reports indicate some success in short-range forecasting of severe tremors. This is done by detecting certain pre-quake seismic waves, which usually precede a major event.

Long-term earthquake forecasting is an entirely different problem. It requires detection of periodicities or trends in the historical record of earthquake activity, which can be extrapolated into the future. The objective is not to foretell individual quakes but to predict those periods when seismic activity will be unusually high.

Table 4.23 gives a 100-year record of an index of worldwide seismic activity, based on the annual incidence of severe quakes (Quenouille, 1952). Examine this record to see if significant trends, periodicities, or other features are contained within. If these exist, how may this information be used in a predictive model?

TABLE 4.23   Index of Worldwide Earthquake Severity, 1770–1869

1770	66	1795	78	1820	90	1845	86
1771	62	1796	110	1821	86	1846	127
1772	66	1797	79	1822	119	1847	201
1773	197	1798	85	1823	82	1848	76
1774	63	1799	113	1824	79	1849	64
1775	0	1800	59	1825	111	1850	31
1776	121	1801	86	1826	60	1851	138
1777	0	1802	199	1827	118	1852	163
1778	113	1803	53	1828	206	1853	98
1779	27	1804	81	1829	122	1854	70
1780	107	1805	81	1830	134	1855	155
1781	50	1806	156	1831	131	1856	97
1782	122	1807	27	1832	84	1857	82
1783	127	1808	81	1833	100	1858	90
1784	152	1809	107	1834	99	1859	122
1785	216	1810	152	1835	99	1860	70
1786	171	1811	99	1836	69	1861	96
1787	70	1812	177	1837	67	1862	111
1788	141	1813	48	1838	26	1863	42
1789	69	1814	70	1839	106	1864	97
1790	160	1815	158	1840	108	1865	91
1791	92	1816	22	1841	155	1866	64
1792	70	1817	43	1842	40	1867	81
1793	46	1818	102	1843	75	1868	162
1794	96	1819	111	1844	99	1869	137

From Quenouille (1952).

Cross-Correlation

If it is possible to compare a time series with itself at successive lags in order to detect dependencies through time, it would seem possible to compare two time series with each other in order to determine positions of pronounced correspondence. Two items of information may emerge from such a comparison: the strength of the relationship between the two series, and the lag or offset in time or distance between them at their position of maximum equivalence. The process of comparing two time series at successive lags is called *cross-correlation*. In many instances, it is not possible to designate a position of "zero lag" because either series may lead the other. Since the two series are not identical, the cross-correlogram is not symmetric about its middle; lags in which series *A* leads series *B* differ from lags in which *B* leads *A* (Fig. 4.34). A complicating factor arises because series *A* may not necessarily be the same length as series *B*. In fact, one approach to "automated correlation" (in the geological sense of equating two stratigraphic sections) consists of moving a short, distinctive part of one stratigraphic interval past another entire stratigraphic section to determine the position of best match.

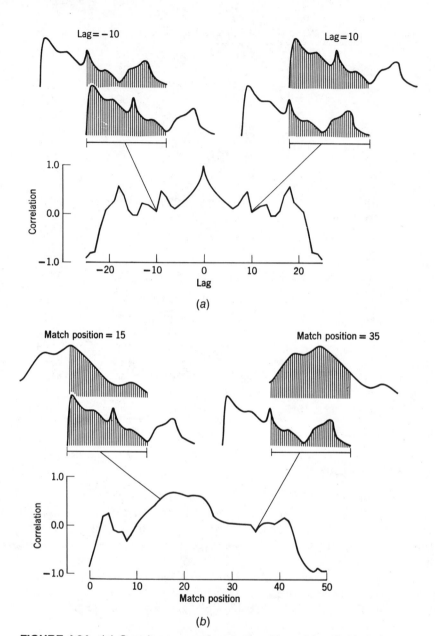

FIGURE 4.34 (*a*) Correlogram produced at positive and negative lags by autocorrelation. Correlogram is symmetric around zero lag. (*b*) Correlogram produced by cross-correlation of two dissimilar sequences. Correlogram is asymmetric unless both sequences are identical.

The equation for cross-correlation is the same as the ordinary linear correlation coefficient, and differs somewhat from the autocorrelation coefficient. If we designate the two series being compared as Y_{1i} and Y_{2i} and define n^* as the number of overlapped positions between the two, the cross-correlation for match position m is

$$r_m = \frac{n^*\Sigma Y_1 Y_2 - \Sigma Y_1 \Sigma Y_2}{\sqrt{[n^*\Sigma Y_1^2 - (\Sigma Y_1)^2][n^*\Sigma Y_2^2 - (\Sigma Y_2)^2]}} \tag{4.66}$$

or, equivalently,

$$r_m = \frac{\text{COV}_{1,2}}{s_1 s_2} \tag{4.67}$$

In this equation $\text{cov}_{1,2}$ is the covariance between the overlapped portions of sequences 1 and 2, and s_1 and s_2 are the corresponding standard deviations. Note that the summations are understood to extend only over the segments of the two sequences which are overlapped at the match position. The match positions are numbered sequentially as shown in Figure 4.34, and the *cross-correlogram* is a plot of match position versus cross-correlation.

Because the summations in eq. (4.66) extend only over the overlapped segment, the denominator of the cross-correlation changes with m. In contrast, the denominator of the autocorrelation coefficient is based on the variance of the entire chain, and is considered to be constant for all lags. A variance derived from an entire sequence is more stable than an estimate derived from a shorter segment, and so is preferable. However, in cross-correlation we cannot expect the variances to be constant through the lengths of both chains, especially when one of the chains is short with respect to the other.

The significance of the cross-correlation coefficient can be assessed by the approximate test

$$t = r_m \sqrt{\frac{n^* - 2}{1 - r_m^2}} \tag{4.68}$$

which has $(n^* - 2)$ degrees of freedom. This test is derived from a test for the significance of the correlation between two samples drawn from normal populations. The null hypothesis states that the correlation between the two sequences at the specified match position is zero, or that which is expected if the two sequences are independent, random series.

Cross-correlation is most appropriately used to compare two series that may have a temporal dependency between them. As an example, we may analyze the data given in Table 4.24. The Rocky Mountain Arsenal is a manufacturing plant for producing various noxious military compounds; it is located at Denver, Colorado, near the Front Range of the Rocky Mountains. Tremendous quantities of contaminated water are produced as a by-product of the weapons industry. In an attempt to dispose of this waste water, a deep injection well was drilled into basement rocks in 1961. Unfortunately, the well penetrated the shear zone of a

TABLE 4.24   Waste Injected by Rocky Mountain Arsenal and Frequency of Recorded Earthquakes in the Denver, Colorado, Area Between March 1962 and October 1965

Month and Year	Gallons of Waste Injected (millions)	Number of Earthquakes
Mar. 1962	4.2	—
Apr.	7.2	2
May	8.4	12
June	8.0	35
July	5.2	23
Aug.	6.0	29
Sept.	5.0	24
Oct.	5.6	8
Nov.	4.0	6
Dec.	3.6	20
Jan. 1963	6.0	25
Feb.	7.6	22
Mar.	7.8	21
Apr.	6.4	42
May	3.6	21
June	4.0	8
July	3.4	6
Aug.	2.4	10
Sept.	3.9	11
Oct.	0	12
Nov.	0	4
Dec.	0	2
Jan. 1964	0	5
Feb.	0	2
Mar.	0	9
Apr.	0	9
May	0	2
June	0	4
July	0	4
Aug.	0	5
Sept.	0.6	2
Oct.	1.8	14
Nov.	2.4	2
Dec.	2.0	7
Jan. 1965	2.0	1
Feb.	1.7	30
Mar.	1.6	9
Apr.	3.6	19
May	4.0	11
June	6.4	38

TABLE 4.24   (*Continued*)

Month and Year	Gallons of Waste Injected (millions)	Number of Earthquakes
July	8.9	62
Aug.	5.4	48
Sept.	6.4	87
Oct.	3.8	5

From Bardwell (1970).

major fault along the Rocky Mountain front, and there is evidence that the high-pressure injection of waste fluids served to lubricate and mobilize the fault. One of the sets of data given in Table 4.24 is the month-by-month record of volume of water injected into the disposal well at the Rocky Mountain Arsenal over the 4-year period the well was in use. The other set gives the number of earthquakes detected in Denver each month. In a study of the statistical relationship between these two time series, Bardwell (1970) plotted each in cumulative form and concluded there was a pronounced 3-month lag between injection and earthquake incidence. Unfortunately, this method of visual comparison between two curves may be misleading because it requires arbitrary scaling to bring the two records into coincidence.

The two records are shown in graphic form in Figures 4.35 and 4.36. The cross-correlogram of these two time series is shown in Figure 4.37. Since in this example both series have a common origin and time scale, it is possible to express match

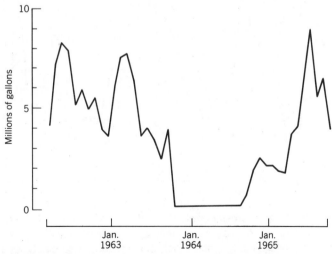

FIGURE 4.35 Quantity of liquid waste injected through Rocky Mountain Arsenal disposal well, in millions of gallons per month.

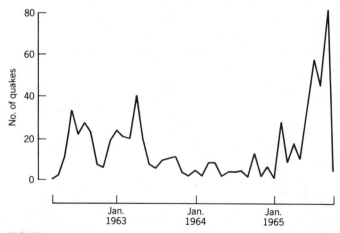

FIGURE 4.36　Numbers of earthquakes detected per month whose epicenters were near Denver, Colorado.

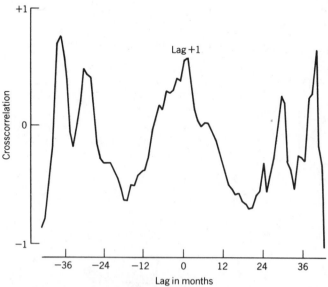

FIGURE 4.37　Cross-correlogram of monthly earthquake incidence and monthly volume of injected waste. Maximum correlation occurs at lag +1 (apparent high correlations at extremely long lags are statistically insignificant because of the low number of observations).

position in terms of "positive" and "negative" lags from the position of initial coincidence. The position of greatest correspondence between the two series occurs at lag $+1$, when the number of earthquakes in a given month is compared to the volume of waste water injected one month previously. The correlation at this match position is $r = 0.60$. The second highest cross-correlation, $r = 0.57$, occurs at lag 0. This indicates that the tectonic response to operation of the injection well began very quickly and extended over a period of about a month.

Sometimes we may want to compare two time series that are periodic over the same interval. We can take advantage of the periodicity and compute a cross-correlogram in "circular" form. In effect, each time series is wrapped in a circle and the start of a series is connected to its end. Rather than imagining the two time series as chains that are moved by each other, we can picture them in the form of wheels. Each wheel has the same number of divisions around its rim, corresponding to the n successive observations. If one wheel is rotated with respect to the other, cross-comparisons can be made between the two at n different positions before repeating.

In Chesapeake Bay, as in many estuaries, there are complex changes in salinity caused by the mingling of fresh water with sea water during the diurnal tidal cycle. Fresh Chesapeake River water floats across the denser brine in the Bay; during periods of low tide, this water moves farther down the estuary. However, there is a counter-flow along the bottom that carries dense marine water up the Bay during the waning tide. Table 4.25 gives salinity measurements (in parts per thousand)

TABLE 4.25 Salinity of Water in Chesapeake Bay at Station 11, Offshore from Annapolis, Maryland, on July 3–4, 1927

Time		State of Tide	Surface Salinity (ppt)	Bottom Salinity (ppt)
July 3	2:30 pm	1/4 ebb	6.97	11.10
	4:00	1/2 ebb	6.20	11.54
	5:30	ebb	5.93	13.12
	7:00	ebb	6.32	13.52
	8:30	1/4 flood	6.36	13.35
	10:00	1/2 flood	6.72	12.83
	11:30	3/4 flood	6.80	13.31
July 4	1:00 am	flood	6.90	13.02
	2:30	1/4 ebb	7.14	12.14
	4:00	1/2 ebb	6.91	12.44
	5:30	ebb	6.76	12.60
	7:00	begin flood	6.74	12.79
	8:30	3/8 flood	7.20	13.46
	10:00	1/2 flood	7.45	12.33
	11:30	3/4 flood	7.47	12.40
	1:00 pm	flood	7.47	12.14

From Wells, Bailey, and Henderson (1928).

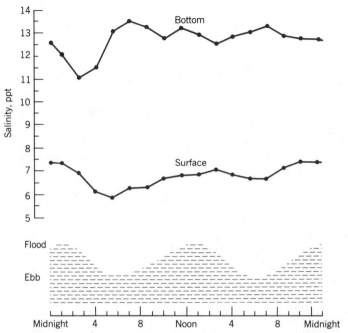

FIGURE 4.38 Salinities of bottom water and surface water in Chesapeake Bay near Annapolis, Maryland. Bottom curve represents state of the tide.

made at 1.5-hour intervals over a 24-hour period, for both surface water and bottom water (11-meter depth average) at a collecting station offshore from Annapolis, Maryland. The records are shown in graphical form in Figure 4.38.

The cross-correlation function is shown in Figure 4.39 and clearly indicates a lag of 2, representing a difference of 3 hours between the crest of salt water invasion along the bottom and the maximum salinity in the surface waters. Because the

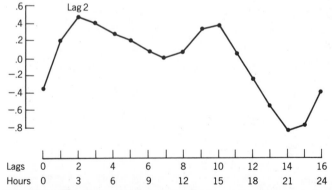

FIGURE 4.39 Cross-correlogram between salinities of bottom water and surface water in Chesapeake Bay. Maximum correspondence occurs at lag 2, representing a 3-hour difference.

records represent a 24-hour tidal cycle that repeats, the cross-correlations can be calculated in circular form. Therefore, cross-correlations can be found for lags up to 24 hours.

Cross-Correlation and Geologic Correlation

Geologists have repeatedly succumbed to the temptation to use cross-correlation procedures to match stratigraphic sections, thereby performing automated geological correlation. Although the literature on this topic is extensive, these efforts have met with a notable lack of success, except in special circumstances. The reasons for failure are not difficult to deduce. Cross-correlation presumes that the two sequences being compared are sampled at discrete, uniformly spaced points, and that the sampling interval in one succession is the same as in the other. Unfortunately, it is difficult to collect stratigraphic measurements that meet these requirements. In general, it is not possible to place sample points so they are equally spaced in geologic time, and if points are equally spaced in distance we are in effect assuming that the rates of sediment accumulation were constant throughout the time of deposition of the two sequences. If we consider stratigraphic sections to be analogous to the records produced by chart recorders, it is as though the recorders ran at different (and unknown) speeds at different times. To make matters worse, it is probable that the recorders were turned off entirely much of the time!

Having said this, we will now consider a problem of geologic correlation in which the special circumstances not only make it appropriate to use cross-correlation, but it is about our only hope for success. Table 4.26 contains measurements of varve thicknesses from a sequence through an Eocene lake deposit in the Rocky Mountains. The deposit, part of the Green River Formation, consists of thinly laminated dolomitic oil shales hundreds of feet thick. A similar section is listed in Table 4.22. The two sections are only 10 miles apart, and presumably the shorter

TABLE 4.26 Thickness of Successive Varves in a Section Through the Green River Oil Shale; This Section is Located 10 Miles North of the Section Listed in Table 4.22

Section B	Thickness (mm)		
(Top of section)	10.8	15.6	9.0
	11.7	15.0	9.2
	11.0	13.4	10.7
	9.9	14.6	10.8
	9.8	13.0	8.9
	9.9	10.3	9.4
	10.0	9.4	10.6
	10.0	9.0	12.6
	10.2	10.1	14.2
	10.8	10.3	12.3
	11.3	9.2	11.1
	12.0	9.1	11.0	(Bottom of section)
	13.5		

sequence is equivalent to some part of the larger. There are no marker beds or distinct features in this part of the monotonous varves, so we must base a correlation on the best match in thickness of individual laminations in the two sections. Compute the cross-correlogram and determine the position of best correlation between the two sections. How do you explain the other positions of lesser, but significant match?

Cross-Association

The methods for comparing two time series we have just considered require that both series consist of measured variables. Unfortunately, many studies, particularly those involving stratigraphy, produce only nominal data. Conventional procedures such as cross-correlation cannot be used because they rely on measures of similarity that presume the data are continuously distributed. The problem can be circumvented by substituting a measure of similarity that does not require measurement data as does the correlation coefficient. One possible approach used by geologists is called *cross-association*. The data consist of a series of states such as lithologic types (i.e., limestone, sandstone, shale, etc.), perhaps encountered in a stratigraphic section. These states are mutually exclusive and cannot be ranked in a meaningful way. With cross-association, two such sequences are moved past one another, and the degree of correspondence between the overlapped segments is calculated. At each match position, the number of matching states and the total number of comparisons are counted. The simple ratio (number of matches)/(number of comparisons) is calculated and can be used as an index of the similarity of the two chains at the overlap position.

Suppose we have measured a section and coded the units by the scheme sandstone = 1, shale = 2, limestone = 3, coal = 4, siltstone = 5. As you can see, the coding system is completely arbitrary. Two stratigraphic sections might be represented by

$$(a) \quad 5\ 1\ 2\ 1\ 2\ 5\ 2\ 3\ 2\ 4\ 2\ 5$$
$$(b) \quad 3\ 2\ 1\ 5\ 2\ 5\ 1\ 2\ 1\ 2\ 3\ 2$$

The second section may be "moved" by the first and compared at each match position as in this sequence:

Match position 1:

$$5\ 1\ 2\ 1\ 2\ 5\ 2\ 3\ 2\ 4\ 2\ 5$$
$$3\ 2\ 1\ 5\ 2\ 5\ 1\ 2\ 1\ 2\ 3\ 2$$

number of comparisons = 1   number of matches = 0   ratio = 0

Match position 2:

$$5\ 1\ 2\ 1\ 2\ 5\ 2\ 3\ 2\ 4\ 2\ 5$$
$$3\ 2\ 1\ 5\ 2\ 5\ 1\ 2\ 1\ 2\ 3\ 2$$

number of comparisons = 2   number of matches = 0   ratio = 0

Match position 3:

$$5\ 1\ 2\ 1\ 2\ 5\ 2\ 3\ 2\ 4\ 2\ 5$$
$$3\ 2\ 1\ 5\ 2\ 5\ 1\ 2\ 1\ 2\ 3\ 2$$

number of comparisons = 3   number of matches = 1   ratio = 0.33

If we plot the match position versus the matching ratio, we obtain a graph that we might call an "associatogram" because it is analogous to the correlogram. The graph for the two sections is shown in Figure 4.40.

The significance of the matching ratio can be determined by a χ^2 test. To perform this test we need to determine the number of matches and the number of mismatches in the overlapping segment. We also need the probability of a number of matches occurring between two totally random sequences. These random sequences contain the same number of observations in each state as our two samples.

We can designate the two sequences as chain 1 and chain 2. There are m possible categories into which observations can be classed. If we denote the number of observations in the kth state of chain 1 as X_{1k}, the total length of chain 1 is $n_1 = \sum_{k=1}^{m} X_{1k}$. In chain 2, there also are m categories, each with X_{2k} observations, giving a total length of chain 2 of $n_2 = \sum_{k=1}^{m} X_{2k}$. To determine the probability that a given number of observations from the two sequences will match, we must find

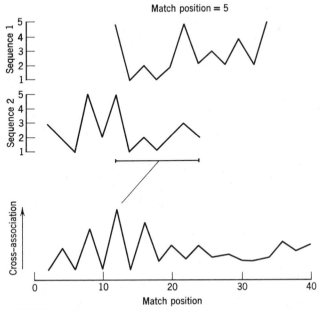

FIGURE 4.40 Comparison of two nonnumeric sequences by cross-association. Two sequences are shown at the position of maximum match.

the sum of the products of each of the m categories in the two chains divided by the product of the chain lengths. That is,

$$P = \frac{\sum_{k=1}^{m} X_{1k}X_{2k}}{n_1 n_2} \tag{4.69}$$

In our measured sections, the number of categories is $m = 5$. The first chain contains two sandstones, five shales, and so forth. The complete count of the number of observations in each category of the two chains is given in Table 4.27. The final column of the table contains the product of the number of observations in each category, and the bottom row contains the column sums. The probability of a match at any position of comparison in random sequences equivalent to our sections is

$$P = \frac{39}{12 \cdot 12} = 0.27$$

The probability of a mismatch is $(1.00 - 0.27)$ or 0.73.

We now have obtained the probability associated with the matching of two randomly ordered chains containing the same distribution of observations as our two sequences. An expected number of matches for an overlapped interval can be determined by P times the number of comparisons. Similarly, the expected number of mismatches is $(1 - P)$ times the number of comparisons. Suppose the two sequences in our example are completely overlapped, giving 12 positions of comparison. The expected number of matches in the sequence is $12 \times 0.27 = 3.2$ and the expected number of mismatches is $12 \times 0.73 = 8.8$. Using these quantities in a χ^2 test, we have

$$\chi^2 = \frac{(O - E)^2}{E} + \frac{(O' - E')^2}{E'} \tag{4.70}$$

where

$$
\begin{aligned}
O &= \text{observed number of matches} \\
O' &= \text{observed number of mismatches} \\
E &= \text{expected number of matches} \\
E' &= \text{expected number of mismatches}
\end{aligned}
$$

TABLE 4.27 Number of Observations in Each Lithologic State in Two Stratigraphic Sequences

Category	Chain 1	Chain 2	$X_{1k}X_{2k}$
1. Sandstone	2	3	6
2. Shale	5	5	25
3. Limestone	1	2	2
4. Coal	1	0	0
5. Siltstone	3	2	6
TOTALS	$n_1 = 12$	$n_2 = 12$	$\sum_{k=1}^{m} X_{1k}X_{2k} = 39$

The χ^2 test statistic has $\nu = 1$ degree of freedom. If the expected number of matches is small, as in match positions near ends of chains, the test is improved by applying Yates' correction (Fisher, 1970, pp. 92–95). This consists of subtracting 1/2 from the absolute difference of the observed and expected values:

$$\chi^2 = \frac{(O - E - 1/2)^2}{E} + \frac{(O' - E' - 1/2)^2}{E'} \tag{4.71}$$

Degrees of freedom remain unchanged, $\nu = 1$.

In our example,

$$O = 2 \qquad O' = 10$$
$$E = 3.2 \qquad E' = 8.8$$

so the χ^2 test has the value

$$\chi^2 = \frac{(2 - 3.2)^2}{3.2} + \frac{(10 - 8.8)^2}{8.8} = 0.61$$

which is not significant. We may conclude on the basis of this test that the observed number of matches between the two successions, at this position, is no greater than that expected from two random sequences of equivalent composition. Yates' correction for small samples does not alter this conclusion.

Because we are dealing with discrete events (matching or mismatching) the frequency distribution of matches is a binomial. The mean number of expected matches from two random sequences that are being compared at n^* overlapped positions is

$$\bar{E} = P(n^*) \tag{4.72}$$

The distance of the observed number of matches from this mean, in units of standard deviation, is given by

$$S = \frac{O - E}{E(1 - P)} \tag{4.73}$$

Unfortunately, we cannot evaluate the probability of obtaining an observation so many standard deviations from the binomial mean unless we have extensive tables of the binomial distribution. These are not readily available for the large values of n likely to be encountered in cross-association problems. However, we can create a normal approximation (Owen, 1962, p. 293) to the binomial deviation by (4.74). Using this normal approximation, we can find the probability of obtaining a given deviation from the mean by use of standardized normal tables.

$$Z = \sqrt{n^*}(2 \arcsin \sqrt{O/n^*} - 2 \arcsin \sqrt{P}) \tag{4.74}$$

We can substitute the appropriate values for our two sequences into (4.74), giving us

$$Z = \sqrt{12}(2 \arcsin \sqrt{2/12} - 2 \arcsin \sqrt{0.27})$$

which becomes

$$Z = 3.46(2 \arcsin 0.41 - 2 \arcsin 0.53) = -0.97$$

We expected three matches when comparing the two random sequences and observed two; this is about one standard deviation away from the expected number. The observed number of matches at this position could well arise by random comparisons.

As an example of the application of cross-association, we will consider the correlation of two measured sections in the Coal Measures of central England. Good outcrops are rare, and electric logs of wells are nonexistent; therefore, most information on the stratigraphic succession comes from deep excavations such as quarries and mines. The lithologies are coded as 1 = sandstone, 2 = siltstone, 3 = nonfossiliferous shale, 4 = underclay, 5 = coal, 6 = fossiliferous shale, and 7 = limestone. The first section was measured in a shaft being sunk for a coal mine. The second, much shorter, is exposed in the wall of an open-pit mine about 6 miles away. Find the position of best match of the shorter sequence with the longer interval. The data are given in Table 4.28.

It should be obvious from the preceding example that cross-association is not restricted to a qualitative equivalent of cross-correlation. In time series, we must assume that the observations are located at points along a scale; this restriction has been relaxed in cross-association. Our data may consist simply of the succession of states listed in the order in which they occur. The distance between successive points in this case is regarded as immaterial, as in the English stratigraphic sections.

In a similar manner, a nonnumeric sequence may be compared to itself; the process is referred to as *autoassociation*. Autoassociation can be a useful procedure in the search for periodicities in the order of a succession of states and has been extensively used in the investigation of cyclothems (Sackin and Merriam, 1969).

In this application, we are not comparing two sequences, but only one with itself. We must therefore adjust the probability of a match (P) to reflect this fact. The binomial probability of a given number of matches occurring when a random sequence is compared to itself is given by

$$P = \frac{\sum_{k=1}^{m} X_k^2 - n}{n^2 - n} \tag{4.75}$$

TABLE 4.28 Two Coded Stratigraphic Sections from Central England[a]

	(Bottom)													
Mine section	2	4	5	6	3	4	5	3	1	4	5	3	4	5
	3	4	5	4	5	3	2	4	5	3	4	5	3	1
	4	5	4	5	6	3	4	5	6	3	4	5	2	1
	3	4	5	3	5	3	2	4	5	3	5	2		
												(Top)		
	(Bottom)													
Quarry section	4	5	3	4	5	4	5	3	2	1	2	4	5	3

[a]Key: 1 = sandstone, 2 = siltstone, 3 = nonfossiliferous shale, 4 = underclay, 5 = coal, 6 = fossiliferous shale, 7 = limestone.

We assume that the sequence is a random arrangement of m states or classes of items, each state containing X_k observations. The total number of observations is $\sum_{k=1}^{m} X_k = n$. This probability can be substituted directly for (4.72) and used in both χ^2 tests and in computation of the standard deviation. The tests are against the null hypothesis that the number of matches is no different from that expected by matching a random sequence with itself.

We may use the data from the mine section (Table 4.27) to illustrate the application of autoassociation. If the mine section contains repetitive elements, these will appear as unusually high match ratios and as significant deviations from the expected mean number of matches. The interpretation of ''associatograms'' proceeds in a manner analogous to the interpretation of correlograms. However, the association coefficient is based on nominal data: therefore, the information content of the sequence is much lower than in an equivalent time series of measured attributes. Because we are using qualitative input, we cannot in general expect the same resolution we would obtain from analyses of a true time series. This factor must be considered in the interpretation of results from cross-association or autoassociation.

Semivariograms

The term *geostatistics* is now widely applied to a special branch of applied statistics originally developed by Georges Matheron of the Centre de Morphologie Mathématique in Fontainebleau, France. Geostatistics was devised to treat problems that arise when conventional statistical theory is used in estimating changes in ore grade within a mine. However, because geostatistics is an abstract theory of statistical behavior, it is applicable to many circumstances in different areas of geology and other natural sciences.

A key concept of geostatistics is that of the *regionalized variable*, which has properties intermediate between a truly random variable and one completely deterministic. Typical regionalized variables are functions describing natural phenomena that have geographic distributions, such as the elevation of the ground surface, changes in grade within an ore body, or the spontaneous electrical potential measured in a well by a logging tool. Unlike random variables, regionalized variables have continuity from point to point, but the changes in the variable are so complex that they cannot be described by any tractable deterministic function.

Even though a regionalized variable is spatially continuous, it is not usually possible to know its value everywhere. Instead, its values are known only through samples, which are taken at specific locations. The size, shape, orientation, and spatial arrangement of these samples constitute the *support* of the regionalized variable, and the regionalized variable will have different characteristics if any of these are changed. For example, suppose we wish to determine the variation in ore grade in a disseminated molybdenum deposit. It seems likely that the answer we might obtain from analysis of 2-inch diamond drill cores could be considerably different from what we might determine if we used the mill runs on muck carsized ore samples. In both instances we might take exactly the same number of samples, and they might come from identical locations within the mine. However, the fact that the samples in one set have volumes of a few cubic inches and the volumes of samples in the second set are measured in cubic yards must inevitably affect the pattern of variation in ore grade we map through the mine. A principal concern of geostatistics is to relate the results obtained from one support (such as core samples) to that obtained from another support (such as stope blocks).

Geostatistics involves estimating the form of a regionalized variable in one, two, or three dimensions. In the next chapter, we will consider the estimation procedure, called kriging, more extensively. Now we will be concerned only with one of the basic statistical measures of geostatistics, the semivariance, which is used to express the rate of change of a regionalized variable along a specific orientation. Estimating the semivariance involves procedures similar to those of time series analysis, hence the introduction of geostatistics at this point.

The semivariance is a measure of the degree of spatial dependence between samples along a specific support. For the sake of simplicity, we will assume the samples are point measurements of a property such as depth to a subsurface horizon. For computational tractability, we will further assume that the support is regular; that is, the samples are uniformly spaced along straight lines. If the spacing between samples along a line is some distance Δ, the semivariance can be estimated for distances that are multiples of Δ:

$$\gamma_h = \sum_i^{n-h} (X_i - X_{i+h})^2 / 2n \qquad (4.76)$$

In this notation, X_i is a measurement of a regionalized variable taken at location i, and X_{i+h} is another measurement taken h intervals away. We are therefore finding the sum of the squared differences between pairs of points separated by the distance Δh. The number of points is n, so the number of comparisons between pairs of points is $n - h$.

If we calculate the semivariances for different values of h, we can plot the results in the form of a *semivariogram*, which is analogous to a correlogram. Figure 4.41 shows a semivariogram for depth to a seismic reflecting horizon, as measured along the seismic profile shown in Figure 4.42. Note that when the distance between sample points is zero, the value at each point is being compared with itself. Hence, all the differences are zero, and the semivariance for γ_0 is zero. If Δh is a small distance, the points being compared tend to be very similar, and the semivariance will be a small value. As the distance Δh is increased, the points being compared are less and less closely related to each other and their differences become larger, resulting in larger values of γ_h. At some distance the points being compared are so far apart that they are not related to each other, and their squared differences become equal in magnitude to the variance around the average value. The semivariance no longer increases and the semivariogram develops a flat region called a *sill*. The distance at which the semivariance approaches the variance is referred to as the *range* or *span* of the regionalized variable, and defines a neighborhood within which all locations are related to one another.

For some arbitrary point in space, we can imagine the neighborhood as a symmetrical interval (or area or volume, depending on the number of dimensions)

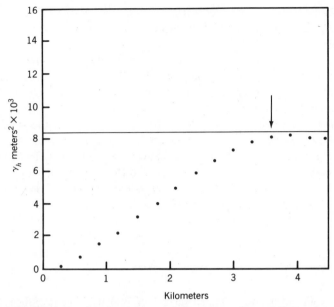

FIGURE 4.41   Semivariogram of elevations of top of Cretaceous Springhill Formation measured along marine seismic traverse in Straits of Magellan, Chile. Semivariance is in meters^2. Light line represents sill, or variance of elevations, and is equal to 8380 meters^2. Range, at arrow, is distance beyond which difference between semivariance and sill is considered negligible and is equal to 3.5 km. From Olea (1977).

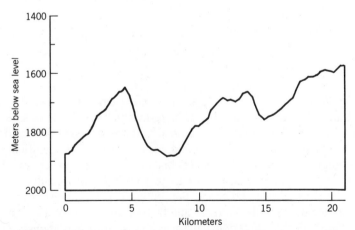

FIGURE 4.42   Subsurface structural elevations of top of Cretaceous Springhill Formation, estimated from reflection seismic measurements along 21-km marine traverse in Straits of Magellan. Depths in meters below sea level. Seismic measurements taken at 300-meter intervals. From Olea (1977).

about the point. If the regionalized variable is stationary, or has the same average value everywhere, any locations outside the interval are completely independent of the central point, and cannot provide information about the value of the regionalized variable at that location. Within the neighborhood, however, the regionalized variable at all observation points is related to the regionalized variable at the central location and hence can be used to estimate its value. If we use a number of measurements made at locations within the neighborhood to estimate the value of the regionalized variable at the central location, the semivariogram provides the proper weightings to be assigned to each of these measurements.

We here digress to demonstrate a point that will be useful later. The semivariance is not only equal to the average of the squared differences between pairs of points spaced a distance Δh apart, it is also equal to the variance of these differences. That is, the semivariance can also be defined as

$$\gamma_h = \frac{\sum \left\{ (X_i - X_{i+h}) - \dfrac{\sum (X_i - X_{i+h})}{n} \right\}^2}{2n} \tag{4.77}$$

Note that the mean of the regionalized variable X_i is also the mean of the regionalized variable X_{i+h}, because these are the same observations, merely taken in a different order. That is,

$$\frac{\sum X_i}{n} = \frac{\sum X_{i+h}}{n}$$

Therefore, their difference must be zero:

$$\frac{\sum X_i}{n} - \frac{\sum X_{i+h}}{n} = 0$$

We can combine the summations

$$\frac{\sum X_i - \sum X_{i+h}}{n} = \frac{\sum (X_i - X_{i+h})}{n} = 0$$

Substituting into eq. (4.77), we see that the second term in the numerator is zero, so the equation is equal to eq. (4.76). Note that this relationship is strictly true only if the regionalized variable is stationary. If the data are not stationary, the mean of the sequence changes with h and eq. (4.77) must be modified.

As you might expect, there are mathematical relationships between the semivariance and other statistics such as the autocovariance and the autocorrelation. If the regionalized variable is stationary, the semivariance for a distance Δh is equal to the difference between the variance and the spatial autocovariance for the same distance (Fig. 4.43). If the regionalized variable is not only stationary but also is standardized to have a mean of zero and variance of 1.0, the semivariogram is a mirror image of the autocorrelation function (Fig. 4.44).

Unfortunately, it often happens that regionalized variables are not stationary,

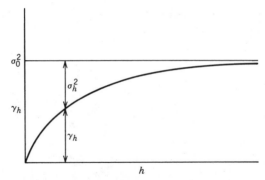

FIGURE 4.43 Relationship between semivariance γ and autocovariance σ^2 for a stationary regionalized variable. σ_0^2 is the variance of the observations, or the autocovariance at lag 0. For values of h beyond the range, $\gamma_h = \sigma_0^2$.

but rather exhibit changes in their average value from place to place. If we attempt to compute a semivariogram for such a variable, we will discover that it may not have the properties we have described. However, if we reexamine the definition of semivariance given in eq. (4.77), we note that it contains two parts, the first being the difference between pairs of points and the second being the average of these differences. If the regionalized variable is stationary, we have shown that the second part vanishes, but if it is not stationary, this average will have some value. In effect, the regionalized variable can be regarded as composed of two parts, called the residual and the drift. The *drift* is the expected value of the regionalized variable at a point i, or computationally, a weighted average of all

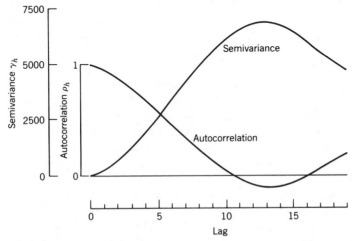

FIGURE 4.44 Relationship between semivariance γ_h and autocorrelation ρ_h for a stationary regionalized variable.

the points within the neighborhood around point i. The drift will have the form of a smooth approximation of the original regionalized variable. If the drift is subtracted from the regionalized variable, the *residuals* $R_i = X_i - \overline{X}_i$ will themselves be a regionalized variable and will have local mean values equal to zero. In other words, the residuals will be stationary and it will be possible to compute their semivariogram.

Here we come to an awkwardly circular problem. The drift could be estimated if we knew the size of the neighborhood and the weights to be assigned points within the neighborhood. However, the weights can be calculated only if we know the semivariances corresponding to the distances between point i, the center of the neighborhood, and the various other points. Having once calculated the drift, it could be subtracted from the observed values to yield stationary residuals, which in turn could be used to estimate the neighborhood size and form of the semivariogram.

At this stage we must relax our definitional rigor and resort to trial and error. We first concede that we cannot determine the neighborhood in the sense that we have been using the term. Instead, the neighborhood is defined as a convenient but arbitrary interval within which we are reasonably confident that all locations are related to one another. Within this arbitrary neighborhood, we assume that the drift can be approximated by a simple expression such as

$$\overline{X}_0 = \Sigma b_1 X_i$$

or

$$\overline{X}_0 = \Sigma(b_1 X_i + b_2 X_i^2)$$

where the first represents a linear drift and the second a quadratic drift. The calculations involve all of the points within the arbitrary neighborhood, so there is an interrelation between neighborhood size, drift, and semivariogram for the residuals. If the neighborhood is large, the drift calculations will involve many points and the drift itself will be very smooth and gentle. Consequently, the residuals will tend to be more variable and their semivariogram will be complicated in form. Conversely, specification of a small neighborhood size will result in a more variable drift estimate, smaller residuals, and a simpler semivariogram.

Determining the b coefficients for the drift requires solving a set of simultaneous equations of somewhat foreboding complexity; they will be left to the section on kriging. The only variables in the equations are the semivariances corresponding to the different distances between point i and the other points within the neighborhood. However, we do not yet have a semivariogram from which to obtain the necessary semivariances. We must assume a reasonable form for the semivariogram and use it as a first approximation. It will be much easier to guess the form of a simple semivariogram, so this is an argument for using as small a neighborhood size as possible.

Next, the experimental estimates of the drift are subtracted from their corresponding observations to yield a set of experimental residuals. A semivariogram can be calculated for these residuals and its form compared to that of the semivariogram that was first assumed. If the assumptions that have been made are appropriate, the two will coincide, and we have successfully deduced the form of the drift and the semivariogram. Most likely they will differ and we must try again.

The process of attempting to simultaneously find satisfactory representations of the semivariogram and drift expression is a major part of "structural analysis." It is to a certain extent an art, requiring experience, patience, and sometimes luck. The process is not altogether satisfying, because the conclusions are not unique; many combinations of drift, neighborhood, and semivariogram model may yield approximately equivalent results. This is especially apt to be true if the regionalized variable is erratic or we have only a short sequence. In such circumstances it may be difficult to tell when (or if) we have arrived at the proper combination of estimates.

The semivariogram expresses the spatial behavior of the regionalized variable or its residual. Some idealized forms of semivariograms are shown in Figure 4.45.

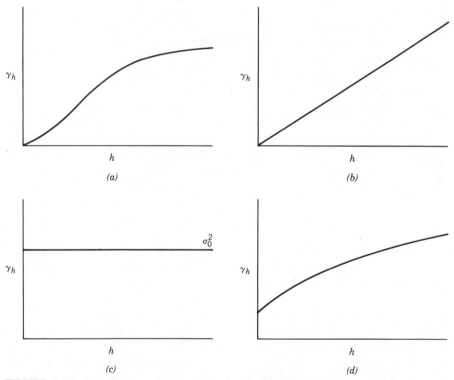

FIGURE 4.45 Idealized semivariograms showing (a) parabolic form, indicative of excellent continuity of the regionalized variable; (b) linear form, indicating moderate continuity; (c) horizontal form at the level of σ_0^2, produced by a random variable having no spatial autocorrelation; (d) nugget effect, or apparent failure of the semivariogram to go through the origin, indicative of a regionalized variable that is highly variable over distances less than the sampling interval.

A semivariogram tangent to the X axis at the origin, as in 4.45a, is described as parabolic and indicates that the regionalized variable is exceptionally continuous. Figure 4.45b is a semivariogram that is linear in form; this indicates moderate continuity of the regionalized variable. A truly random variable will have no continuity and its semivariogram will be a horizontal line equal to the variance (Fig. 4.45c). In some circumstances the semivariogram will appear to not go through the origin but rather will assume some nonzero value. This is referred to as the "nugget effect" and is shown in Figure 4.45d. In theory, γ_0 must equal zero; the nugget effect arises because the regionalized variable is so erratic over a very short distance that the semivariogram goes from zero to the level of the nugget effect in a distance less than the sampling interval.

Modelling the Semivariogram

In principle, the experimental semivariogram could be used directly to provide values for the estimation procedures we will discuss in the next chapter. However, the semivariogram is known only at discrete points representing distances Δh; in practice, semivariances may be required for any distance, whether it is a multiple of Δ or not. For this reason, the discrete experimental semivariogram must be modelled by a continuous function that can be evaluated for any desired distance.

Fitting a model equation to an experimental semivariogram is a trial-and-error process, usually done by eye. Clark (1979) describes and gives examples of the manual process, while Olea (1977) provides a program that computes a linear semivariogram having the same slope at the origin as the experimental semivariogram.

Ideally, the model chosen to represent the semivariogram should begin at the origin, rise smoothly to some upper limit, then continue at a constant level. The *spherical model*, shown in Figure 4.46, has these properties. It is defined as

$$\gamma_h = \sigma_0^2 \left(\frac{3h}{2a} - \frac{h^3}{2a^3} \right) \tag{4.78}$$

for all distances up to the range of the semivariogram, a. Beyond the range, $\gamma_h = \sigma_0^2$. The spherical model usually is described as the ideal form of the semivariogram. Another that is sometimes used is the *exponential model*:

$$\gamma_h = \sigma_0^2 \left(1 - e^{-h/a} \right) \tag{4.79}$$

Figure 4.47 compares the spherical and exponential models. The exponential never quite reaches the limiting value of the sill, but approaches it asymptotically. Also, the semivariance of the exponential model is lower than the spherical for all values of h less than the range.

The *linear model* is simpler than either the spherical or exponential, as it has only one parameter, the slope. The model has the form

$$\gamma_h = \alpha h \tag{4.80}$$

and plots as a straight line through the origin. Obviously, this model cannot have

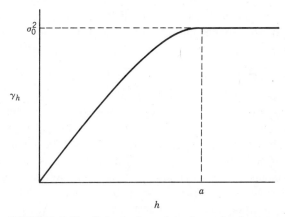

FIGURE 4.46 Spherical model of semivariogram. Range, *a*, is distance at which semivariance becomes equal to variance, σ_0^2.

a sill, as it rises without limit. Sometimes the linear model is arbitrarily modified by inserting a sharp break at the sill value so that

$$\gamma_h = \alpha h \qquad \text{for } h < a$$
$$\gamma_h = \sigma_0^2 \qquad \text{for } h \geq a$$

(4.81)

Armstrong and Jabin (1981) have criticized the use of such a model, because the kriging estimation procedure presumes the semivariogram is a continuous, smoothly varying function. However, for distances much less than the range, the linear model is a perfectly good approximation. This is apparent in Figure 4.47, where both the spherical and exponential models are almost coincident with a straight line near the origin. If the regionalized variable has been sampled at a sufficient density,

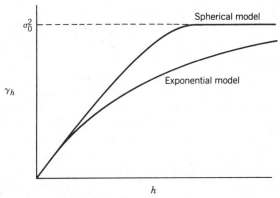

FIGURE 4.47 Exponential and spherical models of semivariogram. Both models have same initial slope and sill. After Clark (1979).

relative to the range, there will be no significant differences between estimates made assuming a linear model and those obtained using a spherical or other model.

Spectral Analysis

Spectral analysis is the partitioning of the variation in a time series into components according to the duration or length of the intervals within which the variation occurs. This is done by considering the time series to be the sum of many simpler time series that have the form of regular sinusoids of differing amplitudes, wavelengths, and starting points. These sinusoids are calculated so they are orthogonal, or statistically independent of one another. Since the sum of all of the sinusoids is equal to the original time series, the sum of the variation in all of the sinusoids must also be equal to the total variation in the series.

Spectral analysis is known, in its various guises, as harmonic analysis, Fourier analysis, and frequency analysis. The roots of the method are interwoven with those of musical theory, reflecting a common concern with vibratory motion. Kepler, in the seventeenth century, applied the harmonic relationships he found in arithmetic, geometry, and music to the positions of the planets and discovered the laws of planetary motion. However, it was Jean Baptiste Fourier (1768–1830) who provided proof that any continuous, single-valued function could be represented by a series of sinusoids, the relationship that now bears Fourier's name. Before examining spectral analysis in detail, a few terms must first be defined. Most of these terms were coined by electrical engineers and are used in the analysis of electrical signals. Although an electrical signal is an energy wave that changes with time, engineers are accustomed to seeing the signal "frozen" on an oscilloscope. Engineering discussions of Fourier analysis always assume that the signal is time-variant; however, the fact that engineers examine the signal as a spatial phenomenon on an oscilloscope indicates that time and space can be considered equivalent. This is, of course, mathematically correct, and the unconscious ease with which electrical engineers make the transition should assure us that the two are conceptually interchangeable as well.

Figure 4.48 shows a regularly repeating signal that can be described as a pure sinusoidal wave. The distance from a point on one wave form to the equivalent point on the next wave form is called the *wavelength*, λ. *Frequency*, f, is the reciprocal of wavelength, or $f = 1/\lambda$. It is the number of wave forms in a unit of length or time. In most engineering problems the signal is expressed in terms of frequency. In geologic problems, we usually are interested in wavelengths. The time required for a regularly repeating signal to exactly repeat itself is called its *period*. The term period is equivalent to wavelength, but the period is measured in units of time such as milliseconds rather than units of distance such as centimeters. With some phenomena, as for example sea waves, it is appropriate to use both wavelength and period. Wavelength in this instance refers to the spatial distance from crest to crest between adjacent wave forms. Period refers to the length of time required for two successive waves to pass a fixed reference point, such as the end of a pier. It is the duration in time between the appearance of one wave

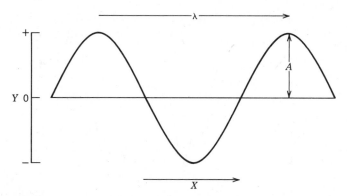

FIGURE 4.48 A regularly repeating sinusoidal wave; λ = wave-length, *A* = amplitude.

and the appearance of the next. (Use of the word period has given rise to the expression *periodic*, which means a signal that repeats at regular intervals.) Half of the height from trough to crest of a wave form is called its *amplitude*, *A*.

Figure 4.49 shows two identical sine waves that have been displaced with respect to one another. The amplitudes and wavelengths of the two signals are the same. The difference in Y_1 and Y_2 between the two signals, at a specified value of x_i, is attributable entirely to the offset between the two wave forms. The offset is described as a difference in the *phase angle*, ϕ, and its derivation can be shown with the help of Figure 4.50. A simple mechanical device that draws a sine wave consists of a disk of radius *r* revolving at constant speed around its center. A pen fastened to a push rod connected to the edge of the disk will trace a line on a sheet of paper being moved at constant speed under the end of the rod. The line is a sine wave having an amplitude $A = r$ and a wavelength that is a function of the rate of rotation of the disk and the velocity of the paper strip. The value of *Y* at any position is equal to $Y = A \sin \phi$.

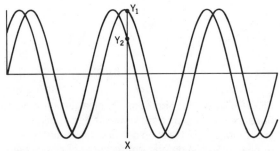

FIGURE 4.49 Two sinusoidal waves that are identical in form. Difference between Y_1 and Y_2 for a specified value *X* is attributable to difference in phase between the two wave forms.

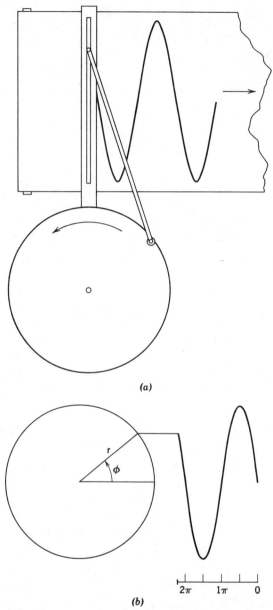

(a)

(b)

FIGURE 4.50   (a) Simple mechanical device to produce a sine wave by transforming rotary motion into linear motion. (b) Relation between angle and radius of disk to amplitude and phase angle of sine wave. Record shown from $\phi = 0$ to slightly more than 2π radians, or over one complete revolution of the disk.

To draw one complete wave form, the pen must revolve completely around the disk, moving through 360° or 2π radians. Suppose we start the device operating with the pen initially resting at an arbitrary location on the paper that we will call 0. The angle α between the pen, the center of the disk, and the center line on the paper is some value ϕ. These are shown in Figure 4.51. If we allow the device to operate for a distance x_i down the record and then stop, the pen will be resting

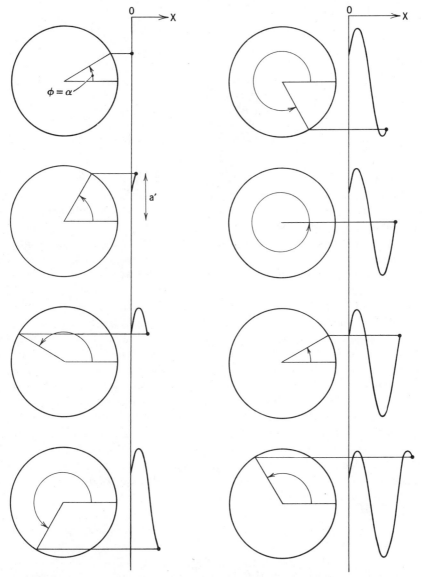

FIGURE 4.51 Progressive changes in phase angle from an initial value ϕ. At successive values of x_i, $Y_i = A \sin (2\pi x_i/X + \phi)$ and $\alpha_i = (2\pi x_i/X + \phi)$.

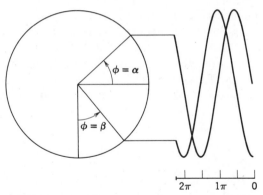

2π 1π 0

FIGURE 4.52   Sine and cosine waves related to rotation of disk. Phase angles for sine and cosine waves are α and β, respectively. Record is shown starting at $\phi = 0$ and continuing for slightly more than 2π radians, or one revolution of the disk.

at a point whose Y coordinate is equal to $A \sin(2\pi x_i/X + \phi)$. The angle α_i will be $(2\pi x_i/X + \phi)$. In these expressions, X is the total length of the record. The important thing to note is that the initial angle $\alpha = \phi$ at the starting point appears as a constant in all subsequent determinations of the Y coordinate and the angle α. The constant ϕ is known as the *phase*, *phase angle*, or *phase constant*. The three parameters of amplitude, wavelength, and phase angle completely describe the wave form. Note that the phase angle of a sine wave is measured from the horizontal center line.

   To construct a series of different wave forms, we may alter the amplitude, the wavelength, or the phase angle. We can also describe a cosine wave, identical in form to a sine wave, but having a phase angle β measured from a vertical line through the origin of the record. That is, a cosine wave is 90° or $\pi/2$ radians out of phase with a sine wave. If we add another pen to our mechanical signal generator, as in Figure 4.52, we can create both a cosine wave and a sine wave simultaneously. Obviously, the two cannot be distinguished except by the phase angle.

   We can summarize the basic trigonometric relationships with the help of Figure 4.53, which shows a sinusoidal curve. This development is given in terms of the cosine, but an equivalent development, leading to the same relationship, can be based on the sine. Assume the wave makes one complete fluctuation in a time or distance X. Any point x within that interval can be expressed in radians by the conversion

$$\theta = \frac{2\pi x}{X} \qquad (4.82)$$

The X coordinates now cover the interval from 0 to 2π radians. For convenience, we will assume that the amplitude of the curve is unity. Then, the equation of the

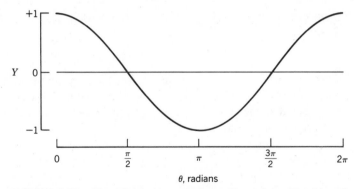

FIGURE 4.53   Sinusoidal curve expressed in radians. Total length of curve, X, is defined as equal to 2π.

curve is

$$Y = \cos \theta \qquad (4.83)$$

The amplitude can be changed to any desired magnitude simply by multiplying by a coefficient A,

$$Y = A \cos \theta \qquad (4.84)$$

The number of cycles that occur within the basic interval (the frequency) may be increased or decreased by multiplying θ by a coefficient k,

$$Y = \cos(k\theta) \qquad (4.85)$$

As drawn, the crest of the wave form occurs at the origin, but it may be shifted to any position by subtracting a phase angle, ϕ.

$$Y = \cos(\theta - \phi) \qquad (4.86)$$

Combining all of these modifications, we find that any curve of regular sinusoidal form may be written as

$$Y_k = A_k \cos(k\theta - \phi_k) \qquad (4.87)$$

The subscript k represents the *harmonic number*, or the number of cycles per basic interval. Since in spectral analysis many different harmonics will be combined to reproduce the original time series, this subscript is necessary to designate a particular wave form.

Next, we may utilize the trigonometric identity for the difference between two angles,

$$\cos(R - S) = \cos S \cos R + \sin S \sin R$$

to rewrite eq. (4.87). The equation then becomes

$$Y_k = A_k \cos \phi_k \cos(k\theta) + A_k \sin \phi_k \sin(k\theta) \qquad (4.88)$$

This expression can be simplified by defining two coefficients, $\alpha_k = A_k \cos \phi_k$ and $\beta_k = A_k \sin \phi_k$, giving

$$Y_k = \alpha_k \cos(k\theta) + \beta_k \sin(k\theta) \tag{4.89}$$

Any time series, regardless of how complex in form (except that it must be continuous or without breaks, and there can only be one value of Y for each value of x), can be represented as the sum of a series of cosine wave forms defined in this manner, as illustrated in Figure 4.54. This is an expression of the Fourier relationship,

$$Y = \sum_{k=0}^{\infty} \alpha_k \cos(k\theta) + \beta_k \sin(k\theta) \tag{4.90}$$

You will note that (4.90) is a linear equation (that is, all of the terms are added together) and resembles in form the polynomial regressions we considered earlier in this chapter. In the Fourier equation, the trigonometric terms $\cos(k\theta)$ and $\sin(k\theta)$ are equivalent to the power terms of a polynomial, such as X^2 and X^3. The α_k and β_k coefficients of these terms can be found by least squares. However, if we attempt

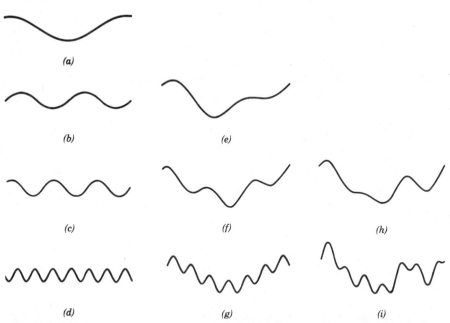

FIGURE 4.54   Effect of adding successive harmonics of a cosine wave. (a) Fundamental or first harmonic, $A = 0.5$, $k = 1$, $\phi = 0°$. (b) Second harmonic, $A = 0.3$, $k = 2$, $\phi = 0°$. (c) Third harmonic, $A = 0.3$, $k = 3$, $\phi = 45°$. (d) Seventh harmonic, $A = 0.25$, $k = 7$, $\phi = 180°$. (e) First plus second harmonic, (a) + (b). (f) First plus third harmonic, (a) + (c). (g) First plus seventh harmonic, (a) + (d). (h) First plus second and third harmonics, (a) + (b) + (c). (i) First plus second, third and seventh harmonics, (a) + (b) + (c) + (d).

to estimate the coefficients by matrix methods, we quickly face insurmountably large computational problems if many harmonics are to be determined.

Harmonic Analysis

If a time series represents a truly periodic phenomenon, we may use a technique called harmonic analysis to decompose the series into its constituent parts. Naturally occurring time series that are periodic include tidal fluctuations, seasonal variation in temperatures, and varve sequences. They also include records generated by instruments such as X-ray diffractometers, although the Fourier transformation of such data is a specialized topic. The time series must be sampled at discrete points which are equally spaced a distance Δ apart. If there are n of these points, one of which is j, the α_k and β_k coefficients of the Fourier equation can be found by the computational formulas:

$$\beta_k = \frac{2}{k} \sum_{j=0}^{n-1} Y_j \sin\left(\frac{2\pi jk}{n}\right) \tag{4.91}$$

$$\alpha_k = \frac{2}{k} \sum_{j=0}^{n-1} Y_j \cos\left(\frac{2\pi jk}{n}\right) \tag{4.92}$$

The equations are not so formidable as they may appear at first glance; the expressions in parentheses simply convert the position of the jth point into radians, defining the total length of the time series as 2π radians. Because of trigonometric relationships, the coefficient β_0 is always zero, and α_0 simplifies to

$$\alpha_0 = \frac{1}{n} \sum_{j=0}^{n-1} Y_j \tag{4.93}$$

which is just the mean of the time series. The subscripts range from 0 at the beginning of the time series to $n - 1$ at the end, rather than from 1 to n, in order to correctly convert the position of the first observation to zero radians.

Once we have determined the α_k and β_k coefficients for the kth harmonic, we can determine the amplitude of the wave form by

$$A_k = \sqrt{\alpha_k^2 + \beta_k^2} \tag{4.94}$$

The phase angle of the kth harmonic is

$$\phi_k = \tan^{-1}\left(\frac{\beta_k}{\alpha_k}\right) \tag{4.95}$$

The expression $\tan^{-1}$ means "find the angle whose tangent is given."

If a regular sinusoid is uniformly sampled at discrete points, the variance of these samples is related to the amplitude of the wave form. In the limit, the variance is simply half the square of the amplitude, or

$$s_k^2 = \frac{A_k^2}{2} = \frac{\alpha_k^2 + \beta_k^2}{2} \tag{4.96}$$

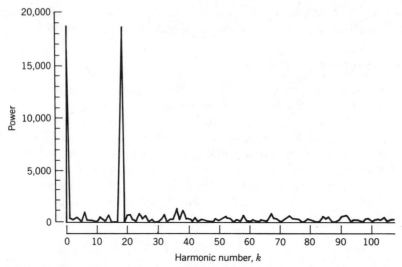

FIGURE 4.55 Periodogram of monthly flow of Cave Creek, Kentucky. Spectral values indicate that flow has a single periodic annual cycle. Power is given as the square of the measurement units, which are hundredths of inches.

Since the Fourier theorem holds that a time series can be regarded as the sum of many sinusoidal functions, or harmonics, the variance of the time series must be composed of the sum of the variances of these same harmonics. We can therefore express the variance of the kth harmonic as a proportion of the total variance of the time series from which it has been derived. If we compute a large series of harmonics, the variances of the successive harmonics can be plotted in a *peri-*

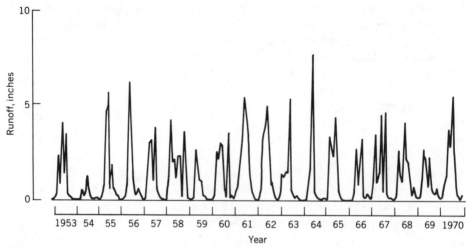

FIGURE 4.56 Monthly flow of Cave Creek, Kentucky, for the period October 1952 to September 1970.

TABLE 4.29 Monthly Runoff of Cave Creek, Kentucky, Given in Hundredths of Inches

Water Year	Oct.	Nov.	Dec.	Jan.	Feb.	Mar.	Apr.	May	Jun.	Jul.	Aug.	Sep.
1953	2	5	19	240	86	416	147	354	31	18	7	1
1954	0	2	4	54	22	40	139	35	8	7	6	14
1955	2	4	30	73	463	579	59	197	55	24	28	3
1956	4	6	13	59	637	469	192	28	32	64	38	8
1957	7	10	172	308	325	103	392	68	24	6	5	2
1958	3	106	432	200	221	117	235	236	19	369	170	12
1959	6	9	17	270	195	112	102	24	24	5	4	2
1960	3	36	269	219	313	291	68	19	364	138	14	30
1961	12	52	79	204	295	532	476	414	159	48	18	4
1962	2	6	76	346	401	508	330	79	96	30	8	7
1963	39	141	124	150	146	548	52	25	14	29	11	3
1964	1	4	3	87	173	788	45	21	11	8	2	16
1965	15	7	347	276	230	449	146	31	8	5	1	2
1966	4	2	2	48	281	79	202	332	25	14	41	11
1967	7	119	357	97	161	466	50	476	33	14	15	7
1968	9	38	271	135	98	425	238	199	91	29	75	16
1969	14	22	112	278	216	73	237	74	40	27	66	17
1970	7	25	91	130	389	291	568	206	38	14	6	27

Data are collected during a "water year" beginning in October of the preceding year and ending in September in order to avoid breaking the sequence in the middle of the winter. From Haan (1977).

odogram, or, as it is sometimes called, a *discrete* or *line power spectrum*. This is simply a plot of power or variance versus harmonic number, k, and traditionally is shown in the form given in Figure 4.55. (The terms *power* and *variance* are synonymous; the former arose in electrical engineering and is now widely used in all fields of engineering, signal analysis, and pattern recognition.) The original time series is shown in Figure 4.56. The time series represents the monthly flow of water in Cave Creek in Kentucky; data are given in Table 4.29, taken from Haan (1977). The periodogram indicates that the variation in monthly flow has an annual (12-month) period, plus a random component.

The spectrum computed in the preceding manner is sometimes called the *raw spectrum*, and is an estimate of the true or population spectrum. These estimates have associated with them very large standard errors, which cannot be reduced by increasing n for reasons that will be explained in the next section. Some of this error may represent *aliasing*, or the incorporation of irresolvable high frequencies into lower frequencies. These high frequencies, whose wavelengths are less than twice the spacing between sample points, cannot be detected. The highest frequency that can be estimated is the *Nyquist frequency*, whose wavelength is exactly 2Δ, where Δ is the distance between successive observations. The reason why the variances of frequencies beyond the Nyquist limit are confounded with those of lower frequencies can be seen readily in Figure 4.57.

Nevertheless, it is possible to statistically test for the existence of the dominant periodic component of the periodogram. The test, devised by Fisher, calculates the probability that a spectral value s_k^2 will exceed the value σ_k^2 of a time series

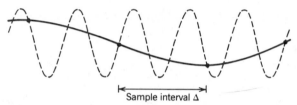

FIGURE 4.57 High-frequency sinusoidal wave (dashed line) sampled at discrete points yields apparent lower frequency wave (heavy line).

composed of independent random points. The test involves calculation of the ratio

$$\hat{g} = \frac{s^2_{max}}{2s^2} \tag{4.97}$$

where s^2_{max} is the largest peak in the periodogram and s^2 is the variance of the entire time series. The critical value of g for a specified probability p is given by

$$g \approx 1 - e^{\frac{\ln p - \ln m}{m - 1}} \tag{4.98}$$

where $m = n/2$ if the time series contains an even number of observations and $m = (n - 1)/2$ if n is odd. If the test statistic $\hat{g}$ exceeds the critical value g, the periodic component may be presumed to exist. If the test value does not exceed the critical value, the observed spectral peak s^2_{max} could have arisen simply by random chance.

The Continuous Spectrum

Harmonic analysis and the construction of the periodogram or line spectrum is appropriate if the time series being investigated is truly periodic. There are few naturally occurring phenomena that possess true periodicity, except for those related to astronomical cycles such as monthly tides or seasonal changes. Most geologic time series are not periodic, but are stochastic. A sequence is *stochastic* if it can be characterized only by its statistical properties; this is in contrast to a deterministic sequence whose state can be predicted exactly from its coefficients.

Even though a time series may not contain truly periodic components, the methods of spectral analysis can provide valuable insights into the behavior of the process that has generated the sequence. Most time series can be regarded as continuous sequences, even though they ordinarily are sampled at discrete points. It is possible to calculate a *continuous spectrum* or *spectral density function* for such sequences, in which the variance of the time series is apportioned among a set of *frequency bands*. The continuous spectrum has the form of a continuous plot of variance versus frequency, and is analogous to a continuous probability distribution, with variances proportional to the areas under the spectral curve between limiting frequencies. The total area under the spectrum is equal to the total variance of the time series from which it is calculated. In contrast, the line spectrum of a periodic time series shows the variances attributable to defined individual frequencies.

The specific sequence of observations we wish to analyze can be considered a random sample from a large, perhaps infinite, set of such time series that could be produced by the process under study. The complete set of time series is called an *ensemble*, and is the population from which our particular sample is taken.

A time series is said to be *stationary* (or time-invariant or nonevolutionary) if its properties do not change with time. A spatial series with the same characteristics is said to be *homogeneous*. If a time series is divided into small segments and the means of all these segments tend to be the same (and the same as the mean of the entire time series), it is referred to as being *first-order stationary* or stationary in the mean. If in addition the autocovariance changes only with lag and not with position along the time series, the series is *second-order stationary*. This is also called *weak stationarity*, or stationarity in the wider sense. If all higher moments are dependent only on lag and not on position, the series is *strongly stationary*, or possesses stationarity in the strict sense.

If a time series is not only strongly stationary, but all statistics are invariant from time series to time series within the ensemble, the ensemble is *ergodic*. Many statistical tests of time series assume ergodicity, just as univariate statistical tests may assume uniformity of variance. We may check for ergodicity if we have several time series that are all realizations of the same stochastic process. Commonly, however, we have only one time series available, and then no check is possible. We can test for stationarity in the mean and variance of a single time series by regression, or by dividing the series into segments and testing to see if the statistics for the segments are the same. If they are, the series is *self-stationary* and ergodicity may be assumed. If the series is not self-stationary, the ensemble from which it is derived cannot be ergodic.

Sometimes a time series can be made stationary by *levelling*, or subtracting a linear trend from the observations. That is, a linear regression of Y_j on j is computed, using least squares as described earlier. Then, a new time series is defined as the deviations,

$$Y_j' = Y_j - \hat{Y}_j \tag{4.99}$$

where $\hat{Y}_j$ are the predicted values given by the regression equation. If the original series was characterized by a slow change in its average value, the new series will have a stationary mean of zero.

For a stationary, stochastic time series that is continuous and sampled at discrete, equally spaced points, the continuous variance (or power) spectrum may be calculated by either of two methods. The newer, and more widely used approach involves calculating many values of the line spectrum by the Fast Fourier Transform (FFT) computer algorithm. These spectral values are then averaged across frequencies to produce a smoothed estimate of the continuous spectrum. The Fast Fourier Transform, as its name implies, is extremely rapid and requires only $n \cdot \log_2 n$ arithmetic operations rather than n^2 operations as do alternative procedures. This approach requires that the Fourier relationship be expressed in complex form, as

$$s_k^2 = \frac{1}{n} \sum_{j=-n/2}^{n/2} Y_j e^{(-i2\pi jk/n)} \tag{4.100}$$

where i is the imaginary number $\sqrt{-1}$. Developing the FFT algorithm using complex arithmetic requires mathematical and computational skills beyond those assumed in this book. Therefore, we will not pursue the topic further, even though this approach is now one of the most widely used in spectral analysis. An excellent introduction to the Fast Fourier Transform, including a discussion of complex arithmetic, is given in Rayner (1971), who also describes applications of the FFT in geography. The first landmark article describing the Fast Fourier Transform algorithm was published by Gentleman and Sande (1966), based upon the mathematics previously introduced by Cooley and Tukey (1965). Modern texts on time series analysis, such as Bloomfield (1976) and Bendat and Piersol (1971), treat the subject extensively.

A somewhat older procedure for calculating the continuous spectrum involves finding the Fourier transform of the autocorrelation function of a time series. Developed originally by Bartlett (1948), this approach achieves the same results as the newer FFT method, and is easier to understand even though not as computationally efficient. It is still widely used, especially when the autocorrelation function is also of interest and when the time series is not extremely long. Jenkins and Watts (1968), for example, discuss this approach at length, and Yevjevich (1972) applies it to problems in hydrology.

The Fourier transform of the autocorrelation function is the power spectrum; for a discretely sampled time series, Jenkins and Watts (1968) give the transform in the following form:

$$s_f^2 = \frac{A_f^2}{2} = \sum_{l=-(L-1)}^{L-1} r_l e^{i2\pi fl} \tag{4.101}$$

Here, s_f^2 is the variance or power in the frequency band centered at frequency f, r_l is the autocorrelation at lag l, the maximum lag in the autocorrelogram is L, and i is the imaginary number $\sqrt{-1}$.

Note that the equation is very similar to the complex Fourier transform given in eq. (4.100). However, eq. (4.101) may be greatly simplified. Although the autocorrelation function can be calculated for negative as well as positive lags, it is symmetrical about lag zero; that is, $r_l = r_{-l}$. Therefore, the calculations need only to be performed over the range of l from zero to $(L - 1)$, and then doubled. Also, because our data consist of a finite set of real numbers that are evenly spaced, we may use a cosine representation rather than complex notation. The computational form of eq. (4.101) is

$$s_f^2 = \frac{A_f^2}{2} = 2 \left(\sum_{l=1}^{L-1} r_l \cos 2\pi fl \right) \tag{4.102}$$

It is necessary to work in frequencies rather than harmonics, because specific lags of the autocorrelation function do not transform into simple multiples of one another in the frequency spectrum. The relationship between lag and frequency is given by

$$f_j = \frac{j}{2\Delta L} \qquad (4.103)$$

where Δ is the spacing between successive observations in the time series, and j represents both the jth lag and the jth spectral band. The frequency of the jth band is f_j and is measured in cycles per unit, where the units are the same as those of Δ, the spacing between data points. If Δ is measured in millimeters, f_j is given in cycles per millimeter.

Use of eq. (4.102) yields what is called the *raw spectrum*. In general, it will not be satisfactory because of the high standard error in each of the spectral bands. This cannot be overcome in the usual manner, by making the time series longer so that n (the total number of observations) is greater. It is true that as n increases for a fixed lag l, there is more information in the form of more crossproducts ($x_j - \overline{X}$)($x_{j+1} - \overline{X}$), and the autocorrelation r_l will have a smaller standard error. However, this is not true for s_f^2. Jenkins and Watts (1968) show that the information contained in each estimate s_f^2 of the spectrum is spread over a band of frequencies whose width is $\pm l/n$ about f. As n increases, the total information contained in the variance spectrum is distributed over an increasing number of bands of decreasing width. As a result, increasing n makes it possible to estimate the variance in narrower frequency bands, yielding a less biased estimate of the true spectrum. However, the standard error of these narrower bands is not improved.

Bartlett (1948) first introduced a method for reducing the standard error in spectral estimates by a process variously referred to as *windowing, filtering, smoothing*, or *spatial averaging*. The autocorrelation function may be weighted by a *lag window* or filter, and the weighted autocorrelogram transformed into the spectral domain. The weights emphasize the shorter lags, which are based on more observations than are longer lags. Alternatively, the raw spectrum may be calculated and then smoothed by weighting and averaging together adjacent spectral values. The set of weights is called a *spectral window* or filter, and is the Fourier transform of the lag window. Both methods produce equivalent results, which may be thought of as a blurring or smoothing of the variances over adjacent frequencies.

Many windows or filters have been suggested by various authors; the design of windows that have superior properties is a fine art among time series analysts. One commonly used window is the Tukey-Hanning filter, which in the spectral domain is

$$s_f^2 = \frac{1}{4} s_{f-1}^2 + \frac{1}{2} s_f^2 + \frac{1}{4} s_{f+1}^2 \qquad (4.104)$$

The Fourier transform of the Tukey-Hanning filter into the lag domain is

$$W_l = (1 + \cos \pi l/L)/2 \qquad (4.105)$$

This defines a set of weights for the autocorrelation function which drops smoothly from $W_0 = 1.0$ at lag zero to $W_L = 0.0$ at the maximum lag, L. Each autocorrelation coefficient is multiplied by its corresponding weight before the Fourier transform is made. Equation (4.102) becomes

$$s_f^2 = 2\left(1 + s \sum_{l=1}^{L-1} W_l r_l \cos 2\pi f l\right) \tag{4.106}$$

Other windows have been proposed, usually either in the lag domain or in the frequency domain because their application is easier in one form or the other. The widely used Parzen window has the form

$$W_l = 1 - \frac{6l^2}{L^2}\left(1 - \frac{l}{L}\right) \tag{4.107}$$

for lags between zero and $L/2$, and

$$W_l = 2\left(1 - \frac{l}{L}\right)^3 \tag{4.108}$$

for lags from $L/2$ to L.

A practical strategy for investigating the spectral characteristics of a stochastic time series includes the following steps.

1. If necessary, de-trend or level the data by computing the coefficients of the regression of Y_j on j, then finding the residuals $Y_j' = Y_j - \hat{Y}_j$.
2. Compute the autocorrelation of the original series (if stationary) or of the series of residuals Y', up to a maximum lag L, which is no more than $n/3$. Some authors recommend more severe limits on the maximum lag to be calculated, such as $n/5$, $n/6$, or even $n/10$.
3. Compute the raw spectral density, s_f^2, using the discrete Fourier transform given in eq. (4.102).
4. Smooth the raw spectrum by using the Tukey-Hanning window (eq. 4.104) as a moving average to produce the smoothed estimate of the variance spectrum, s_f^2.

Obviously, there are many decisions that must be made in spectral analysis that may have a significant effect on the final results. These include choice of n, the length of the sequence to be analyzed (if this is under the analyst's control); L, the maximum lag; and the window to be used for smoothing. The interactions between these variables, and their effect on both the resolution and significance of the smoothed spectrum, are discussed at length by Jenkins and Watts (1968).

In performing Fourier analysis, we have transformed our data from one domain to another. We began with observations in the form of values Y_j at points in space or time, x_j. The succession of points forms a signal or wave form, defined by X and Y coordinates. The data, defined in this manner, are said to be in the time or spatial domain, depending upon whether X denotes points in time or distance. By determining the component frequencies in the signal, we have transformed the data to the frequency domain. A physical analogy can be drawn with the effect of a

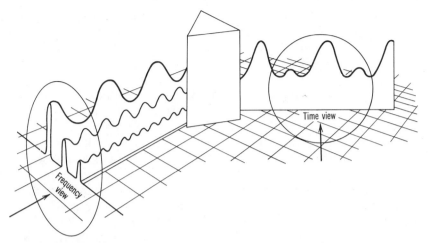

FIGURE 4.58   A prism acts as a frequency analyzer, transforming white light (time or spatial domain) into its constituent spectrum of colors (frequency domain). Courtesy Hewlett-Packard Corp.

glass prism on sunlight, as shown in Figure 4.58. A beam of white light can be regarded as a complex wave form changing with time, and composed of many colors (or wavelengths) of light. A prism acts as a frequency analyzer, and separates the beam into its components, which appear as a rainbow display. Each colored band is separated from its neighbor by an amount proportional to the difference in their wavelengths or frequencies, and the intensity of each band is proportional to the contribution of that particular wavelength to the total intensity of the original beam. Examining the spectrum of a light source may tell us a great many things: the composition of the source, its temperature, nature of the material through which the light passed, and so forth. In a similar fashion, examining the power spectrum of a data sequence may tell us a great deal about its nature and origin, information that may not be apparent in any other way.

Filters

A *filter* is a mathematical operator that changes a time series into another time series having some desired form. Such operators are called "filters" because originally they were electronic filter circuits, consisting of networks of resistors and capacitors, used to selectively suppress or enhance specific frequencies in electronic signals. Now, the same function is performed on digital signals by mathematical filters implemented in a computer. Much of the basis for filtering stems from the work of Norbert Wiener on statistical communication theory, and many practical applications occur in seismic processing and remote sensing. Because of this background, most of the terminology of filtering is derived from the fields of electrical engineering and physics, rather than classical statistics.

The most extensive applications of filtering techniques are associated with the processing of reflection seismic data. Reflection seismology can be regarded as the creation of an input signal (the seismic energy produced by a dynamite shot or a vibrator) which is then "filtered" by absorption of energy as it travels through the Earth. Higher frequencies tend to be preferentially absorbed, so the nature of the seismic pulse changes as it travels. In addition, lower frequency waves travel through rocks more rapidly, so the initial sharp input becomes blurred and attenuated. The final output, which is detected by an array of geophones at the surface, consists of reflections returning from successively deeper horizons. Unfortunately, after their travels, the reflections are severely distorted and difficult to interpret.

The job of the geophysicist is to remove, as much as possible, the deleterious effects of the physical filtering to which the seismic signal has been subjected. This is done by processing the seismic recording with digital filters that remove unwanted frequencies, sharpen diffuse reflections, and subdue the noise in the record until the desired but attenuated signal can be seen.

Because so much of the distortion of seismic signals is frequency-related, geophysicists are especially interested in filters that preferentially pass or remove specific frequencies. Seismic filtering often is done in the frequency domain, by taking the Fourier transform of the seismic record, removing unwanted parts of the spectrum, and then retransforming the filtered spectrum back into the real domain. However, exactly the same process can be performed by transforming the Fourier domain filter into the real world and convolving its transform with the seismic record. Because these filters are designed to pass certain frequencies while suppressing others, they are referred to as high-pass, low-pass, or band-pass filters.

Although in theory it is possible to produce perfect filters that will pass only signals of specified frequencies and no others, such filters usually are not realizable. Typically, these theoretically perfect filters require an infinite number of terms. If they are truncated to manageable size, the filters themselves may introduce spurious features in the output. The design of seismic filters thus is an art of the practical, seeking filters that are sufficiently short so that their use is economically feasible, and that produce good approximations of the desired output with minimum undesirable side effects. An adequate discussion of frequency-specific filters would require much more space than is available for the topic in this text, so interested readers must be referred to the many specialized books on the subject. We will turn instead to conceptually more simple types of filters that also see wide use in many areas of time series analysis.

A time series to be filtered is referred to as the *input*, and the altered time series that emerges from the filter is called the *output*. The filtering operation convolves the input with the filter to produce the output:

$$[C] = [B] * [f] \tag{4.109}$$

If there is some form we want the output to assume, say $[D]$, a filter can be designed so that the actual output $[C]$ is as close to the desired output $[D]$ as possible. To do this, we compute elements of the filter that will minimize the sum of the squared differences between $[C]$ and $[D]$. The mathematical process is very similar to that used to determine the coefficients of a linear regression equation.

Robinson and Treitel (1980) have an extensive and exceptionally lucid discussion of the process, which is called *least squares filter design*.

As a simple example, suppose the input [B] is a vector having two elements, [b_0 b_1]. The filter also contains two elements, [f_0 f_1]. Since the output from the filter is the convolution of the input with the filter, it will consist of three elements.

$$[C] = [B] * [f]$$

$$[C] = \begin{bmatrix} b_0 \\ b_1 \end{bmatrix} \begin{bmatrix} f_0 b_0 & f_1 b_0 \\ f_0 b_1 & f_1 b_1 \end{bmatrix} \qquad \begin{matrix} [\ f_0 & f_1\ ] \end{matrix}$$

(4.110)

$$[C] = [f_0 b_0 \quad f_0 b_1 + f_1 b_0 \quad f_1 b_1]$$

We want the filter output to be equal to some specified output [D], so the two can be equated, defining a set of simultaneous equations.

$$\begin{aligned} f_0 b_0 + \quad 0 \quad &= d_0 \\ f_0 b_1 + f_1 b_1 &= d_1 \\ 0 \quad + f_1 b_1 &= d_2 \end{aligned}$$

(4.111)

In matrix form, this can be written as

$$\begin{bmatrix} b_0 & 0 \\ b_1 & b_0 \\ 0 & b_1 \end{bmatrix} \begin{bmatrix} f_0 \\ f_1 \end{bmatrix} = \begin{bmatrix} d_0 \\ d_1 \\ d_2 \end{bmatrix}$$

(4.112)

We will denote the left-hand matrix as [β].

The matrix equation is overdetermined; that is, there are three equations but only two unknowns. However, if both sides of an equation are multiplied by the same quantity, the equation is unchanged. So, multiplying both sides by the transpose of [β]:

$$[β]' [β] [f] = [β]' [D]$$

or

$$\begin{bmatrix} b_0 & b_1 & 0 \\ 0 & b_0 & b_1 \end{bmatrix} \begin{bmatrix} b_0 & 0 \\ b_1 & b_0 \\ 0 & b_1 \end{bmatrix} \begin{bmatrix} f_0 \\ f_1 \end{bmatrix} = \begin{bmatrix} b_0 & b_1 & 0 \\ 0 & b_0 & b_1 \end{bmatrix} \begin{bmatrix} d_0 \\ d_1 \\ d_2 \end{bmatrix}$$

which becomes

$$\begin{bmatrix} (b_0^2 + b_1^2) & b_0 b_1 \\ b_0 b_1 & (b_0^2 + b_1^2) \end{bmatrix} \begin{bmatrix} f_0 \\ f_1 \end{bmatrix} = \begin{bmatrix} (b_0 d_0 + b_1 d_1) \\ (b_0 d_1 + b_1 d_2) \end{bmatrix}$$

(4.113)

Geophysicists refer to the left-hand matrix as the autocorrelation matrix of the input, and the right-hand matrix as the cross-correlation between the input and output. Actually, the elements of these matrices are not correlation coefficients, but are uncorrected sums of squares and cross-products. However, the denominators of any correlation terms would be identical, and so would cancel out; the solution to the equation set would be the same using either correlations or sums of squares and cross-products. The solution is found by inverting the left-hand matrix and postmultiplying by the right-hand vector. If we call the left-hand matrix $[R]$ and the right-hand vector $[X]$, this is

$$[f] = [R]^{-1} [X]$$

One common use of filtering is to search through a time series for occurrences of a specific wave form, or its digital representation, which is a specific pattern of numbers. The analyst may wish to turn the occurrences of this wave form into something more conspicuous, such as a sharp peak or spike in the output record.

The input wave form $[B]$ will consist of $[2 \quad 1]$ and the desired output will consist of $[D] = [3 \quad 0 \quad 0]$. These are shown as wave forms in Figure 4.59. Substituting into eq. (4.113) yields

$$\begin{bmatrix} 2 & 0 \\ 1 & 2 \\ 0 & 1 \end{bmatrix} \begin{bmatrix} f_0 \\ f_1 \end{bmatrix} = \begin{bmatrix} 3 \\ 0 \\ 0 \end{bmatrix}$$

Premultiplying both sides by $[\beta]'$:

$$\begin{bmatrix} 2 & 1 & 0 \\ 0 & 2 & 1 \end{bmatrix} \begin{bmatrix} 2 & 0 \\ 1 & 2 \\ 0 & 1 \end{bmatrix} \begin{bmatrix} f_0 \\ f_1 \end{bmatrix} = \begin{bmatrix} 2 & 1 & 0 \\ 0 & 2 & 1 \end{bmatrix} \begin{bmatrix} 3 \\ 0 \\ 0 \end{bmatrix}$$

$$\begin{bmatrix} 5 & 2 \\ 2 & 5 \end{bmatrix} \begin{bmatrix} f_0 \\ f_1 \end{bmatrix} = \begin{bmatrix} 6 \\ 0 \end{bmatrix}$$

Inverting the left-hand matrix yields

$$\begin{bmatrix} 5 & 2 \\ 2 & 5 \end{bmatrix}^{-1} = \begin{bmatrix} .2381 & -.0952 \\ -.0952 & .2381 \end{bmatrix}$$

The desired filter coefficients are

$$\begin{bmatrix} f_0 \\ f_1 \end{bmatrix} = \begin{bmatrix} .2381 & -.9852 \\ -.0952 & .2381 \end{bmatrix} \begin{bmatrix} 6 \\ 0 \end{bmatrix}$$
$$= \begin{bmatrix} 1.4286 \\ -.5712 \end{bmatrix}$$

The graphic representation of the filter is shown in Figure 4.59. We can determine its performance by using it to operate on the original input. This is done by inserting the coefficients into the definitional equation of the filtering operation given in eq. (4.109):

$$[C] = [2 \quad 1] * [1.4286 \quad -.5712]$$

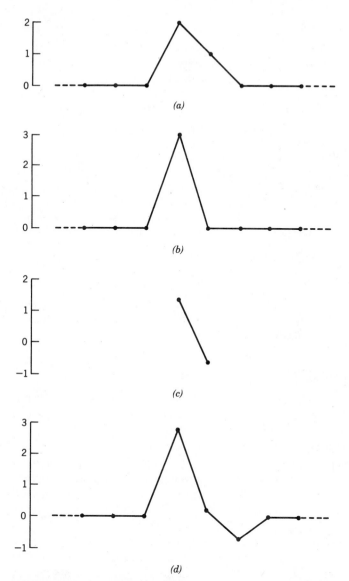

FIGURE 4.59 (a) Arbitrary input with values [2 1]. (b) Desired form of output is [3 0 0]. (c) Least-squares filter [1.43 − .57]. (d) Output from least-squares filter [2.9 .3 − .6].

The convolution is

$$[C] = \begin{bmatrix} 2 \\ 1 \end{bmatrix} * \begin{bmatrix} 1.4286 & -.5712 \\ 2.8572 & -1.1424 \\ 1.4286 & -.5712 \end{bmatrix}$$

$$= [2.8572 \quad (1.4286 - 1.1424) \quad -.5712]$$

$$= [2.8572 \quad .2862 \quad -.5712]$$

The filter output is also shown on Figure 4.59; obviously, we were not successful in creating a pure [3 0 0] spike, but the output is a fairly good approximation. In fact, it is the best approximation possible for this particular input and a filter of only two terms. We can measure just how good the filter is by comparing its output with the desired output. This is done by finding the *error*, defined as the sum of the squared differences between [C] and [D]:

$$\epsilon = (3 - 2.8572)^2 + (0 - .2862)^2 + (0 + .5712)^2$$

$$= .1428$$

Forms of output other than a spike can be specified, and in certain instances will yield a more efficient filter. Common filter designs are referred to as the boxcar (a square wave), Gaussian, and sawtooth.

Smoothing and Time-Trend Analysis

Perhaps the most familiar types of filters used in geology are those designed to reduce the variance in a time series. These are arbitrary filters whose general action is to "smooth" a data sequence; the output from the filter is a subdued approximation of the input. The justification for such a filtering process is that the time series consists of two components—a "long-term" signal or meaningful part, and superimposed random noise. By its nature, such noise must be "short-term." As the signal tends to be the same from point to point and the noise does not, an average of several adjacent points will tend to converge on the value of the signal alone.

One method of smoothing consists of approximating short segments of the original sequence with smooth lines or curves. These curves can be fitted by least squares, using regression techniques discussed earlier. Orthogonal polynomials would seem to be especially convenient for this purpose, because they require only the multiplication of a series of terms with the observations in the segment being fitted, followed by summation and a division, to yield a coefficient of the fitted curve. Once the coefficients have been determined, the equation can be evaluated and the original observations Y_i replaced with the predicted values $\hat{Y}_i$. As long as the order of the fitted curve is less than the number of points in the segment, the curve will be an approximation that is less variable than the original observations, and hence smoother.

Ordinarily, a sequence would not be smoothed by a series of nonoverlapping segments, since these would tend to have abrupt discontinuities at their joins. Instead, the smoothing operation consists of fitting an approximation to a small segment, and determining the predicted value $\hat{Y}_i$ corresponding to the middle observation in the segment. The next segment is chosen so that it overlaps all but one of the observations in the first, and the process is repeated. Eventually, the original data set is replaced by a smoothed series derived from the curves fitted to these overlapping segments.

Since a series of coefficients must be determined and evaluated for each overlapping segment, and there are almost as many segments in a time series as there are original observations, this would be a tedious process even using orthogonal polynomials. However, the coefficients themselves are not needed; all we require are the estimates $\hat{Y}_i$ at the center points of the segments. With this constraint, it is possible to rewrite the orthogonal polynomial equations to yield an alternative set of orthogonal terms that will produce the estimated values of the center points directly. Details of the derivation are given by Savitzky and Golay (1964).

Because we are interested only in estimating the central point in a short sequence of points, n must always be an odd number. Also, it happens that either a quadratic or cubic curve will have the same value at the central point, so one set of weights will do for either. The same is true for the fourth- and fifth-order curves. Of course, if we fit a linear function to the data, the value at the central point will simply be the mean of the points in the segment. This is equivalent to saying that all the terms of a linear orthogonal polynomial equation are equal to 1.

Table 4.30 reflects these considerations; it gives the orthogonal polynomial weighting functions ω for odd numbers of points in a sequence for the second- or third-order and the fourth- or fifth-order curves.

To perform smoothing using orthogonal polynomials, we simply convolve the data sequence with the set of polynomial terms, then divide by the sum of the terms. That is,

$$[C] = [B] * [\omega]/\Sigma\omega$$

Since the set of terms is symmetrical about the center term, the result is the same if we compute a moving average, which is simply a moving cross-product between the set of terms and the time series, divided by $\Sigma\omega$. This process is perhaps more familiar in geology under the name *time-trend analysis*. Among the equations used for this purpose is Shepard's five-term filter, usually written as a moving average of the form

$$\hat{Y}_i = \frac{1}{35} [17Y_i + 12(Y_{i+1} + Y_{i-1}) - 3(Y_{i+2} + Y_{i-2})]$$

If you compare the weights in Shepard's moving-average equation with the orthogonal polynomial terms given in Table 4.30, you will see that it is simply a quadratic equation fitted by least squares to five points. Other smoothing equations used in time-trend analysis are also either identical or very similar to orthogonal polynomials. These smoothing functions are described in the classic work by Whittaker and Robinson (1944, pp. 285–291). Figure 4.60 shows the results of smoothing a drilling-time log by various equations.

It may not be obvious that these moving-average functions are filters, or that the moving-average process is mathematically equivalent to the convolution of two

TABLE 4.30 Orthogonal Polynomial Weighting Functions for Approximation (Smoothing); The Function Directly Estimates the Central Value in a Sequence of n Points, Where n is an Odd Number from 5 to 17

Second-Order (Quadratic) and Third-Order (Cubic) Terms

n	5	7	9	11	13	15	17
	−3	−2	−21	−36	−11	−78	−21
	12	3	14	9	0	−13	−6
	17	6	39	44	9	42	7
	12	7	54	69	16	87	18
	−3	6	59	84	21	122	27
		3	54	89	24	147	34
		−2	39	84	25	162	39
			14	69	24	167	42
			−21	44	21	162	43
				9	16	147	42
				−36	9	122	39
					0	87	34
					−11	42	27
						−13	18
						−78	7
							−6
							−21
$\Sigma\omega$	35	21	231	429	143	1105	323

Fourth-Order (Quartic) and Fifth-Order (Quintic) Terms

n	7	9	11	13	15	17
	5	15	18	110	2145	195
	−30	−55	−45	−198	−2860	−195
	75	30	−10	−160	−2937	−260
	131	135	60	110	−165	−117
	75	179	120	390	3755	135
	−30	135	143	600	7500	415
	5	30	120	677	10125	660
		−55	60	600	11053	825
		15	−10	390	10125	883
			−45	110	7500	825
			18	−160	3755	660
				−198	−165	415
				110	−2937	135
					−2860	−117
					2145	−260
						−195
						195
$\Sigma\omega$	231	429	429	2431	46189	4199

Adapted from Savitsky and Golay (1964).

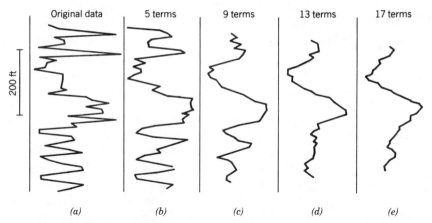

200 ft

(a)   *(b)*   *(c)*   *(d)*   *(e)*

FIGURE 4.60 Digitized drilling time log smoothed by various equations. From Harbaugh and Merriam (1968, Fig. 4.2).

vectors. However, they are entirely equivalent, as can be easily demonstrated using the numerical example worked earlier. There is one caveat: In order to use the filter found by least squares as a weighted moving average rather than as a vector in a convolution, it is necessary to reverse the order of the filter elements. That is, the first element of the filter becomes the last term of the moving average, the second element becomes the next-to-the-last term, and so on until the last element becomes the first term in the moving-average expression.

It is not apparent that the order of the elements in most moving averages has been reversed, because the filters are symmetric. That is, element $f_{i+j} = f_{i-j}$. In geophysical applications, most filters are asymmetric and reversing the elements into proper order for use as a moving average is critical.

We will now apply the two-element asymmetric filter calculated previously to the time series:

$$\ldots \quad 0 \quad 0 \quad 0 \quad 2 \quad 1 \quad 0 \quad 0 \quad \ldots$$

which you will recognize as the wave form of Figure 4.59 embedded within a sequence of zeros. If the filter coefficients $[b_0 \quad b_1] = [1.4286 \quad -.5712]$ are reversed in order and used as weights in a moving average, the result is

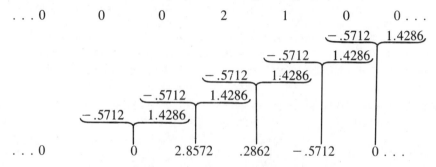

This is exactly the same output we obtained by convolution.

Derivatives

Orthogonal polynomials also can be transformed to yield another useful type of filter output. It is a relatively simple matter to calculate the first derivative of a polynomial curve. The first derivative can be regarded as a measure of the change in slope of the fitted line, or rate of change in the values of $\hat{Y}$. If $\hat{Y}$ increases or decreases at a near-constant rate within an interval, its derivative over that interval will be near zero. A large positive or negative derivative indicates an abrupt change in slope, perhaps caused by a jump in average value or the presence of a sharp peak.

If a fitted curve is defined by an orthogonal polynomial, the derivative can be expressed using orthogonal terms. Because we are interested in computing the derivative of the fitted line only at the central point of the interval, the polynomial equation can be rewritten to yield the derivative directly. The steps are given by Savitzky and Golay (1964), who also provide tables for finding the first and higher derivatives of orthogonal polynomials. Table 4.31 contains terms for finding the first derivative, taken from this source.

The derivative operator is used exactly like a smoothing filter, either by convolving the derivative terms with the original data sequence or using them as weights in a moving average. Either process will yield an output that is equivalent to fitting polynomial curves to successive segments and calculating the derivatives of the fitted curves.

We can see the effect of a derivative operator by using it to process the arbitrary step function shown in Figure 4.61. The steep slope between the two segments of the time series is expressed as a peak in the derivative, while the remainder of the time series has local derivatives of zero.

Polynomial smoothing filters can be used in conjunction with derivative filters if the original time series is so erratic that the derivative is uninterpretable. Table 4.32 lists porosities measured on core sample taken at 10-ft intervals through a producing interval in an oil well drilled on the Arctic slope of Alaska. A plot of these porosity values in shown in Figure 4.62. Also shown on the figure is its first derivative, or rather, the first derivatives of quadratic functions fitted to successive sets of five points. Core porosity measurements are highly variable, in part because of experimental variation but also because the very small core plugs used may not be representative samples of the reservoir rock. Although the first derivative plot clearly shows a maximum change in porosity at sample 16, the erratic nature of the original data is also reflected in the remainder of the plot.

Figure 4.63 shows the measurements of core porosity after smoothing by a five-term quadratic filter. Although the difference in porosity between the upper part of the reservoir and the lower part is preserved, the large point-to-point fluctuations are subdued. If the raw data are indeed composed of a mixture of "true" porosities plus random error, this smoothed approximation may be a more realistic picture of porosity variation than are the original data.

The first derivative of the smoothed data is also shown in Figure 4.63. The pronounced change in porosity between sample 15 and sample 16 is clearly shown. As we expect, derivatives throughout the remainder of the smoothed curve are near zero.

TABLE 4.31 Orthogonal Polynomial Functions for the First Derivative; The Function Directly Estimates the Derivative at the Central Point of a Polynomial Curve Fitted by Least Squares to n Points, where n is an Odd Number from 5 to 17

Derivative of Second-Order (Quadratic) Polynomial

n	5	7	9	11	13	15	17
	−2	−3	−4	−5	−6	−7	−8
	−1	−2	−3	−4	−5	−6	−7
	0	−1	−2	−3	−4	−5	−6
	1	0	−1	−2	−3	−4	−5
	2	1	0	−1	−2	−3	−4
		2	1	0	−1	−2	−3
		3	2	1	0	−1	−2
			3	2	1	0	−1
			4	3	2	1	0
				4	3	2	1
				5	4	3	2
					5	4	3
					6	5	4
						6	5
						7	6
							7
							8
$\Sigma\omega^2$	10	28	60	110	182	280	408

Derivative of Third-Order (Cubic) and Fourth-Order (Quartic) Polynomial

n	5	7	9	11	13	15	17
	1	22	86	300	1133	12922	748
	−8	−67	−142	−294	−660	−4121	−98
	0	−58	−193	−532	−1578	−14150	−643
	8	0	−126	−503	−1796	−18334	−930
	−1	58	0	−296	−1489	−17842	−1002
		67	126	0	−832	−13843	−902
		−22	193	296	0	−7506	−673
			142	503	832	0	−358
			−86	532	1489	7506	0
				294	1796	13843	358
				−300	1578	17842	673
					660	18334	902
					−1133	14150	1002
						4121	930
						−12922	643
							98
							−748
Normalizing term	12	252	1188	5148	24024	334152	23256

Adapted from Savitsky and Golay (1964).

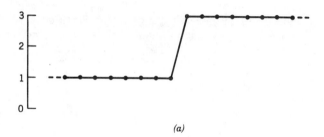

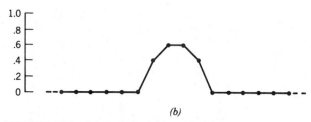

(a)

(b)

FIGURE 4.61   (a) Step function consisting of sequences of 1's and 3's. (b) Plot of first derivative of step function, calculated using five-term quadratic polynomial approximation.

Geophysicists assess the effectiveness of a filter by examining the power spectrum before and after filtering. Changes in the spectrum indicate both how the filter is operating, and how much effect it has. We can perform a simpler analysis that tells us the total effectiveness of a filter by comparing the variance of a time series before filtering to the variance after filtering. The necessary sums of squares to calculate the variances are:

for the original time series

$$SS_o = \Sigma Y^2 - \frac{(\Sigma Y)^2}{n}$$

$$SS_o^* = \Sigma Y^2 - \frac{(\Sigma Y)^2}{n^*}$$

for the filtered series

$$SS_f = \Sigma \hat{Y}^2 - \frac{(\Sigma \hat{Y})^2}{n^*}$$

for the deviations

$$SS_d = \frac{\Sigma (Y - \hat{Y})^2}{n^*}$$

The approximate percentage of the sum of squares preserved is

$$\frac{SS_f}{SS_o^*} \cdot 100\%$$

TABLE 4.32   Porosities Measured on Core Samples from Producing Interval of Oil Well Drilled on Arctic Slope of Alaska; Core Plugs Taken at 10-ft Intervals; Porosities in Percent

(Top) 1	13.2	9	6.7	17	18.9	25	21.2
2	12.6	10	12.8	18	22.5	26	21.7
3	14.3	11	14.2	19	21.6	27	23.0
4	16.2	12	16.3	20	23.2	28	22.1
5	15.2	13	15.4	21	22.8	29	22.4
6	14.1	14	14.0	22	24.7	30	21.8
7	11.2	15	15.0	23	20.0	31	19.8
8	15.3	16	23.2	24	23.4	32	24.7 (Bottom)

In these equations, the summations extend over the n data points in the original sequence, or over the n^* data points in the smoothed sequence. In particular, note that two values for the sum of squares of the original data are computed. The first covers the entire data string from $i = 1$ to n, the other includes only those observations for which estimated values $\hat{Y}_i$ are available. That is, $n^* = n - (m - 1)$, because of loss of data points at the ends of the smoothed string. Although

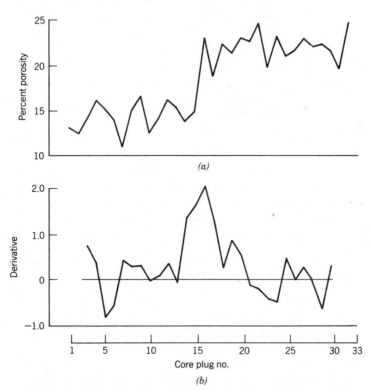

(a)

(b)

FIGURE 4.62   Porosities measured in core plugs taken at 10-ft intervals through oil reservoir in well on Arctic slope of Alaska. (a) Raw data. (b) First derivative of raw data.

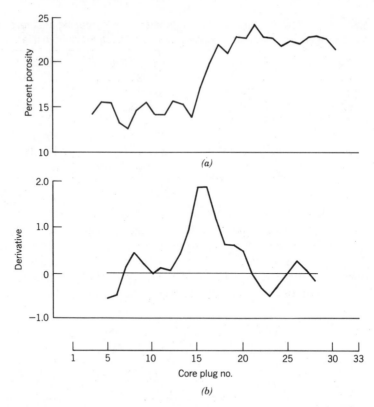

FIGURE 4.63 Smoothed porosities from Arctic well. (*a*) Data smoothed by five-term quadratic polynomial. (*b*) First derivative of smoothed data.

we might expect the equality $SS_o^* = SS_f + SS_d$ as in regression analysis, this relation does not hold for moving averages. This is because values of $\hat{Y}_i$ near the ends of the data sequence are based in part on values not included in the calculation of SS_o^*. The percentage of sum of squares is therefore approximate, but can be used as an index of the effectiveness of the filtering process.

Substitutability Analysis

Part of a hypothetical stratigraphic section is shown in Figure 4.64, accompanied by the coded succession of the various lithologies. If you examine the coded sequence, you will note that the series $A \rightarrow B \rightarrow C$ and $A \rightarrow D \rightarrow C$ occur frequently. It is interesting to speculate that states B and D are somehow related, and one can "proxy" or substitute for the other in a preferred succession. The tendency for two or more states to occur in a common context is called *substitutability*, and an investigation of this phenomenon may reveal groupings of states that are not apparent otherwise. The technique has been applied to such diverse areas as analysis

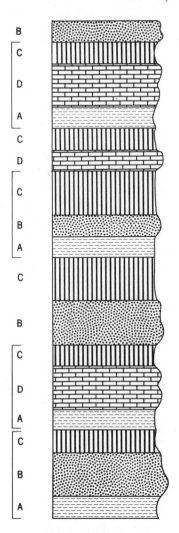

FIGURE 4.64 Part of a four-state stratigraphic succession containing unusually numerous sequences of the form $A \rightarrow B \rightarrow C$ and $A \rightarrow D \rightarrow C$.

of parts of speech and automatic processing of satellite photographs (Rosenfeld and Huang, 1968). The methodology seems equally applicable to geologic data.

From the transition probability matrix, we can develop two quantities, the *first-order left substitutability* and the *first-order right substitutability*. These terms were derived from problems in the analysis of written text, where the states consist of words, and the strings of states proceed from left to right. If two words tend to be followed by the same words, they can be substituted for one another. This is called left substitutability because the word on the left in a sequence is the one that can be replaced by another. Because stratigraphic sequences proceed from bottom to top, the term might more appropriately be "underlying substitutability." The matrix of left substitutability is obtained from the transition probability matrix by computing a cross-product ratio between rows of the matrix.

Suppose we have an $m \times m$ transition probability matrix, where P_{ij} is the probability that state A_j follows state A_i. The cross-product ratio defining the left substitutability between two states, say A_r and A_s, is

$$L_{rs} = \frac{\sum_{j=1}^{m} P_{rj} P_{sj}}{\sqrt{\sum_{j=1}^{m} P_{rj}^2 \sum_{j=1}^{m} P_{sj}^2}} \qquad (4.114)$$

That is, we square every entry in the r and s rows, sum each, multiply the totals together, and find the square root of the product that becomes the denominator of L_{rs}. To find the numerator, we must multiply every element in the rth row by its corresponding element in the sth row and sum these products. The ratio between these two is the cross-product ratio or left substitutability, L_{rs}. Note that if the rows of r and s are identical ($r = s$), the numerator and denominator become identical and the ratio $L_{rs} = 1.0$. Therefore, the substitutability measure is constrained to the range $0 \leqslant L_{rs} \leqslant 1$. Because the P_{ij}'s are found by dividing each entry in a row of the transition frequency matrix by the row totals, the same result can also be achieved directly by computing the cross-product ratio between rows of the transition frequency matrix.

The end result of computations between the m rows of the transition probability matrix is the creation of a symmetrical $m \times m$ matrix of cross-product ratios. The diagonal elements, of course, will be 1.0. The other elements indicate the degree of similarity between one state and another, based on the percentage of times they are followed (or overlain) by a common third state.

The transition frequency and probability matrices for the complete sequence from which the segment in Figure 4.64 was taken are given below.

Transition Frequency Matrix

	A	B	C	D	Row Totals
A	0	11	2	10	23
B	4	0	13	3	20
C	14	6	0	9	29
D	4	3	15	0	22

Upward Transition Probability Matrix

	A	B	C	D	Row Totals
A	0.00	0.48	0.09	0.43	1.00
B	0.20	0.00	0.65	0.15	1.00
C	0.48	0.21	0.00	0.31	1.00
D	0.18	0.14	0.68	0.00	1.00

The left substitutability between A and B is given by the cross-product ratio between

$$\begin{array}{ccccc}
A & [0.00 & 0.48 & 0.09 & 0.43] \\
& \updownarrow & \updownarrow & \updownarrow & \updownarrow \\
B & [0.20 & 0.00 & 0.65 & 0.15]
\end{array}$$

Calculated by (4.114), this value is $L_{AB} = 0.27$ and is the entry in the L_{12} and L_{21} elements of the left substitutability matrix. In a similar manner, the other elements are found. The completed matrix is

$$\begin{bmatrix} 1.00 & 0.27 & 0.60 & 0.25 \\ 0.27 & 1.00 & 0.34 & 0.96 \\ 0.60 & 0.34 & 1.00 & 0.27 \\ 0.25 & 0.96 & 0.27 & 1.00 \end{bmatrix}$$

A high substitutability indicates that there is a strong tendency for the two states to be followed by the same states; that is, they occur in similar contexts. Low substitutabilities, in contrast, indicate that the two states are usually followed by different states.

The matrix of right (or overlying) substitutability is found by computing the cross-product ratios between columns of the downward transition probability matrix, which in turn is found by dividing each element t_{ij} of the transition frequency matrix by the column totals. This probability matrix gives the relative frequency with which a state is underlain or preceded by another. The matrix is exactly the same as we would obtain if we defined the end of the data sequence as the beginning, counted the transitions downward, and then computed the transition probability matrix. The downward probability matrix is shown below.

	Transition Frequency Matrix					Downward Transition Probability Matrix			
	A	B	C	D		A	B	C	D
A	0	11	2	10	A	0.00	0.55	0.07	0.45
B	4	0	13	3	B	0.18	0.00	0.43	0.14
C	14	6	0	9	C	0.64	0.30	0.00	0.41
D	4	3	15	0	D	0.18	0.15	0.50	0.00
Column Totals	22	20	30	22	Column Totals	1.00	1.00	1.00	1.00

In either case, all elements in either a column or row are divided by the total of the column or row, and these cancel out when computing the cross-product ratio between columns or rows. We therefore can obtain the same result by working directly with the transition frequency matrix. The right substitutability of A and B is given by the cross-product ratio between the two columns

$$\begin{array}{cc} A & B \\ \begin{bmatrix} 0.00 \\ 0.18 \\ 0.64 \\ 0.18 \end{bmatrix} & \begin{bmatrix} 0.55 \\ 0.00 \\ 0.30 \\ 0.15 \end{bmatrix} \end{array}$$

with ↔ between corresponding entries.

The right substitutability is $R_{AB} = 0.49$. Other entries are found in the same manner. The completed matrix of right substitutability is

$$\begin{bmatrix} 1.00 & 0.49 & 0.37 & 0.63 \\ 0.49 & 1.00 & 0.27 & 0.94 \\ 0.37 & 0.27 & 1.00 & 0.21 \\ 0.63 & 0.94 & 0.21 & 1.00 \end{bmatrix}$$

Interpretation of this matrix is the same as for left substitutability, except the similarity is based on the tendency for states to be preceded or underlain by the same states.

Finally, we can define a matrix of *mutual substitutability* by forming the products of all pairs of right and left substitutabilities, or

$$C_{ij} = L_{ij}R_{ij} \tag{4.115}$$

This measure describes the similarity of one state to another by the relative frequency with which they appear in a common context; that is, enclosed between similar states. The product of multiplying elements of our two substitutability matrices is

$$\begin{bmatrix} 1.00 & 0.13 & 0.22 & 0.16 \\ 0.13 & 1.00 & 0.09 & 0.90 \\ 0.22 & 0.09 & 1.00 & 0.06 \\ 0.16 & 0.90 & 0.06 & 1.00 \end{bmatrix}$$

We can create a "tree" or hierarchical arrangement that shows the relation of one state to another, based on their mutual substitutability, by connecting the states in the order of greatest substitutability. Methods of constructing these trees by computer will be discussed in Chapter 6; now we will confine ourselves to examining the results of clustering the three substitutability matrices constructed from the stratigraphic section shown in Figure 4.64. The three trees are shown in Figure 4.65. It is obvious that B and D are closely related, both in terms of common preceeding states and common following states.

Few published applications of substitutability analysis to geologic problems have appeared, but this seems to be a measure of the relative obscurity of the technique rather than its potential usefulness. Experiments using data from the Pennsylvanian stratigraphic section in Kansas have been useful in interpreting cyclothems. All beds encountered in a long composite section were classed into 18 lithologic states. Substitutability analysis was then used to combine states to find the minimum number of lithologies necessary to define the sequence. In this way, classification of limestone units by their position in an arbitrary cyclothem was avoided, and an objectively defined cyclic model could be obtained (Davis and Cocke, 1972). Similar applications of the method to other stratigraphic problems seem useful, as do applications to problems of mineral paragenesis.

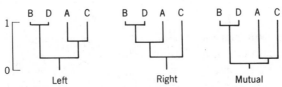

FIGURE 4.65   Clusters of stratigraphic states based on three measures of substitutability. States B and D are similar on the basis of their positions of occurrence within the stratigraphic sequence.

This concludes our discussion of methods for the examination of data in sequences. Those topics covered include the more common techniques in use in geology today, as well as methods that have proven helpful in other fields. These by no means exhaust the list of possibilities, and others may ultimately prove more useful in the solution of problems in the Earth sciences. However, the topics presented should provide a sufficiently broad base for further reading and investigation.

Geologists have not considered the problems of analysis of data in sequences with the same intensity they have given to problems of spatially distributed data. Perhaps map analysis has been emphasized because of the obvious financial stakes involved in prospecting for oil or mineral deposits. Certainly, techniques for analyzing two-dimensionally distributed data are important, and we will consider them at length in Chapter 5. However, the financial rewards for the study of data in sequences are not inconsequential. Consider the potential importance of a method for automatically correlating electric logs of oil wells, or a technique for estimating the duration of time between volcanic eruptions or earthquakes. These problems have not yet been solved, and none of the methods presented here have proven capable of coping with them. However, as we learn more about the nature of geologic data sequences, refined and more powerful methodologies will be devised. The ultimate result will not only be solution of specific problems, but also greater knowledge of the processes that act within the Earth.

SELECTED READINGS

Agterberg, F. P., 1967, Mathematical models in ore evaluation: *Canadian Research Soc. Jour.*, **5,** p. 144–158.

Describes the application of Fourier methods to problems of mineral estimation.

Anderson, R. Y., and L. H. Koopmans, 1963, Harmonic analysis of varve time series: *Jour. Geophysical Research*, **68,** p. 877–893.

A concise treatment of harmonic analysis of a truly periodic geologic phenomenon.

Armstrong, M., and R. Jabin, 1981, Variogram models must be positive-definite: *Jour. Int'l. Assoc. Mathematical Geology*, **13,** no. 5, p. 455–459.

Bardwell, G. E., 1970, Some statistical features of the relationship between Rocky Mountain Arsenal waste disposal and frequency of earthquakes: *Geol. Soc. America, Engineering Geology Case Histories,* No. 8, p. 33–37.

Bartlett, M. S., 1948, Smoothing periodograms from time series with continuous spectra: *Nature*, **161,** p. 686–687.

Bendat, J. S., and A. G. Piersol, 1971, *Random data: Analysis and measurement procedures:* Wiley-Interscience Inc., New York, 407 p.

Chapter 7 of this advanced text is a practical discussion of computer methods in Fourier analysis.

Bloomfield, P., 1976, *Fourier analysis of time series: An Introduction:* Wiley-Interscience Inc., New York, 258 p.

Available as an inexpensive paperback, this text is especially readable; it contains listings of short FORTRAN programs for Fourier analysis.

Bradley, J. V., 1968, *Distribution-free statistical tests:* Prentice-Hall, Inc., Englewood Cliffs, N.J., 388 p.

Runs tests are discussed in Chapter 12.

Clark, I., 1979, *Practical geostatistics:* Applied Science Publishers, London, 129 p.

This slender volume is the best written introduction to geostatistics.

Conover, W. J., 1980, *Practical nonparametric statistics:* John Wiley & Sons, Inc., New York, 493 p.

A textbook with a "how-to-do-it" orientation. Chapter 3 includes runs tests, and Chapter 6 covers K-S statistics.

Cooley, J. W., and J. W. Tukey, 1965, An algorithm for the machine computation of complex Fourier series: *Mathematical Computing,* **19,** p. 297–301.

Cox, D. R., and P. A. W. Lewis, 1966, *The statistical analysis of series of events:* Methuen & Co., Ltd., London, 285 p.

An advanced consideration of all aspects of series of events problems.

Cox, D. R., and H. D. Miller, 1965, *The theory of stochastic processes:* John Wiley & Sons, Inc., New York, 398 p.

Chapters 3–5 contain an advanced treatment of Markov chains and procedures for testing for the Markov property.

Davis, J. C., and J. M. Cocke, 1972, Interpretation of complex lithologic successions by substitutability analysis, *in* D. F. Merriam, ed., *Mathematical models of sedimentary processes:* Plenum Press, New York, p. 27–52.

The application of substitutability analysis in an attempt to interpret Pennsylvanian cyclothems in the American midcontinent.

Doveton, J. H., 1971, An application of Markov chain analysis to the Ayrshire Coal Measures succession: *Scottish Jour. Geology,* **7,** p. 11–27.

One of the clearest and most complete discussions of the problem of embedded Markov chains.

Doveton, J. H., and K. Skipper, 1974, Markov chain and substitutability analysis of turbidite succession, Cloridorme Formation (Middle Ordovician), Gaspé, Quebec: *Canadian Jour. Earth Sciences,* **11,** p. 472–488.

Draper, N. R., and H. Smith, 1981, *Applied regression analysis,* 2nd ed.: John Wiley & Sons, Inc., New York, 709 p.

A detailed consideration of regression analysis, with strong emphasis on computational aspects and applications. Chapters 1, 2, 4, and 5 are especially pertinent to material discussed in this book.

Fisher, R. A., 1970, *Statistical methods for research workers,* 14th ed.: Hafner Publ. Co., New York, 362 p.

A posthumous edition of a classic statistics text, now available as an inexpensive paperback. Section 21 is concerned with Yates' correction, Section 34 with tests of correlation, and Section 37 with cross-correlation.

Fisher, R. A., and F. Yates, 1963, *Statistical tables for biological, agricultural and medical research,* 6th ed.: Oliver and Boyd, London, 126 p.

Gentleman, W. M., and G. Sande, 1966, *Fast Fourier Transforms—for fun and profit:* Bell Telephone Laboratories, Murray Hill, N.J., 65 p.

Gill, D., 1970, Application of a statistical zonation method to reservoir evaluation and digitized-log analysis: *Bull. American Assoc. Petroleum Geologists,* **54,** no. 5, p. 719–729.

Goodman, L. A., 1968, The analysis of cross-classified data: Independence, quasi-independence, and interactions in contingency tables with and without missing entries: *American Statistical Assoc. Jour.,* **63,** p. 1091–1131.

Discusses statistical tests for embedded Markov chains.

Gordon, A. D., and R. A. Reyment, 1979, Slotting of borehole sequences: *Jour. Int'l. Assoc. Mathematical Geology,* **11,** no. 3, p. 309–327.

Haan, C. T., 1977, *Statistical methods in hydrology:* Iowa State Univ. Press, Ames, Ia., 378 p.

An excellent introduction to the time series analysis of surface-water hydrologic data.

Harbaugh, J. W., and G. Bonham-Carter, 1970, *Computer simulation in geology:* John Wiley & Sons, Inc., New York, 575 p.

Chapter 4 is one of the best and most complete discussions of Markov chains available in the geologic literature.

Harbaugh, J. W., and D. F. Merriam, 1968, *Computer applications in stratigraphic analysis:* John Wiley & Sons, Inc., New York, 282 p.

Time-trend analysis is considered in Chapter 4, and Fourier analysis of regularly spaced data is discussed in Chapter 6.

Hawkins, D. M., and D. F. Merriam, 1973, Optimal zonation of digitized sequential data: *Jour. Int'l. Assoc. Mathematical Geology,* **5,** no. 4, p. 389–395.

Hawkins, D. M., and D. F. Merriam, 1974, Zonation of multivatiate sequences of digitized geologic data: *Jour. Int'l. Assoc. Mathematical Geology,* **6,** no. 3, p. 263–269.

Jenkins, G. M., and D. G. Watts, 1968, *Spectral analysis and its applications:* Holden-Day, San Francisco, 525 p.

An advanced treatment of time series, especially Fourier analysis and statistical interpretation of spectra.

Kemeny, J. G., and J. L. Snell, 1960, *Finite Markov chains:* D. Van Nostrand Co., Inc., Princeton, N.J., 210 p.

A basic reference to Markov chain analysis.

Kermack, K. A., and J. B. S. Haldane, 1950, Organic correlation and allometry: *Biometrika,* **37,** p. 30–41.

Krumbein, W. C., 1967, FORTRAN IV computer programs for Markov chain experiments in geology: *Kansas Geological Survey Computer Contribution* 13, 38 p.

Although primarily a program description, this paper contains an excellent introduction to Markov chains.

Kruskal, W., 1953, On the uniqueness of the line of organic correlation: *Biometrics,* **9,** p. 47–58.

Kuno, H., 1962, *Catalogue of the active volcanoes of the world including solfatara fields, part XI, Japan, Taiwan and Marianas:* Inter. Volcanological Assoc., Naples, 332 p.

Li, J. C. R., 1964, *Statistical inference,* v. 1 and 2: Edward Bros., Inc., Ann Arbor, Mich., 658 p. (v. 1), 575 p. (v. 2).

Introductory aspects of linear regression are presented in Chapters 16 and 17 of vol. 1; curvilinear (polynomial) regression is considered in Chapter 30 of vol. 2.

Miller, R. L., and J. S. Kahn, 1962, *Statistical analysis in the geological sciences:* John Wiley & Sons, Inc., New York, 483 p.

Chapters 14 and 15 are concerned with runs tests and elementary time series analysis.

Morrison, D. F., 1983, *Applied linear statistical methods:* Prentice-Hall, Inc., Englewood Cliffs, N.J., 562 p.

Orthogonal polynomials are discussed extensively in Chapter 4.

Olea, R. A., 1977, Measuring spatial dependence with semivariograms: *Kansas Geological Survey Series on Spatial Analysis,* no. 3, Lawrence, Kans., 29 p.

Illustrated with real examples of semivariograms calculated from geologic data, rather than idealized pictures.

Ostle, B., and R. Mensing, 1975, *Statistics in research,* 3rd ed.: Iowa State Univ. Press, Ames, Ia., 612 p.

A basic survey of useful statistical methods. Regression tests are discussed in depth.

Owen, D. B., 1962, *Handbook of statistical tables:* Pergamon Press, London, 580 p.

Panofsky, H. A., and G. W. Brier, 1965, *Some applications of statistics to meteorology:* Pennsylvania State Univ., University Park, Pa., 224 p.

This book contains an easily understood introduction to time series analysis, including harmonic analysis.

Quenouille, M. H., 1952, *Associated measurements:* Butterworths, London, 279 p.

Rayner, J. N., 1971, *An introduction to spectral analysis:* Pion Ltd., London, 174 p.

Written for geographers, this text discusses two-dimensional as well as one-dimensional Fourier analysis and related methods such as Walsh functions.

Robinson, E. A., and S. Treitel, 1980, *Geophysical signal analysis:* Prentice-Hall, Inc., Englewood Cliffs, N.J., 466 p.

A classic text on the analysis of reflection seismic signals.

Rogers, D. F., and J. A. Adams, 1976, *Mathematical elements for computer graphics:* McGraw-Hill, Inc., New York, 239 p.

A compact introduction to the mathematical calculations needed to draw images on a computer terminal or plotter.

Rosenfeld, A., and H. K. Huang, 1968, An application of cluster detection to text and picture processing: *Tech. Rept.* 68–68, Computer Science Center, Univ. of Maryland, College Park, Md., 64 p.

Contains a description of substitutability analysis. Available from the Documents Clearinghouse, Arlington, Va., as document AD670612.

Sackin, M. J., and D. F. Merriam, 1969, Autoassociation, a new geological tool: *Jour. Int'l. Assoc. Mathematical Geology,* **1,** no. 1, p. 7–16.

Although this discussion is concerned primarily with autoassociation, it contains a description of cross-association as well.

Savitzky, A., and M. J. E. Golay, 1964, Smoothing and differentiation of data by simplified least squares procedures: *Analytical Chemistry,* **36,** no. 8, p. 1627–1639.

The authors work for a major instrument maker. Their article discusses the processing of chart recordings, including smoothing, differentiation, and interpolation.

Schwarzacher, W., 1975, *Sedimentation models and quantitative stratigraphy:* Elsevier Publ. Co., Amsterdam, 382 p.

Chapter 5 is an extensive discussion of Markov chains as applied to stratigraphy.

Siegel, S., 1956, *Nonparametric statistics for the behavioral sciences:* McGraw-Hill, Inc., New York, 312 p.

An introductory text on nonparametric tests. Runs tests are discussed in Chapters 4 and 6.

Sneath, P. H. A., and R. R. Sokal, 1973, *Numerical taxonomy: The principles and practice of numerical classification:* W. H. Freeman & Co., San Francisco, 573 p.

A major revision of the seminal book in numerical taxonomy. Required reading for biologists and paleontologists.

Till, R., 1974, *Statistical methods for the Earth sciences:* John Wiley & Sons, Inc., New York, 154 p.

This compact introductory text discusses the RMA method in Chapter 5.

Tipper, J. C., 1979, Surface modelling techniques: *Kansas Geological Survey Series on Spatial Analysis,* no. 4, Lawrence, Kans., 108 p.

A survey of graphics techniques from the field of computer-aided design (CAD) that may be useful in the Earth sciences.

Türk, G., 1979, Transition analysis of structural sequences: Discussion: *Geol. Soc. America Bulletin,* Part I, **90,** p. 989–992.

A criticism of most procedures used by geologists to analyze Markovian sequences.

Vistelius, A. B., 1961, Sedimentation time trend functions and their application for correlation of sedimentary deposits: *Jour. Geology,* **69,** p. 703–728.

This paper is one of the first presentations in English of the application of filtering to geologic sequences.

Webster, R., 1973, Automatic soil-boundary location from transect data: *Jour. Int'l. Assoc. Mathematical Geology,* **5,** no. 1, p. 27–37.

This and the following reference describe the split-moving-window technique of zonation.

Webster, R., 1980, DIVIDE: A FORTRAN IV program for segmenting multivariate one-dimensional spatial series: *Computers & Geosciences,* **6,** no. 1, p. 61–68.

Wells, R. C., R. K. Bailey, and E. P. Henderson, 1928, Salinity of the water of Chesapeake Bay: *U.S. Geological Survey, Prof. Paper* 154, p. 105–152.

Westlake, J. R., 1968, *A handbook of numerical matrix inversion and solution of linear equations:* John Wiley & Sons, Inc., New York, 171 p.

Whittaker, E. T., and G. Robinson, 1944, *The calculus of observations,* 4th ed.: Blackie and Son, Ltd., Glasgow, 395 p.

Contains a detailed discussion of smoothing equations that are closely related to orthogonal polynomials.

Wickman, F. E., 1966, Repose-period patterns of volcanoes: *Arkiv för Mineralogi och Geologi,* Bd. 4, p. 291–366.

Both theoretical and practical aspects of the study of eruption sequences are described in this four-part article.

Wilkes, M. V., 1966, *A short introduction to numerical analysis:* Cambridge Univ. Press, Cambridge, 76 p.

An entire chapter in this small monograph is devoted to interpolation procedures.

Yule, G. U., and M. G. Kendall, 1969, *An introduction to the theory of statistics,* 14th ed.: Hafner Publ. Co., New York, 701 p.

Chapter 25 is concerned with interpolation techniques, Chapter 26 with data smoothing and filtering, and Chapter 27 with autocorrelation and harmonic analysis.

Yevjevich, V., 1972, *Stochastic processes in hydrology:* Water Resources Publications, Fort Collins, Colo., 276 p.

An extensive treatment of periodogram analysis, Fourier analysis, and statistical tests applied to river-level data.

CHAPTER 5

Map Analysis

Geologic Maps—Conventional and Otherwise

Although geologists study a three-dimensional world, their view of it is strongly two dimensional. This reflects in part the fact that the third dimension of depth often is accessible to only a fraction of the extent of the other two spatial dimensions. Also, our thoughts are conditioned by the media in which we express them, and maps, photographs, and cross-sections are printed or drawn on flat sheets of paper. We may be interested in the geologic features exposed in a deep mine with successive levels, adits, and raises creating a complex three-dimensional net, yet we must reduce this network to flat projections in order to express our ideas concerning the relationships we see. Geologists are carefully trained to read, utilize, and create maps; probably no other group of scientists is as adept at expressing and envisioning dimensional relationships. Maps are as important to Earth scientists as the conventions for scales and notes are to the musician, for they are compact and efficient means of expressing relationships and details.

Although maps are a familiar part of every geologist's training and work, surprisingly little thought has gone into the mechanics and philosophy of geologic map-making. Most of the techniques for comparing one map with another in a quantitative way have been developed by geographers, even though geologists are comparing maps constantly and searching for similarities and patterns in them. With the advent of the computer, automatic contouring has become the rage in geologic exploration, and oil companies are among the largest markets for the manufacturers of automatic plotters. Yet almost nothing has been published about the computational algorithms used to drive these plotters, or on the relative merits of the different possible methodologies. Little work has been done on map reliability or efficiency. Trend-surface analysis is probably the single most widely used map analysis technique, but many of those who utilize commercially available programs are unaware of the inherent limitations of the method. As a consequence, some people fortuitously achieve good results and others see their efforts yield nonsense; thus are born advocates and skeptics, both without justification.

A map is a two-dimensional representation of an area. Usually the area is geographical—a quadrangle, mining district, or country—and the map is a method for reducing very large-scale spatial relationships so they can be easily perceived. However, the representation may equally well be a "map" of a thin section or electron photomicrograph, where the relationships between features have been enlarged so they become visible. Maps, in this general definition, include traditional geologic and topographic maps and also aerial photographs, mine plans, peel prints, photomicrographs, and electron micrographs. In fact, any sort of two-dimensional spatial representation is included.

Map relationships are expressed almost always in terms of points located upon the map. We are concerned with distances between points, the density of points, and the values assigned to points. Most maps are estimates of continuous functions, based on discrete observations at control points. An obvious example is the topographic map; although the contour lines are an expression of a continuous and unbroken surface, they are based on measurements made at triangulation and survey control points. Even more obvious is a structural contour map. We do not know that the structural surface is continuous, because we can observe it only at the control points which are the wells that penetrate the surface. Nevertheless, we believe that it is continuous and we estimate its form by the control points, recognizing that our reconstruction is inaccurate and lacking in detail because we have no data between wells.

When mapping the surface geology of a desert region, we can stand at one locality where strike and dip have been measured and extend formation boundaries on our map with great assurance because we can see the contacts across the countryside. In regions of heavy vegetation or deep weathering, however, we must make do with scattered outcrops and poor exposures; the quality of the finished map reflects to a great extent the density of control points. Geologists should be intensely interested in the effects which control-point distributions have on maps, but few studies of this influence have been published. In fact, almost all studies of point distributions have been made by geographers. In this chapter, we will examine some of these procedures and consider their application to maps and also to such problems as the distribution of mineral grains in thin sections.

Geologists exercise their artistic talents as well as their geologic skills when they create contour maps. In some instances, the addition of geologic interpretation to the raw data contained in the observation points is a valuable enhancement of the map. Sometimes, however, geologic judgment becomes biased, and the subtle effects of personal opinion detract rather than add to the utility of a map. Computer contouring methods are totally consistent, and provide a counterbalance to overly interpretative mapping. Of course, a subjective judgment is necessary in choosing an algorithm to perform mapping, but methods are available that allow a choice to be made between competing methods, based upon specified criteria. The principal motive behind the development of automatic contouring is economic, an attempt to utilize the vast investment the petroleum industry has in stratigraphic data banks. However, one of the prime benefits from these methods may come from the attention they have focused on the contouring procedure and the problems they have revealed about map reliability. Contouring methods are the subject of one section in this chapter.

Trend-surface analysis is one of the most widely applied numerical techniques in geology. Because of its popularity, we will examine the method in detail, and we will also consider its variants such as four-dimensional trend surfaces and Fourier surfaces. Although trend-surface analysis is a widely applied technique, it is frequently misused. Therefore, we will discuss the problems of data-point distribution, lack of fit, computational "blowup," and inappropriate applications. Statistical tests are available for trend surfaces if they are used as multiple regressions; we will consider these tests and the assumptions prerequisite to their application. There are conflicting viewpoints about the proper model (and consequently, appropriate application) for trend surfaces.

The exchange between Earth scientists and statisticians has been mostly one-way, with the notable exception of the theory of regionalized variables. Developed originally by Georges Matheron, a French mining engineer, this theory describes the statistical behavior of spatial properties that are intermediate between purely random and completely deterministic phenomena. The most familiar application of the theory is in kriging, an estimation procedure important in mine evaluation, mapping, and other problems where values of a property must be estimated at specific geographic locations.

Two-dimensional methods are, for the most part, direct extensions of techniques discussed in Chapter 4. Trend surfaces are an offshoot of the family of regression techniques, kriging is related to time-series analysis, and contouring is an extension of interpolation methods. We have simply enlarged the dimensionality of our problems by considering a second (and in some cases a third) spatial variable. Of course, there are some problems and some analytical methods that are unique to map analysis. Other methods are a subset of more general multidimensional procedures. It is an indication of the importance of one- and two-dimensional problems in the Earth sciences that they have been included in individual chapters.

Systematic Patterns of Search

Most geologists devote their professional careers to the process of searching for something hidden. Usually the object of a geologist's search is an undiscovered oil field or an ore body, but for some it may be for a flaw in a casting, a primate fossil in an excavation, or a thermal spring on the ocean's floor. Too often the search has been conducted haphazardly, by wandering at random across the area of investigation like an old-time prospector following his burro. Increasingly, however, geologists and other Earth scientists are using systematic procedures of search, particularly when they must rely on instruments to detect their targets.

Most systematic searches are conducted along one or more sets of parallel lines. Ore bodies that are distinctively radioactive or magnetic are sought using airborne instruments carried along equally spaced parallel flight lines. Seismic surveys are laid out in regular sets of traverses. Satellite reconnaissance, by its very nature, must consist of parallel orbital tracks.

The probabilities that targets will be detected by a search along a set of lines can be determined from geometrical considerations. Basically, the probability of discovery is related to the relative size of the target as compared to the spacing of the search pattern. The shape of the target and the arrangement of the lines of search also influence the probability. If the target is assumed to be elliptical and the search consists of parallel lines, the probability that a line will intersect a hidden target of specified size, regardless of where it occurs within the search area, can be calculated. These assumptions do not seem unreasonable for many exploratory surveys. Note that the probabilities relate only to intersecting a target with a line, and do not consider the problem of recognizing a target when it is hit.

McCammon (1977) gives the derivation of the geometric probabilities for circular and linear targets and parallel line searches. His work is based mostly on the mathematical development of Kendall and Moran (1963). An older text by Uspensky (1937) derives the more general elliptical case used here.

Assume the target being sought is an ellipse whose dimensions are given by the major semiaxis a and minor semiaxis b. (If the target is circular, then $a = b = r$, the radius of the circle.) The search pattern consists of a series of parallel traverses spaced a distance D apart (Fig. 5.1). The probability that a target (smaller than the spacing between lines) will be intersected by a line is

$$P = \frac{p}{\pi D} \tag{5.1}$$

where p is the perimeter of the elliptical target. The equation for the perimeter of an ellipse is

$$p = 2\pi \sqrt{\frac{a^2 + b^2}{2}}$$

where a and b are the major and minor semiaxes. Substituting,

$$P = \frac{2\pi \sqrt{\dfrac{a^2 + b^2}{2}}}{\pi D} \tag{5.2}$$

$$= \frac{2 \sqrt{\dfrac{a^2 + b^2}{2}}}{D}$$

Defining Q as

$$Q = 2 \sqrt{\frac{a^2 + b^2}{2}} \tag{5.3}$$

the probability of intersecting an elliptical target with one of a set of parallel lines is

$$P = \frac{Q}{D} \tag{5.4}$$

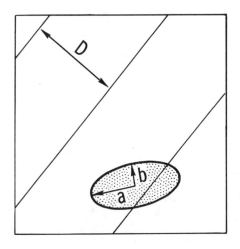

FIGURE 5.1   Search for an elliptical target with major semiaxis *a* and minor semiaxis *b*, using a parallel line search of spacing *D*.

In the specific case of a circular target, *a* and *b* are both equal to the radius, so the equation simplifies to

$$P = \frac{2r}{D} \tag{5.5}$$

At the other extreme, one axis of the ellipse may be so short that the target becomes a randomly oriented line. This geometric relationship is known as *Buffon's problem,* which involves the probability that a needle of length *l* when dropped at random on a set of ruled lines having a spacing *D*, will fall across one of the lines. The probability is

$$P = \frac{2l}{\pi D} \tag{5.6}$$

where *l* is the length of the target.

A similar geometric relationship, known as *Laplace's problem,* also pertains to the probabilities in systematic searches. Laplace's problem requires the probability that a needle of length *l*, when dropped on a board covered with a set of rectangles, will lie entirely within a rectangle. A variant gives the probability that a coin tossed onto a chessboard will fall entirely within a single square. In exploration, the complementary probabilities are of interest, that a randomly located target will be intersected one or more times by a set of lines, such as seismic traverses, arranged in a rectangular grid (Fig. 5.2).

The general equation is

$$P = \frac{Q(D_1 + D_2 - Q)}{D_1 D_2} \tag{5.7}$$

Where D_1 is the spacing between one set of parallel seismic traverses and D_2 is the spacing between the perpendicular set of traverses. In the specific instance of

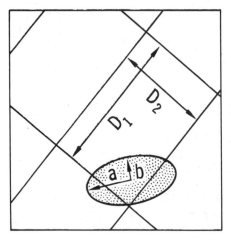

FIGURE 5.2 Search for an elliptical target with major semiaxis *a* and minor semiaxis *b*, using a grid search with spacing D_1 in one direction and D_2 in the other.

a search in the pattern of a square grid, the equation simplifies to

$$P = \frac{Q}{D}\left(2 - \frac{Q}{D}\right) \tag{5.8}$$

Lambie (unpublished report, 1981) has pointed out that these equations for geometric probability are approximations of integral equations. In comparing exact probabilities found by numerical integration with those predicted by the approximation equations, he found that significant differences occur only for very elongate targets that are large with respect to the spacing between the search lines. Then, equations such as 5.4 and 5.7 may seriously overestimate the probabilities of detection.

The probabilities of intersecting a target, as calculated by the approximating equations, can conveniently be shown as graphs. McCammon (1977) has presented such graphs in a particularly useful dimensionless form for various combinations of target shape and size relative to the spacing between the search lines. Figure 5.3 shows the probability of detecting an elliptical target whose shape ranges from a circle to a line, using a search pattern of parallel lines. The size of the target is expressed as the ratio (maximum dimension of target)/(spacing between search lines). Figure 5.4 is the equivalent graph for a search pattern consisting of a square grid of lines.

If the shape of the target is specified, the probabilities of intersection can be graphed for different patterns of search. Figure 5.5, for example, shows the probability of intersecting a circular target with search patterns ranging from a square grid, through rectangular grid patterns, to a parallel line search. Figure 5.6 is the equivalent graph for a line-shaped target. Between the two graphs, all possible shapes of elliptical targets and all possible patterns of search along two perpendicular sets of parallel lines are encompassed.

From the approximation equations, we can calculate specific graphs for special applications. Figure 5.7, for example, gives the probability of detecting a

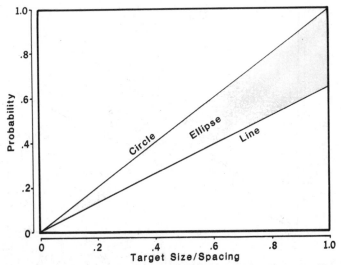

FIGURE 5.3 Probability of intersecting a target with a parallel line search. Shape of target may range from a circle to a line; elliptical targets of various axial ratios fall in the shaded region. Horizontal axis is ratio (major dimension of target)/(spacing between search lines). After McCammon (1977).

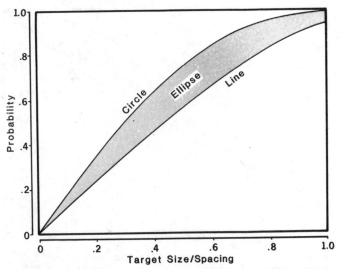

FIGURE 5.4 Probability of intersecting a target with a square grid search. Shape of target may range from a circle to a line; elliptical targets of various axial ratios fall in the shaded region. Horizontal axis is ratio (major dimension of target)/(spacing between search lines). After McCammon (1977).

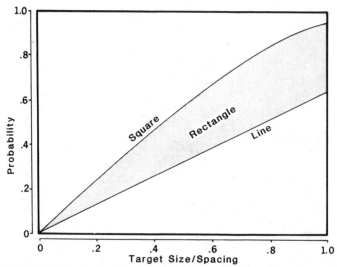

FIGURE 5.5 Probability of intersecting circular target with regular searches, ranging from a square search pattern to a set of parallel search lines. Rectangular search patterns with different ratios of D_1/D_2 fall in the shaded region. Horizontal axis is ratio (major dimension of target)/(minimum spacing between search lines). After McCammon (1977).

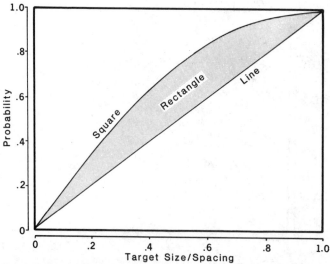

FIGURE 5.6 Probability of intersecting line target with regular searches, ranging from a square search pattern to a set of parallel line searches. Rectangular search patterns with different ratios of D_1/D_2 fall in the shaded region. Horizontal axis is ratio (length of target)/(minimum spacing between search lines). After McCammon (1977).

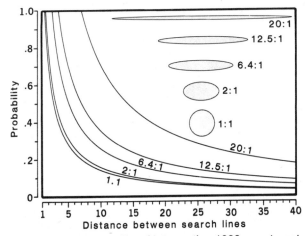

FIGURE 5.7 Probability of intersecting 1000-acre target with parallel line search when spacing between lines is given in miles.

1000-acre elliptical target with a parallel line search when the spacing between lines is given in miles. The graph has been used to estimate the likelihood of detecting a minimum size of seismic anomaly with a reconnaissance seismic survey.

Distribution of Points

Geologists often are interested in the manner in which points are distributed on a two-dimensional surface or a map. The points may represent sample localities, oil wells, control points, or poles and projections on a stereonet. We may be concerned about the uniformity of control-point coverage, the distribution of point density, or the relation of one point to another. These are questions of intense interest to geographers as well as geologists, and the burgeoning field of locational analysis is devoted to these and similar problems. Although much of the attention of the geographer is focused on the distribution of cultural features, the methodology developed is directly applicable to the study of natural phenomena as well.

The patterns of points on maps may be conveniently classified into three categories: regular, random, and clustered. Examples of the three types of distributions are shown in Figure 5.8. Of course, most maps will have patterns intermediate between these extreme values, and the problem becomes one of determining where the observed pattern lies within the spectrum of possible distributions. For example, most people would regard the distribution of points in Figure 5.9 as random. It is not, because the map was first divided into regular cells, and one point then placed at random within each cell. The distribution therefore has both random and regular aspects.

The pattern of points on a map is said to be *uniform* if the density of points in any subarea is equal to the density of points in all other subareas of the same size and shape. The pattern is *regular* if the points are placed on a grid of some sort.

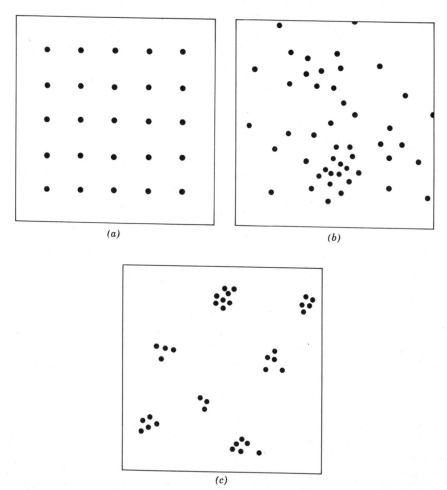

FIGURE 5.8 Possible patterns of points on maps. (*a*) Points regularly spaced on a grid or network. (*b*) Points scattered at random. (*c*) Points grouped in clusters.

That is, the distance between a point i and a point j lying in some specified direction from i is the same for all pairs of points i and j on the map. A *random* pattern can be created if any subarea is as likely to receive a point as any other subarea of the same size, and the placement of a point has no influence on the placement of any other point.

Uniformity of data points is important in many types of analyses, including trend-surface methods, which we will discuss later. The reliability of contour maps is directly dependent upon the density and uniformity of control points. However, most geologic researchers have been content with qualitative assessments of the adequacy of their data distribution. Even though the desirability of a uniform distribution of control points is often cited, the degree of uniformity is seldom measured. The tests necessary to determine uniformity are very simple, and it is

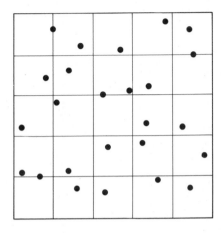

FIGURE 5.9 An apparently random pattern created by assigning one point to a random location within a regular pattern of cells. The distribution of points is more uniform than expected for a completely random pattern.

unfortunate that many geologists seem unaware of their use. They are, however, extensively used by geographers; Haggett, Cliff, and Frey (1977, Chapters 4, 8, and 13), Cole and King (1968, Chapter 4), and King (1969, Chapter 5), can provide an introduction to this literature.

Uniform Patterns

A map area may be divided into a number of equal-sized subareas (sometimes called *quadrats*) such that each subarea contains a number of points. If the data points are distributed uniformly, we expect each subarea to contain the same number of points. This hypothesis of no difference in the number of points per subarea can be tested using a χ^2 method, and is theoretically independent of the shape or orientation of subareas. However, the test is most efficient if the number of subareas is a maximum (this increases the degrees of freedom), subject to the restriction that no subarea contain less than five points. The expected number of points in each subarea is

$$E = \frac{\text{total number of data points}}{\text{number of subareas}} \tag{5.9}$$

A χ^2 test of goodness-of-fit of the expected (uniform) distribution to the observed distribution is given by

$$\chi^2 = \sum \frac{(O - E)^2}{E} \tag{5.10}$$

where O is the observed number of data points in a subarea and E is the expected number. The test has $\nu = (T - 2)$ degrees of freedom, where T is the number of subareas.

As an example of the application of this test, consider the data-point distribution shown in Figure 5.10. These are the locations of 123 oil wells in central Kansas

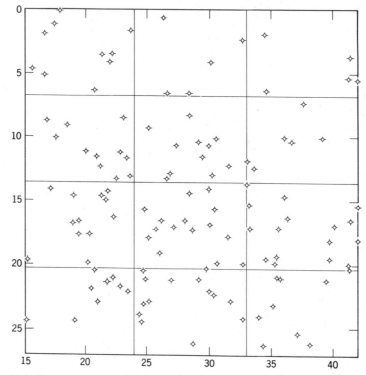

FIGURE 5.10 Locations of 123 wells drilled to the top of Ordovician rocks (Arbuckle Group) in central Kansas. Map has been divided into twelve cells of equal size.

TABLE 5.1 Number of Wells in Twelve Subareas of Central Kansas

Observed Number of Points	$\dfrac{(O - E)^2}{E}$
10	0.00
5	2.60
5	2.60
11	0.06
12	0.32
6	1.73
12	0.32
16	3.30
15	2.26
9	0.14
14	1.42
8	0.48
TOTAL = 123	χ^2 = 15.23[a]

[a]Test value is not significant at the $\alpha = 0.05$ level.

which we will later use in the fitting of a trend surface to the top of the Ordovician stratigraphic succession. In Figure 5.10, the map area has been divided into 12 equal subareas, each of which we expect to contain about 10 points, if the points are uniformly distributed. The observed number of points in each subarea and the computations necessary to find the test value are given in Table 5.1. This test has $\nu = 10$ degrees of freedom, so the critical value of χ^2 at the 5% ($\alpha = 0.05$) significance level is 18.3. The computed test value of $\chi^2 = 15.2$ does not exceed this, so we conclude that there is no evidence suggesting that the quadrats are unevenly populated. Note that the test applies only to the uniformity of point densities between areas of a specified size. It is possible that we could select quadrats of different sizes that might not be uniformly populated, especially if they were smaller than those used in this test.

Random Patterns

Establishing that a pattern is uniform does not specify the nature of the uniformity, for both regular and random patterns are expected to be homogeneous. For many purposes verifying uniformity is sufficient, but if we desire more information about the pattern, we must turn to other tests. Where points are distributed at random across a map area, even though the coverage is uniform, we do not expect exactly the same number of points to lie within each subarea. Rather, there will be some preferred number of points that will occur in most subareas and progressively fewer areas that contain either more or less points than this number. This is apparent in the example we just worked; although our hypothesis of uniformity specified that we expect about 10 observations in each subarea, we actually found some areas that contained more than 10 and some that contained fewer.

You will recall that the Poisson probability distribution is the limiting case of the binomial distribution when p, the probability of a success, is very small and $(1 - p)$ approaches 1.0. The Poisson distribution can be used to model the occurrence of rare, random occurrences in time, as it was used in Chapter 4, or it can be used to model the random placement of points in space. Although the Poisson distribution, like the binomial, uses the numbers of successes, failures, and trials in the calculation of probabilities, it can be rewritten so that neither the number of failures nor the total number of trials are required. Rather, it uses the number of points per quadrat and the density of points in the entire area to predict how many quadrats should contain specified numbers of points. These predicted or expected numbers of quadrats can be used in a χ^2 test to determine if the points are distributed at random within the area.

As an application, we can determine if oil discoveries in a basin occur at random, or are distributed in some other fashion. It is not intuitively obvious that the Poisson distribution can be expressed in a form appropriate for this problem, so we will work through its development.

Assume a basin has an area a, in which m discovery wells are located randomly. The *density* of discovery wells in the basin is designated λ, and is simply

$$\lambda = \frac{m}{a}$$

The basin may be divided into small tracts, each of area A (the term "tract" is equivalent to "quadrat"). In turn, each tract may be divided into n extremely small, equal-sized subareas, which we might regard as potential drilling sites. The probability that any one of these extremely small subareas contains a discovery well tends toward zero as n becomes infinitely large.

The area of each "drilling site" is A/n. The probability that a site contains a discovery well is

$$p = \lambda \frac{A}{n}$$

and the probability that it does not contain a discovery well is

$$1 - p = \left(1 - \lambda \frac{A}{n}\right)$$

We wish to investigate the probability that r of the n "drilling sites" within a tract contain discovery wells, and $n - r$ "drilling sites" do not. The probability of a specific combination of discovery and nondiscovery well sites within a tract is

$$P = \left(\lambda \frac{A}{n}\right)^r \left(1 - \lambda \frac{A}{n}\right)^{n-r}$$

However, there are $\binom{n}{r}$ combinations of the n "drilling sites," of which r contain discovery wells, within a tract and all are equally probable. The probability that a tract will contain exactly r discovery wells is therefore

$$P(r) = \binom{n}{r} \left(\lambda \frac{A}{n}\right)^r \left(1 - \lambda \frac{A}{n}\right)^{n-r}$$

Note that this is simply the binomial probability of r discovery wells on n "drilling sites."

The combinations can be expanded into factorials,

$$P(r) = \frac{n(n-1)(n-2)\cdots(n-r+1)}{r!} \frac{(\lambda A)^r}{n^r} \left(1 - \frac{\lambda A}{n}\right)^n \left(1 - \frac{\lambda A}{n}\right)^{n-r}$$

Rearranging and cancelling terms yields

$$P(r) = \left(1 - \frac{1}{n}\right)\left(1 - \frac{2}{n}\right)\cdots\left(1 - \frac{r-1}{n}\right)\left(1 - \frac{\lambda A}{n}\right)^{n-r}\left[\left(1 - \frac{\lambda A}{n}\right)^n \frac{(\lambda A)^r}{r!}\right]$$

As n becomes infinitely large, all of the fractions that contain n in their denominator become infinitesimally small and vanish, so all terms inside parentheses simply become equal to 1. The terms inside the brackets simplify to

$$P(r) = e^{(-\lambda A)} \frac{(\lambda A)^r}{r!} \tag{5.11}$$

Note that n, the number of "drilling sites," has vanished from the equation leaving only the discovery-well density λ, the number of discovery wells r, and the area A of the tracts. This is an expression of the Poisson distribution, as applied to the probability of rare, random events (discovery wells) occurring within geographic areas. Also note that λA is simply the mean number of wells per tract, because it is the product of the density of discovery wells times the area of a tract. In practice, we estimate λA from the total number of discovery wells, m, and the total number of tracts, T.

$$\lambda A = \frac{m}{T} \tag{5.12}$$

We can now perform a χ^2 test to see if the number of wells per tract matches that expected if the wells are randomly located according to the Poisson model. The number of tracts that contain exactly r discovery wells can be found by

$$n_r = mP_{(r)}$$
$$= me^{(-\lambda A)} \frac{(\lambda A)^r}{r!} \tag{5.13}$$

If λA is estimated by m/T, the equation becomes

$$n_r = me^{(-m/T)} \frac{(m/T)^r}{r!} \tag{5.14}$$

Figure 5.11 shows the locations of discovery wells in part of the Eastern Shelf area of the Permian Basin in Fisher and Noland counties of Texas. The area has been divided into a 10×16 grid of 160 tracts, or quadrats, each containing approximately 10 square miles. Since there are 168 discovery wells in the area, the mean number of wells per tract is

$$\frac{m}{T} = \frac{168}{160} = 1.05$$

We can count the number of tracts in the map that contain no discovery wells, exactly one discovery, two discoveries, and so forth. We also can calculate the expected number of tracts that contain these same numbers of wells, using eq. (5.14). For the Permian Basin area, the expected and observed numbers of tracts are given in Table 5.2.

Table 5.2 contains all of the figures necessary to calculate a χ^2 test of goodness-of-fit, which is essentially a comparison of the two histograms shown in Figure 5.12. The last three categories must be combined so that the observed number of tracts is equal to 5.

$$\chi^2 = \frac{(70 - 56.0)^2}{56.0} + \frac{(42 - 58.8)^2}{58.8} + \frac{(26 - 30.9)^2}{30.9}$$
$$+ \frac{(17 - 10.8)^2}{10.8} + \frac{(5 - 3.5)^2}{3.5} = 13.28$$

The test statistic has $c - 2$ degrees of freedom, where c is the number of categories

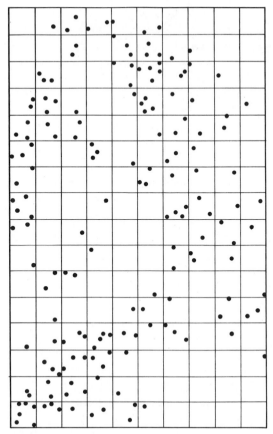

FIGURE 5.11 Locations of oil field discovery wells in part of the Eastern Shelf area of the Permian Basin, Fisher and Noland counties, Texas. Quadrats are approximately 10 square miles in size.

(one degree of freedom is lost because the expected frequencies are constrained to sum to 160, and a second degree of freedom is required for estimation of the parameter λ). For $c = 5$ categories, there are three degrees of freedom. The critical value of χ^2 for $v = 3$ and $\alpha = 0.05$ is 7.81. The test statistic far exceeds this value, so we must reject the hypothesis of equality between the observed and expected distributions and conclude that the Poisson model is not appropriate. Discovery wells are not scattered randomly over this area of the Permian Basin.

In the process of fitting the Poisson model to this data, we have generated some information that may provide additional insight into the nature of the spatial distribution. The mean number of discoveries per tract is estimated by eq. (5.12). The variance in number of discoveries per tract is

$$s^2 = \frac{\sum_{i=1}^{T} (r_i - m/T)^2}{T - 1} \tag{5.15}$$

TABLE 5.2 Calculation of Expected Numbers of Tracts Containing r Discoveries in Eastern Part of Permian Basin, Texas, Assuming a Poisson Distribution

Number of Discoveries Per Tract (r)	Poisson Equation	Probability Tract Contains r Discoveries	Number of Tracts Expected	Number of Tracts Observed
0	$P_{(0)} = e^{(-1.05)} \dfrac{1.05^0}{0!}$	0.3499	56.0	70
1	$P_{(1)} = e^{(-1.05)} \dfrac{1.05^1}{1!}$	0.3674	58.8	42
2	$P_{(2)} = e^{(-1.05)} \dfrac{1.05^2}{2!}$	0.1929	30.9	26
3	$P_{(3)} = e^{(-1.05)} \dfrac{1.05^3}{3!}$	0.0675	10.8	17
4	$P_{(4)} = e^{(-1.05)} \dfrac{1.05^4}{4!}$	0.0177	2.8	3
5	$P_{(5)} = e^{(-1.05)} \dfrac{1.05^5}{5!}$	0.0037	0.6	1
6	$P_{(6)} = e^{(-1.05)} \dfrac{1.05^6}{6!}$	0.0007	0.1	1
TOTALS		0.9998	160.0	160

where r_i is the number of discoveries in the ith tract. The summation extends over all T tracts. The alternative results of comparing the estimated mean and variance are

$m/T > s^2$ Pattern more uniform than random

$m/T = s^2$ Pattern random

$m/T < s^2$ Pattern more clustered than random

Of course, some difference between m/T and s^2 may arise due to random variation in the particular set of tracts chosen. The statistical significance of the observed difference may be tested by a t test based on the standard error of the mean, which is the variance that would be expected in values of m/T if a basin were repeatedly sampled by different sets of tracts of the same size. The standard error in the mean number of discoveries per tract is

$$s_e = \sqrt{2/(T-1)} \tag{5.16}$$

The t test compares the ratio between m/T and s^2, which should be equal to 1.0 if the two statistics are the same:

$$t = \frac{\left(\dfrac{m/T}{s^2}\right) - 1.0}{s_e} \tag{5.17}$$

The test has $T - 1$ degrees of freedom.

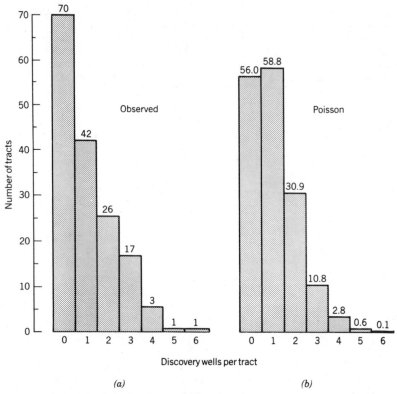

FIGURE 5.12 Histograms showing (a) observed numbers of discovery wells per tract in an area of the Permian Basin, and (b) the number expected if fields are distributed randomly according to a Poisson model.

For the eastern Permian Basin area, the variance in number of wells per tract is

$$s^2 = \frac{231.6}{159} = 1.46$$

The standard error of the mean number of wells per tract can be estimated as

$$s_e = \sqrt{\frac{2}{159}} = 0.112$$

The t statistic for the test of equivalence of the mean and variance is

$$t = \frac{(1.05/1.46) - 1.0}{0.112} = -8.86$$

At a significance level of $\alpha = 0.05$ and 159 degrees of freedom, the critical value of t for a two-tailed test is ± 1.96; the computed statistic far exceeds this and so we may conclude as we did in the χ^2 test that the spatial distribution is not random.

Since the variance is significantly greater than the mean, we must also conclude that discovery wells are areally clustered.

Clustered Patterns

Many naturally occurring spatial distributions show a pronounced tendency toward clustering. This is especially true of certain biological variables, such as presence of specific organisms or occurrence of an infection. The descendants of a sedentary parent, perhaps a coral or a tree, tend to grow nearby, leading to development of densely populated areas surrounded by relatively barren regions. Clustered patterns of points can be modelled by many theoretical distributions, most of which can be regarded as combinations of two or more simpler distributions. One of the distributions describes the locations of the centers of clusters, while the other describes the pattern of individual points around the centers of the clusters.

The negative binomal distribution can be used to model the occurrence of clustered points in space in a manner equivalent to use of the Poisson to model randomly arranged points. An extensive discussion, with citations to studies in many fields, is given by Ripley (1981). Griffiths (1962, 1966) has long advocated the appropriateness of the negative binomial as a model for the occurrence of oil fields and ore bodies.

One derivation of the negative binomial is as a compound Poisson and logarithmic distribution, where clusters of points are randomly located within a region, and individual points within a cluster follow a logarithmic distribution. In the formulation appropriate for describing spatial patterns, the negative binomial is

$$P(r) = \binom{k+r-1}{r} \left(\frac{p}{1+p}\right)^r \left(\frac{1}{1+p}\right)^k \tag{5.18}$$

In terms of the oil-exploration problem we have just considered, r is the number of discovery wells in a tract, p is the probability that a given drilling site contains a discovery well, and k is a measure of the degree of clustering of the discoveries. If k is large, clustering is less pronounced and the spatial distribution approaches the Poisson, or randomness. As k approaches zero, the pattern of clustering becomes more pronounced. The density, λ, is equal to

$$\lambda = kp \tag{5.19}$$

If k is not an integer (and in general it will not be), this combinatorial equation cannot be solved. Then, the following approximation must be used:

$$P(0) = \frac{1}{(1+p)^k} \tag{5.20}$$

$$P(r) = \frac{(k+r-1)(p/1+p)}{r} \cdot P(r-1) \tag{5.21}$$

As with the Poisson distribution, λ is estimated by the average density of discoveries per tract, m/T. The clustering parameter, k, is estimated by

$$k = \frac{(m/T)^2}{s^2 - (m/T)} \tag{5.22}$$

where s^2 is the variance in number of discovery wells per tract. Then, the probability p can be estimated as

$$p = \frac{\lambda}{k} = \frac{(m/T)}{k} \tag{5.23}$$

We can apply the negative binomial model to the data on discovery wells in the eastern part of the Permian Basin (Fig. 5.11) to see if this distribution can adequately describe their spatial distribution. The mean and variance of the number of discovery wells per tract have already been found as $m/T = 1.05$ and $s^2 = 1.46$. The clustering effect can be estimated using eq. (5.22) as

$$k = \frac{1.05^2}{1.46 - 1.05} = 2.69$$

In turn, the probability of a discovery well occurring in a tract is

$$p = \frac{1.05}{2.69} = 0.390$$

Using the approximation equations, the probability that a given tract will contain no discovery wells is

$$P(0) = \frac{1}{(1 + 0.390)^{2.69}} = 0.4124$$

The probability that a tract will contain exactly one discovery well is

$$P(1) = \frac{(2.69 + 1 - 1)(0.390/1.390)}{1} \cdot 0.4124 = 0.3112$$

The probabilities that a tract will contain exactly two, three, or other number of discovery wells can be calculated in a similar fashion. Then, the expected number of tracts containing r discoveries can be determined simply by multiplying these probabilities by 160, the total number of tracts. Table 5.3 gives the expected numbers of tracts for up to six discoveries per tract.

The numbers of tracts containing exactly r discoveries as predicted by the negative binomial model is compared to the corresponding observed numbers of tracts in Figure 5.13. The goodness-of-fit of the negative binomial can be tested by a χ^2 test exactly like that used to check the fit of the Poisson model. Again, it is necessary to combine the final three categories so a frequency of five is obtained. The test statistic is $\chi^2 = 4.82$, with $(5 - 2)$ degrees of freedom. This is less than the critical value of χ^2 for $\alpha = 0.05$ and $\nu = 3$, so we cannot reject the negative binomial as a model of the spatial distribution of discovery wells in the eastern part of the Permian Basin. Keep in mind that this is not equivalent to proof that

TABLE 5.3 Expected Numbers of Tracts Containing r Discoveries in Eastern Part of Permian Basin, Texas, Assuming a Negative Binomial Distribution

Number of Discoveries Per Tract (r)	Probability Tract Contains r Discoveries	Number of Tracts	
		Expected	Observed
0	0.4124	66.0	70
1	0.3112	49.8	42
2	0.1611	25.8	26
3	0.0706	11.3	17
4	0.0281	4.5	3
5	0.0106	1.7	1
6	0.0038	0.6	1
TOTALS	0.9988	159.7	160

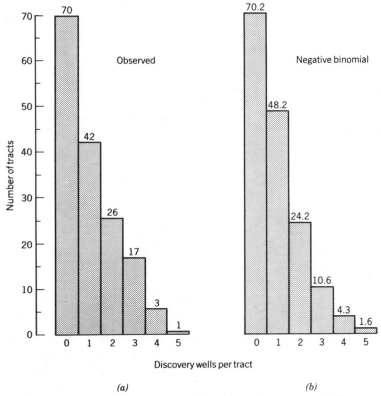

(a)   *(b)*

FIGURE 5.13 Histograms showing (a) observed numbers of discovery wells per tract in an area of the Permian Basin, and (b) the number expected in a clustered (negative binomial) model.

the wells *do* follow a negative binomial model, because it is possible that some other clustered model might provide an even better fit. However, the negative binomial does generate a spatial distribution that is statistically indistinguishable from the one observed.

Nearest-Neighbor Analysis

An alternative to quadrat analysis is *nearest-neighbor analysis*. The data used are not the numbers of points within subareas, but the distances between closest pairs of points. Since it is not necessary to select a quadrat size, nearest-neighbor procedures avoid the possibility of finding that a pattern is random at one scale but not at another. Also, since there usually are many more pairs of nearest neighbors than quadrats, the analysis is more sensitive. A good introduction to nearest-neighbor techniques is given by Getis and Boots (1978). Ripley (1981) provides a review of theory and applications in several fields, as do Cliff and Ord (1981).

Nearest-neighbor analysis compares characteristics of the observed set of distances between pairs of closest points with those that would be expected if the points were randomly placed. The characteristics of a theoretical random pattern can be derived from the Poisson distribution. If we ignore the effect of the edges of our map, the expected mean distance between nearest neighbors is

$$\bar{\delta} = \frac{1}{2} \sqrt{A/n} \qquad (5.24)$$

where A is the area of the map and n is the number of points. You will recall that A/n is the point density, λ. The sampling variance of $\bar{\delta}$ is given by

$$\sigma_{\bar{\delta}}^2 = \frac{(4 - \pi)A}{4\pi n^2} \qquad (5.25)$$

If we work out the constants,

$$\sigma_{\bar{\delta}}^2 = \frac{0.06831A}{n^2} \qquad (5.26)$$

The standard error of the mean distance between nearest neighbors is the square root of $\sigma_{\bar{\delta}}^2$

$$s_e = \frac{0.26136}{\sqrt{A/n^2}} \qquad (5.27)$$

The distribution of $\bar{\delta}$ is normal provided n is greater than 6, so we can use the simple Z test given in Chapter 2 to test the hypothesis that the observed mean distance between nearest neighbors, $\bar{d}$, is equal to the value of $\bar{\delta}$ from a random pattern of points of the same density. The test is

$$Z = \frac{\bar{d} - \bar{\delta}}{s_e} \qquad (5.28)$$

This is the form of the nearest-neighbor test that commonly is presented, but unfortunately it has a serious defect for most practical purposes. The expected value $\bar{\delta}$ assumes that edge effects are not present, which means the observed pattern of points must extend to infinity in all directions if $\bar{d}$ and $\bar{\delta}$ are to be validly compared. Since the map does not extend indefinitely, the nearest neighbors of points near the edges must lie within the body of the map, and so $\bar{d}$ is biased toward a greater value. There are several corrections for this problem. If data are available beyond the limits of the area being analyzed, the map can be surrounded by a *guard region*. Then, nearest-neighbor distances between points inside the map and points in the guard region can be included in the calculation of $\bar{d}$. Alternatively, we can consider our map to be drawn, not on a flat plane, but on a torus. This results in the right edge being adjacent to the left edge and the top adjacent to the bottom. The nearest neighbor of a point along the right edge of the map might lie just inside the left edge (this concept should be familiar to anyone who has contoured point densities on stereonets). Another way of regarding this particular correction is that the pattern of points may be considered as repeating in all directions, like floor tiles. Any point lying adjacent to an edge of the map has the

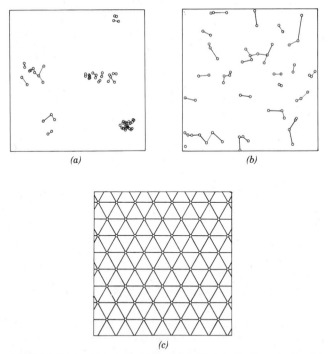

(a)   *(b)*

(c)

FIGURE 5.14 Nearest-neighbor statistics, *R*, for patterns of points on maps. (*a*) Points grouped into five clusters, $R = 0.34$. (*b*) Points scattered at random, $R = 0.91$. (*c*) Points arranged on a hexagonal net, $R = 2.15$. Point density, λ, is the same for all patterns. Adapted from Olea (1984).

opportunity to find a point across the edge that may be a closer neighbor than the nearest point within the map.

A third correction involves adjusting $\bar{\delta}$ so that the boundary effects are included in its expected value. Using numerical simulation, Donnelly (1978) found these alternative expressions for the theoretical mean nearest-neighbor distance and its sampling variance:

$$\bar{\delta} \approx \frac{1}{2} \sqrt{\frac{A}{n}} + \left(0.514 + \frac{0.412}{\sqrt{n}}\right)\frac{p}{n} \tag{5.29}$$

and

$$s_{\bar{\delta}}^2 \approx 0.070 \frac{A}{n^2} + 0.035p \frac{\sqrt{A}}{n^{5/2}} \tag{5.30}$$

In these approximations, p is the perimeter of the rectangular map. Note that if the map has no edges, as when it is considered to be drawn on a torus, p is zero and these equations are identical to eq. (5.24) and (5.26).

The expected and observed mean nearest-neighbor distances can be used to construct an index to the spatial pattern. The ratio

$$R = \bar{d}/\bar{\delta} \tag{5.31}$$

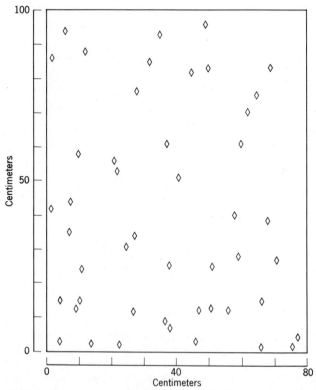

FIGURE 5.15 Representation of a polished slab of anorthosite facing stone, showing locations of magnetite crystals.

TABLE 5.4   Coordinates of Magnetite Grains on Polished Anorthosite Slab

Distance from Lower Left Corner of Slab (cm)			
Horizontal	Vertical	Horizontal	Vertical
1	86	38	25
2	41	38	7
4	3	41	51
4	15	46	2
8	95	47	12
9	13	45	82
7	35	50	83
8	44	49	96
10	58	50	13
12	88	51	25
14	2	56	12
22	2	58	40
21	56	59	28
22	53	60	61
24	31	62	70
27	12	66	0
27	34	66	15
28	76	65	75
37	14	69	38
37	61	69	83
27	85	71	27
11	25	76	1
15	15	77	4
35	93		

is the *nearest-neighbor statistic* and ranges from 0.0 for a distribution where all points coincide and are separated by distances of zero, to 1.0 for a random distribution of points, to a maximum value of 2.15. The latter value characterizes a distribution in which the mean distance to the nearest neighbor is maximized. The distribution has the form of a regular hexagonal pattern where every point is equidistant from six other points. Figure 5.14 shows a series of patterns with different values of the nearest-neighbor statistic, all having the same point density.

We will illustrate the application of the nearest-neighbor method using the "map" shown in Figure 5.15. The "map" actually represents a polished facing stone on the front of a bank in a university town. It provides an interesting subject of study for an igneous petrology class. The stone is black anorthosite and contains small, scattered, euhedral crystals of magnetite. The instructor uses the slab to demonstrate a variety of topics, including examples of numerical techniques in petrography. For pedagogical purposes, it has been decreed that the slab is mounted in its original orientation. That is, it represents a vertical surface; "down" is toward the bottom of the slab. The "map" shows the locations of all visible magnetite grains on the surface. Coordinates of each grain, in centimeters from the lower left corner of the slab, are given in Table 5.4. Are magnetite grains uniformly distributed across the surface, or do they tend to be clustered? Is the density of crystals greater near the bottom of the slab than near the top? These and similar questions are of great importance in determining the petrogenesis of an igneous rock, and can be effectively investigated using the techniques we have discussed. Test the hypothesis of uniform, random distribution of crystals by both quadrat and nearest-neighbor analysis. This problem may be done by hand by measuring distances directly on Figure 5.15, or the distances may be computed using the coordinates in Table 5.4. Ripley (1981, pp. 175–181) gives an exhaustive analysis of these data, using a variety of techniques.

Distribution of Lines

Some naturally occurring patterns are composed of lines, such as lineaments seen on satellite images, the tracery of joints exposed on a weathered granite surface, or the microfractures seen in a thin section of a deformed rock. Just as a set of points can form a pattern that ranges from uniform to tightly clustered, so can sets of lines. Of course, lines are more complex than points, because they possess length and orientation, as well as location. Their analysis is correspondingly more difficult, and statistical methods suitable for the study of patterns of lines seem less well developed than those applied to patterns of points. Few studies have examined the distribution of lengths of lines, except for some work on the lognormal distribution (Aitchison and Brown, 1969). A small number of workers have investigated the spacing between lines in a pattern, a problem analogous to nearest-neighbor analysis of points (Miles, 1964; Dacey, 1967). A much larger body of literature exists on the orientation of lines, a topic we will consider in the next section.

We can define a random pattern of lines as one in which any location is equally likely to be crossed by a line, and any orientation of the crossing line is also equally likely. Such random patterns can be generated in many ways; one procedure consists of choosing two pairs of coordinates from a random number table, then drawing a line through them. Another consists of drawing a radius at a randomly chosen angle, measuring out along the radius a random distance from the center, then constructing a perpendicular to the radial line. Repeating either procedure will result in patterns of lines that are statistically indistinguishable.

We can define a measure of line density that is analogous to λ, the point density:

$$\lambda = \frac{\Sigma l}{A} \tag{5.32}$$

The quantity Σl is simply the total length of lines on the map, which has an area A. λ is the parameter that determines the form of the Poisson distribution; as we would expect, the Poisson model describes the distribution of many properties of a pattern formed by random lines.

The distribution of distances between pairs of lines can be examined by calculating a nearest-neighbor measure. We must first randomly pick a point on each of the lines in the map. From each point, the distance is measured to the nearest line, in a direction perpendicular to that line. The mean nearest-neighbor distance

$\bar{d}$ is the average of these measurements. The procedure is illustrated on Figure 5.16.

Dacey (1967) has determined that the expected nearest-neighbor distance $\bar{\delta}$ for a pattern of random lines is

$$\bar{\delta} = \frac{0.31831}{\lambda} \tag{5.33}$$

and that the expected variance is

$$\sigma_{\bar{\delta}}^2 = \frac{0.10132}{\lambda^2} \tag{5.34}$$

From the expected variance and the number of lines in the pattern, we can find the standard error of our estimate of the mean nearest-neighbor distance. The standard error is

$$s_e = \sqrt{\frac{\sigma_{\bar{\delta}}^2}{n}} \tag{5.35}$$

This allows us to calculate a simple Z statistic for testing the significance of the difference between the expected and observed mean nearest-neighbor distance:

$$Z = \frac{\bar{d} - \bar{\delta}}{s_e} \tag{5.36}$$

The test is two-tailed; if the value of Z is not significant, we conclude that the observed pattern of lines cannot be distinguished from a pattern generated by a random (Poisson) process. We can also create a nearest-neighbor index identical to that used for point patterns, by taking the ratio of the observed and expected mean nearest-neighbor distances, or $\bar{d}/\bar{\delta}$. The index is interpreted exactly like the index for point patterns.

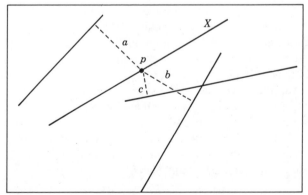

FIGURE 5.16 Calculation of nearest-neighbor distances between lines. Point P is chosen at random on a line X. Dashed lines a, b, and c are perpendiculars drawn from point P to nearby lines. The shortest of these, line c, is the distance to the nearest neighbor of line X. The process is repeated to find the nearest-neighbor distances for all lines.

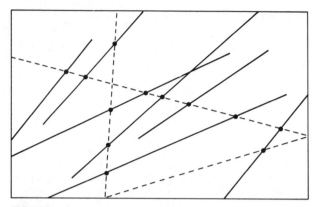

FIGURE 5.17   Random walk sampling line (dashed) drawn across pattern of lines on a map. Intersections along sampling line form a sequence of intervals that can be tested for randomness.

This test will work for sets of lines that are straight or curved, provided the lines do not frequently reverse direction. Also, the lines should be at least one-and-a-half times longer than the average distance between the lines. If the number of lines on the map is small, the estimated density should be adjusted by the factor $(n - 1)/n$, where n is the number of lines in the pattern. The estimate of the line density is, therefore

$$\lambda = \frac{(n - 1) \, \Sigma l}{nA} \tag{5.37}$$

A simple alternative way of investigating the nature of a set of lines on a map involves converting the two-dimensional pattern into a one-dimensional sequence. We can do this by drawing a *sampling line* at random across the map, and noting where the line intersects the lines in the pattern. The distribution of intervals between the points of intersection along the sampling line will provide information about the spatial pattern. We can test this one-dimensional sequence using methods presented in Chapter 4. If a single sampling line does not provide enough inter-sections for a valid test, we can draw a second randomly oriented line continuing from one end of the first sampling line, and a third line continuing from the second, and so on (Fig. 5.17). The zigzag path of the sampling line is a *random walk*, and the succession of intersections can be treated as though they occurred along a single straight line. This and other methods for investigating the density of patterns of lines are reviewed by Getis and Boots (1978).

Analysis of Directional Data

Directional data are an important category of geologic information. Bedding planes, fault surfaces, and joints are all characterized by their attitudes, expressed as strikes and dips. Glacial striations, sole marks, fossil shells, and water-laid pebbles may have preferred orientations. Aerial and satellite photographs may show oriented linear patterns. These features can be measured and treated quantitatively like measurements of other geologic properties, but it is necessary to use special statistics that reflect the circular (or spherical) nature of directional data.

Following the practice of geographers, we can distinguish between *directional* and *oriented* features. Suppose a car is travelling north along a highway; the car's motion has direction, while the highway itself has only a north-south orientation. Strikes of outcrops and the traces of faults are examples of geologic observations that are oriented, while drumlins and certain fossils such as high-spired gastropods have clear directional characteristics.

We may also distinguish observations that are distributed on a circle, such as paleocurrent measurements, and those that are spherically distributed, such as measurements of metamorphic fabric. The former data are conventionally shown as rose diagrams, a form of circular histogram, while the latter are plotted as points on a projection of a hemisphere. Although geologists have plotted directional measurements in these forms for many years, they have not extensively used formal statistical techniques to test the veracity of the conclusions they have drawn from their diagrams. This is doubly unfortunate; not only are these statistical tests useful, but the development of many of the procedures was originally inspired by problems in the Earth sciences.

Figure 5.18 is a map of glacial striations measured in a small area of southern Finland; the measurements are listed in Table 5.5. The directions indicated by the striations can be expressed by plotting them as unit vectors, or on a circle of unit radius, as in Figure 5.19a. If the circle is subdivided into segments and the number

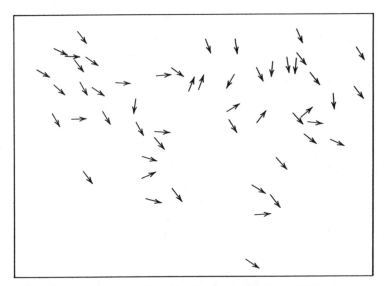

FIGURE 5.18 Map showing location and directions of 51 measurements of glacial striations in a 35 km² area of southern Finland.

TABLE 5.5   Vector Directions of Glacial Striations Measured in an Area of Southern Finland; Measurements are Given in Degrees Clockwise from North

23	123	145
27	125	145
53	126	146
58	126	153
64	126	155
83	127	155
85	127	155
88	128	157
93	128	163
99	129	165
100	132	171
105	132	172
113	132	179
113	134	181
114	135	186
117	137	190
121	144	212

of vectors within each segment counted, the results can be expressed as the *rose diagram* or circular histogram shown as Figure 5.19*b*. However, to compute statistics that describe characteristics of the entire set of vectors, we must work directly with the measurements themselves. (Note that we will follow geologic convention and measure angles clockwise from north, or from the positive end of the Y axis. In most papers on directional statistics, angles are measured counterclockwise from east, or from the positive end of the X axis.)

The dominant direction in a set of vectors can be found by computing the *vector resultant*. The X and Y coordinates of the end point of a unit vector whose direction is given by the angle θ are

$$X_i = \cos \theta_i \qquad (5.38)$$
$$Y_i = \sin \theta_i$$

Three such vectors are shown plotted in Figure 5.20. Also shown is the vector resultant, obtained by summing the sines and cosines of the individual vectors:

$$X_r = \sum_{i=1}^{n} \cos \theta_i \qquad (5.39)$$
$$Y_r = \sum_{i=1}^{n} \sin \theta_i$$

From the resultant, we can obtain the *mean direction*, $\bar{\theta}$, which is the angular average of all of the vectors in a sample. It is directly analogous to the mean value of a set of scalar measurements.

$$\bar{\theta} = \tan^{-1} (Y_r/X_r) \qquad (5.40)$$
$$= \tan^{-1} \left(\sum_{i=1}^{n} \sin \theta_i \Bigg/ \sum_{i=1}^{n} \cos \theta_i \right)$$

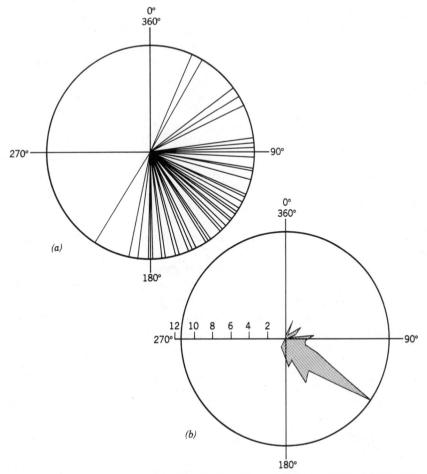

FIGURE 5.19 Directions of glacial striations shown on Figure 5.18. (a) Directions plotted as unit vectors. (b) Directions plotted as a rose diagram, showing numbers of vectors within successive 10° segments.

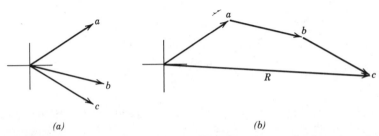

FIGURE 5.20 Determination of mean direction of a set of unit vectors. (a) Three vectors taken from Figure 5.18. (b) Vector resultant, R, obtained by combining the three unit vectors. Order of combination is immaterial.

Obviously, the magnitude or length of the resultant depends in part on the amount of dispersion in the sample of vectors, but also depends upon the number of vectors. In order to compare resultants from samples of different sizes, they must be converted into a standardized form. This is done simply by dividing the coordinates of the resultant by the number of observations, n.

$$\bar{C} = X_r/n$$

$$= \frac{1}{n} \sum_{i=1}^{n} \cos \theta_i$$

$$\bar{S} = Y_r/n \tag{5.41}$$

$$= \frac{1}{n} \sum_{i=1}^{n} \sin \theta_i$$

Note that these coordinates also define the centroid of end points of the individual unit vectors.

The resultant provides information not only about the average direction of a set of vectors, but also on the spread of the vectors about this average. Figure 5.21a shows three vectors that deviate only slightly from the mean direction. The resultant is almost equal in length to the sum of the lengths of the three vectors. In contrast, three vectors in Figure 5.21b are widely dispersed; their resultant is very short. The length of the resultant, designated R, is given by the Pythagorean theorem:

$$R = \sqrt{X_r^2 + Y_r^2} \tag{5.42}$$

$$= \sqrt{\left(\sum_{i=1}^{n} \cos \theta_i\right)^2 + \left(\sum_{i=1}^{n} \sin \theta_i\right)^2}$$

The length of the resultant can be standardized by dividing by the number of

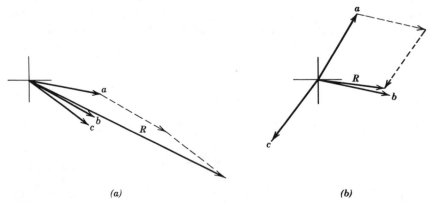

(a)   *(b)*

FIGURE 5.21   Use of the length of the resultant to express dispersion in a collection of unit vectors. (a) Three vectors that are tightly clustered around a common direction. Resultant R is relatively long, approaching the value of n. (b) Three widely dispersed vectors; resultant length is less than one.

observations. The standardized resultant length also can be found from the standardized end points.

$$\overline{R} = \frac{R}{n}$$

$$= \sqrt{\overline{C}^2 + \overline{S}^2}$$

(5.43)

The quantity $\overline{R}$ is called the *mean resultant length* and will range from zero to one. It is a measure of dispersion analogous to the variance, but expressed in the opposite sense. That is, large values of $\overline{R}$ indicate that the observations are tightly bunched together with a small dispersion, while values of $\overline{R}$ near zero indicate that the vectors are widely dispersed. Figure 5.22 shows sets of vectors having different

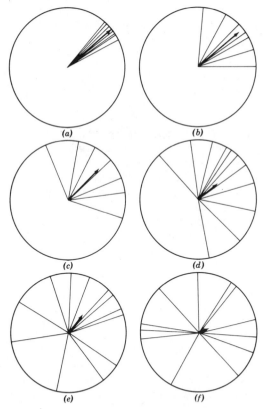

FIGURE 5.22   Sets of unit vectors illustrating the value of $\overline{R}$ produced by different dispersions of the vectors. In all examples, the mean direction is 52°. (a) $\overline{R}$ = 0.997. (b) $\overline{R}$ = 0.90. (c) $\overline{R}$ = 0.75. (d) $\overline{R}$ = 0.55. (e) $\overline{R}$ = 0.40. (f) $\overline{R}$ = 0.10.

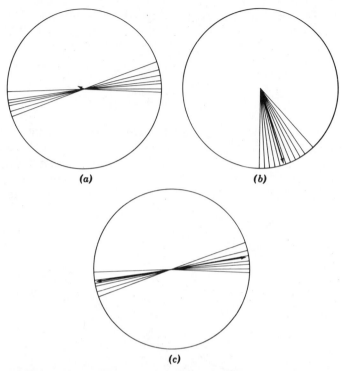

FIGURE 5.23 Effect of doubling angular direction in order to calculate mean orientation. (a) Orientation measurements plotted as
vector directions. Resultant mean direction is 285° and is near zero
in length ($\overline{R}$ = 0.08). (b) Orientation measurements plotted as vector
directions after angles are doubled. Distribution is no longer bimodal.
Resultant reflects correct trend of doubled angles and is near unity
in length (mean direction is 162°; $\overline{R}$ = 0.97). (c) Orientations replotted at original angles and true resultant direction (81°) found by
halving resultant direction in (b).

values of $\overline{R}$. In order to have a measure of dispersion which increases with increasing
scatter, $\overline{R}$ is sometimes expressed as its complement, called the *circular variance*.

$$s_o^2 = 1 - \overline{R} \tag{5.44}$$
$$= (n - R)/n$$

Other directional statistics can be computed, including circular analogs of the
standard deviation, mode, and median. Equations for these are given in a convenient
table by Gaile and Burt (1980).

Orientation data must be modified before mean directions or measures of dispersions can be calculated. Since any oriented feature may be expressed as either
of two opposite directions, some convention must be adopted to avoid inflating
the dispersion of the measurements. Krumbein (1939) hit upon a novel solution to

this problem while studying the orientations of stream pebbles. If all the measured angles are doubled, the same angle will be recorded regardless of which directional sense of the oriented features are used. As an example, consider a fault trace that strikes northeast-southwest. Its orientation could equally well be recorded as 45° or as 225°. If we double the angles, we obtain 45° $\times$ 2 $=$ 90° and 225° $\times$ 2 $=$ 450°, which becomes (450° $-$ 360°) $=$ 90°.

The mean direction, mean resultant length, and circular variance can be found in the usual manner after the orientation angles have been doubled. To recover the true mean orientation, simply divide the calculated mean direction by two. This is illustrated in Figure 5.23.

Testing Hypotheses about Circular Directional Data

In order to test statistical hypotheses about circularly distributed data, we must have a probability model of known characteristics against which we can test. There are circular analogs of univariate distributions, which we discussed in Chapter 2, but the most useful of these is the *von Mises distribution*. It is a circular equivalent of the normal distribution and similarly possesses only two parameters, a mean direction $\bar{\theta}$ and a concentration parameter κ. The von Mises distribution is unimodal and symmetric about the mean direction. As the concentration parameter increases, the likelihood of observing a directional measurement very close to the mean direction increases. If κ is equal to zero, all directions are equally probable, and the distribution becomes a circular uniform. Figure 5.24*a* shows the form of the von Mises distribution for several values of κ. The distribution can also be shown in conventional form as in Figure 5.24*b*; note that the horizontal scale is given in degrees and corresponds to a complete circle.

It is difficult to determine κ directly, but the concentration parameter can be estimated from $\bar{R}$, if we assume that the data are a sample from a population having a von Mises distribution. Table 5.6 gives maximum likelihood estimates of κ for a calculated $\bar{R}$. We will use these estimated values of κ in some subsequent statistical tests.

Test for Randomness   The simplest hypothesis that can be statistically tested is that the directional observations are random. In other words, there is no preferred direction, or the probability of occurrence is the same for all directions. If we assume that the observations come from a von Mises distribution, the hypothesis is equivalent to stating that the concentration parameter κ is equal to zero, because then the distribution becomes a circular uniform. In formal terms, the null hypothesis and alternative are

$$H_0 : \kappa = 0$$

$$H_1 : \kappa > 0$$

The test is extremely simple and involves only the calculation of $\bar{R}$ according to eq. (5.43). This statistic is compared to a critical value of $\bar{R}$ for the desired level of significance. If the observations do come from a circular uniform distribution,

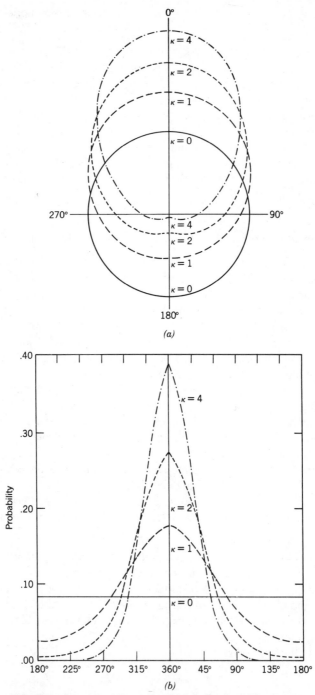

FIGURE 5.24 Von Mises distributions having different concentration parameters. (*a*) Distribution plotted in polar form. (*b*) Distribution plotted as conventional probability distribution. Note that horizontal axis is given in degrees. After Gumbel, Greenwood, and Durand (1953).

TABLE 5.6 Maximum Likelihood Estimates of the Concentration Parameter κ for Calculated Values of $\bar{R}$

$\bar{R}$	κ	$\bar{R}$	κ	$\bar{R}$	κ
0.00	0.00000	0.35	0.74783	0.70	2.01363
.01	.02000	.36	.77241	.71	2.07685
.02	.04001	.37	.79730	.72	2.14359
.03	.06003	.38	.82253	.73	2.21425
.04	.08006	.39	.84812	.74	2.28930
.05	.10013	.40	.87408	.75	2.36930
.06	.12022	.41	.90043	.76	2.45490
.07	.14034	.42	.92720	.77	2.54686
.08	.16051	.43	.95440	.78	2.64613
.09	.18073	.44	.98207	.79	2.75382
.10	.20101	.45	1.01022	.80	2.87129
.11	.22134	.46	1.03889	.81	3.00020
.12	.24175	.47	1.06810	.82	3.14262
.13	.26223	.48	1.09788	.83	3.30114
.14	.28279	.49	1.12828	.84	3.47901
.15	.30344	.50	1.15932	.85	3.68041
.16	.32419	.51	1.19105	.86	3.91072
.17	.34503	.52	1.22350	.87	4.17703
.18	.36599	.53	1.25672	.88	4.48876
.19	.38707	.54	1.29077	.89	4.85871
.20	.40828	.55	1.32570	.90	5.3047
.21	.42962	.56	1.36156	.91	5.8522
.22	.45110	.57	1.39842	.92	6.5394
.23	.47273	.58	1.43635	.93	7.4257
.24	.49453	.59	1.47543	.94	8.6104
.25	.51649	.60	1.51574	.95	10.2716
.26	.53863	.61	1.55738	.96	12.7661
.27	.56097	.62	1.60044	.97	16.9266
.28	.58350	.63	1.64506	.98	25.2522
.29	.60625	.64	1.69134	.99	50.2421
.30	.62922	.65	1.73945	1.00	∞
.31	.65242	.66	1.78953		
.32	.67587	.67	1.84177		
.33	.69958	.68	1.89637		
.34	.72356	.69	1.95357		

Adapted from Batschelet (1965) and Gumbel, Greenwood, and Durand (1953).

TABLE 5.7 Critical Values of $\bar{R}$ for Rayleigh's Test for the Presence of a Preferred Trend

Number of Observations, n	Significance Level, α (%)			
	10	5	2.5	1
4	0.768	0.847	0.905	0.960
5	.677	.754	.816	.879
6	.618	.690	.753	.825
7	.572	.642	.702	.771
8	.535	.602	.660	.725
9	.504	.569	.624	.687
10	.478	.540	.594	.655
11	.456	.516	.567	.627
12	.437	.494	.544	.602
13	.420	.475	.524	.580
14	.405	.458	.505	.560
15	.391	.443	.489	.542
16	.379	.429	.474	.525
17	.367	.417	.460	.510
18	.357	.405	.447	.496
19	.348	.394	.436	.484
20	.339	.385	.425	.472
21	.331	.375	.415	.461
22	.323	.367	.405	.451
23	.316	.359	.397	.441
24	.309	.351	.389	.432
25	.303	.344	.381	.423
30	.277	.315	.348	.387
35	.256	.292	.323	.359
40	.240	.273	.302	.336
45	.226	.257	.285	.318
50	.214	.244	.270	.301

From Mardia (1972).

we would expect that $\bar{R}$ would be small, as in Figure 5.22*f*. However, if the computed statistic is so large that it exceeds the critical value, the null hypothesis must be rejected and the observations may be presumed to come from a population having a preferred orientation. This test was originally developed by Lord Rayleigh at the turn of the century; a modern derivation is given by Mardia (1972). Table 5.7 gives critical values of $\bar{R}$ for various levels of significance and numbers of observations.

Remember that Rayleigh's test presumes that the observed vectors are sampled from a von Mises distribution. That is, the population of vectors is either uniform (if $\kappa = 0$) or has a single mode or preferred direction. If the vectors are actually sampled from a bimodal distribution such as that shown in Figure 5.23, the test will give misleading results.

We will test the measurements of Finnish glacial striations at a 5% level of significance to determine if they have a preferred direction. Since there are 51 observations, Table 5.7 yields a critical value of $\overline{R}_{50,5\%} = 0.244$. The test statistic is simply the normalized resultant $\overline{R}$. The sum of the cosines of the vectors is $X_r = -25.793$ and the sum of the sines is $Y_r = 31.637$. The resultant length is

$$R = \sqrt{(-25.793)^2 + (31.637)^2}$$
$$= 40.819$$

which, when divided by the sample size, yields a mean resultant length of

$$\overline{R} = 40.819/51$$
$$= 0.800$$

Since the computed value of $\overline{R}$ far exceeds the critical value, we reject the null hypothesis that the concentration parameter is equal to zero. The striations must have a preferred trend.

Test for a Specified Trend   On some occasions we may wish to test the hypothesis that the observations correspond to a specified trend. For example, the area of Finland where the measurements of glacial striations were taken occurs within a broad topographic depression aligned northwest-southeast at approximately 105°. Does the mean direction of ice movement, as indicated by the striations, coincide with the axial direction of this depression?

Exact tests of the hypothesis that a sample of vectors has been taken from a population having a specified mean direction require use of extensive charts in order to set the critical value (Stephens, 1969). A simpler alternative is to determine a *confidence angle* around the mean direction of the sample, and see if this angle is sufficiently broad to encompass the hypothetical mean direction. This confidence angle is based on the standard error of the estimate of the mean direction $\overline{\theta}$, and so considers both the size of the sample and its dispersion.

Before computing the confidence angle, the Rayleigh test should be applied to confirm that a statistically significant mean direction does exist. Then, the mean vector length $\overline{R}$ must be computed and the concentration parameter κ estimated using Table 5.6. The approximate standard error of the mean direction, given in radians, is

$$s_e = 1/\sqrt{n\overline{R}\kappa}$$

Since the standard error is a measure of the chance variation expected from sample to sample in estimates of the mean direction, we can use it to define probabilistic limits on the location of the true or population mean direction. Assuming that estimation errors are normally distributed, the interval

$$\overline{\theta} \pm Z_\alpha s_e$$

should capture (or include) the true population mean direction $\alpha\%$ of the time. For example, if we collected 100 random samples of the same size from a population of vectors and computed the mean directions and 95% confidence intervals around each, we would expect that all but about five of those intervals would contain the true mean direction. Of course, we would not know which five of the intervals failed to capture the true direction, so we must assign a probabilistic caveat to all of them. We might, for example, make the statement that "the interval, plus and minus so many degrees around the mean direction of this particular sample, contains the true population mean direction. The probability that this statement is incorrect is 5%."

We have already applied Rayleigh's test and rejected the hypothesis of no trend in the observations of the striations. The approximate standard error of the mean direction can now be found:

$$s_e = \frac{1}{\sqrt{51 \cdot 0.8004 \cdot 2.87129}}$$

$$= \frac{1}{10.826}$$

$$= 0.0924 \text{ radians, or } 3.14°$$

Therefore, the probability is 95% that the interval

$$129.2° \pm 1.96 \cdot 3.14°$$

contains the population mean direction. In other words,

$$126.1° \le \bar{\theta} \le 132.3°$$

Since this interval does not include the direction of alignment of the topographic depression, we must conclude that it does not coincide with the mean direction of the striations.

Test of Goodness-of-Fit   A simple nonparametric alternative to the Rayleigh test of uniformity involves dividing the unit circle into a convenient number of angular segments. If these segments are equal in size and the observed vectors are distributed at random, we should observe approximately equal numbers of vectors in each segment. The number actually observed can be compared to those expected by a χ^2 test. The expected frequency in each segment must be at least 5, and there should be between $n/15$ and $n/5$ segments. The χ^2 is computed in the usual manner (see eq. 2.45) and has $k - 1$ degrees of freedom, where k is the number of segments.

The same procedure can be used to test the goodness-of-fit of the observed vectors to other theoretical models, such as a von Mises distribution with a specified concentration parameter κ greater than zero and a specified mean direction $\bar{\theta}$. Computing the expected frequencies, however, can be complicated. Examples are given by Gumbel, Greenwood, and Durand (1953) and Batschelet (1965).

Testing the Equality of Two Sets of Directional Vectors   We may sometimes wish to test hypotheses about the equivalence of two samples or collections of directional measurements. For example, we may have paleocurrent measurements in two different stratigraphic units, and we want to determine if their mean directions are the same. We may wish to see if the orientations of lineaments seen on a satellite image coincide with the orientations of faults known to exist in the photographed area. At a much smaller scale, we may want to compare the alignment of elongated grains in thin sections from two cored samples of sandstone from a petroleum reservoir.

The equality of two mean directions may be tested by comparing the vector resultants of the two groups to the vector resultant produced when the two sets of measurements are combined or pooled. If the two samples actually are drawn from the same population, the resultant of the pooled samples should be approximately equal to the sum of their two resultants. If the mean directions of the two samples are significantly different, the pooled resultant will be shorter than the sum of their resultants.

If κ is a large value (greater than 10), an F-test statistic can be computed by

$$F_{1,n-2} = \frac{(n - 2)(R_1 + R_2 - R_p)}{(n - R_1 - R_2)} \tag{5.45}$$

where n is the total number of observations, R_1 and R_2 are the resultants of the two samples of vectors, and R_p is the resultant of the set of vectors after the two groups have been pooled.

We can estimate the value of κ from $\overline{R}_p$, the length of the mean resultant of the two pooled samples, by the use of Table 5.6. If κ is smaller than 10 but greater than 2, then a more accurate F test is

$$F_{1,n-2} = \left(1 + \frac{3}{8\kappa}\right) \frac{(n - 2)(R_1 + R_2 - R_p)}{(n - R_1 - R_2)} \tag{5.46}$$

If κ is less than 2, special tables such as those given in Mardia (1972) are necessary.

It is also possible to test the equality of the concentration parameters of two sets of vectors, but the computations are involved. Refer to Mardia (1972) for a detailed discussion, and to Gaile and Burt (1980) for a worked example from geomorphology.

A fold belt, expressed topographically as the Naga Hills and their extensions, occurs at the juncture between the Indian subcontinent and the Indochinese peninsula. Apparently related to compressive movements that created the Himalayas, the fold belt includes a series of subparallel anticlines along the eastern border of Bangladesh. Oil and gas have been found in structural traps in this region, so delineation of the folds is of economic as well as scientific interest. Presumably the folds occur, perhaps with reduced magnitude, to the west of the Naga Hills, but are concealed by modern sediments deposited by the Ganges River and its tributaries. Unfortunately, reflection seismic data that could reveal the buried structures are very sparse.

Interpretations of Landsat satellite images of this region indicate numerous lineations of unknown origin. It is possible that the lineations reflect subsurface folds, and if so, they may provide valuable clues to structural geology and possible petroleum deposits.

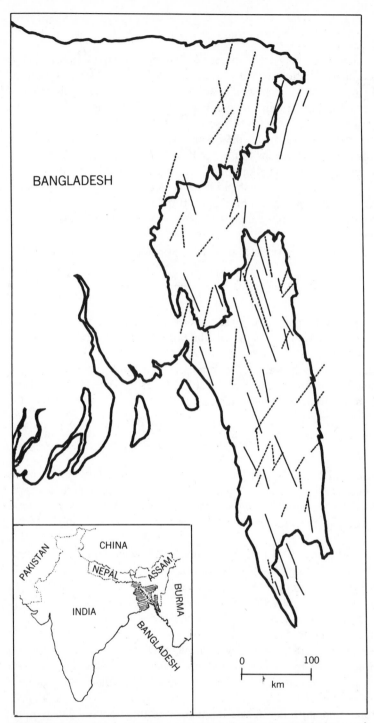

FIGURE 5.25 Map of eastern Bangladesh, showing axial planes of major anticlines (solid lines) and large lineaments interpreted from Landsat images (dashed lines).

Figure 5.25 is a map of eastern Bangladesh, showing the traces of axial planes of major exposed anticlines, and the larger lineations measured on Landsat images. The orientations of these two sets are shown on Figure 5.26. Because the lines have no sense of direction, the plots are bimodal, and we must double the observed angles to obtain the correct distribution of vectors. Table 5.8 lists the orientations of both the axial planes and the lineaments. There is an obvious difference between the two sets, but is this difference statistically significant or could it have arisen through the vagaries of sampling?

To test the hypothesis that the mean directions of the anticlinal axes and the Landsat lineaments are the same, we must first compute the resultants of each of the two groups and the resultant of the two combined. The resultant of the 32 doubled measurements of axial planes is $R_1 = 32.09$, and the resultant of the 40 doubled measurements of the Landsat lineament is $R_2 = 27.81$. The two groups can be combined into a pooled collection of 72 observations that has a resultant of $R_p = 54.00$. The mean resultant of the pooled group is

$$\overline{R}_p = \frac{54}{72} = 0.75$$

and by use of Table 5.6 we can estimate the concentration factor as $\kappa = 2.3693$.

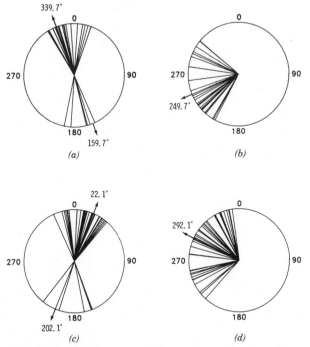

FIGURE 5.26 Circular plots of orientation data from eastern Bangladesh. The mean orientations are indicated. (a) Orientations of axial planes of major anticlines. (b) Angles of axial planes doubled. (c) Orientations of Landsat image lineations. (d) Angles of lineations doubled.

TABLE 5.8 Orientation of Axial Planes of Anticlines and Landsat Lineations in Eastern Bangladesh; Measurements are Given in Degrees Clockwise from North

Anticlinal Axes $n = 32$		Landsat Lineaments $n = 40$	
103.5	294.6	248.9	256.4
288.5	287.2	283.8	232.0
282.2	299.4	247.8	275.8
265.7	290.9	258.7	244.8
256.8	300.1	275.6	254.4
253.6	273.4	238.9	230.5
249.8	294.4	228.7	252.0
250.0	291.6	239.5	257.7
107.9	113.8	277.2	255.3
287.9	290.9	245.5	235.4
291.7	290.1	277.9	281.0
287.1	279.2	236.3	239.4
283.5	265.6	235.9	261.9
301.0	259.7	233.7	257.1
281.3	257.9	293.8	100.8
291.2	104.8	246.8	247.0
		266.4	110.2
		279.4	109.0
		257.4	251.4
		266.1	230.9

Since κ is greater than 2 but less than 10, the appropriate test statistic is given by eq. (5.46). Substituting values we have calculated into that equation gives

$$F = \left(1 + \frac{3}{8(2.3693)}\right)\left(\frac{72 - 2)(32.09 + 27.81 - 54.00)}{(72 - 32.09 - 27.81)}\right)$$

$$= 39.59$$

The test has $\nu_1 = 1$ and $\nu_2 = (72 - 2)$ degrees of freedom. From Table 2.14 we can interpolate to find a critical value for F at the 5% level of significance ($\alpha = 0.05$), with 1 and 70 degrees of freedom; the value is 3.96. Since the test value far exceeds the critical value, we must regretfully conclude that the Landsat lineaments and the fold axes are not drawn from a common population. Although Landsat lineaments may be useful guides for exploration, in this region they apparently do not reflect the trends of structural folds.

Spherical Distributions

Statistical tests of directional data distributed in three dimensions have been developed only in recent years, in part because the mathematics of the distributions are very complicated. However, geologic problems that involve three-dimensional vectors are exceedingly common, and we should not shy away from the use of the available statistical techniques for their interpretation. Some of these methods require matrix algebra, although the matrices are not large, and the extraction of eigenvalues and eigenvectors. The geometric interpretation of eigenvectors presented in Chapter 3 will be of direct application. The mathematics are closely related to multivariate procedures described in Chapter 6. Here we deal with three physical dimensions; later we will apply the same steps to the analysis of multidimensional data in which each "dimension" is a different geologic variable.

Examples of three-dimensional directional data in the Earth sciences include: measurements of strike and dip taken for structural analyses; vectoral measurements of the geomagnetic field; directional permeabilities measured on cores from petroleum reservoirs; measurements of orientation and dip of crossbeds; and determinations of crystallographic axes for petrofabric studies.

As with two-dimensional data, we must first establish a standard method of notation. We can regard three-dimensional directional observations as consisting of vectors; since we are concerned primarily with their angular relationships, these can be considered to be of unit length. If all of the directional measurements from an area are collected together at a common origin, the tips of the unit vectors will lie on the surface of a sphere; hence the term *spherical distribution*.

Some oriented features do not have a sense of direction and can be referred to as *axes*. Examples include the lines of intersections between sets of dipping planes, axes of revolution, and the perpendiculars to planes. In addition, it is sometimes advantageous to disregard the directional aspect of vectors and to treat them as axes.

Standard mathematical notation utilizes three Cartesian coordinates to describe a vector in space (Fig. 5.27a). The direction of the vector OP is specified by the cosines of the angles between the vector and each of the coordinate axes. The coordinates of the point P are equal to

$$X = \cos a, \qquad Y = \cos b, \qquad Z = \cos c$$

Since the vector is considered to have unit length,

$$X^2 + Y^2 + Z^2 = 1$$

Using spherical angles, we can define the direction of vector OP by ϕ, the angle between the X axis and the projection of the vector on the X-Y plane, and by θ, the inclination with respect to the Z axis (Fig. 5.27b). In effect, θ defines the "latitude" of the vector, while ϕ defines its "longitude." The relationship between these spherical polar coordinates and Cartesian coordinates is

$$X = \sin \theta \cos \phi, \qquad Y = \sin \theta \sin \phi, \qquad Z = \cos \theta$$

Measurements of geologic properties are often given in terms of strike and dip, rather than as three-dimensional directional cosines or Cartesian coordinates. Also, the coordinate notation used by geologists differs from that conventionally used in mathematics. If we regard the positive end of the X axis as corresponding to the north, the positive end of the Y axis as corresponding to east, and the positive end

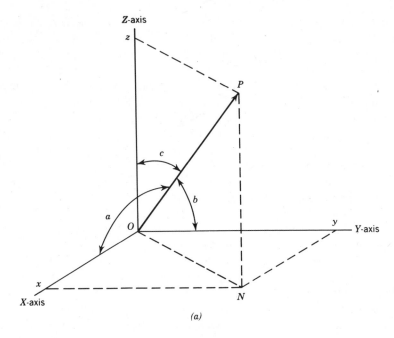

(a)

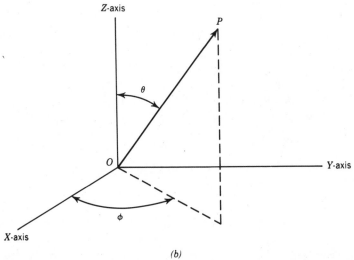

(b)

FIGURE 5.27 Notational system for vectors in three-dimensional space. (a) *OP* is a vector in space defined by Cartesian axes *X, Y,* and *Z.* Angles between *OP* and the axes are *a, b,* and *c.* (b) Vector *OP* in space defined by spherical angles ϕ and θ.

of the Z axis as vertically downward, we have defined a Cartesian system in which dips are expressed as positive angles (Mardia, 1972).

The notation is illustrated in Figure 5.28, for a vector OP defined by the strike and dip of its enclosing plane. The line ON is the *azimuth,* or projection of OP onto the horizontal X-Y plane; it is perpendicular to the line of strike. The angle A is the *angle of strike,* measured clockwise from $0°$ at the north. D is the *dip,* measured as a positive angle from ON downward. The X, Y, and Z coordinates of P are

$$X = -\sin A \sin D, \qquad Y = \cos A \sin D, \qquad Z = \sin D \qquad (5.47)$$

Conversion becomes more complicated if strike is measured in quadrant convention, which requires specifying the direction of dip. Refer to Watson (1970) for a more complete discussion.

Once we have the spherical measurements in terms of X, Y, and Z coordinates of the vector end points, it is a simple matter to compute the mean direction and spherical variance. This is done in a manner analogous to the computation of the circular mean and variance. The mean direction is given by the resultant, R, of the unit vectors. Its length is

$$R = \sqrt{\left(\sum X_i\right)^2 + \left(\sum Y_i\right)^2 + \left(\sum Z_i\right)^2} \qquad (5.48)$$

In normalized form, this is $\overline{R} = R/n$.

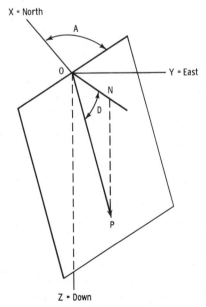

FIGURE 5.28 Strike and dip notation for vectors in three dimensions. Angle *A*, measured clockwise from north, is the strike of the surface containing vector *OP*. The plane *ONP* is perpendicular to the dipping surface. The angle *D* is the dip.

The direction of the resultant with respect to the three coordinate axes is given by the cosines of the angles between the resultant and these axes:

$$\cos \overline{X} = \sum X_i/R, \qquad \cos \overline{Y} = \sum Y_i/R, \qquad \cos \overline{Z} = \sum Z_i/R \quad (5.49)$$

If the observations are tightly clustered around a common direction, the resultant R will be a large number, approaching n. If the observations are scattered, R will be small. As in the case of circular distributions, R can be used as a measure of concentration and can be expressed as the *spherical variance:*

$$s_s^2 = (n - R)/n \qquad (5.50)$$
$$= (1 - \overline{R})$$

These methods for determining the mean direction and spherical variance work well if the vectors are not too widely dispersed. Under certain conditions, however, the mean direction may be misleading. Suppose the dips of nearly flat-lying beds are measured; some dip gently to the west, others a few degrees to the east. Since dip is taken as a vector direction pointing into the lower hemisphere, the vector resultant of the east and west dips will be vertically downward! Of course, the length of the resultant will be near zero so the spherical variance will be large, indicating an extremely high dispersion among the vectors.

If these dips are regarded as nondirectional axes rather than vectors, their two ends project into both the upper and lower hemispheres; it is apparent then that the lines representing the east and west dips are closely related. The mean axis, computed using the eigenvector technique described below, will be horizontal and pass through the bundle of dip axes.

Matrix Representation of Vectors

You will recall from Chapter 3 that the rows of a matrix can be represented graphically by vectors. Conversely, measurements of vector directions can be expressed in matrix form. The eigenvalues and eigenvectors of such a matrix will provide information about the arrangement of the vectors in space. However, in order to express a set of vectors in the appropriate matrix form we must first review a few points of geometry, starting in two-dimensional space.

The geometric relationship that gives the projection of one vector onto another is the scalar product of the two vectors (Fig. 5.29). If the two are assumed to be unit vectors, the Cartesian coordinates of their end points are the same as their directional cosines with respect to the X and Y axes. The projection is

$$l = au + bv \qquad (5.51)$$

where l is the length of the projection of vector a,b onto vector u,v. (It is also the length of the projection of u,v onto a,b).

As can be seen in Figure 5.29, the vector a,b is the hypotenuse of a right triangle whose sides are the projection l on vector u,v and the perpendicular distance d. The Pythagorean theorem defines the relationship between these sides, and can be

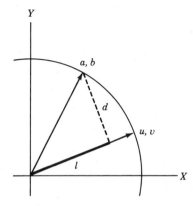

FIGURE 5.29   Projection of vector a,b onto vector u,v. The length of projection of a,b onto u,v is l. The distance of vector end point a,b from the vector u,v is d.

rewritten for this instance as

$$d^2 = 1 - l^2 \qquad (5.52)$$
$$= 1 - (au + bv)^2$$

Any number of vectors can be projected onto the line u,v, by eq. (5.51), and their squared distances from the line u,v determined by eq. (5.52). The sum of these squared distances is

$$M = \sum_{i=1}^{n} d_i^2 \qquad (5.53)$$
$$= n - \sum_{i=1}^{n} (a_i u + b_i v)^2$$

M can be regarded as a *moment of inertia* of the end points of the vectors around the line u,v. This equation can be further generalized to three dimensions by introducing the third spatial coordinate

$$M = \sum_{i=1}^{n} d_i^2 \qquad (5.54)$$
$$= n - \sum_{i=1}^{n} (a_i u + b_i v + c_i w)^2$$

It is possible to express eq. (5.54) in matrix form. First, the coordinates of the line are given as a column vector $[U]$

$$[U] = \begin{bmatrix} u \\ v \\ w \end{bmatrix}$$

We also define the matrix $[B]$

$$[B] = n[I] - [T]$$

where $[T]$ is a 3×3 matrix of the sums of squares and cross-products of the direction cosines of the vectors:

$$[T] = \begin{bmatrix} \Sigma a_i^2 & \Sigma a_i b_i & \Sigma a_i c_i \\ \Sigma b_i a_i & \Sigma b_i^2 & \Sigma b_i c_i \\ \Sigma c_i a_i & \Sigma c_i b_i & \Sigma c_i^2 \end{bmatrix}$$

The matrix $[B]$ therefore has the form

$$[B] = \begin{bmatrix} n - \Sigma a_i^2 & \Sigma a_i b_i & \Sigma a_i c_i \\ \Sigma b_i a_i & n - \Sigma b_i^2 & \Sigma b_i c_i \\ \Sigma c_i a_i & \Sigma c_i b_i & n - \Sigma c_i^2 \end{bmatrix}$$

The moment of inertia of the vectors about the direction $[U]$ is simply

$$M = [U]'[B][U]$$

Rather than determining the moment around an arbitrary line $[U]$, we may find a unique line around which the moment of inertia will be the maximum possible. The coordinates of this line are given by the first eigenvector of the matrix $[B]$. If λ_1 is the first eigenvalue of $[B]$ and $[b_1]$ is its associated eigenvector, then you will recall from Chapter 3 that

$$\lambda_1 = [b_1]'[B][b_1]$$

That is, λ_1 is the moment of inertia of the vectors around the first eigenvector. This means that the sum of the squared distances from the tips of the vectors to the first eigenvector is the maximum possible, or that the eigenvector is simultaneously as nearly perpendicular to all of the vectors as it is possible to be.

The moment of inertia around the second eigenvector is the greatest possible for any line that is orthogonal to the first eigenvector. The third eigenvector must be orthogonal to the other two, and must also account for all of the remaining squared

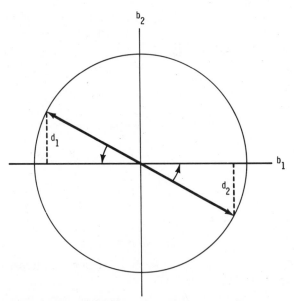

FIGURE 5.30   Projection of two diametrically opposite vectors onto eigenvector $[b_1]$. The distances d_1 and d_2 are identical and act in the same rotational sense.

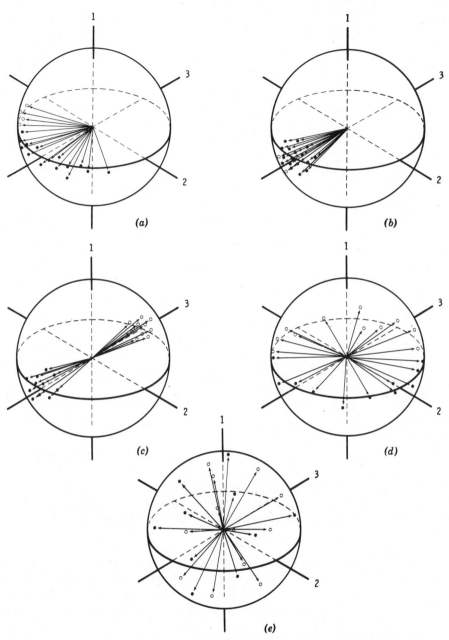

FIGURE 5.31 Patterns of vectors in the unit sphere. (a) Partial girdle pattern in the plane containing eigenvectors 2 and 3. (b) Unimodal distribution of vectors around eigenvector 3. (c) Bimodal distribution of vectors around eigenvector 3. (d) Complete girdle pattern in the plane containing eigenvectors 2 and 3. Their eigenvalues are identical or nearly so. (e) Uniform distribution. Eigenvalues are all approximately equal.

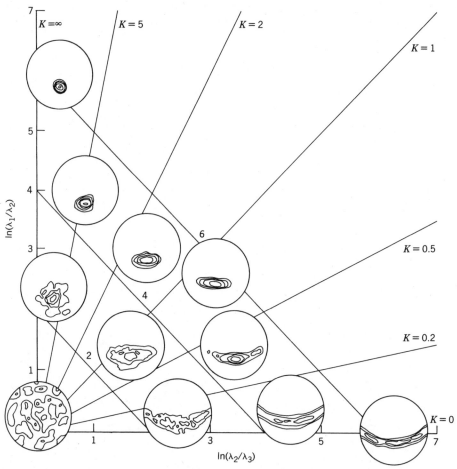

FIGURE 5.32 Classification of patterns of vectors on the unit sphere, according to the logarithms of the ratios of their eigenvalues. Typical petrofabric diagrams are shown for different ratios. K is the ratio $\ln(\lambda_1/\lambda_2)/\ln(\lambda_2/\lambda_3)$. Adapted from Woodcock (1977).

distances to the vector tips. Since the three eigenvectors define an orthogonal framework fully equivalent to the original set of Cartesian axes, the third eigenvector must define a line along which the moment of inertia is a minimum. That is, it will be oriented so as to be simultaneously as close as possible to all of the vectors.

If two vectors are diametrically opposite, as in Figure 5.30, they both will be the same perpendicular distance away from the eigenvector [b] and will have exactly the same influence on the location of the eigenvector. This means that the sense of direction of the vectors is lost; they are indistinguishable from axes. For this reason, eigenvector methods are preferable for the examination of spherically distributed data in those instances where ambiguity may result from an arbitrary distinction between vectors in the upper and lower hemisphere.

The eigenvalues provide direct information about the distribution of the vectors. Mardia (1972) distinguishes four cases.

1. λ_1 is large, while λ_2 and λ_3 are both small. This means that the sum of the squares of the perpendiculars between the vector end points and the axis corresponding to the first eigenvector is very large. Most of the observations must lie in the plane containing eigenvectors 2 and 3, forming a girdle distribution (Fig. 5.31a).
2. λ_1 and λ_2 are both large, while λ_3 is small. The perpendicular distances from the end points to the first and second eigenvectors must be very large, but the distances to the third eigenvector must be small. The observations are clustered around the end of the third eigenvector (Fig. 5.31b,c). Either a bimodal or unimodal distribution will yield the same result; they can be distinguished by the value of $\bar{R}$, which will be large for the unimodal case.
3. Two eigenvalues are identical. This is actually a variation of case 1. The observations form a symmetrical girdle around the axis corresponding to the unique eigenvalue (Fig. 5.31d).
4. All three eigenvalues are identical. The distribution is uniform, as the perpendicular directions from the points are the same for all three orthogonal axes. There is no preferred arrangement of the points on the unit sphere (Fig. 5.31e).

Woodcock (1977) has generalized this classification by graphing logs of ratios of the eigenvalues [ln (λ_1/λ_2) versus ln (λ_2/λ_3)]. On his diagrams, all possible patterns of points on a sphere fall into specific regions. This form of graphical analysis may be especially useful with petrofabric data. Figure 5.32 shows one of Woodcock's diagrams.

Displaying Spherical Data

Although perspective drawings of unit spheres such as those in Figure 5.31 are useful for illustrative purposes, they cannot convey detailed information about the distribution of the vectors. Conventionally, three-dimensional vectors are shown by projecting their end points onto a plane. Since the points actually lie on the surface of a sphere, portraying them in two dimensions requires use of a projection equation. Geologists traditionally have used the equal-area polar Lambert projection, which is referred to as a "Schmidt net." Crystallographers have preferred the equal-angle polar stereographic projection, or "Wulff net."

Figure 5.33 shows a set of vectors within the unit sphere and their projection onto an equal-area diagram. It is necessary to distinguish between vectors that go into the lower hemisphere and those that go into the upper. Because geologists often describe vectors in terms of their "plunge," a word with "downward" connotations, they conventionally plot the lower hemisphere of the unit sphere.

In addition to vectors, it is sometimes necessary to plot the three-dimensional orientations of planes such as faults and fracture surfaces. If a plane is placed through the center of the unit sphere, its intersection with the sphere will form a great circle (Fig. 5.34a). However, it is easier to represent the plane by an axis

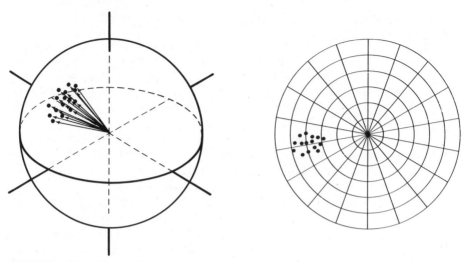

FIGURE 5.33 Vectors within the unit sphere and their projection onto an equal-area polar diagram.

(called a "pole") that is perpendicular to the plane at the origin. Conventionally, geologists plot the intersection of a pole with the lower hemisphere, although it seems more logical to plot its intersection with the upper hemisphere. Then, the projection of the pole to a plane that dips to the west, for example, will plot on the left-hand or "western" side of a diagram (Fig. 5.34b).

Very large sets of three-dimensional data can result in diagrams that contain so

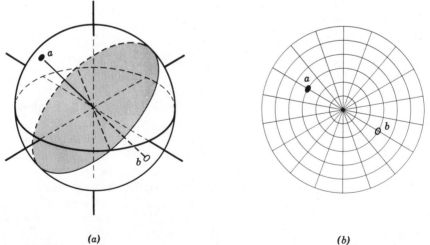

(a)   *(b)*

FIGURE 5.34 Plotting of poles to planes. (a) Plane and its poles in the unit sphere. (b) Poles to a plane projected onto an equal-area diagram. Point a is the projection onto the upper hemisphere, point b is the projection onto the lower hemisphere.

many points that their general pattern cannot be seen. In such instances the local density of points can be contoured, by counting the number of points that lie within a small area of the diagram. This can be done conveniently only if an equal-area projection is used. The projection is covered with a regular array of grid nodes and the number of points within a fixed radius of each node is counted. Usually the radius is set so that it describes a circle containing 10% of the total area. Since the grid nodes are closer together than the radius, the successive areas overlap and there tends to be a gradual change in density from one part of the diagram to another. The small diagrams on Figure 5.32 show typical contoured patterns obtained in petrofabric studies.

Testing Hypotheses about Spherical Directional Data

The simpler tests of three-dimensional orientation are extensions of those used with circular data. As in that instance, we need a probability model of known characteristics against which we can test. A widely used model is the *Fisher distribution*—an extension of the von Mises distribution and a spherical equivalent of the normal curve. The Fisher distribution is characterized by two parameters: a mean vector direction $\bar{\theta}$ and a dispersion κ. Because we are dealing with three dimensions, the mean vector has three elements, each a direction cosine with respect to the three coordinate axes.

The mean vector is estimated by the direction cosines of the resultant (eq. 5.49). The dispersion can be approximated by

$$\kappa = (n - 2)/(n - R) \qquad (5.55)$$

which is sufficiently accurate if κ is "large," or greater than about 10. Mardia (1972) gives a table of more exact estimates based on the size of the normalized resultant, $\bar{R}$.

A Test of Randomness   As in the circular case, the simplest hypothesis that can be tested is that the data are distributed uniformly in all directions, which is equivalent to stating that the concentration parameter κ is zero.

The null hypothesis and alternative are

$$H_0 : \kappa = 0$$

$$H_1 : \kappa > 0$$

The test statistic is calculated in the same manner as with circular data; it consists of the normalized resultant $\bar{R}$. This statistic is then compared to a critical value of $\bar{R}$ for the desired level of significance. A table of critical values is given in Table 5.9. If the calculated value of $\bar{R}$ exceeds the table value, the hypothesis that the observations are uniformly distributed is rejected at the specified level of significance.

It is also possible to test hypotheses about the specific orientation of the mean vector, and to construct a "cone of confidence" around the vector. These tests, however, require extensive tables such as those published by Stephens (1967) and

TABLE 5.9 Critical Values of $\overline{R}$ for the Test of Uniformity of a Spherical Distribution

Number of Observations, n	Significance Level, α (%)			
	10	5	2	1
5	0.637	0.700	0.765	0.805
6	.583	.642	.707	.747
7	.541	.597	.659	.698
8	.506	.560	.619	.658
9	.478	.529	.586	.624
10	.454	.503	.558	.594
11	.433	.480	.533	.568
12	.415	.460	.512	.546
13	.398	.442	.492	.526
14	.384	.427	.475	.507
15	.371	.413	.460	.491
16	.359	.400	.446	.476
17	.349	.388	.443	.463
18	.339	.377	.421	.450
19	.330	.367	.410	.438
20	.322	.358	.399	.428
21	.314	.350	.390	.418
22	.307	.342	.382	.408
23	.300	.334	.374	.400
24	.294	.328	.366	.392
25	.288	.321	.359	.384
30	.26	.29	.33	.36
35	.24	.27	.31	.33
40	.23	.26	.29	.31
45	.22	.24	.27	.29
50	.20	.23	.26	.28
100	.14	.16	.18	.19

From Mardia (1972).

Mardia (1972). Two-sample tests for the equivalency of the mean directions of two sets of observations can also be performed. The necessary tables are also given in Stephens (1969) and Mardia (1972).

Shape

Shape is an extremely difficult property to measure, or even to define in a precise manner. Perhaps this is why there are so many proposed shape measures, none of which have proved entirely satisfactory. A shape measure should possess several desirable properties. Obviously, objects with different shapes should yield different measures, and similar shapes should yield similar values regardless of the size or orientation of the objects. Unfortunately, a shape measure possessing these properties may be a chimera; it has been proven mathematically that no single measure can be unique to only one shape (Lee and Sallee, 1970).

Earth scientists have attempted to characterize a broad spectrum of shapes, ranging from relatively simple forms such as the projected outlines of sand grains to the more complex patterns presented by fossil organisms. Geomorphologists have been especially prolific in the creation of shape measures, applying them to the study of drainage basins, drumlins, and coral atolls, among other landforms. The shapes of oil fields, as defined by their axial ratios, have been investigated, as have the shapes of certain types of structural traps. Extensive literature reviews are given by Moellering and Rayner (1979) and by Clark (1981); both articles also discuss the theoretical aspects of shape measurement.

Table 5.10 contains a selection of single-value measures of shape, gleaned from the geologic and geographic literature. This list is by no means exhaustive. Most of these are calculated from a few basic measurements, such as axial lengths, perimeters, and areas. Some involve comparisons with standard forms, such as a circle.

Any of these shape measurements can be used in the same manner as any other descriptor. Summary statistics, such as means and variances, can be calculated for the shape measurements from a collection of objects, although there is no guarantee that the measurements will follow a normal distribution.

Fourier Measurements of Shape

The relative merits of alternative shape measurements have been argued with some heat, and perhaps still provide a fruitful area for a master's thesis or two. However, we will turn our attention to the more promising shape descriptors which are multivalued. Among these are various modifications of the Fourier transformation used in Chapter 4 to analyze time series.

The coordinates of a closed line, such as an outline of the projection of a sand grain or a fossil shell, can be expressed in polar form as in Figure 5.35. One of the two coordinates is the angular orientation of a radius extending from a point within the outline; the other is the distance along this radius from the central point to the outline.

This immediately introduces a problem, because we must be able to specify the placement of the point within the object's outline that defines the center of the polar coordinate system. If the center is moved, the distances along all the radii will change, and the Fourier transform will be different as well. If we wish to compare several different shapes, we must identify equivalent points within each of them as the center of the coordinate system. If this is not done, we will not be able to tell if the differences we see between Fourier spectra result from differences in shape, or differences in our choice of the origin.

For some applications, unique points can be identified within each shape that can serve as the origin of the polar coordinate system. This was done, for example, by Kaesler and Waters (1972), who measured radii extending from a characteristic muscle scar in ostracodes to the outlines of their shells. Unfortunately, most shapes

TABLE 5.10 Measures of Shape Used in the Geologic and Geographic Literature; Only Nondimensional Measures are Listed

---

1. Measures based on axial ratios

   Form $\qquad\qquad F = \dfrac{l}{w}$

   Elongation $\qquad\quad E = \dfrac{w}{l}$

   Circularity $\qquad\quad C_1 = \sqrt{\dfrac{lw}{l^2}}$

2. Measures based on perimeters

   Grain Shape Index $\qquad GSI = \dfrac{p}{l}$

   Shape Factor $\qquad SF_1 = \dfrac{p_c}{p}$

   $\qquad\qquad\qquad\quad SF_2 = \dfrac{p}{p_c} \times 100$

3. Measures including both perimeters and areas

   Circularity $\qquad\quad C_2 = \dfrac{4A}{p^2}$

   $\qquad\qquad\qquad\quad C_3 = \dfrac{4A}{lp}$

   Compactness $\qquad K_1 = \dfrac{2\sqrt{\pi A}}{p}$

   $\qquad\qquad\qquad\quad K_2 = \dfrac{p^2}{4\pi A}$

   Thinness Ratio $\qquad TR = 4\pi \left(\dfrac{A}{p^2}\right)$

4. Measures based on areas

   Circularity $\qquad\quad C_4 = \sqrt{\dfrac{A}{A_c}}$

   Shape Factor $\qquad SF_3 = \dfrac{A_i}{A}$

   $\qquad\qquad\qquad\quad SF_4 = \dfrac{A_c - A_i}{A}$

   $\qquad\qquad\qquad\quad SF_5 = \dfrac{A}{A_c} \times 100$

5. Measures based on areas and areal lengths

   Form Ratio $\qquad\quad FR = \dfrac{A}{l^2}$

   Ellipticity Index $\qquad EI = \dfrac{\pi(1/2l)l}{A}$

TABLE 5.10 (*Continued*)

6. Other measures

Circularity	$C_s = \sqrt{\dfrac{D_i}{D_c}}$
Mean Radius	$\bar{R} = \dfrac{\Sigma R_j}{n}$
Radial Variance	$s_R^2 = \dfrac{\Sigma(R_j - \bar{R})^2}{n}$
Mean Side	$\bar{S} = \dfrac{\Sigma S_j}{n}$
Variance of Side	$s_s^2 = \dfrac{\Sigma(S_j - \bar{S})^2}{n}$

Original sources for these measures are given in Folk (1968), and Moellering and Rayner (1979).
Key:
A—Area of object
A_c—Area of smallest enclosing circle
A_i—Area of largest inscribed circle
D_c—Diameter of smallest enclosing circle
D_i—Diameter of largest inscribed circle
l—Length of long axis
p—Perimeter of object
p_c—Perimeter of a circle having same area as object
R_j—jth radius of object, measured from centroid to edge
S_j—Length of jth side of object, considered as a polygon
n—Number of sides, considered as a polygon
w—Width of object perpendicular to long axis

such as sand grains, pebbles, or the outlines of salt domes lack distinctive points that could serve as common centers of reference. We can, however, compute a unique point within any closed shape that can serve as the origin of the coordinate system.

The *centroid* represents the center of gravity of a form, and is unique for each object. If the periphery of an object is represented by a series of Cartesian coordinates, such as those generated by a digitizer, the centroid can be found by integrating the X and Y coordinates. A simple procedure for integration uses trapezoidal approximation.

A series of points have been placed around the periphery of an object in Figure 5.36. Pairs of these points can be used in combination with the axes to define trapezoids. From an arbitrary starting point, we construct a series of trapezoids, moving counterclockwise around the figure. The center of gravity of each trapezoid can be found, and these combined to yield the center of gravity of the figure. The coordinates of the centroid of the figure are

$$\bar{X} = \frac{\Sigma \bar{X}_t A_t}{\Sigma A_t} \tag{5.56}$$

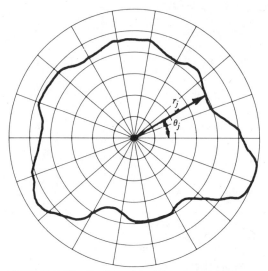

FIGURE 5.35 Projected outline of sand grain expressed in polar coordinates. Coordinate pairs are formed as the length r_j and angle θ_j of radii drawn from the centroid to the edge.

and

$$\overline{Y} = \frac{\Sigma \overline{Y}_t A_t}{\Sigma A_t} \tag{5.57}$$

where $\overline{X}_t$ is the X coordinate of the centroid of the tth trapezoid, $\overline{Y}_t$ is its Y coordinate, and A_t is the area of the trapezoid.

The centroids of the trapezoidal areas can be found in turn by simple geometry.

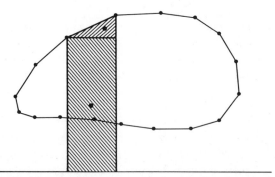

FIGURE 5.36 Determination of centroid by trapezoidal approximation. Digitized form is divided into trapezoids consisting of a rectangle and a right triangle. Points indicate their centroids.

Each trapezoid can be broken down into a rectangle and a right triangle. The rectangular portion has its centroid located at a point midway between each of the four sides. The centroid of the triangular portion is located at coordinates one-third of the way between the right angle and the two acute angles. The centroids of the two portions are combined by weighting each according to their respective areas. These operations can be combined and simplified into the following expressions which provide the terms needed in eqs. (5.56) and (5.57):

$$A_t = \frac{1}{2} (Y_{i+1} + Y_i)(X_i - X_{i+1})$$

$$\overline{X}_t A_t = \frac{1}{6} (X_{i+1}^2 + X_{i+1}X_i + X_i^2)(Y_{i+1} - Y_i)$$

$$\overline{Y}_t A_t = \frac{1}{6} (Y_{i+1}^2 + Y_{i+1}Y_i + Y_i^2)(X_i - X_{i+1})$$

The procedure can be performed rapidly by computer, yielding the location of the centroid of a complicated figure. The accuracy of the location depends upon the number of points placed around the perimeter.

We have now established the centroid of the figure, and can proceed to draw radii from the centroid to its perimeter. The lengths of these radii, together with their angular orientation, provide the polar coordinate pairs which can be analyzed by the Fourier techniques introduced in the preceding chapter. The analysis will yield the spectrum of the closed figure, and from the spectrum we may deduce many interesting things about the shape of the figure. The circular Fourier spectrum has all of the desirable properties of the ordinary Fourier spectrum: it contains all of the information contained in the original figure, the successive harmonics are independent of one another, and each spectral value is a measure of the "contribution," or amount of the total variance added, by the corresponding harmonic wave form.

In order to make the circular Fourier transformation computationally tractable, the radii must be spaced at equal angular increments, just as in ordinary Fourier analysis samples must be equally spaced in time or distance. Unfortunately, it is unlikely that the original coordinates placed around the perimeter of the figure and used to determine the centroid will be in the proper locations to form equal angles when these points are connected to the centroid. We must find a new set of points along the perimeter that do define a set of equal angles with the centroid, either by remeasuring or by interpolation between the existing points.

However, either procedure introduces a new complication, because a centroid calculated from the new set of points is unlikely to coincide exactly with the centroid determined from the original set of points. Unless the radii are measured from the true centroid, a distortion will be introduced into the Fourier spectrum. The effect is analogous to an off-center wheel, which will "wobble" up and down once every revolution. This wobble will contribute to the first harmonic, which otherwise will be zero. Since spectra usually are standardized for comparative purposes, the presence of a spurious first harmonic will reduce the relative magnitudes of successive harmonics. Boon, Evans, and Hennigar (1982) describe an iterative procedure that produces successively closer approximations to the true centroid of a set of coordinates spaced at equal angles around the perimeter of an object.

Converting the coordinate system from rectangular to polar form has the effect of "unrolling" the closed outline (Fig. 5.37). The techniques of ordinary Fourier analysis are obviously applicable to the unrolled figure.

You will recall from Chapter 4 that the Fourier equation can be written as

$$Y = \sum_{k=0}^{\infty} \alpha_k \cos(k\theta) + \beta_k \sin(k\theta) \tag{5.58}$$

and its coefficients estimated as

$$\alpha_k = \frac{2}{k} \sum_{j=1}^{n} Y_j \cos\left(\frac{2\pi jk}{n}\right) \tag{5.59}$$

and

$$\beta_k = \frac{2}{k} \sum_{j=1}^{n} Y_j \sin\left(\frac{2\pi jk}{n}\right) \tag{5.60}$$

These same equations can be modified to find the polar Fourier transform of a closed shape. If we place n radii at equal angles around a complete circle, the angle between each pair of radii is $2\pi/n$. Therefore, the angle from the origin to the jth radius is $2\pi j/n$. Designating the angular direction of the jth radius as θ_j, the length of the jth radius as r_j, and substituting into eqs. (5.59) and (5.60):

$$\alpha_k = \frac{2}{k} \sum_{j=1}^{n} r_j \cos k\theta_j \tag{5.61}$$

$$\beta_k = \frac{2}{k} \sum_{j=1}^{n} r_j \sin k\theta_j \tag{5.62}$$

As in single Fourier analysis, some of the terms can be simplified. The coefficient

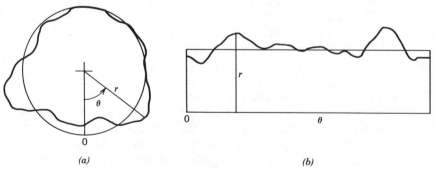

(a)   *(b)*

FIGURE 5.37 Equivalency between polar and Cartesian representations. (a) Grain outline shown in polar notation. (b) Polar coordinates plotted as r versus θ.

β_0 is equal to zero, and α_0 is equal to the mean radius

$$\bar{r} = \frac{\sum_{j=1}^{n} r_j}{n} \tag{5.63}$$

The effect of size can be eliminated from an analysis by dividing all radii by the mean radius; then, α_0 will always be equal to one. For the reason discussed earlier, the coefficients of the first harmonic, α_1 and β_1, should be zero if the radial measurements are taken from the true centroid.

Once the α and β coefficients have been determined for a series of harmonics, the interpretation of the circular Fourier spectrum proceeds in a manner directly analogous to the interpretation of a conventional Fourier spectrum. The amplitudes and phase angles of the polar harmonics can be found from the Fourier coefficients:

$$A_k = \sqrt{\alpha_k^2 + \beta_k^2} \tag{5.64}$$

$$\phi_k = \tan^{-1}\left(\frac{\beta_k}{\alpha_k}\right) \tag{5.65}$$

These relationships are exactly the same as those given in an earlier section for the spectrum of a discrete time series.

It is usually most convenient to combine the α and β coefficients to form the power or variance spectrum. This directly expresses the contribution made by each harmonic to the shape of the figure. Successive approximations of the analyzed form can be constructed by inserting the α and β coefficients into the Fourier equation. As harmonics are added, the reconstruction will become increasingly detailed until it exactly matches the original digitized outline. (This will occur at the Nyquist frequency, when $k = (n - 1)/2$.) Figure 5.38 shows the reconstruction of the shape of a sand grain. The zeroth harmonic produces a circle whose diameter is equal to the mean diameter of the original outline. There is no first harmonic because the grain has been properly centered around its centroid. The second harmonic changes the form of the reconstruction to an ellipse, the third harmonic adds a "triangular" component, the fourth a "square" component, and so on. Most simple closed figures can be closely replicated with fewer than ten harmonics (Fig. 5.39).

The proportion of the variation in shape accounted for by the successive harmonics is given by the power spectrum. Usually, a small number of the lowest harmonics will account for almost all of the variation in simple forms such as the projections of sand grains (Fig. 5.40). The higher harmonics reflect smaller and smaller details of the outline, and have been used by sedimentologists as measures of "surface roughness." The higher harmonics of the outlines of quartz sand grains are said to reflect provenience or transport history (many studies are cited in Ehrlich, Brown, and Yarus, 1980), but caution should be used in making such interpretations from parts of the Fourier spectrum. The standard errors of the estimates of Fourier coefficients are very large, being of the same order of magnitude as the estimates themselves. The higher harmonics of grain outlines typically have extremely low

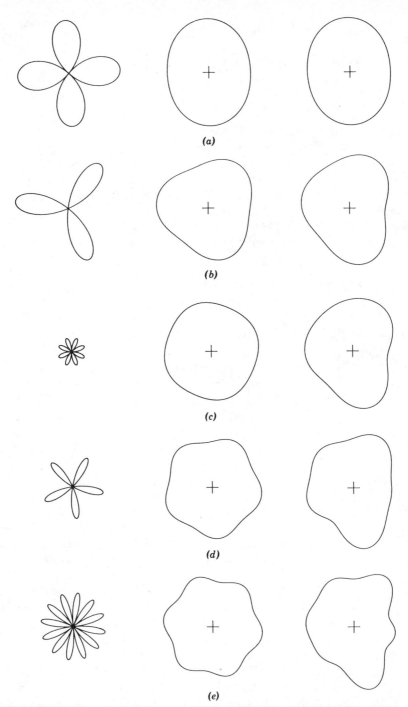

FIGURE 5.38 Reconstruction of the outline of a sand grain by polar Fourier series. (*a*) Plot of second harmonic (left), second harmonic plus circle corresponding to mean radius or zeroth harmonic (center), cumulative sum of harmonics (right). (*b*) Third harmonic. (*c*) Fourth harmonic. (*d*) Fifth harmonic. (*e*) Sixth harmonic.

350

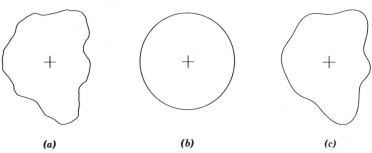

FIGURE 5.39 Comparison between digitized grain outline (*a*), zeroth harmonic or circle of mean radius (*b*), and reconstruction based on six harmonics (*c*).

power, perhaps five or more orders of magnitude below the power contained in the low harmonics. Considering their standard errors, it seems difficult to attach much significance to such frequencies. In addition, the effects of aliasing are most pronounced at the high frequencies, making determination of true spectral values even more difficult.

Polar Fourier transformation has certain limitations, the most conspicuous of which is that only single-valued outlines can be analyzed. That is, a radius drawn

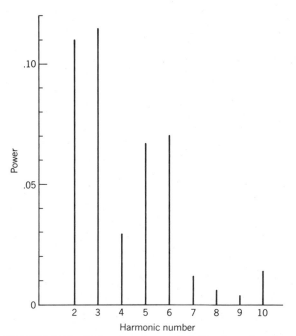

FIGURE 5.40 Power spectrum of digitized sand grain shown in Figure 5.39.

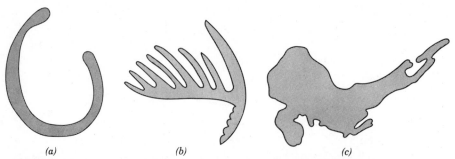

(a)   *(b)*   *(c)*

FIGURE 5.41   Examples of shapes which cannot be analyzed by polar Fourier analysis because of double-valued radii. (*a*) Shoreline of a mid-Pacific atoll. (*b*) Projection of the conodont *Ligonodina*. (*c*) Outline of a granite pluton in southern Ontario.

from the centroid must intercept the perimeter only once. This means that the procedure cannot be used for extremely convoluted shapes, such as those shown in Figure 5.41.

Other Fourier procedures permit the analysis of more complicated forms, which may be double-valued in a polar representation. One method converts the original outline into a series of angular deviations. First, equally spaced points are placed

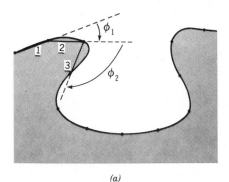

(a)

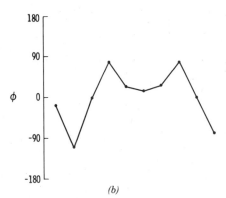

(b)

FIGURE 5.42   Conversion of digitized coordinates into angular deviations. (*a*) The straight lines are the approximation formed by digitizing original shape at equally spaced points. Angular change between segment 1 and 2 is given by ϕ_1; change between segment 2 and 3 is given by ϕ_2. (*b*) Plot of angular change ϕ between successive segments. This is the record that is analyzed by Fourier methods.

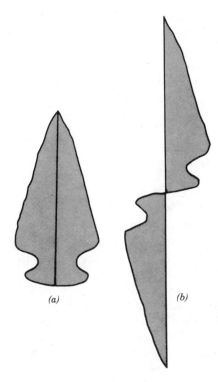

(a)

(b)

FIGURE 5.43 Outline of Late Paleolithic flint arrowhead. (a) Shape shown in conventional representation. (b) Transformation of outline by reversing one-half of figure along axis of symmetry.

around the perimeter of the object. The direction from the first to the second point is found, and then the angular difference between this direction and the direction from the second and third points (Fig. 5.42). As this process is repeated, the original series of X and Y coordinates defining the form is replaced with a series of angular deviations between the successive points. This new series can be analyzed by conventional Fourier procedures, although the spectrum may be difficult to interpret because the Y variable is not a distance but an angular change. This and similar transforms are discussed by Clark (1981).

Some closed forms are bilaterally symmetrical, or possess a distinctive line or fold that allows them to be oriented along an axis. The form can then be "split" and one-half "folded over" by reversing the signs of one coordinate along the symmetry axis for half of the form (Fig. 5.43). The object will then appear as a sinusoidal line that can be analyzed by ordinary Fourier analysis. This approach may be useful for the study of the shapes of certain invertebrate fossils and also has been used to characterize the shapes of arrowheads.

Computer Contouring

The objective of contour mapping is to portray the form of a surface. A contour map is a type of three-dimensional graph or diagram, compressed onto a flat, two-dimensional piece of paper. X_1, the axis running across the page, and X_2, the axis running up and down the page, correspond to the geographic coordinates east-west and north-south. Y, the vertical dimension that may represent elevations above sea level, thickness, or some other quantity, is shown by means of *isolines*, or lines of equal value. (Many authors denote the three axes as X, Y, and Z, corresponding to the two geographic coordinates and the mapped variable. To maintain consistency with later topics, we will continue to use Y as a variable that is expressed as a function of other variables X_i.) *Contour lines*, strictly speaking, are isolines of elevation, whether representing the ground's surface or the top of a subsurface formation. Geologists are casual in their use of terminology, however, and usually call any isoline a contour, whether it depicts elevation, thickness, porosity, composition, or some other property.

The contour lines on a map connect points of equal value, and the space between two successive contour lines contains only points whose values fall within the interval defined by the contour lines. In most circumstances the value of the surface cannot be determined at every possible location, nor can its value be measured at any specific point we might choose. Usually, only scattered measurements of the surface are made at a relatively few number of *control points*, such as well locations, seismic shot points, or the sites where assay samples have been taken.

The drawing of a contour map may be done by hand, as it has been for hundreds of years, or it may be done by computer. Drawing a map by computer usually involves an intermediate step—the construction of a mathematical model of the surface—that must be performed before the contour lines themselves can be constructed.

A computer contouring program traces out contour lines by a precise mathematical relationship based on the geometry of the control points. A geologist, however, contours not only the control points but also his concepts and ideas about what the surface should look like. If these preconceived ideas are indeed correct, a competent geologist may be able to create a map superior to a computer-made product. On the other hand, if the geologist's preconceptions are erroneous, his finished map is likely to be seriously in error.

Experiments have offered some reassuring insight into the nature of computer-produced maps (Dahlberg, 1975). In an experiment, experienced petroleum geologists were pitted against a widely used automatic contouring program. The test data consisted of structural elevations from a collection of wells drilled into and around a Devonian reef in Alberta. Information from only a small number of the wells actually available was presented to the participants, and the objective was to assess the relative capabilities of men and machine in creating a realistic structural contour map.

All the maps tended to be very much alike at and near control points, but differed radically in uncontrolled areas. Some geologists produced "better" structural maps than the computer program, in the sense that their representations were closer to the structural configuration revealed by the complete data set. Other geologists, however, were seriously in error. Most interestingly, the computer-contoured map coincided almost exactly with the average of the manually produced maps. That is, between control points some geologists tended to bend their contour lines in one direction while others bent them in the opposite direction. Most drew their lines through a common middle ground, and only a few seriously deflected their contour lines one way or another. The computer-drawn contours passed almost precisely through the middle of this bundle of lines. In this instance, the computer program behaved like an "average" geologist.

However, one of the strongest arguments for computer contouring is that it creates a mathematical model of the mapped surface that also can be used for other purposes. Among the operations that can be performed are the mapping of derivatives of the surface, calculating volumes beneath the surface, and various surface-to-surface operations such as isopaching (subtracting one surface from another).

It should be emphasized that most computer contouring procedures are ad hoc; there is little theoretical basis for the various methodologies that have been employed. Rather, they are founded on commonsense assumptions about the way surfaces behave and the results of practical experience.

Several assumptions must be embodied in any computer algorithm used to create the mathematical model from which a contour map is constructed. The completed map reflects these assumptions; the model is reasonable and the map is a realistic representation of the surface only if these assumptions are valid. In general, a contouring program is designed to map a surface that is (a) single-valued at a point, (b) continuous everywhere within the map area, and (c) autocorrelated over a distance greater than the typical spacing between control points.

If there is only one possible value that a mapped property can have at a specific geographic location, that property is *single-valued*. An example is the elevation of a stratigraphic horizon, as measured in a well, to be used to construct a structural contour map. It is presumed that there is no uncertainty associated with the measurement except for that arising from instrument error. Only in very unusual geological circumstances, such as an overturned fold, can a stratigraphic unit assume more than one elevation at a single location.

Some important geological variables are not so obviously single-valued. Measurements of porosity or chemical composition, for example, are statistical in nature, and repeated sampling and analysis at a single location may result in a suite of values. This results both from errors in measurement and from random variations in the small samples of rock that are analyzed. Most automatic contouring programs cannot accommodate such repeated data, although it is possible to reduce multiple observations to a single, representative value such as the average or mean, which then can be mapped.

Automatic contouring procedures involve interpolation between control points and extrapolation beyond the control points. Because of the mathematical methods involved, all the values obtained by interpolation lie on a continuous, sloping surface between the control points. If the real surface contains discontinuities such as faults, these will not be recognized by the contouring program but will be mapped simply as areas of very steep slope. Faults or other discontinuities that are known in advance of mapping can be accommodated by procedures that in effect insert boundaries into the map. The mapping program will draw the surfaces on opposite sides of boundaries as though they were entirely separate maps. The corresponding mathematical model will have an abrupt change in numerical values along the boundaries. However, it has not been possible to create a contouring program that automatically recognizes unidentified faults or breaks in a surface.

Contouring programs incorporate the commonsense assumption that the value of a surface at one point is closely related to values at nearby points and less closely related to values at more distant locations. This assumption that the variable being mapped is positively autocorrelated over at least short distances is expressed in some procedures by selecting all the nearest control points around a location to be evaluated, and then estimating the surface at that location as some type of average. If the surface is highly autocorrelated, all of these neighboring control points will have approximately the same value, and their average will be a reasonable estimate for the intermediate location. In contrast, if the surface is poorly autocorrelated, neighboring control points will have little relation to one another, nor will they be related to the value at the location to be estimated. Under such conditions it may be impossible to make reasonable interpretations about the nature of the surface between control points.

Contouring by Triangulation

The first computer programs for contouring were direct implementations of methods used by surveyors for the hand mapping of topography (IBM, 1965). Control points, assumed to be located without any particular regularity, are first connected by straight lines. This forms a mesh of triangles that covers the map (Fig. 5.44). By interpolating down the sides of the triangles, locations can be found where the ground elevation is a constant, specified value. Connecting these points of equal elevation produces a contour line. In effect, the surface is modeled as a series of flat, triangular plates, each held at its corners by a control point. Contouring consists of drawing horizontal lines across these tilted plates. Almost every undergraduate geologist and engineer has experienced a manual equivalent of this process as an exercise in plane-table mapping or surveying.

It is obvious that if the control points are connected in a different manner, a different set of triangular plates will be defined and a different set of contour lines will result. In some early contouring programs, simply entering the data points in a different sequence could result in conspicuously different-appearing contour lines on a map. To avoid this problem attempts were made to select a unique, "optimal" set of triangles for mapping. Usually this meant the individual triangles should be as near to equilateral as possible, or that the triangles should have the minimum possible height, or that the longest leg of each triangle should be the shortest possible (Gold, Charters, and Ramsden, 1977). Unfortunately, no algorithm was available that would insure the construction of a triangular mesh having any of these properties. This was a serious impediment to early contouring programs, because it meant starting with an arbitrary arrangement of triangles, and through iteration trying to adjust the triangular mesh until an optimum configuration was obtained. This often resulted in exorbitant run times, and led to the almost complete abandonment of triangulation algorithms. Their place was taken by procedures that

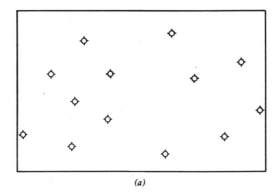

(a)

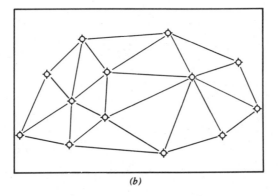

(b)

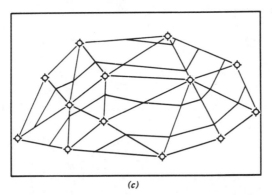

(c)

FIGURE 5.44 Triangulation method of estimating positions of contour lines. (a) Irregularly spaced data points. (b) Triangles formed across map area with data points as vertices. (c) Contour lines drawn through sides of triangles where points of specified elevation have been found by linear interpolation (after IBM, 1965).

used the control points only to make estimates of the surface at the nodes of a regular grid, and then contoured this grid rather than the control points themselves.

In recent years triangulation procedures have made a strong comeback, thanks to the development of algorithms that produce almost optimal networks on the first pass (Gold, Charters, and Ramsden, 1977; McCullagh and Ross, 1980). These networks, referred to as *Delaunay triangulations,* are uniquely defined for a given set of points. In addition, the triangles formed are as nearly equiangular as possible, and the longest sides of the triangles are as short as possible. This means that the greatest distances over which interpolations must be made to find contour levels are smaller than in other triangular nets.

In a field of scattered points, we can imagine that each is surrounded by an irregular polygon so that every location within a polygon is closer to the enclosed point than it is to any other (Fig. 5.45). Conversely, every location outside a particular polygon is closer to some other point than it is to the point within the polygon. This is the most compact division of space possible. Sets of polygons having these properties are called Thiessen, Dirichlet, or Voronoi polygons, and they arise in many diverse fields.

Geographers use Thiessen polygons to model the zones of influence around competing cities. Metallurgists model the growth of crystals in a cooling melt by Voronoi polyhedra, the three-dimensional equivalent of polygons. A mass of soap bubbles forms an easily observed network of Voronoi polyhedra.

Immediately surrounding the Thiessen polygon enclosing a specific point A are other Thiessen polygons, each of which also encloses a single point. These points are called the Thiessen neighbors of point A. If these points are connected by straight lines, the result is a Delaunay triangular network (Fig. 5.46). Both the sets of Thiessen polygons and the Delaunay triangles are unique for any arrangement of points.

The process of triangulation consists of determining the Thiessen neighbors of successive points on a map. In Figure 5.47a the neighbors of point A are to be found. We first assume that a nearby point B is a neighbor and construct a circle whose diameter is defined by the line AB. If no other points occur within the circle, B is indeed a neighbor of A. If a point is found in the circle, it replaces point B.

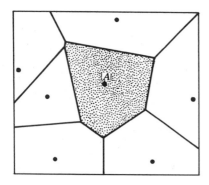

FIGURE 5.45 Thiessen polygon (shaded) around point A. All locations within polygon are closer to A than to any other point.

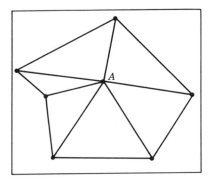

FIGURE 5.46   Delaunay triangle network around point A. All connected points are Thiessen neighbors of A.

The search for the next neighbor proceeds in a clockwise manner around point A. The circle is expanded as shown so that points A and B lie on its perimeter. The interior of the circle is then checked to see if any points are enclosed. If one point is found, it is the second Thiessen neighbor. If two or more points are found, the correct second neighbor must be determined. This is done by computing the angle formed by point B, the candidate point, and point A. The true Thiessen neighbor will form the largest angle (Fig. 5.47b).

The search for the third Thiessen neighbor is conducted by drawing a circle such that point A and point C, the second neighbor, lie on its perimeter. The interior of this circle is checked for an enclosed point, which becomes the third neighbor, D, when found (Fig. 5.47c). Next, a circle is constructed that includes point A and point D on its perimeter, and a search is made for the fourth Thiessen neighbor in its interior. Eventually, point B will be rediscovered as a Thiessen neighbor; then, all of the neighbors of A have been identified (Fig. 5.47d). Connecting these neighbors will form the triangular network around A (Fig. 5.48).

One of the Thiessen neighbors is now designated as a new point A around which a search will be conducted, and the entire process begins again. The network grows, spreading like a wave across the map until every point is included. To achieve computational efficiency, elaborate prior sorting of the point coordinates is done so that the searches first consider only the most likely candidates as neighbors. Although the process is complicated to describe, McCullagh and Ross (1980) state that the number of necessary steps to triangulate a set of n points is proportional to $n \log n$. In contrast, the number of operations required by algorithms that adjust the triangular net by trial and error may be proportional to n^3, and the number of steps in a gridding procedure is proportional to n^2.

The assumption that the triangles represent tilted flat plates obviously results in a very crude approximation of a surface. A better approximation can be achieved using curved or bent triangular plates, particularly if these can be made to join smoothly across the edges of the triangles. Several procedures have been used for this purpose. One of the earliest involved finding the three neighbors closest to the faces of a triangle, then fitting a second-degree polynomial trend surface to these and to the points at the vertices of the triangle. A second-degree trend surface

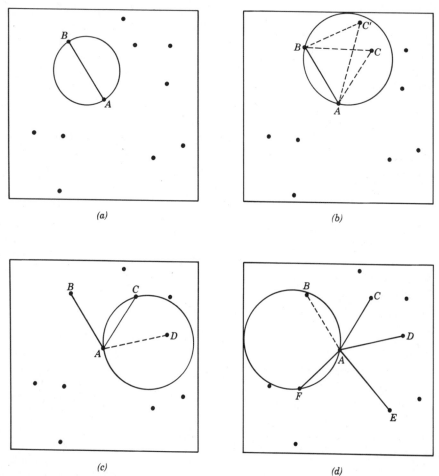

FIGURE 5.47 Determining the Thiessen neighbors of point *A*. (*a*) Nearby point *B* is selected as possible neighbor and circle with diameter *AB* constructed. If circle contains no other points, *B* is a neighbor. (*b*) Larger circle with *A* and *B* on perimeter is used to search for nearest neighbor in clockwise direction. *C* is neighbor because angle *BCA* exceeds angle *BC'A*. (*c*) Search circle from *AC* finds neighbor *D*. (*d*) Final search circle finds point *B* again.

is dome- or basin-shaped, and is defined by six coefficients. This means that the fitted surface will pass exactly through all six points. The equation can then be evaluated to find a series of locations having a specified elevation. These are connected to form the contours, which will be curved rather than straight lines.

Even though adjacent plates are fitted using common points, their trend surfaces will not coincide exactly along the line of overlap. This means there may be abrupt changes in direction where contour lines cross from one triangular plate to another. One way of correcting this problem is to "blend" the contour lines from the two surfaces by averaging them.

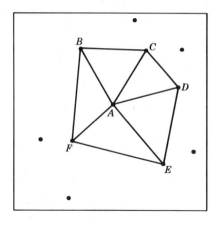

FIGURE 5.48 Connecting Thiessen neighbors of point *A* forms Delaunay triangular network around *A*. Procedure then moves to one of the points *B–F* and begins again.

A more elegant procedure uses three-dimensional equivalents of the spline functions introduced in the previous chapter. These are described in detail by Tipper (1979). Surface interpolation equations used in finite element analysis also can be adapted to contouring (Gold, Charters, and Ramsden, 1977; McCullagh, 1981). The procedure used by McCullagh to model the form of the surface within each triangular plate is too complicated to describe in detail here; those interested can

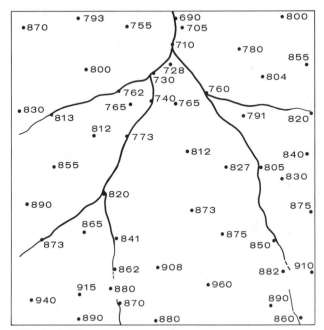

FIGURE 5.49 Survey control points for topographic mapping problem. Coordinates expressed in 50-ft units from origin in southwest corner of map. Elevations in feet above sea level.

TABLE 5.11 Geographic Coordinates and Elevations of Control Points in Surveying Problem; Coordinates Expressed in 50-ft Units Measured from Origin in Southwest Corner

East-West Coordinate (units)	North-South Coordinate (units)	Elevation above Sea Level (ft)
0.3	6.1	870.0
1.4	6.2	793.0
2.4	6.1	755.0
3.6	6.2	690.0
5.7	6.2	800.0
1.6	5.2	800.0
2.9	5.1	730.0
3.4	5.3	728.0
3.4	5.7	710.0
4.8	5.6	780.0
5.3	5.0	804.0
6.2	5.2	855.0
0.2	4.3	830.0
0.9	4.2	813.0
2.3	4.8	762.0
2.5	4.5	765.0
3.0	4.5	740.0
3.5	4.5	765.0
4.1	4.6	760.0
4.9	4.2	790.0
6.3	4.3	820.0
0.9	3.2	855.0
1.7	3.8	812.0
2.4	3.8	773.0
3.7	3.5	812.0
4.5	3.2	827.0
5.2	3.2	805.0
6.3	3.4	840.0
0.3	2.4	890.0
2.0	2.7	820.0
3.8	2.3	873.0
6.3	2.2	875.0
0.6	1.7	873.0
1.5	1.8	865.0
2.1	1.8	841.0
2.1	1.1	862.0
3.1	1.1	908.0
4.5	1.8	855.0
5.5	1.7	850.0
5.7	1.0	882.0

TABLE 5.11   (*Continued*)

East-West Coordinate (units)	North-South Coordinate (units)	Elevation above Sea Level (ft)
6.2	1.0	910.0
0.4	0.5	940.0
1.4	0.6	915.0
1.4	0.1	890.0
2.1	0.7	880.0
2.3	0.3	870.0
3.1	0.0	880.0
4.1	0.8	960.0
5.4	0.4	890.0
6.0	0.1	860.0
5.7	3.0	830.0
3.6	6.0	705.0

refer to his article and to the complete mathematical derivation in Birkhoff and Mansfield (1974). The interpolation equation is referred to as a "tricubic polynomial," and in the form used for contouring requires nine parameters to estimate each point within a triangle. The first three of these parameters are in effect the cross-products of the lengths of the sides of the triangle. The second set of three parameters essentially represents the coordinates of the location where the surface is to be estimated, expressed with respect to each of the three vertices. The final set of three coordinates is the first derivatives of the surface at each vertex. The derivative at a vertex is estimated by fitting a plane by least squares to its Thiessen neighbors. The plane is constrained to pass exactly through the value of Y at the vertex. The X_1 and X_2 coordinates of the plane are then combined to form a general slope.

These nine parameters are combined in a linear equation that yields an estimate, $\hat{Y}$, at a specified location. All estimates within a triangle lie on a smoothly curving surface, which merges continuously with the curved surfaces in adjacent triangles. Continuity is assured across the sides of two adjoining triangles because they have two vertices in common and share two generalized slopes.

In contour mapping, the estimation equation is reversed. The value of $\hat{Y}$ is set at the desired contour level, and for several selected X_1 coordinates the corresponding X_2 coordinates are found (or vice-versa). The result is a series of points having constant values of $\hat{Y}$. A contour line can be drawn merely by connecting the points.

A set of values to be contoured is entered into the computer as a·3 × n array, where each entry corresponds to an X_1 coordinate, an X_2 coordinate, and a Y or dependent value which will be contoured. Figure 5.49 shows a typical set of data points, in this instance representing topographic elevations measured in a small area by a surveying class. These data were obtained by plane table and alidade, and are uniformly distributed across the area in terms of the map scale. The data for each control point, in the form of an east-west coordinate, a north-south co-ordinate, and elevation of the ground surface relative to sea level, are listed in Table 5.11. The point coordinates are expressed in arbitrary units (1 map unit = 50 ft) from an origin in the southwest corner of the map, but this is done only for convenience. The control locations could equally well be expressed by any numerical Cartesian coordinate system.

Figure 5.50 shows the set of Delaunay triangles constructed by McCullagh's contouring program. In order to draw contours beyond the bounding polygon that encloses the outermost control points, a series of "pseudopoints" are placed along the boundary of the map. In this example, one pseudopoint is placed in each corner and two more are placed along each side of the map.

The finished map, contoured by triangulation, is given in Figure 5.51. Although very similar to a map produced using a gridding algorithm, shown in Figure 5.58, there are differences in detail. The most obvious of these are narrow regions where the surface slope abruptly changes, as in the southwest corner and in the north-central part of the map. These are areas where the Delaunay triangles are extremely acute. Along the margins of the map, these features can be corrected by judicious insertion of pseudopoints, but they cannot be altered within the interior of the map unless more data are available.

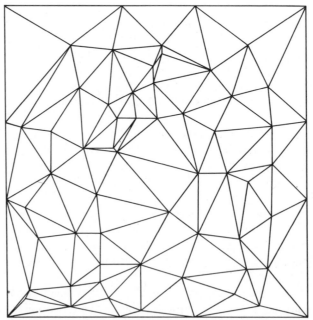

FIGURE 5.50   Delaunay triangular network for survey control points shown in Figure 5.49. Pseudopoints have been added along margin of map.

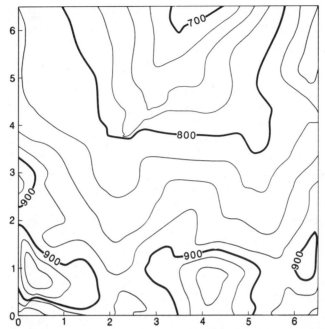

FIGURE 5.51 Contour map of topographic data produced by a contouring program that uses a triangular mesh for the mathematical model of the surface. Contour interval is 25 ft.

Contouring by Gridding

Gridding is the process of determining values of the surface at a set of locations that are arranged in a regular pattern, usually square, which completely covers the mapped area. In general, values of the surface are not known at these uniformly spaced points and so they must be estimated from the irregularly located control points where values of the surface are known. The locations where estimates are made are referred to as "grid points" or "grid nodes."

In this procedure, we first construct a mathematical model of the surface which has the form of a pattern of tilted square plates. In simple algorithms, these plates are flat. In more complex procedures they are curved, and each blends smoothly into the adjacent plates. The mathematical model is constructed purely for practical reasons. It is much easier to draw contour lines through an array of regularly spaced grid nodes than it is to draw them through the irregular pattern of the original points. All the possible ways for a contour line to enter and leave the square defined by four uniformly spaced grid nodes are known, and a line-drawing algorithm can easily be written to handle all of these possibilities. A contour line can be drawn simply by tracing the path of the line from one square of the grid to the next. Determining the path of a contour line through the irregular pattern of control points, as is done in triangulation algorithms, is much more difficult. Since the individual points cannot be connected to form a regular array, the possible paths that a contour line might follow cannot be known in advance. Also, the explicit X_1 and X_2 coordinates of all intermediate points involved in the calculation of the path of the contour lines must be retained in computer memory. In a grid or regular array of estimated values, the X_1 and X_2 coordinates are implied by position in the array. This saves memory and speeds computation.

The grid nodes, or intermediate locations where values of the surface must be estimated, usually are arranged in a square pattern so the distance between nodes in one direction is the same as the distance between them in the perpendicular direction. In most contouring programs, the spacing is under user control, and is one of many parameters that must be chosen before a surface can be gridded and mapped. The area enclosed by four grid nodes is called a *grid cell;* if a large size is chosen for the grid cells, the resulting map will have low resolution and a coarse appearance, but can be computed quickly. Conversely, if the grid cells are small in size, the contour map will have a finer appearance, but will be more expensive to produce.

Since gridding algorithms estimate only a single location from a collection of nearby control points, the estimation procedure is repeatedly applied across the map area until the entire map is covered by a regular grid or mesh of estimated values. Once the regular grid of estimates has been constructed, contour lines can be laced rapidly through this numerical array.

In some computer contouring packages, the initial grid estimation step may be followed by one or more additional steps in which the grid estimates are "refined." Typically, the grid nodes in the immediate vicinity of each control point are re-calculated, using both original control points and the initial estimates at the nearby grid nodes. This may result in a map surface that comes nearer to passing exactly through the control points than would be possible otherwise.

Gridding, or the calculation of the regular array of estimated values, involves three essential steps. First, the control points must be sorted according to their geographic coordinates. Second, from the sorted files, the control points surrounding a grid node to be estimated must be searched out. Third, the program must estimate the value of that grid node by some mathematical function of the values in these neighboring wells. Sorting greatly affects the speed of operation, and hence the cost of using a contouring program. However, it has no effect on the accuracy of the estimates, so we will not consider it further. Both the search procedure and the mathematical function do have significant effects on the form of the final map.

The most obvious function that could be used to estimate the value of a surface at a specific location on a map is simply to calculate an average of the known values of the surface at nearby control points. In effect, this projects all of these surrounding known values horizontally to the location to be estimated (Fig. 5.52). Then, a composite estimate is made by averaging these, usually weighting the closest points more heavily than distant points. If this is done on a regular grid over the entire map area, the resulting map will have certain characteristics. The highest and lowest areas on the surface will contain control points, and most interpolated grid nodes will lie at intervening values, since an average cannot be

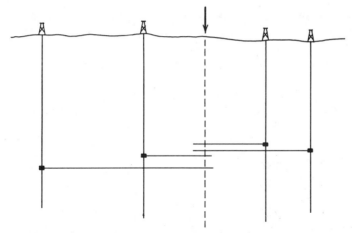

FIGURE 5.52 Cross-section view showing control points (wells) where value of surface (elevation of formation top) is known. Estimate at grid node (arrow) is made by projecting values horizontally, then averaging.

outside the range of the values from which it was calculated. At grid nodes beyond the outermost known points, the estimates are extrapolations and will be close in value to the nearest control points.

Figure 5.53a shows a series of observations on a map; each point is characterized by its X_1 coordinate (east-west or across the page), its X_2 coordinate (north-south or down the page), and its Y coordinate or elevation. On the figure, values of Y are noted beside each point. The observations may be identified by numbering them sequentially as they are read in, from 1 to i. Therefore, an original data point i has coordinates X_{1i} in the east-west direction, X_{2i} in the north-south direction, and elevation Y_i. In Figure 5.53b, we have superimposed a regular grid of nodes on the map. These grid nodes also are numbered sequentially, from 1 to k. Grid node k has coordinates X_{1k} and X_{2k} and has an estimated elevation of $\hat{Y}_k$. We are going to estimate $\hat{Y}_k$ from the nearest n data points; therefore, we must be able to search out the points nearest to each grid node and calculate the distances from these points to the nodes. The search procedure may be simple or elaborate; we will consider different alternatives later. Now we will assume we have by some method located the n nearest data points to grid node k. The distance D_{ik} from observation point i to grid node k is found by the Pythagorean equation:

$$D_{ik} = \sqrt{(X_{1k} - X_{1i})^2 + (X_{2k} - X_{2i})^2} \qquad (5.66)$$

Having found the distances D_{ik} to the n nearest data points, we now can estimate the grid node elevation $\hat{Y}_k$ from these. The estimate is

$$\hat{Y}_k = \frac{\sum_{i=1}^{n} (Y_i/D_{ik})}{\sum_{i=1}^{n} (1/D_{ik})} \qquad (5.67)$$

We can illustrate this process by computing the value of $\hat{Y}_i$ for the grid intersection shown on Figure 5.53c. We will arbitrarily use $n = 4$ nearest observations to calculate the estimate. The four nearest observations are labeled A, B, C, and D. We must first determine the distance of each of these from our grid node.

For A,

$$D_{ik} = \sqrt{(2.0 - 1.5)^2 + (3.0 - 3.6)^2} = \sqrt{0.61} = 0.78$$

For B,

$$D_{ik} = \sqrt{(2.0 - 3.0)^2 + (3.0 - 3.0)^2} = \sqrt{1.00} = 1.00$$

For C,

$$D_{ik} = \sqrt{(2.0 - 2.0)^2 + (3.0 - 2.4)^2} = \sqrt{0.36} = 0.60$$

For D,

$$D_{ik} = \sqrt{(2.0 - 1.0)^2 + (3.0 - 2.9)^2} = \sqrt{1.01} = 1.00$$

Using these distances, we can now find $\hat{Y}_k$. The numerator of (5.67) is

$$\frac{6.0}{0.78} + \frac{6.0}{1.00} + \frac{7.0}{0.60} + \frac{7.0}{1.00} = 32.36$$

The denominator is

$$\frac{1}{0.78} + \frac{1}{1.00} + \frac{1}{0.60} + \frac{1}{1.00} = 4.95$$

so the estimated value of $\hat{Y}_k$ is

$$\frac{32.36}{4.95} = 6.54$$

In a similar manner we can evaluate the remaining grid nodes on the map. The completed grid with all values of $\hat{Y}_k$ posted is shown in Figure 5.53d.

This type of algorithm is sometimes called a "moving average," because each node in the grid is estimated as the average of values at control points within a neighborhood that is "moved" from grid node to grid node. Such algorithms can be considered to be special cases of a more general set of procedures that involve the fitting of planes or curved surfaces to control points within a neighborhood. First, all control points within a specified neighborhood around a grid node to be estimated are found. We can then imagine that the values at these points approximately define a sloping plane. This plane can be expressed as a first-degree trend surface, whose coefficients can be found by least squares. The coefficients of the plane are calculated in exactly the same manner as the coefficients of a trend surface, except, of course, only points within the neighborhood are used.

Once the equation of the plane has been found, values for X_{1k} and X_{2k} that correspond to the location of the grid node can be inserted and the equation evaluated. This will yield $\hat{Y}_k$, which is the estimate of the surface for that grid

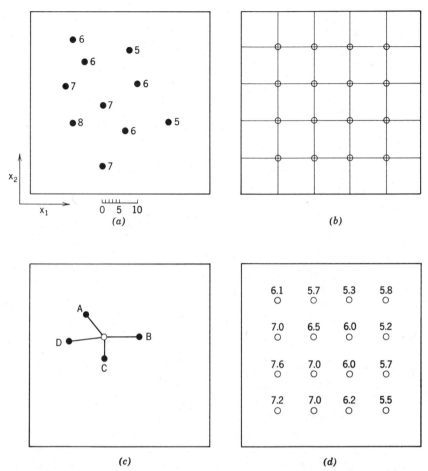

FIGURE 5.53 Steps in computation of grid value for contouring. (a) Original set of irregularly spaced control points on a map. Numbers are "elevations." (b) Regularly spaced grid network, with nodes whose values are to be calculated. (c) Location of four nearest control points to one grid node. These four nearest values will be used to calculate grid value. (d) Completed grid with estimated elevations at every node.

node. The process of fitting a plane and evaluating it to estimate the surface is repeated for every node in the grid. The plane represents a "general slope" of the surface around the grid node. In effect, the values of the surface at control points within the neighborhood are projected parallel to this sloping plane, and then averaged at the grid node (Fig. 5.54). If the fitted plane does not slope but rather is completely flat, an estimate made by this method will be the same as that found by the moving average method.

Gridding procedures that fit planes in this manner are sometimes called "piecewise linear least squares." A variant, called "piecewise quadratic least squares,"

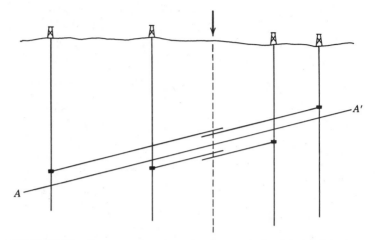

FIGURE 5.54 Cross-section view showing control points (wells) where value of surface (elevation of formation top) is known. Estimate at grid node (arrow) is made by fitting least-squares plane A-A′ to known values, then evaluating equation of plane at grid node location.

differs only by the fitting of a curved surface rather than a plane. A quadratic or second-degree trend surface is dome-, bowl-, or saddle-shaped and is defined by an equation that contains squares and cross-products of the X_1 and X_2 coordinates.

Since these algorithms consider the slope of the surface in a neighborhood, they may perform better than simple moving-average methods when used for interpolation between control points. Grid values can be estimated that are either higher or lower than the range of values at the control points, so the uppermost and lowermost areas of the surface need not correspond exactly with the known values. However, beyond the limits of control, extrapolation may create extreme values that are completely unwarranted. This occurs because any slopes that exist near the margins of the controlled part of the map are continued without constraint beyond the data. Use of a fitted quadratic surface makes this problem even worse.

A somewhat more complicated type of algorithm computes the local dip or slope of the mapped surface at every control point. It does this in almost exactly the same way that the linear least-squares algorithm fits a plane to control points around a grid node. That is, it uses a least-squares method, and determines the coefficients of a plane that comes as close as possible to the values of the surface at the surrounding control points.

This algorithm must proceed in two steps. First, a neighborhood is defined around each control point and all the other points within the neighborhood are found. Then, a plane is fitted to the known values of the surface at these points using least squares. However, the plane is constrained so that it must pass exactly through the value at the central control point. The coefficients of this plane, which define the slope of the surface at the central point, are stored along with the Y value for that point.

In the second step, a neighborhood is defined around each grid node to be estimated. The control points within this neighborhood are found, and the equations for the planes at each of the control points are evaluated for the location of the grid node. The different estimates from the planes are then weighted and combined. In effect, the slopes of the surface at the control points are projected to the grid node where they are averaged (Fig. 5.55).

Variants of this algorithm, sometimes referred to as "linear projection," are among the most popular of those used in commercial contouring packages. Some programs incorporate modifications of this procedure in which quadratic surfaces rather than planes are fitted to the control points. These algorithms are especially good within areas that are densely controlled by uniformly spaced data points. Like the piecewise linear least-squares methods, they have the distressing habit of creating extreme projections when used to estimate grid nodes beyond the geographic limits of the data.

The control points used in estimating a grid node, whether they are projected or not, ordinarily are weighted. The weightings vary according to the distances between the grid node being estimated and the control points. Figure 5.56 shows graphs for a number of commonly used weighting functions. Most contouring programs allow the user to select from among a variety of such functions.

The weights that are assigned to the control points according to the weighting function are adjusted to sum to 1.0. Therefore, the weighting function actually assigns proportional weights, and expresses the relative influence of each control point. A widely used version of the weighting process assigns a function whose exact form depends upon the distance from the location being estimated and the most distant point used in the estimation, or in one variant, over the distance to the outer limits of the neighborhood. The inverse distance-squared weighting function is then scaled so that it extends from one to zero over this distance. The process can be expressed in a single equation,

$$\omega_D = \left(1 - \frac{D}{D_{max}}\right)^2 \bigg/ \left(\frac{D}{D_{max}}\right) \tag{5.68}$$

Like the other weighting functions, the sum of the weights is set equal to 1.0.

The most obvious differences between various contouring programs are in the search methods employed. These are the algorithms used to select the data points within the local neighborhood around the grid location to be estimated. The simplest selection technique is called a nearest-neighbor search (Fig. 5.57a), which locates some specified number of control points or well locations that are closest to the grid node being estimated. A set of possible nearest-neighbor control points are selected from the complete data collection by sorting on the X_1 and X_2 coordinates of the points. The Euclidean distances from the grid node to each of these points are then calculated, and a specified number of the closest points are found.

An objection to a simple nearest-neighbor search is that it may find that all nearby points lie in a narrow wedge on one side of the grid node that is to be estimated. The resulting estimate is essentially unconstrained, except in one di-

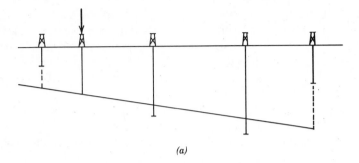

(a)

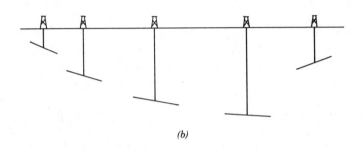

(b)

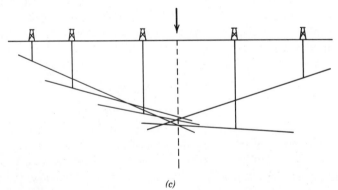

(c)

FIGURE 5.55 Interpolation from irregularly spaced data points
to grid by dip-projection method requires calculation of local dip
at every data point, shown here in cross-section form. (a) Local
dip of surface at data point (shown by arrow) is found by fitting
least squares plane to surrounding points. Fitted plane is constrained to pass through the data point. (b) Local dips become
part of data for each control point. (c) Local dips at control points
are projected to grid node (arrow). Value assigned to node is
weighted average of these projections. After Sampson (1975).

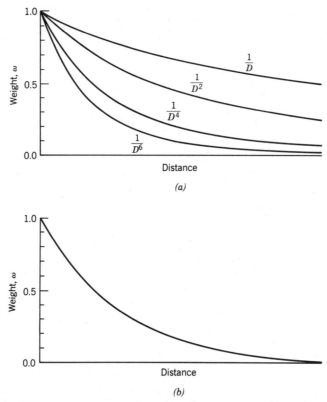

FIGURE 5.56 Distance weighting functions used in contouring programs. (*a*) Inverse distance-powered functions. (*b*) Scaled inverse distance-squared function.

rection. This may be avoided by restricting the search in some way that ensures that the control points are equitably distributed about the location to be estimated. Figure 5.57*b* illustrates one mode of radial constraint, called a quadrant search. Some minimum number of control points must be taken from each of the four quadrants around the grid node being calculated. An elaboration on the quadrant search is an octant search (Fig. 5.57*c*), which introduces a further constraint on the radial distribution of the points used in the estimating equation. A specified number of control points must be found in each of the octants surrounding the grid node being estimated. This search method is one of the more elegant procedures currently employed and is widely used in commercial programs.

Any constraints on the search for nearest control points, such as a quadrant or octant requirement, will obviously expand the size of the neighborhood around the grid node being estimated. This is because some nearby control points are likely to be passed over in favor of more distant points in order to satisfy the requirement that only a few points may be taken from a single sector. Unfortunately, the autocorrelation of a typical geological surface decreases with increasing distance,

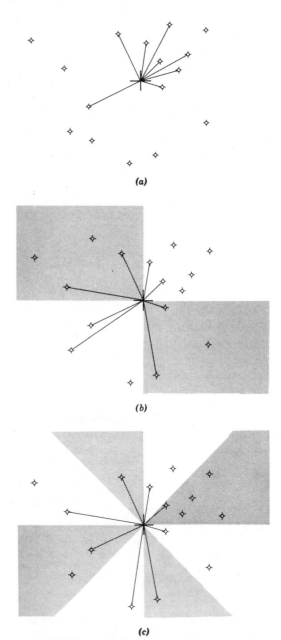

FIGURE 5.57 (a) Search technique that locates *n* nearest neighbors around grid node being estimated. No constraints are placed on radial distribution of control points. (b) Quadrant search pattern used to insure equitable distribution of control points around grid node being estimated. (c) Octant constraint on search pattern around grid node being estimated.

so these more remote control points are less closely related to the location being estimated. This means the estimate may be poorer than if a simple nearest-neighbor search procedure were used.

A matter of major concern to those who use and evaluate contouring programs is how well they "honor the data points." That is, how closely does the mathematical model accord with the control points that were used in its construction? Since the contour lines are drawn based on relationships between the grid nodes, rather than between the control points themselves, it is possible for a contour line to be drawn correctly with respect to the grid values but incorrectly with respect to the original data points. This is especially apt to happen if the grid is relatively coarse. Usually the errors are very small and are easily overlooked in areas where the surface is complex in shape or where slopes on the surface are steep so that the contour lines are closely spaced. In areas of very gentle slope, however, a small discrepancy between the grid node values and a control point may be sufficient to displace a contour line some distance from the point that it is supposed to honor. This may result in a conspicuous instance of a contour line passing by a control point on the wrong side. This problem, of course, does not occur with triangulation procedures because the model of the surface is formed by the control points themselves.

Appearance of a contour map is not a reliable guide to how well the underlying mathematical model represents the original control points. There is no formal statistical theory that allows us to predict, on theoretical grounds alone, which contouring procedure might be superior. In any given situation, the performance of a particular algorithm is determined by the complexity of the surface being mapped, the density and arrangement of the control points, the size of the grid, and, of course, the algorithm itself. Empirical tests of how well various gridding algorithms perform, using typical subsurface data, have been published by Davis (1976).

However, representing the known data as precisely as possible is not the real objective of most contour mapping exercises. Rather, we wish to estimate, with the smallest possible error, values of the surface at locations where measurements have not yet been made. The ability of various algorithms to produce accurate estimates at locations where no control exists can be checked by empirical tests in which a small proportion of the available control points are removed from the data set prior to mapping. Comparisons can then be made between the true values at these "blind" locations and the estimates made by the mapping program. Then, the omitted points can be returned to the data set, another random set selected for removal, and the process repeated over and over (Davis, 1976).

The distressing (although not surprising) conclusion from such an empirical study is that the different objectives we might set for a contouring procedure may not be mutually obtainable. In order to faithfully reproduce or honor the original control points with a gridding-type algorithm, it is necessary to utilize a weighting function that drops off extremely rapidly with distance and that uses only a few nearest neighbors. Such an algorithm will produce poor predictions or estimates at locations where no control is available. To make the best prediction of values of a surface

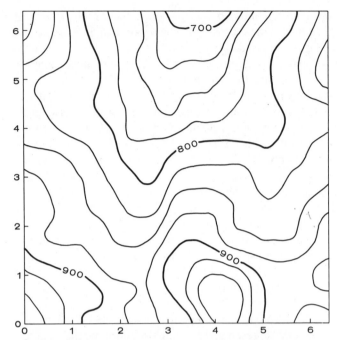

FIGURE 5.58 Contour map of topographic data produced by a contouring program that uses a regular grid for the mathematical model of the surface. Contour interval is 25 ft.

at unsampled locations, we must use many control points in each calculation of a grid node, and weight distant points relatively heavily. Unfortunately, the smooth, generalized surface created in this manner performs poorly in honoring the original control points. Some commercial contouring programs attempt to circumvent this impasse by first gridding in a manner designed to achieve good predictions in areas of no control, then regridding in the immediate vicinity of control points. The resulting composite surface does at least partially satisfy the different objectives of contouring. However, it may be marked by peculiar-appearing small features that surround each control point and that occur nowhere else on the map.

A map of the topographic data given in Table 5.11 is shown in Figure 5.58, contoured by a program that computes a regular grid. In this instance, the mathematical model of the surface contains 63 rows and 63 columns; each grid node was estimated by a linear projection algorithm that used the nearest 16 control points. Control points were weighted according to their distance from the grid node by the function given in eq. (5.68). This map should be compared to both Figure 5.51, contoured by a triangulation algorithm, and Figure 5.59, contoured manually by a member of the surveying class. There are differences between the two computer-drawn maps, and also between them and the hand-drawn map. These differences reflect the underlying mathematical models in the two methods, and also the fact that the human could incorporate information about the effects of streams

FIGURE 5.59 Contour map of topographic data produced by manual contouring. Note the inferred effect of streams on the form of contour lines.

on topography. This additional information is not available to either contouring program. Ripley (1981) provides several additional examples of these data mapped by other procedures.

Moving Averages

Although some of the mapping techniques we have discussed have been applied to problems in mining geology, the estimation of ore grade and control of quality in a mine present special problems. In response to these, mathematical and statistical techniques for mine evaluation have been developed somewhat separately from the growth of mathematical geology in general. In the United States, mining geology statistics have generally followed the lines of traditional statistical analysis, especially analysis of variance (several examples and an extensive bibliography are given in Koch and Link, 1980, Chapter 5). In South Africa and France, however, ore evaluation and prediction have proceeded along an independent path which has led to regionalized variable theory and the estimation method called kriging. Precursors of this approach were based on spatial moving averages, which we will now consider.

In a bedded sedimentary ore body, the quality or value of ore may be distributed normally. That is, the deposit may have a mean grade and the grades of individual samples may be distributed more or less symmetrically about this mean, with declining abundances of extreme values. Conventional parametric statistics may be applied appropriately to the analysis of samples from these deposits, and regression techniques may be useful for prediction and reserve estimation. However, deposits of valuable minerals or precious metals typically do not follow such distribution patterns. Drill-hole or hand samples are characterized by extreme values which are erratic in their spatial distribution. In general, there is a notable absence of the behavior that characterizes "good" statistical variables.

Figure 5.60a shows values obtained from samples along a drift in a Mexican silver mine. Two characteristics of the data are apparent at once; the values increase in a greater than linear manner (rich zones are many times richer than poor zones), and variation in grade increases as the grade itself increases. In the rich part of the drift, changes in value are more extreme than the entire range of values in poorer parts of the same drift. These two characteristics may be summarized by stating that changes in ore grade occur exponentially with distance, and the variance is heteroscedastic.

In such a case, we can transform the data by taking the logarithms of the ore

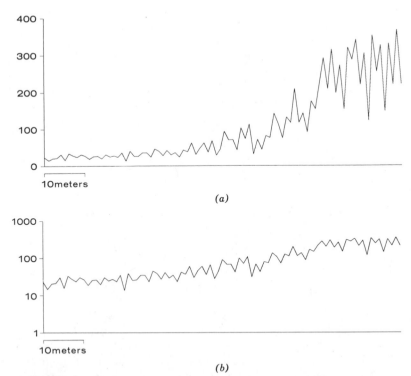

FIGURE 5.60 Variations in ore grade with distance along a drift in a silver mine in Mexico. (a) Ore grade plotted on an arithmetic scale. (b) Ore grade plotted on a logarithmic scale.

values. The exponential curve of the data then becomes a linear trend, and the heteroscedastic variance becomes constant about the trend (Fig. 5.60*b*).

Another characteristic of many ore deposits is that the mean grade of a block of ore is largely independent of the size of the block, once some minimum volume is exceeded. However, it has often been stated that the variance of ore samples is inversely related to block size, declining as the volume of an ore sample is increased. This could conceivably lead to problems in the estimation of reserves, because the volumes of core samples from which the estimates must be made are many orders of magnitude smaller than the blocks of ore that actually will be removed. (See Koch and Link, 1980, pp. 280–284, for a discussion of this point.) Therefore, the variance in the grade of core samples may be so high that a realistic estimate of the true grade of the ore blocks cannot be made. To produce estimates that have less variance, we may utilize moving averages of the drill samples, hoping to average out the extreme variation associated with individual samples.

Two-dimensional moving averages are an extension of data-smoothing techniques considered in Chapter 4. Generally, a variable must be estimated at a series of points on a grid, or values are to be assigned to successive adjacent squares or rectangles on a map. The data on which the estimates are based are scattered through the map area, and may or may not lie on a grid. A pattern, analogous to the smoothing interval of time-trend analysis, is centered on the first point to be estimated. All data points within the pattern, perhaps a square or circle, are weighted in some manner and used to estimate the central point. In the simplest type of moving average, the mean of all the observations within the pattern is applied to the estimated point. The pattern is then moved to the next grid intersection and the process is repeated. When one row or column of the grid has been evaluated, the next row or column is taken, until the map area has been completely covered.

Any moving average is an expression of the general model

$$\hat{Y}_{ij} = \sum_{k=1}^{n} W_k Y_k \tag{5.69}$$

That is, an estimated grid value $\hat{Y}_{ij}$ is based on the weighted sum of n adjacent observations Y_k. The nature of the weighting function varies from one moving-average scheme to another. For example, in a contour mapping program, a moving average based on points weighted inversely to their distance from $\hat{Y}_{ij}$ may be used. It is possible to use a simple plane as a moving average, analogous to smoothing functions discussed in Chapter 4. Other weighting equations are equally applicable.

Most moving-average techniques consider distance from the estimated point to the estimating points in some manner. In a contouring program, the distances to n nearest points are measured directly, and each point is weighted accordingly. Methods analogous to one-dimensional smoothing equations require that the data be located upon a grid. Then, the spatial relationship between $\hat{Y}_{ij}$ and each value of Y within the moving-average interval is known. In this case, weights remain constant for equivalent points of Y_k as the moving-average surface estimates successive values of $\hat{Y}_{ij}$. In a third technique, a series of zones or blocks adjacent to the point being estimated may be established. All observations Y_k within each of these are averaged, and the block averages $\overline{Y}$ are weighted and used to estimate $\hat{Y}_{ij}$. If many observations Y_k occur within each block, the mean can be considered to be located at the center of the block with negligible error. Because the block centers always are located a fixed distance from the estimated point, constant weighting functions can be used. This method has significant advantages when estimating points from highly irregular control values. We will examine in detail a moving-average method of the third type, because it is one of the most advanced techniques and has been extensively used for ore evaluation in some large mines in North America.

Moving Weighted Averages of Block Means

Even if a mineralized vein is very small, a sizable block of rock must be removed during mining. The minimum size of this block is determined by the method of extraction, the size of equipment, and the nature of the rock. Although removal of ore in blocks may reduce the grade by dilution with country rock, the mechanics of mining make this inevitable. All of these various physical factors set a lower limit to the size of the block whose grade we wish to estimate. It is senseless, for example, to estimate the grade of ore in a 1-ft cube if the smallest mass of rock that will actually be removed measures 10 ft on each side. In the mines where weighted block averaging is used, the smallest practical mining unit is about 100 ft on each side. The smallest practical mining unit varies from one ore deposit to another, and probably is largest in disseminated ore bodies mined by stoping and smallest in very rich hydrothermal vein deposits.

Fortunately, large blocks are less subject to extreme grade fluctuation than are small samples. In fact, a fundamental assumption in a moving weighted block method is that the minimum mining unit is so large compared to the distance over which rapid grade changes occur that the variance is not altered by increasing the size of the blocks. Therefore, the smallest practical mining units may be combined into successively larger blocks. If variation in assay values occurs mostly between units much smaller than the minimum practical block, these small blocks will have no greater variance than the larger combined blocks. A theoretical plot of variance versus volume of sample is shown in Figure 5.61; this illustrates the assumption that variance is stable once a critical minimum size of sample is exceeded.

To establish a moving average based on block means, we must first determine the weightings to be attached to the blocks. The arrangement of the blocks themselves also must be established, but this is more or less arbitrary. Assume that we have decided on a moving-average design such as that shown in Figure 5.62. We wish to create an estimate, $\hat{Y}$, of the grade of ore that will be produced from block 1 from the means of exploration samples taken in blocks 1–9. (Note that $\hat{Y}$ is the estimated grade of ore to be produced from the center block, and $\overline{Y}_1$ is the mean assay value of exploration cores in the same block.) Once we have determined an estimating equation for $\hat{Y}$, we will move the sample design over the unknown region, estimating the grade of successive adjacent blocks until the entire area is evaluated.

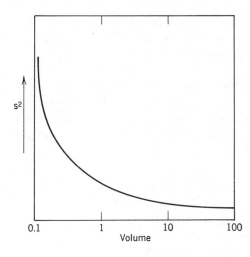

FIGURE 5.61 Theoretical change in variance of ore grade with changes in volume of ore sample or mined block.

The moving-average equation is determined from records of the grade of ore removed from developed regions of the mine. Suppose we place our moving-average design over a map of an area of the ore body that already has been mined out. From the exploration cores and assays in each of the blocks, we can compute block means, which become our variables $\overline{Y}_1, \overline{Y}_2, \ldots, \overline{Y}_9$. Production records give the value of ore actually produced from the block of interest. If we denote this as Y, we can relate production ore grade to exploration assays by a linear equation

$$Y = \beta_0 + \sum_{i=1}^{9} \beta_i \overline{Y}_i + \epsilon \tag{5.70}$$

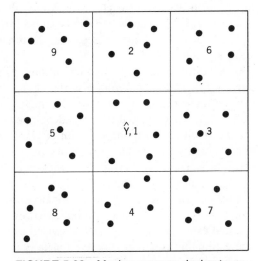

FIGURE 5.62 Moving-average design to estimate mean grade of ore in center block, $\hat{Y}$, from averages of drill-holes in blocks 1–9.

which you will recognize as a regression of the $\overline{Y}$'s on **Y**. Placing our design over many successive blocks will give us a set of observations of the $\overline{Y}$'s, and we can solve (5.70) by least-squares methods almost identical to those used to compute trend surfaces.

We now have an estimator, $\hat{Y}$, of the development grade of ore in the central block, which is based on the means of exploratory cores in surrounding blocks. The estimating equation contains a constant term and nine weighting coefficients. However, we can define the constant, β_0, as equal to a new constant times the mean of all exploration samples in all blocks:

$$\beta_0 = \beta_0'\overline{\overline{Y}} \quad \text{or} \quad \beta_0' = \frac{\beta_0}{\overline{\overline{Y}}} \tag{5.71}$$

The new constant, β_0', is found simply by dividing β_0 by the grand mean of exploration samples (the term grand mean refers to the overall mean, or $\overline{\overline{Y}} = \sum_{i=1}^{9}\overline{Y}_i/9$. Thus, the regression equation can be written as

$$\hat{Y} = \beta_0'\overline{\overline{Y}} + \sum_{i=1}^{9}\beta_i\overline{Y}_i + \epsilon \tag{5.72}$$

All coefficients in the regression equation, including the constant term, can now be regarded as weighting functions. By equating β_0 to $\beta_0'\overline{\overline{Y}}$, we have made the regression independent of any trend that may exist in the data, and the equation may be applied to other areas even though the mean value in these new areas is higher or lower than in the region where the equation was determined.

We may now use the regression equation as a moving average in an undeveloped region of the mine. Placing the design on the exploration map, we need only to select the exploration cores that lie within each of the nine blocks. From the assay values, we can compute mean values for each block, which then become the $\overline{Y}$'s in (5.72) and are used to estimate $\hat{Y}$. The design can be moved from block to block through an undeveloped part of the mine, successively estimating development grade from exploration or evaluation data.

The success of the method depends upon two conditions. First, the regression equation must adequately fit development grades in the already developed part of the mine. If the fit of the linear regression is poor, the moving average $\hat{Y}$ will be an ineffective estimator of **Y** when applied to the undeveloped parts of the mine. However, because of the stability of variances within the large blocks, a good fit usually will be obtained. A low fit indicates variability on a scale comparable in size to the blocks themselves. The second critical assumption is that the distribution of values in the developed part of the mine is closely similar to that in the undeveloped region. Nothing is accomplished by computing an estimating equation and then applying it in an area which is basically different. However, mine evaluation is a dynamic process, and estimates are continually revised as mining progresses. The area in which the estimating equation is derived usually is as close to current mining activity as possible.

As you might surmise, weighted moving block designs are "custom-tailored" to individual mines and even to parts of mines. The design of the moving-average pattern is influenced by the presence of a strong trend in values, by spacing and distribution of exploratory holes, by the mining method employed, and by many other factors. The development of this method and an example of its application to a large mine are given by Krige (1966, pp. 17–21).

Kriging

The concept of a *regionalized variable* was introduced in Chapter 4 as a naturally occurring property that has characteristics intermediate between a truly random variable and one that is completely deterministic. Many geological surfaces, both real and conceptual, can be regarded as regionalized variables. These surfaces are continuous from place to place and hence must be spatially correlated over short distances. However, points on an irregular surface that are widely separated tend to be statistically independent. The degree of spatial continuity of a regionalized variable can be expressed by a semivariogram, as discussed in Chapter 4. If measurements have been made at scattered sampling points and the form of the semivariogram is known, it is possible to estimate the value of the surface at any unsampled location. The estimation procedure is called *kriging*, named after D.G. Krige, a South African mining engineer and pioneer in the application of statistical techniques to mine evaluation.

Kriging can be used to make contour maps, but unlike conventional contouring algorithms, it has certain statistically optimal properties. Perhaps most importantly, the method provides measures of the error or uncertainty of the contoured surface. Kriging uses the information from the semivariogram to find an optimal set of weights that are used in the estimation of the surface at unsampled locations. Since the semivariogram is a function of distance, the weights change according to the geographic arrangement of the samples.

Punctual Kriging

Punctual kriging is the simplest form of kriging, in which the observations consist of measurements taken at dimensionless points, and the estimates are made at other locations that are themselves dimensionless points. Punctual kriging is used, for example, in contour mapping where the observations may be elevations of the top of a formation, as measured in a set of exploratory drill holes. Constructing a structural contour map requires that estimations of the elevation of the formation top be made at closely spaced locations over the map area. Once made, contour lines can be drawn through these estimates in the manner described in the previous section.

To simplify the problem, we may assume that the variable being mapped is statistically stationary, or free from drift. The value at an unsampled location may be estimated as a weighted average of the known observations. That is, the value at point p is based on a small set of nearby known control points:

$$\hat{Y}_p = \Sigma \, W_i Y_i$$

We expect that the estimate $\hat{Y}_p$ will differ somewhat from the true (but unknown) value Y_p by an amount we may call the *estimation error*:

$$\epsilon_p = (\hat{Y}_p - Y_p) \qquad (5.73)$$

If the weights used in the estimation equation sum to one, the resulting estimates are *unbiased* provided there is no drift. This means that, over a great many estimations, the average error will be zero, as overestimates and underestimates will tend to cancel one another. However, even though the average estimation error may be zero, the estimates may scatter widely about the correct values. This scatter can be expressed as the *error variance*,

$$s_\epsilon^2 = \frac{\Sigma(\hat{Y}_p - Y_p)^2}{n} \qquad (5.74)$$

or as its square root, the *standard error of the estimate*:

$$s_\epsilon = \sqrt{s_\epsilon^2} \qquad (5.75)$$

As noted in the section on contour mapping, it seems intuitively reasonable that nearby control points should be most influential in estimating the value at an unsampled location on a surface, and more distant control points should be less influential. It also seems reasonable to expect that the weights used in the estimation process, and the error in the estimate, should be related in some way to the semivariogram of the surface. In a simple example, Clark (1979) demonstrates that this is so.

Suppose we wish to estimate the value of Y at a point p from three nearby points, using as our estimator a weighted average of the three known values:

$$\hat{Y}_p = W_1Y_1 + W_2Y_2 + W_3Y_3$$

The weights are constrained to sum to one, so the estimate is unbiased if there is no trend. Suppose that weight W_1 is chosen to be equal to 1.0. Then, weights W_2 and W_3 must be zero and the estimate at p is

$$\hat{Y}_p = 1.0Y_1 + 0.0Y_2 + 0.0Y_3$$

or

$$\hat{Y}_p = Y_1$$

Obviously, the estimation error is simply $\epsilon = Y_p - Y_1$, since Y_1 *is* the estimate $\hat{Y}_p$. If many other locations like Y_p are estimated from points arranged in a manner spatially similar to Y_1, the estimation variance can be calculated as the average squared difference between these pairs of points. For convenience, we may call these other estimated locations Y_{pi} and the other estimating points Y_{1i}. Then,

$$s_\epsilon^2 = \frac{1}{n} \Sigma_{i=1}^n (Y_{pi} - Y_{1i})^2$$

If the equation is compared to eq. (4.71), you will see that the estimation variance is equal to twice the semivariance for a distance equal to the separation between points Y_{pi} and Y_{1i}.

We have chosen one particular combination of weights to arrive at an estimate $\hat{Y}_p$ and to determine the estimation error. There are an infinity of other possible combinations of weights that could be chosen, each of which will give a different estimate and a different estimation error. There is, however, only one combination that will give a *minimum* estimation error. It is this unique combination of that kriging attempts to find.

Deriving the kriging equations requires calculus and so will not be considered here. A simple discussion is contained in Clark (1979) and a complete derivation is provided by Olea (1975) for the case of punctual kriging. Optimum values for the weights can be found by solving a set of simultaneous equations, which includes values from a semivariogram of the variable being estimated. The weights are optimal in the sense that the resulting estimates are unbiased and have minimum estimation variance. No other linear combination of the observations can yield estimates that have a smaller scatter around their true values.

In the simplest possible situation, we may wish to make a kriged estimate of the value $\hat{Y}$ at a point p from three known observations, Y_1, Y_2, and Y_3. Three weights, W_1, W_2, and W_3 must be found for the kriging equation. To find these requires the solution to a system of three simultaneous equations:

$$W_1\gamma(h_{11}) + W_2\gamma(h_{12}) + W_3\gamma(h_{13}) = \gamma(h_{1p})$$
$$W_1\gamma(h_{12}) + W_2\gamma(h_{22}) + W_3\gamma(h_{23}) = \gamma(h_{2p})$$
$$W_1\gamma(h_{13}) + W_2\gamma(h_{23}) + W_3\gamma(h_{33}) = \gamma(h_{3p})$$

In this notation, $\gamma(h_{ij})$ is the semivariance over a distance h corresponding to the separation between control points i and j. For example, $\gamma(h_{13})$ is the semivariance for a distance equal to that between known points 1 and 3; $\gamma(h_{1p})$ is the semivariance for a distance equal to that between known point 1 and the location p where the estimate is to be made. The left-hand matrix is symmetrical because $h_{ij} = h_{ji}$. It has zeroes along the main diagonal because h_{ii} represents the distance from a point to itself, which is zero. Assuming the semivariogram goes through the origin, the semivariance for zero distance is zero. Values of the semivariance are taken from the semivariogram, which must be known (or estimated) prior to kriging.

However, a fourth equation is needed to insure that the solution is unbiased, by constraining the weights to sum to one. The fourth equation is

$$W_1 + W_2 + W_3 = 1.0$$

This gives a set of four equations but only three unknowns. Since we have more equations than unknowns, we can use the extra degree of freedom to assure that the solution will have the minimum possible estimation error. This is done by adding a slack variable, called a Lagrange multiplier, λ, to the equation set. The complete set of simultaneous equations has the following appearance:

$$W_1\gamma(h_{11}) + W_2\gamma(h_{12}) + W_3\gamma(h_{13}) + \lambda = \gamma(h_{1p})$$
$$W_1\gamma(h_{12}) + W_2\gamma(h_{22}) + W_3\gamma(h_{23}) + \lambda = \gamma(h_{2p})$$
$$W_1\gamma(h_{13}) + W_2\gamma(h_{23}) + W_3\gamma(h_{33}) + \lambda = \gamma(h_{3p})$$
$$W_1 + W_2 + W_3 + 0 = 1$$

$$(5.76)$$

Rearranging in matrix form,

$$\begin{bmatrix} \gamma(h_{11}) & \gamma(h_{12}) & \gamma(h_{13}) & 1 \\ \gamma(h_{12}) & \gamma(h_{22}) & \gamma(h_{23}) & 1 \\ \gamma(h_{13}) & \gamma(h_{23}) & \gamma(h_{33}) & 1 \\ 1 & 1 & 1 & 0 \end{bmatrix} \cdot \begin{bmatrix} W_1 \\ W_2 \\ W_3 \\ \lambda \end{bmatrix} = \begin{bmatrix} \gamma(h_{1p}) \\ \gamma(h_{2p}) \\ \gamma(h_{3p}) \\ 1 \end{bmatrix} \qquad (5.77)$$

In general terms, we must solve the matrix equation

$$[A] \cdot [W] = [B]$$

for the vector of unknown coefficients, $[W]$. The terms in matrix $[A]$ and vector $[B]$ are taken directly from the semivariogram or from the mathematical function that describes its form. Once the unknown weights have been determined, the variable at location p is estimated by

$$\hat{Y}_p = W_1 Y_1 + W_2 Y_2 + W_3 Y_3 \qquad (5.78)$$

The estimation variance is

$$s_\epsilon^2 = W_1 \gamma(h_{1p}) + W_2 \gamma(h_{2p}) + W_3 \gamma(h_{3p}) + \lambda \qquad (5.79)$$

That is, the variance of the estimate is essentially the weighted sum of the semi-variances for the distances to the points used in the estimation, plus a contribution from the λ coefficient that is equivalent to a constant term. Kriging has two powerful advantages over conventional estimation procedures such as those used for contour mapping. Kriging produces estimates that, on the average, have the smallest possible error, and also produces an explicit statement of the magnitude of this error.

To demonstrate punctual kriging, we will estimate the elevation of the water table at point p on the map shown in Figure 5.63. The estimate will be made from known elevations measured in three observation wells. The map coordinates of the wells and the distances between them are given in Table 5.12. A prior structural analysis has produced the semivariogram in Figure 5.64, which is linear with a slope of 4.0 m²/km within a neighborhood of 20 km. Values of the semivariance corresponding to distances between the wells are also given in Table 5.12; these may be read directly off the semivariogram or calculated from the slope.

The equations that must be solved to find the weights W_i in this example are

$$W_1(0) + W_2(12.2) + W_3(11.5) + \lambda = 4.0$$
$$W_1(12.2) + W_2(0) + W_3(18.1) + \lambda = 12.1$$
$$W_1(11.5) + W_2(18.1) + W_3(0) + \lambda = 7.9$$
$$W_1 + W_2 + W_3 + 0 = 1.0$$

not correct

Set in matrix form, this is

$$\begin{bmatrix} 0 & 12.2 & 11.5 & 1.0 \\ 12.2 & 0 & 18.1 & 1.0 \\ 11.5 & 18.1 & 0 & 1.0 \\ 1.0 & 1.0 & 1.0 & 0 \end{bmatrix} \cdot \begin{bmatrix} W_1 \\ W_2 \\ W_3 \\ \lambda \end{bmatrix} = \begin{bmatrix} 4.0 \\ 12.1 \\ 7.9 \\ 1.0 \end{bmatrix}$$

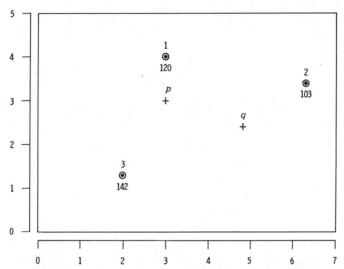

FIGURE 5.63 Map showing water table elevations (in meters) at three observation wells. Estimates of the water table elevation will be made at locations p and q. Coordinates are given in kilometers from an arbitrary origin.

TABLE 5.12 Observation Wells Used to Estimate Water Table Elevation at p

	X_1 Coordinate	X_2 Coordinate	Water Table Elevation
Well 1	3.0	4.0	120
2	6.3	3.4	103
3	2.0	1.3	142
Location p	3.0	3.0	

Distances between Wells and Location p

	1	2	3	p
Well 1	0	3.35	2.88	1.00
2		0	4.79	3.32
3			0	1.97

Semivariances for Distances between Wells and Location p

	1	2	3	p
Well 1	0	13.42	11.52	4.00
2		0	19.14	13.30
3			0	7.89

The inverse of the left-hand matrix can be found using the procedure described in Chapter 3, although it may be necessary to rearrange the order of the equations to avoid zeroes along the main diagonal. The inverse is

$$
\begin{bmatrix}
-0.0680 & 0.0326 & 0.0354 & 0.1932 \\
0.0326 & -0.0433 & 0.0106 & 0.4072 \\
0.0354 & 0.0106 & -0.0461 & 0.3995 \\
0.1932 & 0.4072 & 0.3995 & -9.5851
\end{bmatrix}
$$

The unknown weights can now be found by post-multiplying the transpose by the right-hand vector of the semivariances, yielding

$$
\begin{bmatrix}
W_1 \\
W_2 \\
W_3 \\
\lambda
\end{bmatrix}
=
\begin{bmatrix}
0.5954 \\
0.0975 \\
0.3071 \\
-0.7298
\end{bmatrix}
$$

The estimate of the elevation of the water table at location p is found by inserting the appropriate weights in the linear equation (5.78):

$$
\hat{Y}_p = 0.5954(120) + 0.0975(103) + 0.3071(142)
$$

$$
= 125.1 \text{ meters}
$$

Similarly, the error variance is the weighted sum of the semivariances for the distances from the control points to the location of the estimate. In matrix terms, $s_\epsilon^2 = [W]^T[B]$:

$$
s_\epsilon^2 = 0.5954(4) + 0.0975(12.1) + 0.3071(7.9) - 0.7298(1)
$$

$$
= 5.25 \text{ meters}^2
$$

The standard error of the estimate is simply the square root of the estimation variance, or

$$
s_\epsilon = \sqrt{5.25} = 2.3 \text{ meters}
$$

If we assume the errors of estimation are normally distributed about the true value, we can use the standard error as a confidence band around the estimates. The probability that the true elevation of the water table at point p is within one standard error above or below the value estimated is 68%, and the probability is 95% that the true elevation lies within two standard errors. That is, the water table elevation at this location must be

$$
Y_p = 125.1 \pm 4.6 \text{ meters, with 95\% probability}
$$

At every point on this map we can estimate the elevation of the water table and can also determine the standard errors of these estimates. From these we can construct two maps; the first is based on the estimates themselves and is a "best guess" of the configuration of the mapped variable. The second is an *error map* showing the confidence envelope that surrounds this estimated surface; it expresses the relative reliability of the first map. In areas of poor control, the error map will show large values, indicating that the estimates are subject to high variability. In areas of dense control the error map will show low values, and at the control points themselves the estimation error will be zero.

The system of equations used to find the kriging weights must be solved for every estimated location, unless the samples are arranged in a regular pattern so the distances between points remain the same. If we wish to estimate the elevation of the water table at point q on Figure 5.63, the distances between q and the three observation wells must be considered. These distances are

$$\text{from well} \quad \begin{matrix} 1 \\ 2 \\ 3 \end{matrix} \quad \begin{bmatrix} 2.4 \\ 1.6 \\ 3.0 \end{bmatrix}$$

The corresponding semivariances, taken from Figure 5.64, are

$$\begin{bmatrix} 9.6 \\ 6.2 \\ 12.0 \end{bmatrix}$$

Since the arrangement of the observation wells remains the same, all distances between them are the same and the left-hand side of the set of simultaneous equations is unchanged. The inverse is likewise unchanged, so multiplying it by the new vector of semivariances will yield weights for estimating the elevation of the water table at point q. This new set of weights is

$$\begin{bmatrix} W_1 \\ W_2 \\ W_3 \\ \lambda \end{bmatrix} = \begin{bmatrix} 0.1676 \\ 0.5796 \\ 0.2528 \\ -0.3711 \end{bmatrix}$$

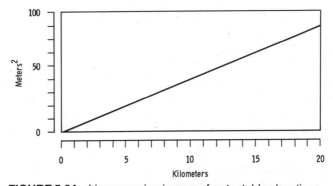

FIGURE 5.64 Linear semivariogram of water table elevations in an area which includes the map in Figure 5.63. Semivariogram has a slope of 4.0 m²/km within a 20-km neighborhood.

The estimate of the water table elevation is

$$\hat{Y}_q = 0.1676(120) + 0.5796(103) + 0.2528(142)$$
$$= 115.7 \text{ meters}$$

and the estimation variance is

$$s_\epsilon^2 = 0.1676(9.6) + 0.5796(6.3) + 0.2528(12.0) - 0.3711(1)$$
$$= 7.91 \text{ meters}^2$$

The standard error of the estimate at q is

$$s_\epsilon = \sqrt{7.91} = 2.8 \text{ meters}$$

so the elevation of the water table at this new location can be expressed as

$$Y_q = 115.7 \pm 5.6 \text{ meters, with 95\% probability}$$

The ground water surface is lower at point q than at p, and the standard error is greater, reflecting the greater total distance to the control wells.

If one of the control points is changed, some of the distances are also changed, and the system of equations must be solved anew. In Figure 5.65, observation well 2A has been drilled at a site closer to location p, and a water table elevation of 115 m measured for the regionalized variable. The new interpoint distances and corresponding semivariances are given in Table 5.13. The set of simultaneous equations is now

$$\begin{bmatrix} 0 & 7.2 & 11.5 & 1.0 \\ 7.2 & 0 & 8.4 & 1.0 \\ 11.5 & 8.4 & 0 & 1.0 \\ 1.0 & 1.0 & 1.0 & 0 \end{bmatrix} \cdot \begin{bmatrix} W_1 \\ W_2 \\ W_3 \\ \lambda \end{bmatrix} = \begin{bmatrix} 4.0 \\ 4.0 \\ 7.9 \\ 1.0 \end{bmatrix}$$

whose solution is

$$\begin{bmatrix} W_1 \\ W_2 \\ W_3 \\ \lambda \end{bmatrix} = \begin{bmatrix} 0.4545 \\ 0.3858 \\ 0.1598 \\ -0.6001 \end{bmatrix}$$

The new estimate of the water table elevation at point p, using information from observation well 2A, is

$$\hat{Y}_p = 0.4545(120) + 0.3858(115) + 0.1598(142)$$
$$= 121.6 \text{ meters}$$

The variance of this new estimate is

$$s_\epsilon^2 = 0.4545(4.0) + 0.3858(4.0) + 0.1598(7.9) - 0.6001(1)$$
$$= 4.02 \text{ meters}^2$$

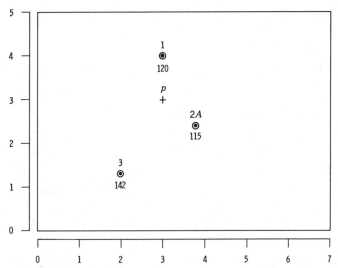

FIGURE 5.65 Map showing water table elevations (in meters) at three observation wells. Well 2A is closer to the location p being estimated than is well 2 in Figure 5.63.

TABLE 5.13 Second Set of Observation Wells Used to Estimate Water Table Elevation at p

	X_1 Coordinate	X_2 Coordinate	Water Table Elevation
Well 1	3.0	4.0	120
2A	3.8	2.4	115
3	2.0	1.3	142
Location p	3.0	3.0	

	Distances between Wells and Location p			
	1	2A	3	p
Well 1	0	1.79	2.88	1.00
2A		0	2.11	1.00
3			0	1.97

	Semivariances for Distances between Wells and Location p			
	1	2A	3	p
Well 1	0	7.16	11.52	4.00
2A		0	8.44	4.00
3			0	7.89

The standard error of the estimate at p is now

$$s_\epsilon = \sqrt{4.02} = 2.0 \text{ meters}$$

which is a somewhat lower value than was found when using observation well 2 rather than well 2A. This illustrates the fact that the estimation errors are reduced if the control points are closer to the location where the estimate is to be made.

Suppose one of the control points coincides with the location to be estimated. Then, one of the values on the right-hand side of the matrix equation becomes zero, and the remaining values become equal to some of the values in the left-hand matrix. We can determine the effect of this change in our example by assuming that an observation well 2B is drilled at location p, and the water level is measured at 125 m. The distance between any point i and point 2B is now the same as the distance between any point i and location p. The semivariances are likewise the same, so the set of simultaneous equations becomes

$$\begin{bmatrix} 0 & 4.0 & 11.5 & 1.0 \\ 4.0 & 0 & 7.9 & 1.0 \\ 11.5 & 7.9 & 0 & 1.0 \\ 1.0 & 1.0 & 1.0 & 0 \end{bmatrix} \cdot \begin{bmatrix} W_1 \\ W_2 \\ W_3 \\ \lambda \end{bmatrix} = \begin{bmatrix} 4.0 \\ 0 \\ 7.9 \\ 1.0 \end{bmatrix}$$

The vector of weights can be calculated and is, as we should expect,

$$\begin{bmatrix} W_1 \\ W_2 \\ W_3 \\ \lambda \end{bmatrix} = \begin{bmatrix} 0.0000 \\ 1.0000 \\ 0.0000 \\ 0.0000 \end{bmatrix}$$

If these weights are used to estimate p, we see that the estimated elevation is exactly equal to the measured value of the water level in well 2B.

$$\hat{Y}_p = 0.0000(120) + 1.0000(125) + 0.000(142)$$

$$= 125.0 \text{ meters}$$

Also, as we should expect, the estimation variance is

$$s_\epsilon^2 = 0.0000(4.0) + 1.0000(0) + 0.0000(7.9) + 0.0000(1)$$

$$= 0.00 \text{ meters}^2$$

This demonstrates what is meant by the oft-heard statement that kriging is an "exact interpolator"; it predicts the actual values measured at the known points, and does this with zero error. Of course, we do not ordinarily produce estimates for locations already known, but this does occasionally occur when using punctual kriging for contouring. If any of the control points happen to coincide with grid nodes, kriging will produce the correct, error-free values. We also can be assured that the estimated surface must pass exactly through all control points, and that the confidence bands around the estimated surface go to zero at the control points.

In these examples, we have assumed that each estimate is made using only three control points in order to simplify the mathematics as much as possible. In actual practice, we would expect to use more points, perhaps many more, in making each estimate. Every control point used in an estimate must be weighted, and finding each weight requires another equation. Most contouring routines use 16 or more control points to estimate every grid intersection, which means a set of at least 17 simultaneous equations must be solved for each location. When used in this manner as a contouring procedure, kriging is computationally expensive.

In theory, the number of points needed to estimate a location varies with the local density of control. All control points within the neighborhood around the location to be estimated provide information and should be considered. In practice, many of these points may be redundant, and their use will improve the estimate only slightly. Practical rules-of-thumb have been developed for contour mapping by kriging, which limit the number of control points actually needed to a subset of the points within the zone of influence or neighborhood. The optimum number of control points is determined by the semivariogram and the spatial pattern of the points (Olea, 1982). The structural analysis thus plays a doubly critical role in kriging; it provides the semivariogram necessary to solve the kriging equations, and also determines the neighborhood size within which the control points are selected for each estimate.

Universal Kriging

A significant problem with punctual kriging is that it will not work unless the regionalized variable being mapped is stationary. In the presence of a trend, or slow change in average value, a linear estimator is no longer unbiased. The computed estimates will be systematically shifted upwards or downwards from the true values depending upon the arrangement of the control points and the direction of dip of the surface.

In the parlance of geostatistics, a nonstationary regionalized variable can be regarded as having two components. The *drift* consists of the average or expected value of the regionalized variable within a neighborhood, and is a slowly varying, nonstationary part of the surface. The *residual* is the difference between the actual measurements and the drift. Obviously, if the drift is removed from a regionalized variable, the residuals must be stationary and kriging can be applied to them. Universal kriging can thus be regarded as consisting of three operations: First, the drift must be estimated and removed. Then, the stationary residuals are kriged to obtain the needed estimates. Finally, the estimated residuals are combined with the drift to obtain estimates of the actual surface.

The drift is analogous to a trend surface, except that the drift estimate is based only on the control points within the neighborhood around the location being evaluated. In general, a different arrangement of control points will be found within the neighborhood around each separate location to be estimated, so the equation that defines the drift must be solved as many times as there are locations. This is not as onerous as it sounds, because the kriging equations must also be solved for each of these locations, and it is possible to combine the two operations.

We may define the drift as an arbitrary function of the control point coordinates, such as a low-order polynomial. The drift M at point p might be defined as either a first-order (eq. 5.80) or second-order (eq. 5.81) polynomial:

$$M_p = \alpha_1 X_{1i} + \alpha_2 X_{2i} \tag{5.80}$$

or

$$M_p = \alpha_1 X_{1i} + \alpha_2 X_{2i} + \alpha_3 X_{1i}^2 + \alpha_4 X_{1i} X_{2i} + \alpha_5 X_{2i}^2 \tag{5.81}$$

Here, X_{1i} and X_{2i} are the geographic coordinates of the ith control point within the neighborhood, and the α's are unknown drift coefficients that must be found. Before this can be done, however, a structural analysis must be performed to determine the best combination of neighborhood size and expression for the drift. As noted in the section on semivariograms, this is not a trivial undertaking, because the drift model and neighborhood size are interdependent.

The expressions for the drift may be incorporated into the system of simultaneous equations used to find the kriging weights as additional constraints. Solving this expanded set of equations will produce a set of weights for the kriged estimate, which will include the effect of the specified drift within the local neighborhood. The drift expressions relate the geographic coordinates of each control point to the geographic coordinates of the location being kriged.

The complexity of the drift model, the size of the neighborhood, and the form of the semivariogram of the residuals from the drift are interrelated. This means that the variance of the residuals depends in part on the somewhat arbitrary specification of the drift. In universal kriging, the weights to be applied to control points must be determined, and so must the coefficients of the drift. Since more

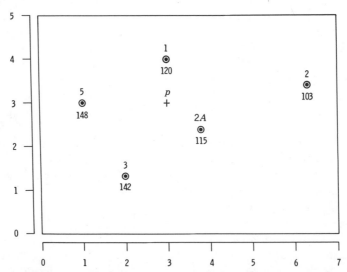

FIGURE 5.66 Map showing water table elevations (in meters) at five observation wells. Estimates of the water table elevation will be made by universal kriging at location p and at the southwest corner of the map.

terms are being estimated, a larger sample of control points within the neighborhood must be used. The simplest example consists of kriging a point, assuming the drift is linear and that the semivariogram of the residuals from the drift is also linear. The linear drift model, given in eq. (5.80), has two coefficients, so a minimum of three points must be used in the drift estimation process, or we will run out of degrees of freedom.

If we wish to estimate both the drift and the regionalized variable by universal kriging, additional control points are required in order to provide degrees of freedom for the kriging coefficients. Otherwise, the kriging process will produce the same estimate for both the drift and the kriged surface. A reasonable choice might be five control points, giving three degrees of freedom for the determination of the drift, plus two additional degrees of freedom for estimation of the surface itself.

The east-west coordinate of control point i may be indicated by X_{1i} and the north-south coordinate as X_{2i}. The coordinates of the kriged location are X_{1p} and X_{2p}. We must determine a set of five weights, plus a coefficient for the unity constraint, plus two more constraints for the linear drift. This requires a set of eight simultaneous equations:

$$W_1\gamma(h_{11}) + W_2\gamma(h_{12}) + W_3\gamma(h_{13}) + W_4\gamma(h_{14}) + W_5\gamma(h_{15}) + \lambda + \alpha_1 X_{11} + \alpha_2 X_{21} = \gamma(h_{1p})$$

$$W_1\gamma(h_{12}) + W_2\gamma(h_{22}) + W_3\gamma(h_{23}) + W_4\gamma(h_{24}) + W_5\gamma(h_{25}) + \lambda + \alpha_1 X_{12} + \alpha_2 X_{22} = \gamma(h_{2p})$$

$$W_1\gamma(h_{13}) + W_2\gamma(h_{23}) + W_3\gamma(h_{33}) + W_4\gamma(h_{34}) + W_5\gamma(h_{35}) + \lambda + \alpha_1 X_{13} + \alpha_2 X_{23} = \gamma(h_{3p})$$

$$W_1\gamma(h_{14}) + W_2\gamma(h_{24}) + W_3\gamma(h_{34}) + W_4\gamma(h_{44}) + W_5\gamma(h_{45}) + \lambda + \alpha_1 X_{14} + \alpha_2 X_{24} = \gamma(h_{4p})$$

$$W_1\gamma(h_{15}) + W_2\gamma(h_{25}) + W_3\gamma(h_{35}) + W_4\gamma(h_{45}) + W_5\gamma(h_{55}) + \lambda + \alpha_1 X_{15} + \alpha_2 X_{25} = \gamma(h_{5p})$$

$$W_1 + W_2 + W_3 + W_4 + W_5 + 0 + 0 + 0 = 1$$

$$W_1 X_{11} + W_2 X_{12} + W_3 X_{13} + W_4 X_{14} + W_5 X_{15} + 0 + 0 + 0 = X_{1p}$$

$$W_2 X_{21} + W_2 X_{22} + W_3 X_{23} + W_4 X_{24} + W_5 X_{25} + 0 + 0 + 0 = X_{2p}$$

In matrix form, this becomes

$$
\begin{bmatrix}
\gamma(h_{11}) & \gamma(h_{12}) & \gamma(h_{13}) & \gamma(h_{14}) & \gamma(h_{15}) & 1 & X_{11} & X_{21} \\
\gamma(h_{12}) & \gamma(h_{22}) & \gamma(h_{23}) & \gamma(h_{24}) & \gamma(h_{25}) & 1 & X_{12} & X_{22} \\
\gamma(h_{13}) & \gamma(h_{23}) & \gamma(h_{33}) & \gamma(h_{34}) & \gamma(h_{35}) & 1 & X_{13} & X_{23} \\
\gamma(h_{14}) & \gamma(h_{24}) & \gamma(h_{34}) & \gamma(h_{44}) & \gamma(h_{45}) & 1 & X_{14} & X_{24} \\
\gamma(h_{15}) & \gamma(h_{25}) & \gamma(h_{35}) & \gamma(h_{45}) & \gamma(h_{55}) & 1 & X_{15} & X_{25} \\
1 & 1 & 1 & 1 & 1 & 0 & 0 & 0 \\
X_{11} & X_{12} & X_{13} & X_{14} & X_{15} & 0 & 0 & 0 \\
X_{21} & X_{22} & X_{23} & X_{24} & X_{25} & 0 & 0 & 0
\end{bmatrix}
\cdot
\begin{bmatrix}
W_1 \\ W_2 \\ W_3 \\ W_4 \\ W_5 \\ \lambda \\ \alpha_1 \\ \alpha_2
\end{bmatrix}
=
\begin{bmatrix}
\gamma(h_{1p}) \\ \gamma(h_{2p}) \\ \gamma(h_{3p}) \\ \gamma(h_{4p}) \\ \gamma(h_{5p}) \\ 1 \\ X_{1p} \\ X_{2p}
\end{bmatrix}
$$

An additional step that can simplify calculations is to shift the origin of the coordinate system to the location being kriged. The coordinates X_{1p} and X_{2p} then become zero. This alters all of the X_{1i} and X_{2i} coordinates but not the interpoint distances, so the kriging weights are unchanged.

We may extend our example problem, which is based on data from a western Kansas aquifer, to demonstrate the steps in universal kriging. Figure 5.66 shows the locations of five observation wells that will be used to estimate the drift and kriged value of the water table elevation at location p. We will assume that Figure 5.64 now represents the estimated semivariogram for the residuals and that it is linear in form with a slope of 4.0 m²/km. All of the basic information required is given in Table 5.14, which also includes the necessary semivariances. The equation that must be solved to estimate the water table elevation at point p is:

$$
\begin{bmatrix}
0 & 13.4 & 11.5 & 7.2 & 8.9 & 1 & 0 & 1.0 \\
13.4 & 0 & 19.1 & 10.8 & 21.3 & 1 & 3.3 & 0.4 \\
11.5 & 19.1 & 0 & 8.4 & 7.9 & 1 & -1.0 & -1.7 \\
7.2 & 10.8 & 8.4 & 0 & 11.5 & 1 & 0.8 & -0.6 \\
8.9 & 21.3 & 7.9 & 11.5 & 0 & 1 & -2.0 & 0 \\
1 & 1 & 1 & 1 & 1 & 0 & 0 & 0 \\
0 & 3.3 & -1.0 & 0.8 & -2.0 & 0 & 0 & 0 \\
1.0 & 0.4 & -1.7 & -0.6 & 0 & 0 & 0 & 0
\end{bmatrix}
\cdot
\begin{bmatrix}
W_1 \\ W_2 \\ W_3 \\ W_4 \\ W_5 \\ \lambda \\ \alpha_1 \\ \alpha_2
\end{bmatrix}
=
\begin{bmatrix}
4.0 \\ 13.3 \\ 7.9 \\ 4.0 \\ 8.0 \\ 1 \\ 0 \\ 0
\end{bmatrix}
$$

TABLE 5.14 Observation Wells Used to Estimate Water Table Elevation and Drift at p

	X_1 Coordinate	X_2 Coordinate	Water Table Elevation
Well 1	3.0	4.0	120
2	6.3	3.4	103
3	2.0	1.3	142
2A	3.8	2.4	115
5	1.0	3.0	148
Location p	3.0	3.0	

Distances between Wells and Location p

	1	2	3	2A	5	p
Well 1	0	3.35	2.88	1.79	2.24	1.00
2		0	4.79	2.69	5.32	3.32
3			0	2.11	1.97	1.97
2A				0	2.86	1.00
5					0	2.00

Semivariances for Distances Between Wells and Location p

	1	2	3	2A	5	p
Well 1	0	13.42	11.52	7.16	8.94	4.00
2		0	19.14	10.77	21.26	13.30
3			0	8.44	7.89	7.89
2A				0	11.45	4.00
5					0	8.00

Solving the equation gives a set of eight coefficients, the first five of which are the kriging weights.

$$
\begin{bmatrix} W_1 \\ W_2 \\ W_3 \\ W_4 \\ W_5 \\ \lambda \\ \alpha_1 \\ \alpha_2 \end{bmatrix} = \begin{bmatrix} 0.4119 \\ -0.0137 \\ 0.0934 \\ 0.4126 \\ 0.0957 \\ -0.7245 \\ 0.0660 \\ 0.0229 \end{bmatrix}
$$

The estimate of the water table elevation at location p is

$$
\begin{aligned}
\hat{Y}_p &= 0.4119(120) - 0.0137(103) + 0.0934(142) \\
&\quad + 0.4126(115) + 0.0957(148) \\
&= 122.9 \text{ meters}
\end{aligned}
$$

which is only slightly different than the results obtained from three observations without assuming a drift. The estimation error variance can also be calculated in exactly the same manner as without a drift, by pre-multiplying the vector of right-hand terms $[B]$ by the transpose of the solution vector, $[W]$. The estimation error variance is 8.1 m^2.

This example does not illustrate a major distinction between simple punctual kriging and universal kriging with drift, because in this instance the two procedures yield almost identical estimates. However, punctual kriging, in common with other weighted averaging methods, cannot extrapolate much beyond the range of the control points. That is, most estimated values will lie on the slopes of the surface, and the highest and lowest points on the surface usually will be defined by control points. Suppose we estimate the water table elevation at a location where it seems obvious that the surface should be outside the interval which the observation wells define. The water table appears to dip from west to east, dropping almost 40 m between observation well 2 and well 3. If this dip continues, we would expect water levels higher than 142 m at locations west of observation well 3, and levels below 103 m at locations east of observation well 2.

We will estimate the water table elevation in the extreme southwest corner of the map, at coordinates $X_1 = 0$, $X_2 = 0$. We will first use simple punctual kriging and the set of five observation wells. This will yield the following set of weights:

$$
\begin{bmatrix} W_1 \\ W_2 \\ W_3 \\ W_4 \\ W_5 \end{bmatrix} = \begin{bmatrix} -0.1221 \\ 0.0110 \\ 0.7523 \\ -0.0307 \\ 0.3895 \end{bmatrix}
$$

The estimate of the water level is

$$\hat{Y} = -0.1221(120) + 0.0110(103) + 0.7523(142)$$
$$- 0.0307(115) + 0.3895(148)$$
$$= 147.4 \text{ meters}$$

As we would expect, the estimate is based almost entirely on the closest observation wells, but is within the interval defined by the highest and lowest observed values. The estimation error variance is $s_\epsilon^2 = 17.3 \text{ m}^2$.

If a first-degree drift is assumed, the universal kriging coefficients are

$$\begin{bmatrix} W_1 \\ W_2 \\ W_3 \\ W_4 \\ W_5 \\ \lambda \\ \alpha_1 \\ \alpha_2 \end{bmatrix} = \begin{bmatrix} -0.5594 \\ -0.3020 \\ 1.3133 \\ 0.1451 \\ 0.4030 \\ 26.3832 \\ -1.7940 \\ -4.1795 \end{bmatrix}$$

Using these weights yields an estimate of the water table of

$$\hat{Y} = -0.5594(120) - 0.3020(103) + 1.3133(142)$$
$$+ 0.1451(115) + 0.4030(148)$$
$$= 164.6 \text{ meters}$$

which is much higher than the highest control elevation. Universal kriging has considered the dip of the water table, or drift, within the local neighborhood, and has projected this to the location being kriged. The estimation error variance is $s_\epsilon^2 = 26.8 \text{ m}^2$, a much larger value that incorporates the uncertainty in both the drift estimation and the estimation of the regionalized variable itself.

Calculating the Drift

The observations selected to estimate the kriged value at some location p are chosen from within the range or zone of influence around p. Kriging finds a set of weights that equates the sum of the semivariances between the observations to the semi-variances between the observations and location p. Suppose, however, that all of the available observations were beyond the range. The semivariance γ_{ip}^2 between location p and a distant observation i would be identical for all observations, and equal to the variance s_0^2 of the regionalized variable (or if a drift is present, equal to the variance of the residuals from the drift). That is, the first n elements of the right-hand vector of the kriging equation would contain values of s_0^2 rather than γ_{ip}^2. Since the regionalized variable at location p would be statistically independent of its value at each of the observation points, we could not predict the local value of the surface. Instead, the estimate we would obtain by solving the kriging system of equations would be based on the global or average properties of the regionalized variable. In other words, we would be estimating the drift itself.

We can compute the drift at a location p even if the observations used are within a neighborhood smaller than the range around p. All that is necessary is to replace the semivariances on the right-hand side of the kriging equations with the semivariances that would be observed if location p were so far from the control points that it was independent. Since the semivariance for all distances beyond the range is equal to the variance of the residuals, the terms for the right-hand side can be set equal to the variance of the residuals, s_0^2. Unfortunately, we again arrive at a circular impasse, because we cannot *know* s_0^2 until the drift has been calculated. Fortunately, the structural analysis allows us to make an a priori estimate of its value, because it is equal to the semivariance at the sill, or beyond the range.

It should be emphasized that the drift is an arbitrary but convenient construct, necessary to satisfy the requirement for stationarity in the regionalized variable. There may be many alternative combinations of drift model, neighborhood size, and estimated semivariogram that will satisfactorily represent the structure of a regionalized variable. The choice of a specific combination depends upon the availability of data, computational convenience, and other considerations. We will continue our simple example, assuming that the drift in water table elevations is linear, and that the semivariogram for the residuals is also linear. From a structural analysis, we may conclude that the range of the regionalized variable extends up to 30 km. Since the slope of the semivariogram is 4 m²/km, the variance beyond the range should be about 4 × 30, or 120 m².

The right-hand part of the kriging matrix will have the following appearance when calculating the drift at point p:

$$[C] = \begin{bmatrix} 120 \\ 120 \\ 120 \\ 120 \\ 120 \\ 1 \\ 0 \\ 0 \end{bmatrix}$$

The left-hand side, $[A]$, is unchanged from the kriging calculation, so all that is necessary to estimate the drift is to multiply the inverse of $[A]$ by $[C]$. This yields a set of five weights, M_i, which are used to compute the drift, plus three constant terms that contribute to the estimation error variance for the drift. The solution vector, $[M]$, is

$$[M] = \begin{bmatrix} 0.1311 \\ 0.3702 \\ 0.2202 \\ -0.1587 \\ 0.4372 \\ 109.2283 \\ -0.9048 \\ 0.4935 \end{bmatrix}$$

The drift at location p is found by multiplying the elevations at the observation wells by the appropriate drift coefficients and summing.

$$\overline{M} = 0.1311(120) + 0.3702(103) + 0.2202(142)$$
$$- 0.1587(115) + 0.4372(148)$$
$$= 131.6 \text{ meters}$$

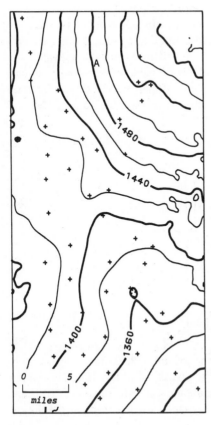

FIGURE 5.67 Map showing elevation of the water table in the Equus Beds, a major aquifer in south-central Kansas. Map produced by universal kriging, assuming a first-order drift. Contours are in feet above sea level. Crosses indicate observation wells.

FIGURE 5.68 Map showing standard error of the estimates of water table elevations in the Equus Beds. Contour interval is 1 ft.

The estimation error is found as a variance by pre-multipying [C] by the transpose of [M]. For this example,

$$s_m^2 = 229.3 \text{ meters}^2$$

As before, the standard error of the estimate is the square root of the estimation variance, or

$$s_m = \sqrt{229.35} = 15.1 \text{ meters}$$

Again, by assuming normality, we can make probabilistic statements in the form of a confidence interval about the true value of the drift. For example, the probability is 95% that the true linear drift lies within the interval 131.6 ± 30.2 m, or between 199.0 and 259.4 m. (We must remember that there is nothing "real" about the drift in a physical sense. The confidence interval merely indicates the sampling error and gives the likely value for the drift if the regionalized variable were to be repeatedly sampled and the drift recalculated.)

In order to keep the numerical examples in this section computationally tractable, many simplifications have been made. For example, the number of observations used is the smallest possible. In actual applications, many more control points

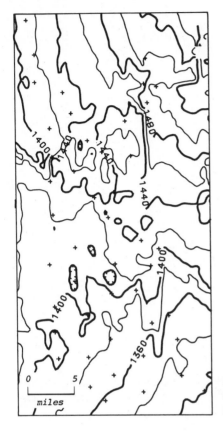

FIGURE 5.69 Map showing first-degree drift of the water table in the Equus Beds. Contours are in feet above sea level.

should be considered because this will improve the accuracy of the kriged estimate and reduce the estimation error. Also, the semivariogram is assumed to be linear because this is the simplest model possible, having only one parameter. A strictly linear semivariogram would not have a sill, but rather would continue on to an infinite variance. Here, we have in effect assumed that the semivariogram is linear up to some value, then breaks and becomes a constant. Armstrong and Jabin (1981) have pointed out that semivariograms possessing sudden changes in slope may lead to unstable solutions and the calculation of negative variances. It may be better to use a continuous function to represent the semivariogram, such as the spherical model, although this complicates calculations somewhat. The possible severity of problems resulting from use of a linear semivariogram is not known but apparently is negligible in most instances, as linear semivariograms are widely used in practice. The uncertainty in specifying a correct a priori estimate of the variance of the residuals is not too troublesome because only the estimation error for the drift depends upon this parameter. The drift itself will be calculated correctly regardless of the value chosen as s_0^2.

An Example

You may now appreciate that kriging, even the highly simplified variations we have considered, may be arithmetically tedious. The practical application of kriging to a real problem is only possible by using a computer, because the estimations must be made repeatedly for a large number of locations in order to characterize the changes in a regionalized variable throughout an area. As an example, consider the map shown in Figure 5.67, adapted from a study by Olea (1982). This represents the elevation of the water table of the Equus Beds, an aquifer in south-central Kansas. A structural analysis, made over a much wider area, shows that the water table can be regarded as a nonstationary regionalized variable having a first-order drift. The range of the semivariogram for the residuals from the drift is 28 mi. The semivariogram can be modelled as a linear function with a slope of 60 ft²/mi.

In order to map the water table elevation by universal kriging, a contouring program was used to generate estimates of the water table at locations spaced at equal intervals across the map area. The map in Figure 5.67 represents $31 \times 61 = 1891$ kriging estimates, each based on the eight nearest control wells

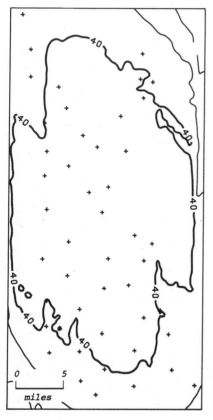

FIGURE 5.70 Map showing standard error of the first-degree drift of the water table elevation in the Equus Beds. Contour interval is 10 ft.

selected by an octant search around the location being estimated. Each estimate in turn required the solution of a set of eleven simultaneous equations.

In addition to the map of the water table itself, kriging also was used to produce a map of the standard error of the estimates, shown in Figure 5.68. The standard error is zero at each of the 47 observation wells, but increases with distance away from the known control points (Fig. 5.68). With 95% probability, the true surface of the ground water table lies within an interval defined by plus or minus twice the indicated value. For example, at point A the water table is estimated to be at about 1480 ft. Because there are relatively few observation wells near this location, the map of standard error indicates a value of over 6 ft. Therefore, the true elevation at this point must be 1480 ± 12 ft, or between 1468 and 1492 ft, with 95% confidence.

In this application there is little geologic significance to the drift itself, but it can be mapped if desired. Figure 5.69 shows the first-order drift of the water table elevation; Figure 5.70 is the standard error map for the drift. The residuals from the drift, found by subtracting the drift map from the kriged map, are shown in Figure 5.71. Areas of negative residuals where the water table is lower than the drift are shown by shading.

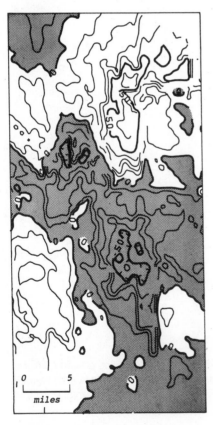

FIGURE 5.71 Map showing residuals from the first-degree drift of water table elevations in the Equus Beds. Contour interval is 10 ft. Areas of negative residuals where the water table is below the drift are shaded.

The estimation of the value of a variable at a point from observations which are themselves points is only one application of kriging. The method can be extended to the estimation of the value of an area from samples that consist of areas, or to the estimation of the value contained within a volume from samples that are volumes. The latter application is especially important in mining, where the estimated quantities may be the grade of ore in a stope block, and the observations are the assay grades of diamond-drill core samples. The estimation procedure is essentially the same as that presented here, but there are additional complexities that arise because of variation within the areas or volumes. An excellent introduction to the use of kriging for mine evaluation is given in the slim volume by Clark (1979). More extensive discussions, including considerations of more advanced topics in geostatistics, are presented by David (1977) and Journel and Huijbregts (1978).

Trend Surfaces

Trend analysis is the geology profession's name for a mathematical method of separating map data into two components—that of a regional nature, and local fluctuations. This has been done intuitively or graphically by geologists for years. Petroleum geologists, for example, refer to "regional dip" or "basinal configuration" as opposed to "local structures." Petrologists may speak of the "regional grain" of a metamorphic terrain. Geophysicists have long been accustomed to the concept of "regional trends" and "local anomalies." All of these expressions imply a belief that any given observation is the outcome of two interacting geologic forces or sets of forces—that which shaped the region or general geologic setting, and that which caused small areas to deviate from the regional pattern. Obvious examples may be drawn from structural geology. The Tertiary basins of Wyoming are the result of major movement along faults that extend deep into the crust. Within basins, folded structures have developed as the result of such diverse local agents as gravity sliding, minor antithetic faulting, failure of incompetent beds in areas of high dip, and so forth. If we regard the geometric shape of a basin as a regional structure, these small structures represent local deviations.

What we consider to be "regional" and "local" is largely subjective. It depends in part upon the size of the region being examined. If the buried Precambrian surface of the United States is under consideration, the basins and intervening mountain ranges of Wyoming will appear as anomalies or local deviations, as will the Black Hills, Ozark Dome, Michigan Basin, and other major features of the basement. Within an individual Wyoming basin, as we have seen, "regional" and "local" take on quite different meanings.

The availability of data exerts a very real influence on the nature of regional trends and local deviations. It is fruitless, for example, to search for meaningful local features whose suspected size approaches the spacing between sample points. Such features, whether they exist or not, simply cannot be detected. The relationship between the size of detectable features and the spacing of control points can be calculated for regular nets of points (Singer and Wickman, 1969, pp. 2–19), but not for the less tractable situation of irregularly distributed points.

Even the purpose of the geologic examination exerts an influence on our concept of the two spatial components. In a South African gold mine, for example, the only "deviations" of interest may be those which exceed a certain arbitrary value determined by the economics of the operation. On the other hand, a petroleum exploration group reexamining an area may be searching for small structural anomalies, knowing that the larger structures in the region have been adequately tested. The larger features are then considered to be part of the "regional trend."

For purposes of illustration, we can consider the set of observations in X and Y in Figure 5.72a. These may be separated into "regional" and "local" components in a variety of ways. Suppose we decide that the regional trend has the form of a straight line placed through the points in some manner. The data may then be separated into a major linear trend, shown by the heavy line, and three large anomalies, with a smaller anomaly at one extreme of the sample set (Fig. 5.72b). However, we may decide that a parabolic function would be more representative of the regional trend than a straight line. In that case, we might arrive at the separation shown in Figure 5.72c. The parabolic trend is very much different from that shown in Figure 5.72b, and the distribution of deviations consequently is different. We could postulate still more complex forms for the regional trend, resulting in smaller and smaller deviations, as in the cubic function fit to the data in Figure 5.72d. Eventually, our trend and samples would coincide, at which point there would be no residuals. Then, of course, there would be no separation of the data into components, and the purpose of the exercise would be defeated.

An obvious question at this point is how can the data be objectively separated into two components if the definition of the components is entirely subjective? This may be done if we use, instead of a geologic definition of trend and deviation, an operational definition which specifies the way in which our data are to be treated. A trend may then be defined as a linear function of the geographic coordinates of a set of observations so constructed that the squared deviations from the trend are minimized. Let us examine the three parts of this definition carefully.

1. It is based on the *geographic coordinates*. This means that an observation, whether it is the elevation of a horizon or the amount of gold in a vein, is considered to be in part a function of the location of the observation.
2. The trend is a *linear function*. That is, it has the form $Y = b_1X_1 + b_2X_2 + \cdots$, where the b's are coefficients and the X's are some combination of the geographic coordinates. The equation will yield values of $\hat{Y}$ that are the trend components of an observation. Note that the terms of the equation are added together.
3. The specific linear function chosen for the trend must *minimize the squared deviations* from the trend. The appearance of squared deviations at this point may bring to mind some material from Chapter 2. The sum of the squared deviations from the mean defines the variance of the sample. If we substitute "line" or "plane" for mean in this sentence, we see that the trend can be regarded as a function having the smallest variance about it. You may also

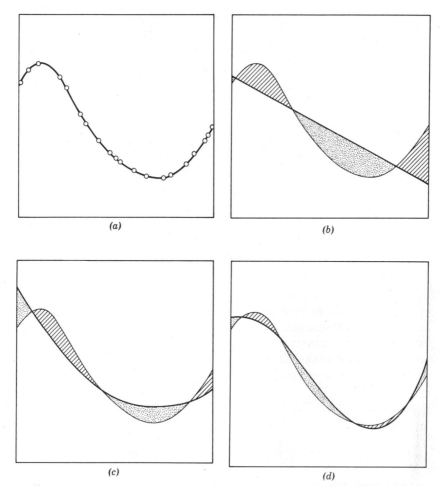

FIGURE 5.72 Concept of trend illustrated in two dimensions. (a) Collection of original data points and the line on which they lie. (b) Straight-line trend fit to the observations. (c) Parabolic trend. (d) Cubic trend. Shadings represent positive and negative residuals from the trends.

recall that we defined a line of regression in a similar fashion. Indeed, trend analysis is an adaptation of the statistical field of multiple regression, and the techniques have been borrowed directly from the discipline. In some cases, we can even use the powerful tests of hypotheses of multiple regression on geologic problems.

In Chapter 4, we computed a line of regression of Y on X, which was the line of best estimate of values of Y for any specified value of X. The equation of the line, $Y = b_0 + b_1X$ was found by solving the set of so-called *normal equations*

$$\Sigma Y = b_0 n + b_1 \Sigma X$$

$$\Sigma XY = b_0 \Sigma X + b_1 \Sigma X^2$$

(5.82)

for the unknown coefficients b_0 and b_1. The summations in this and subsequent equations extend from $i = 1$ to n. They have been omitted to simplify the written equations. We can easily expand this set of equations to the case where there are two independent variables, such as mutually perpendicular geographic coordinates. A linear trend surface is an equation of this type:

$$Y = b_0 + b_1X_1 + b_2X_2 \tag{5.83}$$

That is, a geologic observation, Y, may be regarded as a linear function of some constant value (b_0) related to the mean of the observations, plus an east-west (b_1) coordinate component and a north-south (b_2) component. Because the equation contains three unknowns, we need three normal equations to find the solution:

$$\Sigma Y = b_0 n + b_1\Sigma X_1 + b_2\Sigma X_2$$

$$\Sigma X_1 Y = b_0\Sigma X_1 + b_1\Sigma X_1^2 + b_2\Sigma X_1 X_2 \tag{5.84}$$

$$\Sigma X_2 Y = b_0\Sigma X_2 + b_1\Sigma X_1 X_2 + b_2\Sigma X_2^2$$

Solving this series of simultaneous equations will give the coefficients of the best-fitting linear trend surface, "best-fitting" being defined by the least-squares criterion.

The equations can be rewritten into matrix form as

$$\begin{bmatrix} n & \Sigma X_1 & \Sigma X_2 \\ \Sigma X_1 & \Sigma X_1^2 & \Sigma X_1 X_2 \\ \Sigma X_2 & \Sigma X_1 X_2 & \Sigma X_2^2 \end{bmatrix} \cdot \begin{bmatrix} b_0 \\ b_1 \\ b_2 \end{bmatrix} = \begin{bmatrix} \Sigma Y \\ \Sigma X_1 Y \\ \Sigma X_2 Y \end{bmatrix} \tag{5.85}$$

The similarity between this matrix equation and (4.33) should be obvious. Both may be regarded as curve fitting using two independent variables. Here, our two variables are X_1 and X_2, mutually perpendicular geographic coordinates. In fitting a quadratic curve to data along a line, the two variables were X and X^2. However, there basically is no difference between the two procedures. As an example of the fitting of a linear trend surface, we will consider the following problem.

The Anglo-Barren Oil Company (ABOC) has been given a concession in a remote part of northeastern Africa. The area is extremely inhospitable, nearly inaccessible, and almost completely unknown geologically. Under the terms of the concession, ABOC must drill ten wells during the year or forfeit their rights. The company management has decided to drill a series of widely spaced exploratory holes to develop the geologic background necessary for continued prospecting. Well locations within the 100-km² concession are shown in Figure 5.73 and are listed in Table 5.15 (coordinates are given in kilometers from the southwest corner of the concession) along with the elevations of the bottom of the Cretaceous. We wish to determine the linear trend, and from an examination of residuals, pick areas that seem the most promising for additional exploration.

We must first accumulate the sums, sums of powers, and sums of cross-products required in (5.84). The necessary entries are

$\Sigma X_1 = 539$	$\Sigma X_2 = 482$	$\Sigma Y = -4579$
$\Sigma X_1^2 = 36{,}934$	$\Sigma X_2^2 = 31{,}692$	$\Sigma X_1 Y = -211{,}098$
$\Sigma X_1 X_2 = 27{,}030$		$\Sigma X_2 Y = -232{,}342$

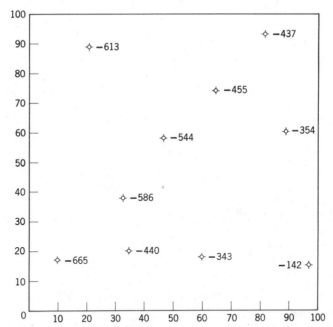

FIGURE 5.73 Map of well locations and elevations of the base of Cretaceous units in the Anglo-Barren Oil Company concession in northeastern Africa. Elevations are in meters below sea level. Coordinates of the map are in kilometers from southwest corner of concession.

Substituting these values into (5.85) gives

$$\begin{bmatrix} 10 & 539 & 482 \\ 539 & 36,943 & 27,030 \\ 482 & 27,030 & 31,692 \end{bmatrix} \cdot \begin{bmatrix} b_0 \\ b_1 \\ b_2 \end{bmatrix} = \begin{bmatrix} -4,579 \\ -211,098 \\ -232,342 \end{bmatrix}$$

This matrix equation can be solved by methods described in Chapter 3. The solutions

TABLE 5.15 Well Coordinates and Elevations of the Base of the Cretaceous within the Anglo-Barren Oil Company Concession

X_1 (km)	X_2 (km)	Y (m)
10.0	17.0	−665.0
21.0	89.0	−613.0
33.0	38.0	−586.0
35.0	20.0	−440.0
47.0	58.0	−544.0
60.0	18.0	−343.0
65.0	74.0	−455.0
82.0	93.0	−437.0
89.0	60.0	−354.0
97.0	15.0	−142.0

are

$$b_0 = -621.0, \quad b_1 = 4.8, \quad b_2 = -2.0$$

Having obtained the coefficients of the linear equation $Y = b_0 + b_1X_1 + b_2X_2$, we can calculate the expected or trend values, $\hat{Y}$, of the base of the Cretaceous at each of the ten wells. These, with the deviations $(Y_i - \hat{Y}_i)$, are listed in Table 5.16. We also can calculate measures of how well the trend surface corresponds to the observations by using (4.18)–(4.24), which we developed for the fitting of a line. In particular, we can compute the total variation as the sum of squares of the dependent variable, depth:

$$SS_T = \Sigma Y^2 - \frac{(\Sigma Y)^2}{n} = 215,324.9$$

For the estimated values of $\hat{Y}$ in Table 5.16, we can calculate the sum of squares due to the trend or regression:

$$SS_R = \Sigma \hat{Y}^2 - \frac{(\Sigma \hat{Y})^2}{n} = 193,861.4$$

The difference between these gives the sum of squares due to residuals or deviations from the trend and is

$$SS_D = SS_T - SS_R = 21,463.5$$

Now, the percentage of goodness-of-fit of the trend surface is

$$100\% \cdot R^2 = \frac{SS_R}{SS_T} = 90.0\%$$

The coefficient of multiple correlation is

$$R = \sqrt{R^2} = 0.95$$

From these extremely high values, we would conclude that the base of the Cretaceous in this area is an almost smooth, uniformly dipping plane. Deviations from the surface are relatively small, as a glance at Table 5.16 will confirm. Apparently,

TABLE 5.16   Geographic Coordinates, Elevation of Base of Cretaceous, Trend-Surface Estimate of Cretaceous Base, and Residual $(Y - \hat{Y})$

X_1 (km)	X_2 (km)	Y (m)	$\hat{Y}$ (m)	$(Y - \hat{Y})$ (m)
10.0	17.0	−665.0	−606.6	−58.3
21.0	89.0	−613.0	−695.7	82.7
33.0	38.0	−586.0	−537.8	−48.1
35.0	20.0	−440.0	−492.8	52.8
47.0	58.0	−544.0	−510.2	−33.7
60.0	18.0	−343.0	−369.2	26.2
65.0	74.0	−455.0	−455.5	0.5
82.0	93.0	−437.0	−411.5	−25.4
89.0	60.0	−354.0	−313.0	−40.9
97.0	15.0	−142.0	−186.1	44.1

the basal Cretaceous contact is a gently dipping surface of low relief, as shown in Figure 5.74. Although this simple analysis seems sufficient for this example, we have already discussed the possibility that a geologic trend may not be a plane, and may be extremely complex. Furthermore, we very rarely have any prior knowledge about what the functional form of the trend should be. Physicists can state that the path of a falling projectile should be a parabola, because they know something of the controlling forces—that is, the acceleration of gravity and the conservation of momentum. Geologists can seldom speak with any authority about what form a geologic surface or distribution ''should'' take. Instead, they do the next best thing, and approximate the unknown function with one of arbitrary nature. In particular, they use a polynomial expansion of the linear trend surface, introducing powers and cross-products of the geographic coordinates. Polynomials are extremely flexible, and if expanded to sufficiently high orders, can conform to very complex surfaces.

A note of caution should be injected at this point. Polynomial functions are used for geologic trend analysis merely as a matter of convenience. The equations that are necessary to find the coefficients of the trend may easily be established and solved by computer. Use of polynomials in no way intimates a belief that geologic processes are polynomial functions or even that they are linear. Their nature, unknown and perhaps ultimately unknowable, can only be approximated by a polynomial expansion. Other approximations (or in rare cases, model equations) may be more appropriate in specific instances; some of these will be discussed in subsequent parts of this section.

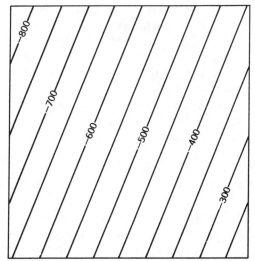

FIGURE 5.74 Trend-surface map of Anglo-Barren Oil Company concession. Contour interval is 50 m.

As we saw in Chapter 4, a least-squares line may be expanded to a second-order curve (parabola) by adding a squared term to the linear equation:

$$Y = b_0 + b_1X_1 + b_2X_1^2 \tag{5.86}$$

An equivalent expansion of (5.83) gives a second-degree trend surface:

$$Y = b_0 + b_1X_1 + b_2X_2 + b_3X_1^2 + b_4X_2^2 + b_5X_1X_2 \tag{5.87}$$

Note that the equation contains terms which are the squares of the two geographic coordinates and a cross-product term, X_1X_2. The expansion of this trend-surface equation to even higher powers should be relatively obvious. Each geographic variable is simply raised to a higher power, creating two new variables, then the appropriate cross-products of the two coordinates are calculated to give other new variables. For example,

$$Y = b_0 + \overbrace{b_1X_1 + b_2X_2}^{1} + \overbrace{b_3X_1^2 + b_4X_2^2 + b_5X_1X_2}^{2} \tag{5.88}$$
$$+ \overbrace{b_6X_1^3 + b_7X_2^3 + b_8X_1^2X_2 + b_9X_1X_2^2}^{3}$$

is a third-degree trend surface. The first-degree coefficients are b_1 and b_2. The coefficients b_3, b_4, and b_5 are second-degree, because the variables in these terms are of the form $X_3 = (X_1 \cdot X_1)$, $X_4 = (X_2 \cdot X_2)$, and $X_5 = (X_1 \cdot X_2)$. That is, the variables are the products of multiplying two original variables together. Similarly, b_6, b_7, b_8, and b_9 are third-degree coefficients, as the variables in these terms result from the multiplication of three original variables together; that is, $X_6 = (X_1 \cdot X_1 \cdot X_1)$, $X_7 = (X_2 \cdot X_2 \cdot X_2)$, $X_8 = (X_1 \cdot X_1 \cdot X_2)$, and $X_9 = (X_1 \cdot X_2 \cdot X_2)$.

Petroleum explorationists have made good use of trend surface analysis in the search for oil and gas in central Alberta, Canada. The Alberta Basin is extremely petroliferous, with production coming primarily from Lower Cretaceous sands and from much deeper carbonate reefs of Upper Devonian age. Many more wells penetrate the shallower Cretaceous horizons than the Devonian, especially near the Rocky Mountain front where reef prospects may be 15,000 ft or more deep. The Devonian reefs are thick accumulations of carbonates in the Leduc Formation, overlain and surrounded by shales that represent lagoonal and open marine clastic sediments. The carbonate reefs were rigid and incompressible, while the enclosing fine-grained clastic sediments of the Ireton Shale were compacted to a fraction of their original thickness as the basin subsided and deposition continued. This differential compaction created drape structures over the buried reefs; these structures persist in the overlying rocks, although their magnitudes become less on shallower horizons.

Deep-seated compaction features are not readily apparent on the Cretaceous horizons because of the overriding effect of the strong regional dip (Fig. 5.75). Closure at depth is expressed only as slight changes in local gradient that appear on structural maps of Cretaceous horizons as subtle variations in the spacing of contour lines. However, if the strong regional dip component could be removed

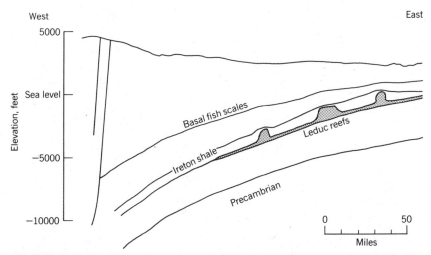

FIGURE 5.75 Diagrammatic cross-section across western margins of Alberta Basin. Leduc reefs and Ireton Shale are Upper Devonian; Basal Fish Scales is Lower Cretaceous.

from these maps, presumably the underlying drape features would appear as closed structures. Since the density of wells that bottom in the Cretaceous is relatively great, their analysis could provide valuable information about possible prospects in the Devonian even though few wells penetrate the deeper horizon.

The Basal Fish Scales is a black shale occurring near the boundary between the Lower and Upper Cretaceous. The unit derives its name from the abundant fish scales it contains. It is also characterized by numerous bentonite beds, which produce conspicuous spikes on gamma ray logs because of their high radioactivity. Bentonites are synchronous and so form excellent markers for regional correlation. The top of the Basal Fish Scales can be picked with exceptional consistency on log traces because of the pronounced response from a thick, persistent bentonite. The map shown in Figure 5.76 includes an area of about 3500 square miles in west-central Alberta, and was constructed using picks of this bentonite at the top of the Basal Fish Scales in 360 exploratory holes. The map area lies on the western margin of the Alberta Basin, immediately in front of the overthrust zone marking the Rocky Mountain front. Beds dip downward to the southwest, with increasing dip as the western edge of the basin is approached. Within the map area, depths of the Basal Fish Scales range from slightly in excess of 1000 ft below sea level in the northeast, to almost 5000 ft below sea level in the southwest.

Figures 5.77 and 5.78 are first- and second-degree polynomial trend surfaces of the Basal Fish Scales; statistics relating to the fit of the trend surfaces are given in Table 5.17. Both trend surfaces provide an extremely high fit to the observations, although the second degree is significantly better than the first degree. In this application, statistical tests of significance are not an appropriate guide for the selection of degree of trend surface, because the problem is not one of statistical

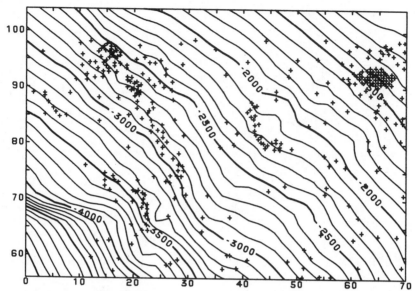

FIGURE 5.76 Structure contour map of part of Alberta Basin, drawn on top of Basal Fish Scales. Contours in feet below sea level. Well control shown by crosses.

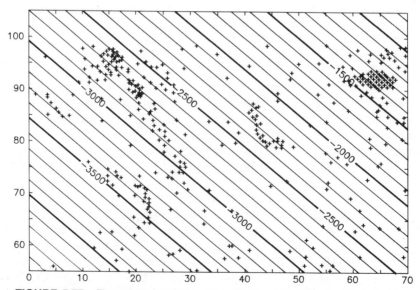

FIGURE 5.77 First-degree trend surface of Basal Fish Scales. Contour interval in feet below sea level.

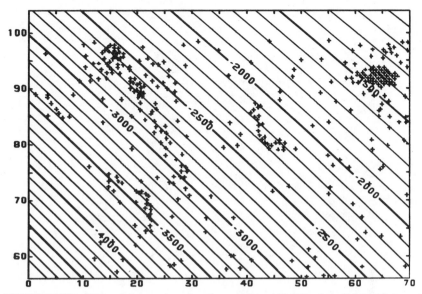

FIGURE 5.78 Second-degree trend surface of Basal Fish Scales. Contour interval in feet below sea level.

estimation. Rather, the objective is to closely simulate the regional features of the structure contour map, so subtraction of the trend will remove the regional component of structure. In effect, the trend surface is used as a high-pass filter, removing the large-scale structural variation from the map and leaving behind small-scale features.

Figure 5.79 shows residuals from the second-degree trend surface. Figure 5.80 is a paleogeographic map representing the Upper Devonian, reconstructed from well and seismic information. Note the strong coincidence between positive trend-surface residuals on the Basal Fish Scales, and the locations of major Devonian

TABLE 5.17 Statistics of First- and Second-Degree Polynomial Trend Surfaces Fitted to Elevation of Basal Fish Scales in West-Central Alberta

First-Degree Trend Surface
  Percent goodness-of-fit (R^2)................................... 98.6%
  Correlation coefficient (R)...................................... 0.993
  Trend surface equation:
    $Y = -6351.4 + 29.3X_1 + 33.8X_2$

Second-Degree Trend surface
  Percent goodness-of-fit (R^2)................................... 99.7%
  Correlation coefficient (R)...................................... 0.999
  Trend surface equation:
    $Y = -7993.3 + 63.4X_1 + 59.2X_2$
    $-0.1X_1^2 - 0.3X_2^2 - 0.1X_1X_2$

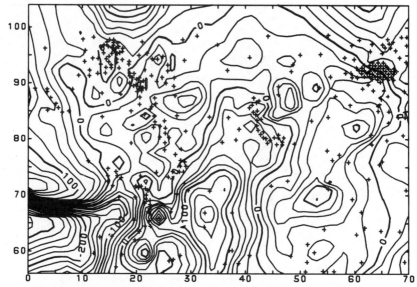

FIGURE 5.79   Residual map constructed by subtracting second-degree trend surface (Fig. 5.78) from structural contour map of Basal Fish Scales (Fig. 5.75). Positive residuals are shaded. Contour interval is 20 ft.

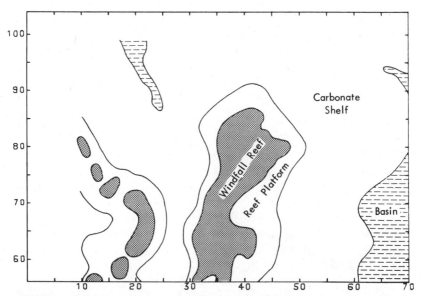

FIGURE 5.80   Paleogeographic map of Upper Devonian showing Leduc Reefs growing on carbonate platform along west margin of Devonian sea.

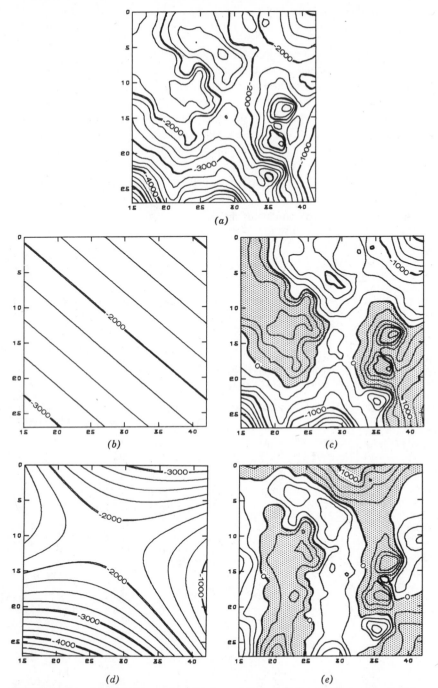

FIGURE 5.81 (a) Contour map of the top of Ordovician rocks (Arbuckle Gp.) in central Kansas. Contours in feet below sea level. Original well control shown in Figure 5.10. (b) First-degree trend surface. (c) Contoured residuals from first-degree trend. (d) Second-degree trend surface. (e) Contoured residuals from second-degree trend. Areas of positive residuals shown by shading.

reefs, particularly the Windfall Reef. Removal of the regional trend has successfully isolated minor components of structure representing drape over these reefs. Several major oil fields were discovered in this area using this exploration technique.

Still unanswered is the question, "what is a high (or low) goodness-of-fit or correlation in a trend surface?" In geologic studies, we often must rely on experience and intuition to provide an answer. For example, in structural analyses performed on data from Kansas, Oklahoma, Texas, Wyoming, California, England, and other localities, we have judged first- and second-degree correlations of less than 0.3 as providing a "poor fit." Correlations of 0.4–0.6 yield interpretable

TABLE 5.18 Geographic Coordinates (in Arbitrary Units) and Elevations of the Top of the Ordovician in Central Kansas

X_1 (units)	X_2 (units)	Y (ft)	X_1 (units)	X_2 (units)	Y (ft)	X_1 (units)	X_2 (units)	Y (ft)
23.36	22.10	−2961.0	35.90	9.90	−1537.0	26.86	21.17	−2951.0
29.80	10.58	−2240.0	21.54	14.90	−1667.0	29.93	16.80	−2435.0
22.18	16.24	−1872.0	33.46	12.31	−1694.0	24.62	23.12	−3177.0
22.91	3.36	−2584.0	19.40	16.57	−2300.0	26.04	16.52	−2241.0
39.33	21.14	−1119.0	35.18	19.79	−1465.0	32.86	13.52	−1775.0
15.10	18.50	−3062.0	21.20	3.50	−2349.0	15.10	25.85	−5400.0
28.96	10.30	−2540.0	23.50	13.02	−1564.0	36.99	25.28	−1852.0
18.90	16.69	−2300.0	19.09	24.40	−3657.0	22.40	13.23	−1500.0
21.11	12.26	−1505.0	34.45	19.45	−1257.0	27.91	16.48	−2481.0
29.92	4.00	−1921.0	25.60	17.18	−2337.0	19.92	11.10	−1599.0
41.99	5.31	−2056.0	28.60	26.15	−4373.0	29.62	20.27	−2875.0
21.86	4.01	−2466.0	29.32	11.48	−2109.0	36.50	10.21	−1353.0
41.30	16.38	−1077.0	32.60	24.15	−2923.0	16.58	5.12	−1608.0
22.75	21.67	−2780.0	25.90	19.05	−2607.0	34.45	6.22	−1499.0
35.93	14.53	−707.0	20.12	19.80	−2751.0	30.26	22.27	−3029.0
21.59	21.16	−2677.0	33.87	23.97	−1626.0	35.60	20.95	−1860.0
28.20	6.40	−2801.0	34.28	1.78	−2305.0	23.23	11.59	−1412.0
30.50	19.83	−2678.0	39.60	19.40	−1135.0	27.02	17.01	−2407.0
41.20	3.50	−2586.0	39.01	9.93	−1971.0	20.94	22.90	−3044.0
31.38	17.74	−2190.0	24.60	20.45	−2483.0	17.53	10.02	−1657.0
24.30	23.90	−3367.0	25.10	22.84	−3095.0	28.27	8.16	−2540.0
24.80	21.10	−2959.0	24.48	24.45	−3589.0	17.83	0.10	−1647.0
16.75	8.66	−1709.0	26.47	13.19	−1490.0	32.58	19.86	−2140.0
22.72	11.15	−1431.0	20.23	17.58	−2307.0	35.45	17.03	−883.0
20.40	21.82	−3022.0	35.27	19.23	−1037.0	21.21	14.57	−1695.0
26.49	6.43	−2431.0	25.00	9.20	−1407.0	36.20	16.25	−1746.0
33.12	17.04	−1792.0	22.95	8.42	−2133.0	28.52	17.20	−2440.0
24.67	15.61	−2146.0	35.00	23.13	−3090.0	29.80	13.96	−2346.0
30.10	12.92	−2131.0	30.07	15.18	−1890.0	30.40	10.02	−2182.0
26.12	0.50	−2295.0	40.00	16.82	−1366.0	37.45	7.18	−1934.0
28.25	14.35	−2421.0	18.45	9.01	−1651.0	25.02	17.82	−2063.0
27.21	10.58	−2204.0	38.01	26.05	−1857.0	34.20	26.20	−2834.0
19.38	17.60	−2348.0	29.05	21.08	−2998.0	16.60	1.85	−1583.0
42.00	17.90	−885.0	35.40	20.90	−1902.0	17.40	1.10	−1521.0
32.48	2.24	−1618.0	41.18	20.14	−998.0	41.12	19.86	−835.0
18.96	14.54	−1834.0	26.70	12.80	−1785.0	29.92	22.02	−3061.0
32.97	11.79	−1884.0	15.58	4.63	−1593.0	17.10	14.05	−2402.0
23.50	1.57	−2308.0	20.70	20.39	−2722.0	21.70	14.21	−1608.0
41.89	15.25	−1108.0	20.80	11.50	−1477.0	20.60	6.28	−1613.0
31.40	12.20	−2058.0	41.11	5.22	−2274.0	30.32	15.55	−2195.0
31.60	22.82	−2711.0	22.11	21.00	−2598.0	39.62	18.03	−1421.0

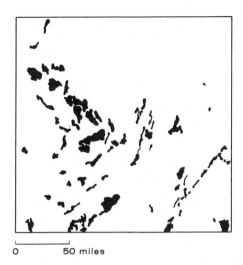

0   50 miles

FIGURE 5.82 Locations of major oil and gas fields in central Kansas.

trend and residual maps, and those higher than 0.7 are regarded as conforming closely to the original data points. Again, the purpose of the investigation must be kept firmly in mind while interpreting goodness-of-fit. In all of these structural studies, we were examining basins having relatively simple shapes and were searching for relatively small residuals compared to the total size of the basin. Third- and fourth-degree polynomials usually provided very high fits (over 0.8). As a reference value, randomly generated data sets having the same range of values as actual data yielded correlations of up to 0.3 for fourth-degree polynomials. Realistic-appearing trends can be extracted from random data, so geologists should proceed cautiously in the interpretation of trend and residual maps having low goodness-of-fit.

Figure 5.81*a* is a contour map of the data given in Table 5.18, which are elevations of the top of the uppermost formation of Ordovician age in central Kansas. Figure 5.81*b* is a first-degree trend surface fit to this same data; a map of residuals from the surface is shown as Figure 5.81*c*. A second-degree trend surface and residuals from the same data are shown as Figures 5.81*d* and 5.81*e*.

As an exercise, using the data from Table 5.18, fit a third-degree polynomial trend surface to the Ordovician top. Compare your map to the second-degree trend. Contour the residuals from the surface and compare this map with the second-degree residuals. Does the pattern of residuals change? Figure 5.82 is a generalized map of the major oil fields discovered in central Kansas. It may be instructive to compare the field distributions with the distribution of positive residuals from the three trends.

Statistical Tests of Trends

The goodness-of-fit of a trend surface may be tested statistically, by comparing the variance due to regression or trend to the variance due to deviations from the trend. You will recall from Chapter 2 that tests of equality of variances involve the F distribution and are valid only if the data satisfy certain conditions. If these assumptions can justifiably be made, we may regard the b_i coefficients found by least squares as estimates of the true population regression coefficients, β_i, and test hypotheses about their nature. We must assume that the population of the dependent variable is normally distributed about the regression, and that its variance does not change with changes in the independent variables. In addition, the samples must be drawn without bias from this population. The last condition often is the most difficult to satisfy, especially in a structural analysis based on oil wells, as the wells presumably were not located by random selection! Testing of statistical hypotheses about trend surfaces is perhaps most easily defended when dealing with data such as geochemical assays on samples collected according to a planned experimental design.

The significance of a trend or regression may be tested by performing an analysis of variance, which is the process of separating the total variation of a set of observations into components associated with defined sources of variation. This, or course, has been done by dividing the total variation of Y into two components, the trend (or regression) and the residuals (or deviations). The degrees of freedom associated with total variation in a trend analysis are $(n - 1)$, where n is the number of observations. The degrees of freedom associated with regression are determined by the number of terms or coefficients in the polynomial equation fit to the data. Degrees of freedom for deviation are the number of degrees of freedom associated with total variation minus those that are accounted for by regression, or $\nu_D = \nu_T - \nu_R$. A formal analysis of variance table is shown in Table 5.19.

The mean squares are found by dividing the various sums of squares by the appropriate degrees of freedom. By reducing sums of squares to mean squares, they have been converted to estimates of variance and may be compared using an F probability distribution. The MS_D is the variance about the regression line; MS_R is the variance of the regression line about its mean. If the regression is significant, the deviation about the regression will be small compared to the variance of the regression itself.

In a general test of a trend-surface equation, the ratio of interest is that between

TABLE 5.19 General ANOVA for Significance of Regression of kth-Degree Polynomial Trend Surface; Number of Coefficients in Trend-Surface Equation not Counting the b_0 Coefficient is m, Number of Data Points is n

Source of Variation	Sums of Squares	Degrees of Freedom	Mean Square	F Test
Polynomial Regression	SS_R	m	MS_R	MS_R/MS_D
Deviation from Polynomial	SS_D	$n - m - 1$	MS_D	
Total Variation	SS_T	$n - 1$		

variance due to regression and variance due to deviation. The F test gives a probabilistic answer to the question of whether the variances being examined have been obtained by random sampling from the same population. Or, is the regression effect not significantly different from the random effect? An affirmative answer may be interpreted as meaning that (a) the distribution of Y is random and independent of values of $X_1, \ldots, X_m$, or (b) the distribution of Y may be in part a function of $X_1, \ldots, X_m$, but the wrong functional model has been fit to the data.

In more formal terms, the F test for significance of fit is a test of the hypothesis and alternative

$$H_0 : \beta_1 = \beta_2 = \cdots = \beta_m = 0 \qquad (5.89)$$
$$H_1 : \beta_1, \beta_2, \ldots, \beta_m \neq 0$$

The hypothesis to be tested is that the partial regression coefficients are equal to zero, or in other words, there is no regression. If the computed value of F exceeds the table value of F, this hypothesis is rejected and the alternative, H_1, is accepted.

In polynomial trend-surface analysis, it is customary for some investigators to fit a series of equations of successively higher degrees to the data. In such an analysis, a number of regression sums of squares will be produced, each larger than the preceding sum. The analysis of variance table may be expanded to analyze the contribution of the additional partial regression coefficients and give a measure of the appropriateness of increasing the order of the equations. The test is developed by finding the difference in sums of squares due to regression of the higher polynomial equation minus the regression sums of squares due to fitting the lower order polynomial. This difference is divided by the difference in regression degrees of freedom, giving the mean square of regression due to increasing the degree of the polynomial. This mean square is then divided by the mean square due to deviation from the higher polynomial. If the resulting F value is significant, the deleted order was contributing to the regression and should be retained. If the value is not significant, nothing has been gained by fitting the higher degree polynomial. An ANOVA table for testing the significance of a higher degree polynomial trend surface is given in Table 5.20.

The F test for significance of added terms is a test of the hypothesis and alternative

$$H_0 : \beta_{k+1} = \beta_{k+2} = \cdots = \beta_m = 0 \qquad (5.90)$$
$$H_1 : \beta_{k+1}, \beta_{k+2}, \ldots, \beta_m \neq 0$$

The null hypothesis states that partial regression coefficients after the kth term are all equal to zero, or, they do not contribute to the regression caused by the first through kth term. (Remember that the polynomial trend surface of degree p contains k coefficients, whereas the polynomial equation of the $(p + 1)$ trend contains m coefficients.) Again, if the computed value of F exceeds the table value, the hypothesis is rejected. The test procedure is given for an equivalent curvilinear model by Li (1964, pp. 176–181).

In some instances we may be interested in assessing the effect of a single partial regression coefficient. This may be done simply by omitting that term from the polynomial equation and recomputing the sums of squares due to regression and

TABLE 5.20 General ANOVA for the Significance of Increasing the Degree of a Polynomial Trend from p to $(p + 1)$ Degree; Polynomial Equation of Degree p Has k Coefficients, not Counting the b_0 Term; Equation of Degree $(p + 1)$ Has m Coefficients, not Counting the b_0 Term; Number of Observations is n

Source of Variation	Sum of Squares	Degrees of Freedom	Mean Squares	F Test
Regression of Degree $(p + 1)$	SS_{RP+1}	m	MS_{RP+1}	MS_{RP+1}/MS_{DP+1} [a]
Deviation from Degree $(p + 1)$	SS_{DP+1}	$n - m - 1$	MS_{DP+1}	
Regression of Degree p	SS_{RP}	k	MS_{RP}	MS_{RP}/MS_{DP} [b]
Deviation from Degree p	SS_{DP}	$n - k - 1$	MS_{DP}	
Regression Due to Increase from p to $(p + 1)$ Degree	$SS_{RI} = SS_{RP+1} - SS_{RP}$	$m - k$	MS_{RI}	MS_{RI}/MS_{DP+1} [c]
Total Variation	SS_T	$n - 1$		

[a] Test of significance of the $(p + 1)$-degree trend surface.
[b] Test of significance of the p-degree trend surface.
[c] Test of significance of increase in fit of the $(p + 1)$ degree over p degree.

deviation. The contribution of the deleted term is the difference in the two sums of squares due to regression. Significance of this term may be tested by computing the ratio between the mean square due to the deleted term and the mean square due to the complete regression. The F ratio has one and $(n - m - 1)$ degrees of freedom. The ANOVA for the significance of a single deleted coefficient is given in Table 5.21.

Individual terms in the regression equation also may be tested by calculating the increase in SS_R when a new variable is added. However, this is not an advisable practice, because the tendency may be to regard all further terms as nonsignificant after encountering several successive nonsignificant coefficients. This may not be the case. In trend-surface analysis, the complete set of terms for the next higher order should be added, and then the added individual terms tested one by one by elimination. Abbreviated sets of higher degree terms should not be added blindly, unless there are compelling reasons to do so. In one example, because of computer limitations, third-order "hypersurfaces" were fit to oil-gravity data with an equation that contained no cubic cross-product terms. The resulting regression was considerably different from that produced from the same data when they were rerun on a larger computer using a program that fitted a complete cubic equation. Additional terms are especially apt to be significant if the correlation coefficients of lower orders are small.

TABLE 5.21 ANOVA for Testing the Significance of a Single Deleted Coefficient; the Complete Polynomial Regression Equation Contains m Coefficients, not Counting the b_0 Term; the Regression Equation Contains Only $(m - 1)$ Coefficients after Deleting the kth Coefficient; Number of Observations is n

Source of Variation	Sum of Squares	Degrees of Freedom	Mean Squares	F Test
Regression of All Terms	SS_R	m	MS_R	MS_R/MS_D[a]
Deviation	SS_D	$n - m - 1$	MS_D	
Regression Omitting kth Term	SS_{R-1}	$m - 1$	MS_{R-1}	MS_{R-1}/MS_{D-1}[b]
Deviation	SS_{D-1}	$n - m - 2$	MS_{D-1}	
Regression of kth Term Only	$SS_{RK} = SS_R - SS_{R-1}$	1	MS_{RK}	MS_{RK}/MS_D[c]
Total	SS_T	$n - 1$		

[a]Test of significance of the p-degree trend surface.
[b]Test of significance of the p-degree trend surface fitted without the kth coefficient.
[c]Test of significance of the kth coefficient alone.

The two data sets presented previously (Tables 5.15 and 5.18) are representative of problems in structural trend-surface analysis. The purpose of both investigations was to seek areas where the structural surface departed from a polynomial model equation. In these problems, the error distribution is such that the appropriateness of tests of the significance of the regressions is suspect. However, in the following example experimental and sampling conditions seem adequate to justify application of regression tests.

Figure 5.83 shows a planar view of single crystal of sphalerite collected from a mine in northern Mexico. Investigators are interested in the distribution of iron through the crystal, so it has been carefully cleaved through its center and the surface has been polished. The iron content at 10 Å spots spaced 1 mm apart has been determined by electron microprobe. The sample points are shown as dots on Figure 5.83a; a contour map of the values is shown as 5.83b; and the data are listed in Table 5.22. Although use of iron content in sphalerite as a temperature indicator has been criticized because of the possibility of nonequilibrium conditions at the time of crystallization, the researchers postulate that the growing shell of the crystal was in equilibrium with the ore solution at all times. Therefore, the average composition of the crystal may be inadequate as a temperature indicator, but the composition of successive thin shells of the crystal may define a temperature-change curve. Excluding the possibility of zoning, the simplest model for the distribution of iron in the crystal would be a gradual increase or decrease symmetrically outward from the center. The fitting of a second-degree polynomial

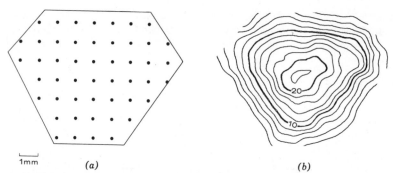

FIGURE 5.83 (a) Locations of sampling points in an electron microprobe study of the iron content of a sphalerite crystal. (b) Contours showing iron concentration, derived from data taken at points shown in part a.

TABLE 5.22   Iron Content of 10 Å Spots Spaced on a 1-mm Grid Across Cleavage Surface of Sphalerite Crystal

		Iron Content (%)			
X_1	X_2	Y	X_1	X_2	Y
2.0	1.0	3.1	2.0	4.0	6.4
3.0	1.0	4.6	3.0	4.0	14.6
4.0	1.0	5.8	4.0	4.0	17.6
5.0	1.0	7.2	5.0	4.0	21.2
6.0	1.0	8.4	6.0	4.0	21.0
7.0	1.0	6.3	7.0	4.0	13.4
8.0	1.0	2.4	8.0	4.0	7.5
1.0	2.0	2.5	9.0	4.0	0.4
2.0	2.0	10.2	2.0	5.0	3.1
3.0	2.0	12.8	3.0	5.0	8.6
4.0	2.0	16.1	4.0	5.0	15.0
5.0	2.0	14.2	5.0	5.0	16.2
6.0	2.0	15.1	6.0	5.0	14.8
7.0	2.0	12.8	7.0	5.0	9.8
8.0	2.0	9.0	8.0	5.0	3.1
9.0	2.0	5.3	3.0	6.0	5.0
1.0	3.0	4.3	4.0	6.0	7.2
2.0	3.0	14.1	5.0	6.0	12.3
3.0	3.0	15.6	6.0	6.0	10.6
4.0	3.0	20.2	7.0	6.0	4.5
5.0	3.0	20.6	3.0	7.0	0.6
6.0	3.0	18.5	4.0	7.0	2.4
7.0	3.0	16.2	5.0	7.0	3.5
8.0	3.0	10.2	6.0	7.0	4.7
9.0	3.0	4.6			

regression of iron content on the coordinates of the sample points seems an appropriate way of testing this hypothesis. Using the data from Table 5.22, fit a series of trend surfaces to the iron content. From the various sums of squares, construct an ANOVA similar to Table 5.20 and test the significance of the polynomial regressions.

Two Trend-Surface Models

You may have noted that the preceding discussions imply that two basically different types of geologic problems are analyzed using trend-surface methods. On one hand, trend surfaces are fit to structural data in an attempt to isolate "local structures." It has been demonstrated empirically that in a sedimentary basin these residuals may be associated with structurally or hydrodynamically trapped oil. Alternatively, trend surfaces have been used to define regional trends in petrographic and geochemical data. These two applications differ in objectives and in underlying assumptions, but their common methodology has obscured these differences.

Trend surfaces fit to structural data can be represented by the model equation

$$Y_i = \beta_0 + \beta_1 X_1 + \beta_2 X_2 + \cdots + \beta_m X_2^p + (\gamma_i + \epsilon_i) \qquad (5.91)$$

The equation states that a given observation (elevation of a formation top) is equal to the sum of a constant term related to the means of the geographic coordinates, plus a polynomial expansion of degree p of the geographic coordinates, plus a local component, plus a randomly distributed measurement error. The latter two terms are considered to be confounded, and are the parameters of interest.

In contrast, a trend surface fit to petrographic or similar data usually is represented by an ordinary response-surface model:

$$Y_i = \beta_0 + \beta_1 X_1 + \beta_2 X_2 + \cdots + \beta_m X_2^p + \epsilon_i \qquad (5.92)$$

The response-surface model equation is similar in all respects to the trend-surface equation except that the local component γ_i is not considered to be present. Interest is centered upon the nature of the trend, as expressed in the estimates of the β's or polynomial coefficients

Petrographic and geochemical variates usually are characterized by high variance among replicates. This variance arises because of inhomogeneities within the samples analyzed, local or small-scale variation in composition (on a scale larger than the sample but smaller than the interval between samples), and analytical or instrumental errors. In a typical study, these are confounded and produce a normal distribution of error about each observation. Although each source of variance could be isolated and measured by replication at several levels within the experiment, this may not be done because of economic or other factors.

A trend surface or regression on geographic variables can be fitted appropriately to such data if certain basic assumptions seem reasonable. These require that the errors ϵ_i be randomly and normally distributed about the regression, that they have zero mean, and that they have constant variance. This in turn means that the errors are independent of one another. If these conditions are fulfilled, the regression may be tested for significance and inferences drawn about the trend. The appropriate statistical tests are those we have presented in Tables 5.19–5.21. Many other special statistical designs have been created and are widely used in such fields as agriculture and chemical engineering; an introduction to these is given in Mendenhall (1968, Chapter 10). Koch and Link (1980, pp. 217–222) discuss a geologic application of one such design. The important aspect of such inferences is that they pertain to the trend. This is emphasized in Figure 5.84, where they observed values of Y_i can be seen to fall within the distribution of errors around the regression.

In the fitting of trend surfaces to structural data, observations (typically elevations of a geologic horizon) are not replicated. Although it is possible to relog a well, it usually is not possible to drill a second, nearby well which would constitute a replicate sample. Repeated down-hole measurements may vary by a few feet depending upon depth and amount of stretch in the cable, but this source of experimental error is always one or more orders of magnitude smaller than deviations in trend-surface analysis. Lack of replicates means that local variation cannot be assessed. However, is may be rationalized that this source of error is also inconsequential, as the drill-hole is not sampling from a population of surfaces. Only one formation top exists and the only variance is due to the negligible measurement error. Thus, all residual variance in trend-surface analysis can be ascribed to lack of fit.

In terms of the model equation, this is equivalent to saying that ϵ_i is negligibly small compared to γ_i. Although the random error ϵ_i has zero mean and is independent for all values of Y_i, it cannot be isolated because replicates have not been taken. The confounded term $(\gamma_i + \epsilon_i)$ also has a mean of zero, but it is not in general independent for values of Y_i. In fact, the purpose of the analysis is to define regions of specified size in X_1 and X_2 over which $(\gamma_i + \epsilon_i)$ are correlated. Figure 5.85 shows a theoretical distribution of ϵ_i about the structural surface. For the most part, deviations of the surface from the polynomial model do not reflect the magnitude of the error term but rather the local component γ_i.

The difference between these two model equations for trend surfaces is reflected in the way in which autocorrelation among residuals is regarded. In polynomial regression analysis, autocorrelation is considered to be a violation of model as-

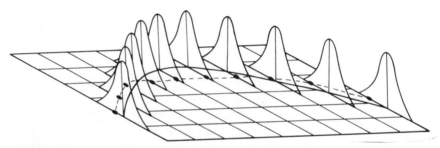

FIGURE 5.84 Distribution of error variance around the regression line according to polynomial regression model. Observations, indicated by dots, are assumed to lie within the variance distribution around the model regression line.

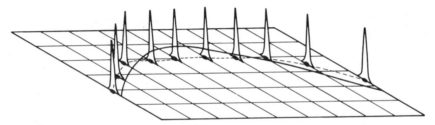

FIGURE 5.85 Distribution of error variance around observed surface according to structural model of trend surfaces. Errors in replications of points are centered about some mean value for each point, rather than around the polynomial model equation.

sumptions and is assumed to invalidate (or seriously weaken) inferences drawn from the analysis. Such an attitude may be appropriate for petrographic and geochemical data, because replication tends to reveal a normal distribution of error around relatively simple regressions. Although analyses for lack-of-fit may suggest more complex regressions, in general the error term is sufficiently large to account for all deviations.

In contrast, geologists using trend-surface analysis on structural data are seeking areas of autocorrelated residuals. As stated before, almost all structural deviations are ascribable to lack-of-fit, and the presence of autocorrelated residuals indicates a region larger than the sampling interval where the surface deviates from the polynomial model in a consistent direction. Either large areas of autocorrelated residuals or single points of large deviations are of interest in petroleum exploration, because they may indicate regions where local structures (γ_i) have a strong influence. Because the deviations are not randomly distributed, standard tests of significance of the regression cannot be used. This generally is not regarded as a drawback, as the regression itself is only of incidental interest.

Pitfalls

At this point, it may be appropriate to emphasize some factors which can adversely affect trend-surface analyses, or indeed, any type of map analysis. These cautionary statements have been repeated many times in the literature, but they seem to be ignored more times than not (Chayes, 1970). The effect of these factors can range from a slight distortion of the trend (a form of bias) to total invalidation of the results.

The first and most obvious point is that adequate control must be present. At an absolute minimum, the number of data points must exceed the number of coefficients in the polynomial equation or the results of the regression are invalid. If statistical tests are to be run, the number of control points determines the degree of freedom and these must be sufficiently large so a meaningful F test can be performed. If the degrees of freedom for deviation are small (a consequence of the number of polynomial coefficients approaching the number of data points), only extremely high correlations will be found significant. Furthermore, the power of the test (the probability of not committing a type II error) decreases drastically with small sample sizes. Of course, the number and spacing of control points has a direct influence on the size of local deviations that can be detected in trend-surface analysis of structural data, and relates to the resolution with which these can be defined.

Ordinarily, we do not consider control points beyond the boundaries of our map. Often, the map area may even extend slightly beyond the actual lateral extent of the data points, there being few (or no) control points actually on the map boundaries. In such circumstances, there are almost no constraints on the form of the trend surface near the edges of the map. Whatever slope exists in the region of control is extrapolated without limits along the map boundaries. This creates what are called "edge effects." If a high-order trend is being fitted to data, extrapolated values near the edges of the map may reach astronomical proportions. Minor edge effects will exist even if the entire map area is uniformly covered with control points up to the boundary. Therefore, it is good practice to have data over an area in excess of the size of the area to be mapped. This forms a "buffer region" around the map in which edge effects are concentrated; the control points in this region constrain the form of the trend surface within the map area proper. The necessary width of the buffer depends primarily on the density of available control. If the map contains many control points, a narrow bordering strip will suffice. If control density is low, a much wider belt around the map will be necessary to absorb edge effects. Incidentally, it should be pointed out that edge effects are not unique to trend surfaces, but also occur in contour maps, moving-average surfaces, and other forms of fitted surfaces.

The arrangement of the data points within the map also has a pronounced effect on the form of the regression. The illustrations in Figure 5.86 are taken from an evaluation of the effect of data distributions on polynomial trend surfaces (Doveton and Parsley, 1970, pp. B198–B202). A series of points were placed at random locations on a basin-shaped surface, and a second-degree trend was calculated.

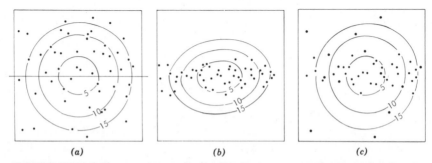

(a)   *(b)*   *(c)*

FIGURE 5.86   Influence of control-point distribution on trend surfaces. (a) Original test surface with randomly placed control points. (b) Distorted trend produced by sampling original test surface along a narrow band. (c) Nearly correct trend produced by adding a few outlying points to the narrow band of observations.

The regression equation was then used to calculate values of the dependent variable at points placed according to various sampling patterns. Ideally, surfaces fit to these points would be identical to the original form from which the data were derived. Figure 5.86a shows the surface produced by randomly distributed points; the fit is over 95% and the trend is essentially identical to the original model. In Figure 5.86b, however, the sample points are distributed along a narrow band. The fit of the surface is still good (93%), but the form of the regression has become badly distorted parallel to the sample pattern. Only a few control points outside the band are necessary to correct this bias, however, as can be seen in Figure 5.86c (also 93% fit). These tests demonstrate that the shape of the map area can seriously affect the form of a polynomial fit. If the data do not occupy an approximately equidimensional area, the trend surface will be elongated parellel to the pattern of points. Remember that these examples are for idealized models having no local or random component; distortions are more severe if small-scale "noise" is present.

Cautionary remarks about the deleterious effect of clustered data points on trend surfaces also have been published. Clustering or bunching of control points is especially apt to be troublesome in petroleum exploration by trend-surface methods, because wells are most abundant on and around known oil fields. These areas may exert an undue influence on the regional trend, although there is evidence that this effect is not as severe as sometimes feared (Doveton and Parsley, 1970, pp. B200–B201). Figure 5.87a shows a cubic trend-surface model used to generate data points for analysis of the clustering effect. Points were taken in various clustered patterns and were used in an attempt to recreate the original surface. Figure 5.87b shows a strongly clustered data-point distribution and the trend surface created.

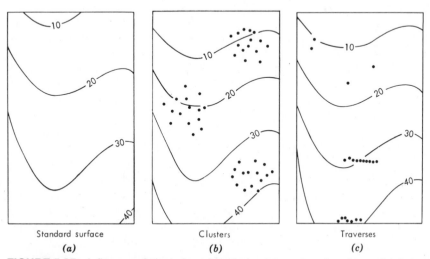

Standard surface

(a)

Clusters

(b)

Traverses

(c)

FIGURE 5.87 Influence of clustering of control points on trend surfaces. (a) Original third-degree test surface. (b) Trend surface obtained from control points placed in clusters on the test surface. (c) Trend surface obtained from control points located along traverses across parts of the test surface.

The surface accounts for 99% of the original variance. An even more radically clustered distribution is shown in Figure 5.87c, but the cubic surface accounts for 100% of the total variance. Both Figures 5.87b and 5.87c are essentially identical to the original surface. These experiments suggest that trend-surface methods may be more robust against effects of clustering than is commonly supposed. Again, it should be remembered that these experimental tests are essentially noise-free, and more severe distortions are to be expected in the presence of local variation.

Four-Dimensional Trend Surfaces

A logical extension of polynomial trend-surface analysis is the inclusion of the third geographic coordinate (and its powers and cross-products with the other spatial coordinates) as an independent variable. The resulting regression is variously referred to as a "hypersurface," an "isopleth envelope," a "U-V-W trend," or simply as a "four-dimensional trend surface." In this technique, a dependent variable, usually percentage composition of some constituent, is regressed upon east-west, north-south, and elevation coordinates. Contour lines become contour envelopes in the higher dimensionality of this method. A completed analysis has the form of a solid (representing the space defined by the three coordinates) containing nested contour sheets or envelopes. These sheets are interpreted in the same manner as contour lines in a conventional map; they enclose volumes of equal composition. Just as the area between two successive contour lines on a topographic map is occupied by points which have the same range in elevations, so the volume between two successive contour sheets in a four-dimensional trend surface is occupied by points which have the same range in compositions.

As a simple example, we can consider the data listed in Table 5.23. These represent the uranium oxide content, in percent, of a small carnotite body in Jurassic sediments in the Colorado Plateau. Such small but rich ore bodies were developed by replacement of logs or organic "trash pockets" in the sediment by uranium and vanadium minerals. The ore bodies have the general form of an ellipsoid with increasing concentration of uranium toward the center. The amount of elongation in the ore body is roughly determined by the length of the log or organic trash accumulation that forms its core. This particular body was carefully sampled and analyzed during mining, and provides an excellent example of the type of data most amenable to four-dimensional trend surfacing.

As you can imagine, the construction of either a stereoscopic or perspective map, or a solid model of a variable that changes in three-dimensional space is difficult. This difficulty is compounded when the values at control points contain a random component. A traditional method of construction is to create a series of maps at different levels or a set of cross-sections, contour these, and then assemble the whole by stacking the sheets together. Unfortunately, the contouring method considers gradients between points in only one, or at most two planes. Assembling and smoothing the representation is almost completely subjective, and a model

TABLE 5.23 Uranium Oxide Content (Y) of Carnotite Roll in Colorado Plateau; Coordinates Given Are Distances from Arbitrary Origin in Upper Northwest Corner of Ore Block (X_1 = North-South, X_2 = Depth, X_3 = East-West)

X_1 (ft)	X_2 (ft)	X_3 (ft)	Y (%)	X_1 (ft)	X_2 (ft)	X_3 (ft)	Y (%)
4.0	3.0	5.0	1.2	24.0	5.0	11.0	0.8
5.0	3.0	8.0	12.4	25.0	3.0	6.5	23.2
5.0	5.0	8.0	0.6	26.0	2.5	9.0	12.6
6.0	2.0	9.0	1.1	27.0	1.5	6.5	1.5
8.0	3.0	10.5	11.0	28.0	1.5	7.0	3.1
9.0	2.0	8.0	6.7	29.0	1.5	11.0	4.0
10.0	3.0	5.5	12.4	29.0	3.0	9.0	17.2
10.0	5.0	8.5	1.4	30.0	4.0	13.0	6.0
13.0	1.0	6.0	3.2	31.0	4.5	7.0	4.8
14.0	3.0	8.5	17.4	34.0	1.0	10.5	0.3
15.0	3.0	6.0	9.8	35.0	4.0	8.0	4.9
16.0	4.5	9.0	3.3	36.0	3.0	10.5	17.7
17.0	2.0	7.5	1.8	37.0	3.0	13.0	8.1
19.0	3.0	4.5	4.7	38.0	1.5	10.0	1.6
20.0	3.0	7.0	21.4	40.0	2.0	14.5	4.1
20.0	3.0	10.0	7.6	40.0	4.0	15.5	2.3
22.0	5.0	8.0	2.9	42.0	4.0	13.0	8.7

made from contoured levels will probably differ significantly from one made from contoured cross-sections in all but the simplest examples.

Four-dimensional trend-surface analysis is especially useful in the construction of these models. The points may be collected irregularly through space and do not need to be projected onto planes. Gradients between points in all directions are considered simultaneously. Because the surface is a least-squares fit, a smoothed representation or simple form is created which can be displayed in a number of ways. Computer programs are available which produce "slice maps" (sections perpendicular to any one of the three geographic coordinates) which can be assembled into "egg-crate" models. More sophisticated programs create perspective, isometric, or stereoscopic drawings of the contour envelopes using a plotter.

Because we have added another independent variable, we must solve four simultaneous equations to compute the first-degree trend surface. The normal equations are listed below. Again, the limits of summation have been omitted to simplify the notation. Summation extends over the observations from $i = 1$ to n.

$$\Sigma Y = b_0 n + b_1 \Sigma X_1 + b_2 \Sigma X_2 + b_3 \Sigma X_3$$
$$\Sigma X_1 Y = b_0 \Sigma X_1 + b_1 \Sigma X_1^2 + b_2 \Sigma X_1 X_2 + b_3 \Sigma X_1 X_3 \qquad (5.93)$$
$$\Sigma X_2 Y = b_0 \Sigma X_2 + b_1 \Sigma X_1 X_2 + b_2 \Sigma X_2^2 + b_3 \Sigma X_2 X_3$$
$$\Sigma X_3 Y = b_0 \Sigma X_3 + b_1 \Sigma X_1 X_3 + b_2 \Sigma X_2 X_3 + b_3 \Sigma X_3^2$$

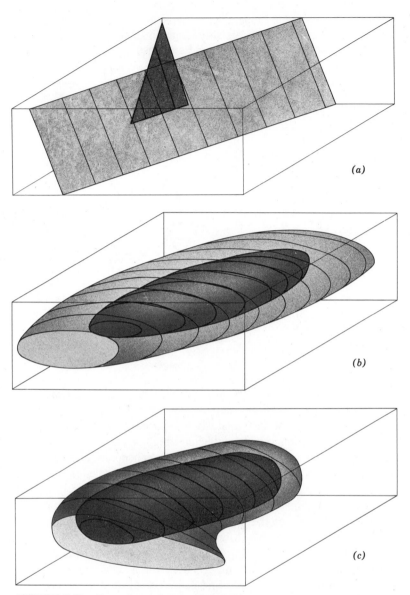

FIGURE 5.88 Four-dimensional polynomial trend surfaces of uranium content. Darker, innermost envelopes contain ore having over 10% uranium oxide. Between inner and outer envelopes, ore contains 5–10% uranium oxide. Outside outer envelope, ore contains less than 5% uranium oxide. (*a*) First-degree surface. (*b*) Second-degree surface. (*c*) Third-degree surface.

If you compare this equation set with (5.82), you will see that it is a direct extension, which is in turn an extension of (4.13). The simultaneous equations can be rewritten into matrix form as

$$
\begin{bmatrix}
n & \Sigma X_1 & \Sigma X_2 & \Sigma X_3 \\
\Sigma X_1 & \Sigma X_1^2 & \Sigma X_1 X_2 & \Sigma X_1 X_3 \\
\Sigma X_2 & \Sigma X_1 X_2 & \Sigma X_2^2 & \Sigma X_2 X_3 \\
\Sigma X_3 & \Sigma X_1 X_3 & \Sigma X_2 X_3 & \Sigma X_3^2
\end{bmatrix}
\cdot
\begin{bmatrix}
b_0 \\ b_1 \\ b_2 \\ b_3
\end{bmatrix}
=
\begin{bmatrix}
\Sigma Y \\ \Sigma X_1 Y \\ \Sigma X_2 Y \\ \Sigma X_3 Y
\end{bmatrix}
\tag{5.94}
$$

The expansion of this matrix expression to higher orders follows the same pattern as in trend-surface analysis with the addition of an extra coordinate and its cross-products. As an exercise, develop the matrix equation for a second-degree polynomial trend surface with three independent coordinates. How many terms are there in the matrix of simultaneous equations? This should suggest one of the major limitations of this technique, because at higher powers the matrix equations become very large. Consequently, solutions tend to become unstable and the researcher must resort to standardizing the data to scale the matrix of coefficients into a small range, and must utilize highly efficient and sophisticated inversion routines. Even with these precautions, it often is impossible to obtain reliable high-order solutions from real data sets.

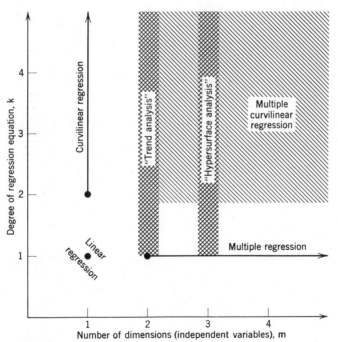

FIGURE 5.89 Relation between curvilinear regression, trend-surface analysis, four-dimensional trend surfaces, and multiple regression.

Figure 5.88*a* shows a first-degree polynomial trend fitted to the uranium content data. The trend has the form of parallel plates enclosing volumes of equal composition. As would be expected, the fit of the linear model is poor, only 2.3%. A second-degree surface is shown in Figure 5.88*b*; the fit of this regression is dramatically improved over the simple linear surface and accounts for 59.9% of the total variance in uranium content. Figure 5.88*c* is the third-degree regression and has approximately the same form as the second-degree surface. Total variance accounted for, however, is increased to 79.3%.

The problems of data-point distribution which affect trend-surface analyses apply as well to four-dimensional trend surfaces. However, they are apt to be exaggerated, because of the difficulty in many instances of obtaining samples through an adequate range of depths. Ideally, samples should be uniformly dispersed through the space within a cube, or at worst, through a thick prism similar to the data distribution in our example. Although it is sometimes possible to design equidimensional sampling schemes, often it is impractical because the region of interest has an areal extent many times in excess of its depth. This is the situation, for example, in

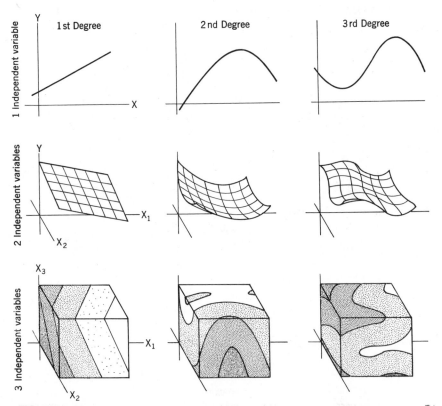

FIGURE 5.90 Trends of one, two, and three independent variables for polynomial equations of the first, second, and third degrees. After Harbaugh (1964).

problems concerning distribution of constituents through stratigraphic intervals. Even if the area of research interest extends to great depths, the availability of samples often limits us to a consideration of a thin plate. In either case, the effect is a foreshortening of contour intervals in the depth dimension. This is a three-dimensional analog to the problem illustrated in Figure 5.86*b*.

In analyses of constituents in stratigraphic units such as formations, this problem is overcome by a change in scale; vertical dimensions commonly are measured in feet, while horizontal distances are measured in miles. This has the effect of expanding the vertical dimension, and distortions due to the sampling pattern are not apparent on the exaggerated vertical scale. Of course, if the final regression is converted to true scale representation, the distortion immediately reappears. One compensating factor is that changes in composition within sedimentary units tend to be more pronounced perpendicular to stratification, so that changes near the top and bottom margin may overshadow the distortion effect. However, in studies of the composition of bodies that extend to great depths such as granite plutons, the distortion created by a limited sample space may be severe. In these instances, rapid gradients near the surface and near the bottom limits of data control may be entirely due to distortions created by the sampling pattern.

The relationships between least-squares or regression methods involving geographic coordinates and their powers and cross-products are shown in Figure 5.89. Typical trends from each category are shown in Figure 5.90. In the next chapter, we will again increase the generality of linear regression and consider regressions having k independent variables. Of course, these are not geographic variables, although coordinates may be included.

Double Fourier Series

As we stated earlier, the use of a polynomial expansion as an approximating function for trend surfaces is due primarily to the ease with which the power series can be computed. Experience has shown that relatively low-order trend surfaces correspond well to the distribution of many geologic variables, especially the elevation of structural horizons. However, no significance can be attached to the fact that a polynomial series is used to fit the data. Other approximating functions are available, and in some instances their use may be more appropriate for trend analysis. The most widely used of these alternative functions is the double Fourier series.

The terminology associated with double Fourier series is derived largely from electrical engineering and time-series analysis. The necessary definitions and nomenclature have been presented in Chapter 4 when we discussed single Fourier series, and will be used here with appropriate extensions. You will recall that a complex oscillating or repeating pattern, such as an electrical signal, can be considered to be the sum of a large number of simple sinusoidal wave forms. The amplitudes and phase angles of these simple wave forms can be determined by fitting a series of harmonics of sine and cosine waves to the data. In an analogous manner, a complex surface can be considered to be the sum of two interacting sets of two-dimensional sinusoidal wave forms, each containing many harmonics of differing amplitudes and phase angles. In the simplest example, all harmonics in one direction have zero amplitude, and only one harmonic in the other direction has an amplitude greater than zero. The resulting surface resembles a sheet of corrugated iron roofing, or a set of parallel ripples on a still body of water (Fig. 5.91a). Next, we might consider a surface in which waves of two harmonics are present in one direction, and none in the other (Fig. 5.91b). These surfaces obviously are direct extensions of one-dimensional signals such as those shown in Figure 4.53.

A more complex surface can be produced by a single harmonic in one direction and a single harmonic in the other. If the two waves have the same amplitude and wavelength, the surface pattern resembles a waffle or molded egg carton (Fig. 5.91c). A still more complex surface is shown in Figure 5.91d. It is obvious that surfaces of great intricacy can be created with double Fourier series.

If the fundamental wavelengths in the two mutually perpendicular directions are shorter than the dimensions of map area, the Fourier surface will repeat with the map area. This feature may be desirable in rare circumstances such as the study of harmonic folds, but ordinarily the fundamental wavelength is chosen to exceed the map length, so forced repetition does not occur. Choice of the two fundamental wavelengths, λ_1 and λ_2, and location of the origin of the two-dimensional series usually must be made arbitrarily. Identical double Fourier trend maps can be prepared using different fundamental wavelengths and origins, if a sufficient number of harmonics are created. The contributions made by specific harmonics vary from map to map, but the form of the Fourier trend will remain essentially the same.

Essentially the same model is used for double Fourier analysis as is used for trend-surface analysis. The mapped variable, Y_{ij}, is considered to be a function of a linear trend in the mean value of Y_{ij} across the map area, a regional component, and a local component confounded with a randomly distributed error. This is the model that seems to be tacitly assumed in most double Fouier analyses, especially those of structural deformation of layered sedimentary rock (Harbaugh and Merriam, 1968, p. 125). Other studies, such as those of the distribution of ore bodies in a region or the chemical or mineralogical composition of a body of rock, assume a more conventional time-series model (Agterberg, 1967, pp. 62–67). That is, the distribution of Y_{ij} is regarded as a function of a linear trend, various periodic components, and a random error. As in the first model, the linear trend must be removed prior to double Fourier analysis. The two approaches differ in that the first simply attempts to separate the map variation into large- and small-scale components, whereas the second attempts to ascribe a physical interpretation to the more significant harmonics. Consequently, the time-series approach is the more ambitious of the two, and because it attempts to extract more information, it requires correspondingly more refined data.

In either model, we assume that the distribution of Y_{ij} across the map can be represented by a double Fourier series. This series is a direct expansion of (4.87),

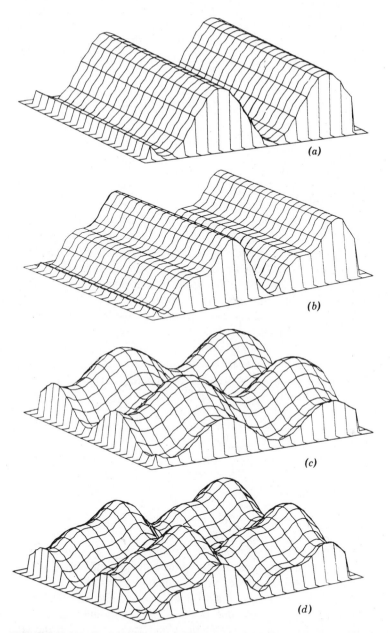

FIGURE 5.91 Two-dimensional sine waves. (a) Single harmonic in X_1 direction. (b) Two harmonics in X_1 direction. (c) Single harmonics in both X_1 and X_2 directions. (d) Two harmonics in both directions.

and has the form

$$
\begin{aligned}
Y_{ij} = &\sum_{i=0}^{\infty} \sum_{j=0}^{\infty} \alpha_{nm} \cos \frac{2k_1 \pi X_{1i}}{n_1} \cos \frac{2k_2 \pi X_{2j}}{n_2} \\
&+ \sum_{i=0}^{\infty} \sum_{j=0}^{\infty} \beta_{nm} \cos \frac{2k_1 \pi X_{1i}}{n_1} \sin \frac{2k_2 \pi X_{2j}}{n_2} \\
&+ \sum_{i=0}^{\infty} \sum_{j=0}^{\infty} \gamma_{nm} \sin \frac{2k_1 \pi X_{1i}}{n_1} \cos \frac{2k_2 \pi X_{2j}}{n_2} \\
&+ \sum_{i=0}^{\infty} \sum_{j=0}^{\infty} \delta_{nm} \sin \frac{2k_1 \pi X_{1i}}{n_1} \sin \frac{2k_2 \pi X_{2j}}{n_2}
\end{aligned}
\tag{5.95}
$$

The symbology is an extension of that used in eqs. (4.88) and (4.89); k_1 and k_2 are the harmonic numbers in directions X_1 and X_2, n_1 and n_2 are the number of equally spaced observations in each of the two directions, and X_{1i} and X_{2j} refer to the observation at location $X_1 = i$ and $X_2 = j$. As in single Fourier series, if we assume that the observations extending from 0 to n_1 and from 0 to n_2 constitute the entire available sample, the upper limits of the summations become n_1 and n_2. In effect, this defines the fundamental harmonic to be equal to the length of the sequence in each direction. Although the equation seems formidable, much of the fractional part of each term is simply converting location X_{ij} into radians.

The expression of the double Fourier series can be simplified if we use the same type of abbreviations for terms as we used in Chapter 4. That is,

$$
C_{k_1} = \cos \frac{2k_1 \pi X_{1i}}{n_1}, \qquad C_{k_2}^* = \cos \frac{2k_2 \pi X_{2j}}{n_2}
$$

$$
S_{k_1} = \sin \frac{2k_1 \pi X_{1i}}{n_1}, \qquad S_{k_2}^* = \sin \frac{2k_2 \pi X_{2j}}{n_2}
$$

Equation (5.95) can then be rewritten as

$$
Y_{ij} = \sum_{i=0}^{\infty} \sum_{j=0}^{\infty} (\alpha_{k_1 k_2} C_{k_1} C_{k_2}^* + \beta_{k_1 k_2} C_{k_1} S_{k_2}^* + \gamma_{k_1 k_2} S_{k_1} C_{k_2}^* + \delta_{k_1 k_2} S_{k_1} S_{k_2}^*)
\tag{5.96}
$$

If the double Fourier series is written in this manner, we can expand it into a series of simultaneous normal equations and solve these for the unknown coefficients in exactly the same manner as trend-surface equations (James, 1966, pp. 3, 4). However, because the equations contain doubly subscripted coefficients and the products of two harmonic series (one in the X_1 direction and the other in the X_2 direction) enter into each term, the equations are complex even in the simplified notation. The organization of the matrix of sums and products is clarified if we label the rows and columns as we did for the single Fourier series matrix in Chapter 4. The matrix of sums and products is the matrix $[A]$ in the equation

$$
[A] \cdot [\beta] = [C]
$$

where $[\beta]$ are the unknown coefficients and the matrix $[C]$ contains the sums of cross-products between Y and the various harmonics. We will find the unknown matrix $[\beta]$ by inversion and multiplication:

$$[\beta] = [A]^{-1} \cdot [C]$$

The matrix of sums and cross-products is developed by expanding the Fourier series to the desired number of harmonics, say k_1 in the X_1 direction and k_2 in the X_2 direction. The rows and columns of the sums and products matrix then have the form

$$
\begin{array}{c}
\\
C_0 C_0^* \\
C_1 C_0^* \\
\vdots \\
C_3 S_1^* \\
\vdots \\
S_{k_1} S_{k_2}^*
\end{array}
\begin{array}{ccccc}
C_0 C_0^* & C_1 C_0^* & \cdots & C_3 S_1^* & \cdots & S_{k_1} S_{k_2}^* \\
\left[\begin{matrix}
\Sigma(C_0 C_0^*)^2 & \Sigma C_0 C_0^* C_1 C_0^* & \cdots \Sigma C_0 C_0^* C_3 S_1^* & \cdots \Sigma C_0 C_0^* S_{k_1} S_{k_2}^* \\
\Sigma C_1 C_0^* C_0 C_0^* & \Sigma(C_1 C_0^*)^2 & \cdots \Sigma C_1 C_0^* C_3 S_1^* & \cdots \Sigma C_1 C_0^* S_{k_1} S_{k_2} \\
\vdots & \vdots & \vdots & \vdots \\
\Sigma C_3 S_1^* C_0 C_0^* & \Sigma C_3 S_1^* C_1 C_0^* & \cdots \Sigma(C_3 S_1^*)^2 & \cdots \Sigma C_3 S_1^* S_{k_1} S_{k_2}^* \\
\vdots & \vdots & \vdots & \vdots \\
\Sigma S_{k_1} S_{k_2}^* C_0 C_0^* & \Sigma S_{k_1} S_{k_2}^* C_1 C_0^* & \cdots \Sigma S_{k_1} S_{k_2}^* C_3 S_1^* & \cdots \Sigma(S_{k_1} S_{k_2}^*)^2
\end{matrix}\right]
\end{array}
$$

$$
\begin{bmatrix}
\alpha_{00} \\
\alpha_{10} \\
\vdots \\
\beta_{31} \\
\vdots \\
\delta_{k_1 k_2}
\end{bmatrix}
=
\begin{bmatrix}
\Sigma Y C_0 C_0^* \\
\Sigma Y C_1 C_0^* \\
\vdots \\
\Sigma Y C_3 S_1^* \\
\vdots \\
\Sigma Y S_{k_1} S_{k_2}^*
\end{bmatrix}
\qquad (5.97)
$$

If many harmonics are to be computed, the matrix equation becomes extremely large, because four coefficients must be calculated for each harmonic. As an example, computation of five harmonics in both directions requires the determination of approximately one hundred coefficients.

You will recall from the discussion of one-dimensional Fourier series in Chapter 4 that the sine of zero degrees is zero; this condition arises in the terms

$$\sin \frac{2k_1 \pi X_{1i}}{n_1} \quad \text{and} \quad \sin \frac{2k_2 \pi X_{2j}}{n_2}$$

if either k_1 or k_2 are equal to zero. Therefore, all terms containing sines of zeroth harmonics are zero, and one row and one column of the matrix of sums and cross-products vanishes. The coefficients that vanish are shown by shading in Figure 5.92. Also, C_0 and C_0^* are equal to 1.0, because the cosine of zero degrees is one. This simplifies the expression of one row and one column of the matrix which contains these terms. In the arrangement of coefficients shown in this figure, block 0 contains only one term, which produces a horizontal plane at the value of its coefficient. Block 1 contains eight terms that represent the fundamental wavelength surfaces. Blocks 0 and 1 together represent the coefficients of the first harmonic trend surface. Block 2 contains sixteen additional terms that represent the second harmonic surface, having a wavelength equal to one-half of the fundamental

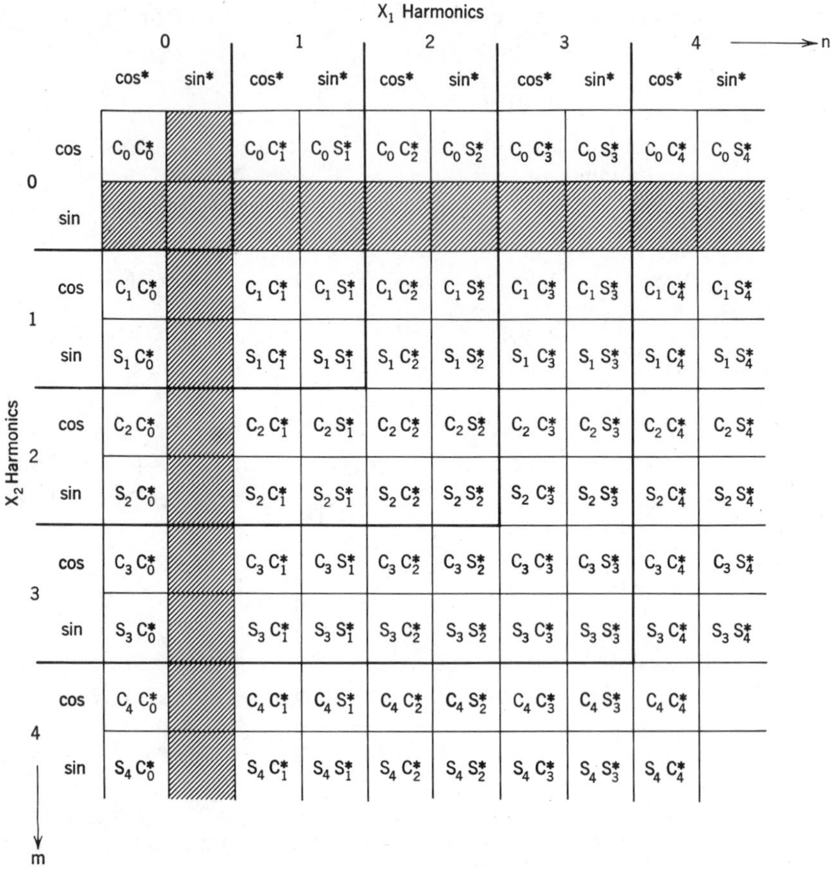

X₁ Harmonics ($\to n$)

X₂ Harmonics		0 cos*	0 sin*	1 cos*	1 sin*	2 cos*	2 sin*	3 cos*	3 sin*	4 cos*	4 sin*
0	cos	$C_0 C_0^*$	▨	$C_0 C_1^*$	$C_0 S_1^*$	$C_0 C_2^*$	$C_0 S_2^*$	$C_0 C_3^*$	$C_0 S_3^*$	$C_0 C_4^*$	$C_0 S_4^*$
0	sin	▨	▨	▨	▨	▨	▨	▨	▨	▨	▨
1	cos	$C_1 C_0^*$	▨	$C_1 C_1^*$	$C_1 S_1^*$	$C_1 C_2^*$	$C_1 S_2^*$	$C_1 C_3^*$	$C_1 S_3^*$	$C_1 C_4^*$	$C_1 S_4^*$
1	sin	$S_1 C_0^*$	▨	$S_1 C_1^*$	$S_1 S_1^*$	$S_1 C_2^*$	$S_1 S_2^*$	$S_1 C_3^*$	$S_1 S_3^*$	$S_1 C_4^*$	$S_1 S_4^*$
2	cos	$C_2 C_0^*$	▨	$C_2 C_1^*$	$C_2 S_1^*$	$C_2 C_2^*$	$C_2 S_2^*$	$C_2 C_3^*$	$C_2 S_3^*$	$C_2 C_4^*$	$C_2 S_4^*$
2	sin	$S_2 C_0^*$	▨	$S_2 C_1^*$	$S_2 S_1^*$	$S_2 C_2^*$	$S_2 S_2^*$	$S_2 C_3^*$	$S_2 S_3^*$	$S_2 C_4^*$	$S_2 S_4^*$
3	cos	$C_3 C_0^*$	▨	$C_3 C_1^*$	$C_3 S_1^*$	$C_3 C_2^*$	$C_3 S_2^*$	$C_3 C_3^*$	$C_3 S_3^*$	$C_3 C_4^*$	$C_3 S_4^*$
3	sin	$S_3 C_0^*$	▨	$S_3 C_1^*$	$S_3 S_1^*$	$S_3 C_2^*$	$S_3 S_2^*$	$S_3 C_3^*$	$S_3 S_3^*$	$S_3 C_4^*$	$S_3 S_4^*$
4	cos	$C_4 C_0^*$	▨	$C_4 C_1^*$	$C_4 S_1^*$	$C_4 C_2^*$	$C_4 S_2^*$	$C_4 C_3^*$	$C_4 S_3^*$	$C_4 C_4^*$	
4	sin	$S_4 C_0^*$	▨	$S_4 C_1^*$	$S_4 S_1^*$	$S_4 C_2^*$	$S_4 S_2^*$	$S_4 C_3^*$	$S_4 S_3^*$	$S_4 C_4^*$	

($\downarrow m$)

FIGURE 5.92 Double Fourier series coefficients arranged according to wavelength. Shaded coefficients are equal to zero. After James (1966).

wavelength. The complete second harmonic surface is composed of coefficients of blocks 0, 1, and 2. Each successive harmonic surface is constructed by adding the terms in the next block.

Double Fourier analysis usually is applied to a class of problems similar to those for which polynomial trend surfaces are used. Irregularly spaced map data are fit by an approximating function based on the geographic coordinates of the data points. The generalized representation created of the approximating function is used to separate the variability of the data into two components; a regional trend represented by the function, and local residuals represented by deviations. Fourier series may be more appropriate approximating functions than power series polynomials if the data seem to contain spatially repetitive elements. As it is not necessary (nor desirable) to fit the data exactly, a limited number of harmonics are sufficient for the analysis. Data in trend-surface problems usually are located irregularly, so it is necessary to compute the Fourier coefficients by solving the general set of normal equations (5.97). Because of the extreme size of the matrices that must be handled, only a few harmonics can be calculated by a small computer. For example, we can see from Figure 5.92 that 49 coefficients are necessary in a double Fourier series having only three harmonics. The matrix that must be inverted to find these coefficients contains 2450 terms! Limitations are imposed not only by the amount of computer memory available, but also by round-off errors that become increasingly troublesome during the inversion of very large matrices.

Fortunately, if our observations can be collected on a regular grid, we can find a way out of this computational impasse. All off-diagonal (cross-product) terms in the normal equations become zero. The matrix we must invert is then a diagonal matrix and the solution of the matrix equation is direct. Furthermore, many harmonics may be computed, rather than the limited number possible with the general solution to the double Fourier equations. This opens up the possibility of performing two-dimensional harmonic analysis, and developing meaningful studies of the power spectra of maps, photographs, and images in general.

Two general groups of geologic problems are currently being investigated by double Fourier analysis. One consists of the search for significant periodicities in ore enrichment within mines and in the placement of ore bodies within a metallogenic province. Many mineral deposits are intimately associated with major fracture systems, and these in turn have developed in response to regional deformation. The mechanical response of large areas of the crust may be such that periodicities, or a form of regularity, are inherent in the location of these major fracture systems. If this is true, and some estimate of the spectrum of the fractures can be obtained, a significant prospecting guide may result. From an analysis of the spatial wavelengths between discovered ore deposits, we may be led to locations where others are likely. The same idea has been applied on a smaller scale to the analysis of individual ore deposits in an attempt to guide further mine development.

A significant application in petroleum geology combines double Fourier analysis with two-dimensional filtering (Robinson, Charlesworth, and Ellis, 1969; Robinson, 1982). The two-dimensional power spectrum of a structural contour map is computed, and from the spectrum the significant spatial wavelengths are determined. Then, filters (which are nothing more than elegant forms of moving averages) are designed to isolate these specific wavelengths. The filters are then used to extract structural features of specified size and/or orientation from the structural contour map. Isolated, positive anomalies may be associated with oil. Experienced researchers have been able to detect and isolate structures in sedimentary beds which reflect deformation in basement rocks many thousands of feet below. These periodic components generally are subdued and obscure in a conventional structural map, but where they can be isolated they have proved to be a significant exploration aid in the prospecting of deeper beds.

At the opposite end of the spatial scale, two-dimensional Fourier analysis has been applied to the study of the arrangement of mineral grains in thin sections and in the examination of pore systems within oil reservoir rocks (Davis and Preston, 1972). In the latter study, researchers are attempting to obtain simple numerical descriptors of the exceedingly complex pattern which pores form in sandstones and limestones, and to use these descriptors to estimate the fluid flow characteristics of the rock. The descriptors are obtained from the power spectrum of an image of the rock. Data consist of the optical density of points measured on a photograph of a thin section, in which the grains of the rock are black and the pores appear as clear areas. The rock photographs are digitized by an electronic device that measures the density of the film at thousands of points on a regularly spaced grid. These data are then processed to obtain the Fourier transform which in turn yields the two-dimensional power spectrum. Further manipulation extracts the desired parameters from the spectra, and these are used in models giving predicted reservoir fluid behavior.

A significant recent development in Fourier analysis of both single and double series is the Fast Fourier Transform (FFT), a computer algorithm that results in a manyfold reduction of computation time and makes the analysis of large data sets practical. Coupled with array processors, this algorithm allows digital processing of seismic records and has resulted in a fundamental advance in geophysical exploration. Fast Fourier Transform methods are being incorporated into other areas of time-series analysis and spatial analysis, even though they necessitate further restrictions on the size and structure of the data set. FFT achieves its great computational speed by a series of special matrix operations that might be compared to constructing a "circular" matrix in which the origin of a series is placed at the center and the data sequence is "wound around" the center. The operation then proceeds by shifting multiplications and additions between adjacent entries. The computational skills necessary to program the FFT are beyond anything presumed for this book, so we will not consider its possibilities further. A good review of the FFT algorithm is given in Cochran and others (1967).

You will recall from Chapter 4 that if our data were spaced at regular intervals, we could develop a set of equations which would yield the Fourier coefficients directly, without the necessity of a matrix operation. These equations can be expanded to the two-dimensional case, except that we must determine four coefficients rather than two for each harmonic. Nonetheless, this is a significant advantage over the general method, because the computation of a large number of harmonics is now possible. This means that we can arbitrarily set fundamental wavelengths and then examine a large suite of harmonics in a search for spatial periodicities. An introduction to double Fourier analysis of gridded geologic data is given in Harbaugh and Merriam (1968, pp. 125–148). The mathematics of two-dimensional Fourier transforms are developed in Cote and others (1960, pp. 18–29) and Rayner (1971, pp. 102–126).

If the data are collected on a regular grid, we can arbitrarily define the fundamental wavelength in the X_1 direction as being equal to the length of the data set in that direction, or n_1. The fundamental wavelength in the X_2 direction can be similarly defined as n_2. We then can compute the coefficients of any harmonic by the equations

$$\alpha_{k_1 k_2} = \frac{\kappa}{n_1 n_2} \sum_{i=0}^{n_1-1} \sum_{j=0}^{n_2-1} Y_{ij} \cos \frac{2\pi k_1 X_{1i}}{n_1} \cos \frac{2\pi k_2 X_{2j}}{n_2}$$

$$\beta_{k_1 k_2} = \frac{\kappa}{n_1 n_2} \sum_{i=0}^{n_1-1} \sum_{j=0}^{n_2-1} Y_{ij} \cos \frac{2\pi k_1 X_{1i}}{n_1} \sin \frac{2\pi k_2 X_{2j}}{n_2} \qquad (5.98)$$

$$\gamma_{k_1 k_2} = \frac{\kappa}{n_1 n_2} \sum_{i=0}^{n_1-1} \sum_{j=0}^{n_2-1} Y_{ij} \sin \frac{2\pi k_1 X_{1i}}{n_1} \cos \frac{2\pi k_2 X_{2j}}{n_2}$$

$$\delta_{k_1 k_2} = \frac{\kappa}{n_1 n_2} \sum_{i=0}^{n_1-1} \sum_{j=0}^{n_2-1} Y_{ij} \sin \frac{2\pi k_1 X_{1i}}{n_1} \sin \frac{2\pi k_2 X_{2j}}{n_2}$$

where

$$\kappa = 1 \text{ if } k_1 = 0 \text{ and } k_2 = 0$$
$$= 2 \text{ if } k_1 = 0 \text{ or } k_2 = 0, \text{ but not both}$$
$$= 4 \text{ if } k_1 > 0 \text{ and } k_2 > 0$$
$$k_1 = \text{harmonic number in direction } X_1$$
$$k_2 = \text{harmonic number in direction } X_2$$
$$n_1 = \text{number of grid points in direction } X_1$$
$$n_2 = \text{number of grid points in direction } X_2$$

This expression of the Fourier transform presumes that the origin of the harmonic series is at the origin of the coordinate system, or where X_1 and X_2 equal zero. Harmonics can be calculated down to $k_1 = n_1/2$ and $k_2 = n_2/2$, but higher harmonic numbers are estimated with fewer points per wave form and so are less reliable estimates than are lower harmonics.

Once the Fourier coefficients have been calculated, the two-dimensional raw power spectrum can be calculated by an extension of (4.96):

$$s_{k_1 k_2}^2 = (\alpha_{k_1 k_2}^2 + \beta_{k_1 k_2}^2 + \gamma_{k_1 k_2}^2 + \delta_{k_1 k_2}^2)/4 \qquad (5.99)$$

You will recall that the power is an expression of variance and measures the variance contributed by the k_1th and k_2th harmonics to the total variance of Y. If we denote the total variance in Y as s_y^2, the percentage of contribution of any harmonic pair is

$$s_{k_1 k_2}^2 / s_2^2 \cdot 100\% = \text{percentage of contribution} \qquad (5.100)$$

However, the raw power spectrum is only an estimate of the contributions to the variance by the harmonics, and like all sample estimates these may deviate markedly from the true population power spectrum. The raw spectrum becomes more stable and a better estimate of the population spectrum as the sample size tends toward infinity, but we cannot continue sampling indefinitely and must end the analysis at some point. Our estimates of the sample power spectrum can be improved by smoothing, because adjacent harmonics tend to be similar and an

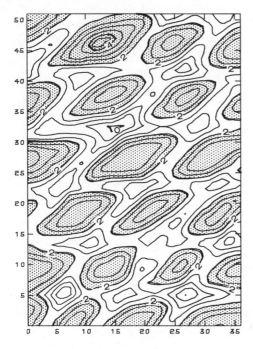

FIGURE 5.93 Contour diagram of interference ripple marks on slab of Precambrian sandstone from eastern Idaho. Depressions deeper than 2 cm are shaded. Coordinate scale in 2-cm units.

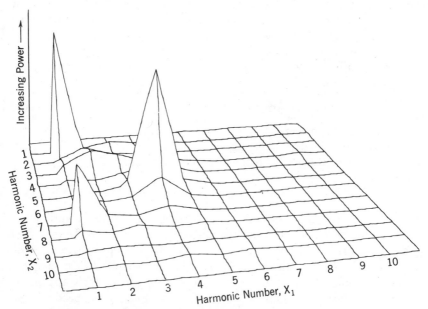

FIGURE 5.94 Two-dimensional power spectrum of ripple marks. Note strong contribution by two dominant sets of wavelengths.

averaging process forces them to converge on a common value. Smoothing is accomplished using techniques discussed in the section on two-dimensional moving averages. A common weighting scheme for producing a smoothed estimate of a spectral value $\hat{s}^2_{k_1 k_2}$ is

$$\hat{s}^2_{k_1 k_2} = \frac{1}{16} (s^2_{k_1-1,k_2-1} + s^2_{k_1-1,k_2+1} + s^2_{k_1+1,k_2-1} + s^2_{k_1+1,k_2+1})$$

$$+ \frac{1}{8} (s^2_{k_1-1,k_2} + s^2_{k_1+1,k_2} + s^2_{k_1,k_2-1} + s^2_{k_1,k_2+1}) + \frac{1}{4} s^2_{k_1,k_2}$$

(5.101)

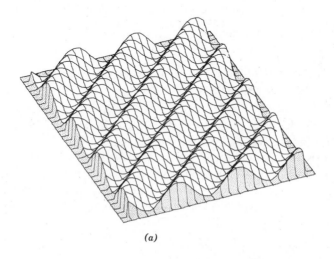

(a)

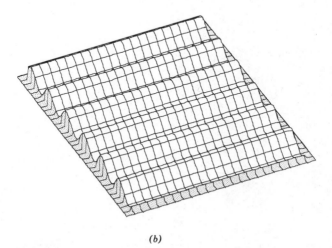

(b)

FIGURE 5.95 Reconstructions of wave forms present in ripple-marked sandstone. (a) Perspective diagram of largest spectral component. (b) Perspective diagram of second major waveform.

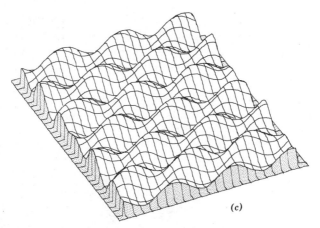

(c)

FIGURE 5.95 (*Continued*)   (c) Surface formed by combination of two dominant waveforms.

We will consider an application of double Fourier analysis in which both the Fourier trend and the power spectrum are of interest. Ripple marks preserved on bedding planes of sandstones may contain valuable paleogeographic information. It is common practice in stratigraphic investigations to measure the dominant orientation, wavelength, and amplitude of ripple marks. Sometimes measurements taken at many different outcrops can be combined to produce maps of paleocurrent

FIGURE 5.96 Contoured reconstruction of two dominant wave forms extracted from rippled sandstone. Compare with original data shown in Figure 5.93.

directions to draw inferences about strand lines. However, few attempts have been made to analyze in detail the pattern of ripple marks exposed over a wide extent of a single outcrop. Figure 5.93 is a diagram of the rippled pattern on a slab of Late Precambrian sandstone exposed in eastern Idaho. A typical ripple mark on this slab has an amplitude of about 3 cm and a wavelength of about 15 cm. The diagram was prepared by contouring measurements collected on a 2-cm grid. A wooden frame was first placed on the rock surface, and then a straightedge was laid across the frame to define successive traverses. At 2-cm increments, the distance from the bottom of the straightedge to the surface of the rock was measured. Because the resulting data are so voluminous, they are not presented here except in the form of the contour diagram.

A double Fourier analysis of this data yields the power spectrum shown in Figure 5.94. This spectrum has been produced as a display that is a two-dimensional extension of power spectra diagrams in Chapter 4. Two major peaks are apparent on the diagram; these represent two significant harmonics.

From the information in the power spectrum, we can recompute the individual wave forms which are the major contributors to the rippled pattern. Figure 5.95a is a computer-drawn perspective diagram of the largest spectral component; Figure 5.95b is a diagram of the second major wave form. These two are combined in Figure 5.95c. A contour map of the surface defined by these two harmonics is shown in Figure 5.96. If you compare this pattern to the original contour map (Fig. 5.93), you will note that the essential features of the original are recaptured by a very simple model using only two harmonics.

Comparing Maps

A common problem in many geologic studies is the comparison of two or more maps of an area with one another. Typically, these may be maps of different chemical or mineralogical constituents, different grain-size parameters, or structural maps on different geologic horizons. Direct comparisons of trend surfaces of different horizons have been made in attempts to date periods of structural deformation. A related but more difficult problem is the comparison of maps of a single parameter for two or more regions. Geologists are continually comparing and contrasting geologic features in one region with those in another. Indeed, much of the interpretative skill of a geologist is based on his prior experiences and his ability to mentally compare a new region with those he has studied in the past. Although quantitative comparisons of mapped variables seem potentially useful to geologists, only a few attempts have been made to measure similarity between two maps. Fortunately, this is a problem that has interested geographers for many years, and they have devised several useful techniques for comparing spatial distributions, giving either an overall, average measure of similarity or a map of the deviations between the two surfaces.

The subject of map comparisons will become increasingly important in the future, because interpreting the voluminous data from Earth-sensing satellites will require development of automatic pattern recognizers and map analyzers. The algorithms that control these machines must be developed by geologists and other Earth scientists, who alone have the knowledge of the Earth necessary to interpret the data. In turn, geologists must learn to quantify and systematize their mental recognition skills so the machines can be taught to assume some of the burden for them. If this is not done, we will be literally buried under the reams of charts, maps, and photographs returned from the resources survey satellites, orbiting geophysical platforms, and other exotic tools of the future.

Overall Similarity

Possibly the simplest way to compare two maps is to compute the correlation between the mapped variables. If the two maps are prepared from variables measured at common points, this consists simply of computing the correlation coefficient between mapped variable 1 and mapped variable 2 with no consideration of the sample locations at all. The correlation coefficient is calculated by (2.24), and results in a measure of overall correspondence between the two variables.

During a study of an estuary on the North Carolina coast, shown on the map in Figure 5.97a, a graduate student collected a series of bottom sediment samples. These were analyzed to determine the grain-size distributions, and the customary size-distribution statistics (mean, standard deviation, skewness, and kurtosis) were calculated. It has been suggested by many authors that various combinations of moment statistics are effective for distinguishing sedimentary environments. The student is interested in interrelations between the statistics to see if they can define areas within the estuary where conditions of sedimentation are markedly different.

Figure 5.97b is a contour map of the standard deviation of sediment grain size within the estuary. Figure 5.97c is a map of grain-size skewness of the same samples. It has been suggested that skewness and standard deviation are effective for classifying sediments when used jointly, so it is important that the relationship between these two variables be assessed. Because both variables were computed from raw data collected at common points, the correlation can be calculated directly from the data. The overall correlation between the two maps is $r = 0.52$. Note that the geographic coordinates of the points are not considered. This reflects a major drawback in the method of overall comparison, because a given correlation may reflect the degree of correspondence over the entire map area, or may be the result of a large deviation in a small region of the map.

We may wish to compare two maps where the variables are not measured at the same points. One possible way of obtaining a measure of overall similarity is to estimate the values of both variables at a set of grid points common to both maps. Because the estimation procedure introduces errors we cannot assess, it is desirable to place the estimated points as close to original sample points as possible. In fact, the most effective scheme might be to estimate one variable from its contour map at every sampling location of the other variable. We then can compute a correlation between the two variables. However, no statistical significance can be attached to the correlation because it is based entirely on interpolated values. Obviously, the reliability of the correlation is improved with increased density of control points.

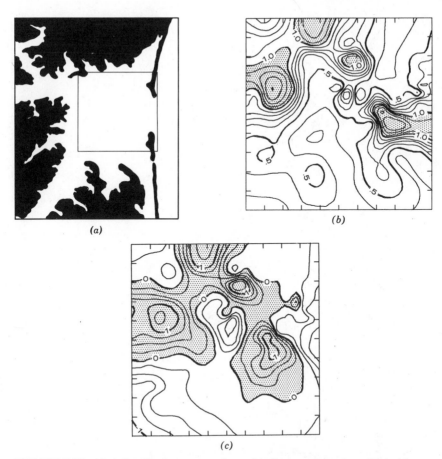

FIGURE 5.97   Variation in size parameters of bottom sediments collected in an estuary on the North Carolina coast. (*a*) Index map showing relation of study area to the estuary shore. (*b*) Standard deviation of grain size, in phi units. Areas of sediment having grain-size standard deviations greater than 1.0 are shaded. (*c*) Skewness of grain size. Areas of sediment having positive skewness are shaded.

Similarity Maps

The simplest possible map comparison that can be made is between two maps of the same type of variable in the same map region, as, for example, in the construction of an isopach map from two structural contour maps. An isopach map is a difference map, in this case of elevation between two surfaces. Similar maps can be constructed showing the differences in variables such as grain size or percentage of a constituent measured in two different horizons. A simple example is shown in Figures 5.98*a* and 5.98*b*, which are contoured maps of the groundwater table of an area in Nebraska. Map *a* was prepared from data gathered in 1950 and shows the water table undisturbed except for the effects of local pumping. Map *b* was prepared from data collected 10 years later, after construction of a large dam

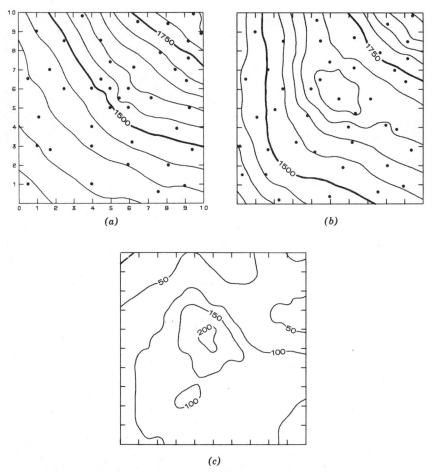

FIGURE 5.98 Elevation of the water table in an area in Nebraska, as measured in observation wells. (*a*) Water table elevation in 1950. (*b*) Water table elevation in 1960, after construction of a small reservoir. (*c*) Change in water table elevation between 1950 and 1960, produced by subtracting contour map grid *a* from grid *b*. Observation wells shown by dots. Elevations in feet above sea level.

near the middle of the area. The contribution of the reservoir to the groundwater level has been significant, and can be shown by producing a difference map (Fig. 5.98*c*). If the original maps are prepared by an automatic contouring program that constructs an intermediate grid, as these were, the difference map is constructed by subtracting the two matrices of grid values from each other and then contouring the resulting matrix. Of course, if the same observation wells have been used for both maps, the difference between the two sets of raw data could be found and then contoured. In this example, you will note that not all points are common to the two maps.

In the example, both maps are expressed in the same units, and a comparison may be made simply by determining the difference between the two. However, the problem of comparison is more complex if two different variables have been mapped. We will now examine a problem in which the two maps to be compared are expressed in different units and were collected at different control points.

Reflection seismic techniques are among the most successful methods of prospecting for geologic structures that may contain oil. A series of seismic shocks are set off along a traverse. The length of time required for seismic waves to travel from the surface to reflecting horizons and back to geophones at the surface can be precisely measured. From these measurements, a seismic cross-section may be produced that gives the configuration of the reflecting horizons along the line of geophones. A series of seismic profiles across a suspected structure can be combined into a seismic contour map, as in Figure 5.99. Although the data have been corrected for all the geometric factors that influence return time, no precise way exists to turn these measurements into depth estimates. This is because the velocity with which seismic energy travels through rocks varies with composition, depth, and a host of other variables. However, a contour map of return times, such as Figure 5.99, should correspond closely in form to a structural map of the reflecting horizon.

Figure 5.100 is a structural contour map of the same area, based on measurements at drill-holes that penetrated the reflecting horizon. This provides a somewhat more refined picture of the structural configuration than does the seismic map, although the two closely resemble each other. We are interested in comparing the two, to determine where our seismic estimate of the structure departs most severely from the picture revealed by drilling. However, seismic observations were not made at the drill-holes where elevations to the top of the reflecting horizon were measured, nor are the two maps expressed in the same units. In order to directly compare these two maps, we must express one in terms of units of the other, or must convert both to a standardized, unitless form.

Expressing one variable in terms of the other has a certain advantage, because the comparison is in units of one of the original maps, allowing us to perceive

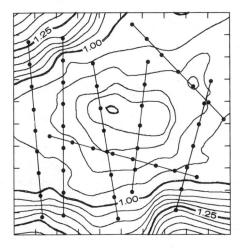

FIGURE 5.99 Structural configuration of a seismic reflecting surface, as defined by seismic return times. Lines are seismic traverses. Contours are in seconds of return time, after corrections.

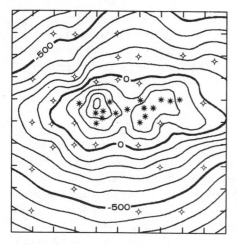

FIGURE 5.100 Structural contour map of top of reflecting horizon, as determined from drilling data. Elevations in feet below sea level.

areas where the mapped variable is "greater than it should be" or "smaller than predicted" on the basis of the other variable. The operation is performed by estimating one variable by the other, using least-squares regression methods, and then mapping the deviation of the predicted variable from the actual values. The map variable to be estimated may be denoted Y and the other map variable X. If we have observed X and Y at common points, we can compute the regression of Y on X from these observations. From the regression equation, we then can predict values of $\hat{Y}$ at every point, in effect mapping X in terms of Y. That is, we compute

$$\hat{Y} = \beta_0 + \beta_1 X_1 \tag{5.102}$$

for all points on the map; $\hat{Y}$ is a linear transformation of X into units of Y, based on a least-squares estimate of the relation between the two variables. Although typically we will compute a linear regression of Y on X to find values of $\hat{Y}$, a low-order polynomial regression also may be used. The procedure is exactly the same as that described in (4.13)–(4.16), except that the independent variable is one of our two mapped variables rather than a spatial coordinate. That is, (5.102) can be solved for the two coefficients by the normal equations

$$\sum_{i=1}^{n} Y_i = b_0 + b_1 \sum_{i=1}^{n} X_i$$
$$\sum_{i=1}^{n} X_i Y_i = b_0 \sum_{i=1}^{n} X_i + b_1 \sum_{i=1}^{n} X_i^2 \tag{5.103}$$

or the equivalent matrix equation

$$\begin{bmatrix} n & \Sigma X \\ \Sigma X & \Sigma X^2 \end{bmatrix} \cdot \begin{bmatrix} b_0 \\ b_1 \end{bmatrix} = \begin{bmatrix} \Sigma Y \\ \Sigma XY \end{bmatrix} \tag{5.104}$$

Occasionally, a better estimate of $\hat{Y}$ can be obtained using a polynomial expansion of (5.102); the appropriate normal equations are given in (4.32) and the accompanying section.

Having found an equation that gives one variable in terms of the other, we then can create a map of predicted variables of $\hat{Y}$. Because the $\hat{Y}$'s are based solely on values of the second variable X and the relationship we have determined between Y and X, the map of $\hat{Y}$ may be regarded as a map of X expressed in units of Y. The difference map $(Y - \hat{Y})$ may be regarded as a map of the difference between X and Y, expressed in units of Y.

In this particular problem, we are interested in how well seismic data predicts structure. Therefore, we can define Y as structural depth and X as seismic return time. By computing the regression

$$\text{structural depth} = b_0 + b_1 \text{ (seismic return time)}$$

we can obtain an equation that will give the best estimate of the structure from the seismic data. Because the two variables were not measured at common points, we must base our equation on the estimates of X and Y produced between the control points by the contouring program. No statistical assessment of the regression of Y on X is possible for this reason.

Figure 5.101a is the map of estimated structural elevation $\hat{Y}$ produced from the regression. Because the regression equation expresses a direct linear relationship, the form of the estimated structure is identical to the form of the seismic return time map. However, the scale has been transformed from one of time to one of elevation. The equation relating structural depth and seismic return may be regarded as a way of estimating seismic velocity. The coefficients found are

$$\text{structural depth} = 1{,}389.6 - 1{,}692.5 \text{ (seismic return time)}$$

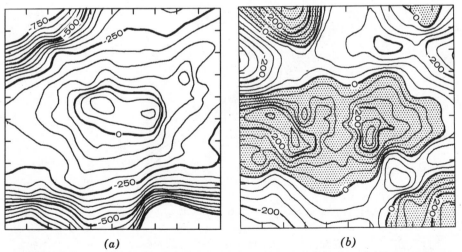

(a)   (b)

FIGURE 5.101 Structural configuration estimated from seismic data. (a) Structural contour map based on regression of seismic return times. Elevations in feet below sea level. (b) Difference map obtained by subtracting measured structural surface (Fig. 5.100) from estimated structure. Positive deviations (shaded) indicate areas where seismic prediction underestimates elevation.

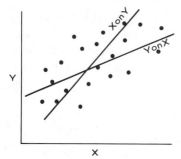

FIGURE 5.102 Two lines of regression, Y on X and X on Y, through a set of sample points.

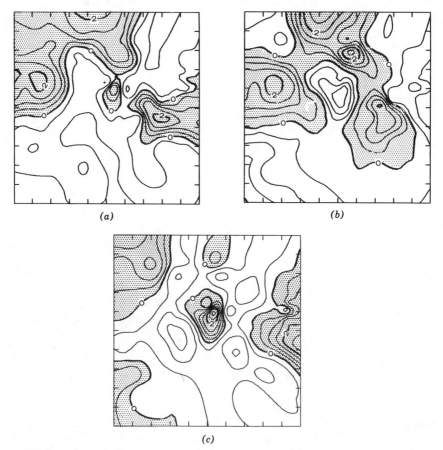

(a)   (b)

(c)

FIGURE 5.103 Variation in standardized size parameters of bottom sediments collected in a North Carolina estuary. Maps of raw parameters shown in Figure 5.97. (a) Standardized standard deviation. (b) Standardized skewness. (c) Difference map, produced by subtracting map b from map a. Maps are expressed in units of standard deviations away from the mean value of the raw parameters. Positive areas are shaded.

Figure 5.101*b* is a difference map found by subtracting Y from $\hat{Y}$. Areas in which seismic data underestimated the depth to the structural horizon appear as positive deviations. Where seismic methods predicted greater depth than that actually encountered, the deviations will be negative.

Several obvious hazards exist in the application of this method. First, the estimates $\hat{Y}$ account for only a part of the variation in Y. Unless the correlation between X and Y is high, serious errors may be introduced by the substitution of the estimate $\hat{Y}$ for X. In this example, $r = 0.87$, so the estimation process seems reasonable.

Although it seems reasonable in the present example to express the difference map in units of structural elevation, it may not always be possible to decide which variable should be used as the estimator. In this case, two lines of regression are possible, one of Y on X and another of X on Y, as shown in Figure 5.102. If the correlation between X and Y is high, the two lines will nearly coincide, but if the correlation is not pronounced, the two estimating lines may produce radically different results (Mills, 1955, pp. 283–289). A possible way out of this impasse is to use the reduced major axis to express one variable in terms of the other, as

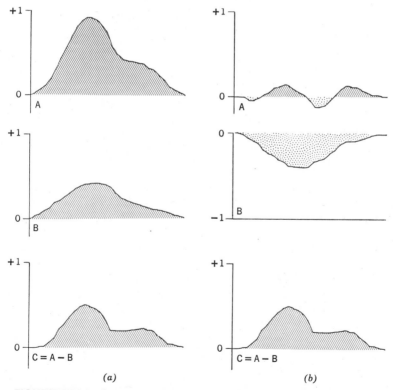

FIGURE 5.104 Cross-sections of maps, showing special cases that can arise if difference maps are created by subtraction of one map from another. (*a*) Subtraction of low positive area from high positive area. (*b*) Subtraction of low negative area from low positive area. Result is identical to *a*.

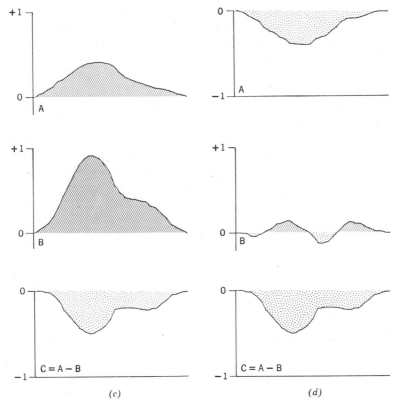

FIGURE 5.104 (*Continued*)   (*c*) Subtraction of high positive areas from low positive area. (*d*) Subtraction of low positive area from low negative area. Result is identical to *c*.

discussed in Chapter 4. Of course, it also can be argued that in the absence of a high correlation between two variables it is pointless to attempt to compare the two by a predictive method.

The problems inherent in difference maps based on estimated or predicted variables can be avoided if the two original maps are converted to standardized form. To do this, the variable at each control point on a map is subtracted from the mean of that variable and divided by the standard deviation. In other words, we perform the same transformation that we used to convert data to standard normal form in Chapter 2:

$$Z_i = \frac{X_i - \overline{X}}{s} \tag{5.105}$$

After the data on each map have been standardized, they can be contoured in the conventional manner. However, the contour values will be in units of standard deviation above or below the mean. This may impose some barrier to interpretation, because the map units are unfamiliar. However, those who use statistical procedures should have little difficulty adjusting to the standardized scale.

As an example, we will convert the maps of standard deviation and skewness of sediments in an estuary (Figs. 5.97*b* and 5.97*c*) to standardized form. Figure 5.103*a* is standardized grain-size standard deviation. Figure 5.103*b* is standardized grain-size skewness.

Both maps are now drawn on the same contour scale, so they can be compared by subtracting one from another to create a difference map similar to an isopach map. The map of standard deviation minus skewness is shown in Figure 5.103*c*. However, some problems may arise, because the difference map may contain ambiguous areas. For example, consider the series of cross-sections of two map surfaces shown in Figure 5.104. In these illustrations, map B has been subtracted from map A. In part *a,* an expected positive difference results from subtracting a low positive area from a high positive area. In part *b,* however, it is apparent that a positive difference also can result by subtracting a large negative area from a low positive area (or even from a zero area or low negative area). Similar ambiguous cases that result in negative differences are shown in parts *c* and *d.* Although the differences indicated are mathematically correct, we would regard the two surfaces in Figures 5.104*a* and 5.104*c* as being more alike than the pairs of surfaces in Figures 5.104*b* and 5.104*d.* In parts *a* and *c,* both surfaces deviate from the mean

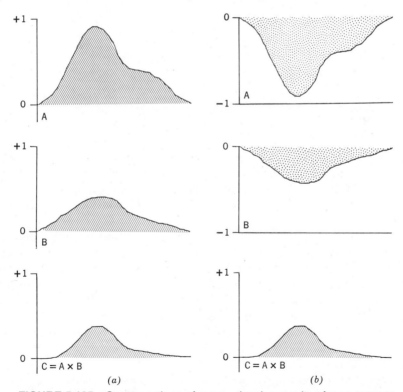

FIGURE 5.105 Cross-sections of maps, showing results of map comparison by cross-multiplication. (*a*) Multiplication of two positive areas. (*b*) Multiplication of two negative areas.

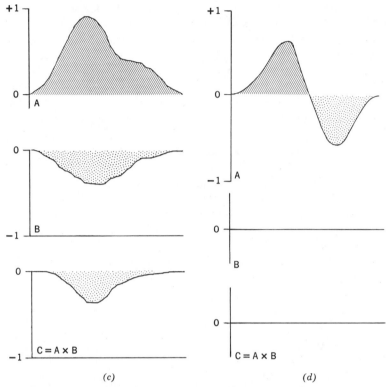

FIGURE 5.105 (*Continued*)   (*c*) Multiplication of positive by negative area.
(*d*) Multiplication of either positive or negative areas by zero area.

in the same direction and are subparallel. The correlation between a series of points common to both surfaces would be positive. The more closely the two surfaces correspond, the higher the correlation between them. In contrast, the pairs of surfaces shown in parts *b* and *d* would be negatively correlated, because they slope in opposite directions.

Because the surfaces are standardized, it is not really necessary to compute a correlation to obtain a measure of similarity between the two surfaces. Instead, the surfaces can be multiplied together. If both surfaces deviate from the mean in the same direction, their product will be positive. If they deviate in opposite directions, their product will be negative. Examples of these two cases are shown in Figure 5.105. The only ambiguous case arises when one of the surfaces is a flat plane, at the mean value. After standardization, all points on such a surface will be zero, so the product map will also be zero regardless of the form of the second map. This case is shown in Figure 5.105*d*.

The map of the product of the two standardized grain-size statistics is shown in Figure 5.106. Areas of high coincidence of the two surfaces are clearly shown, as are areas where they depart markedly from one another.

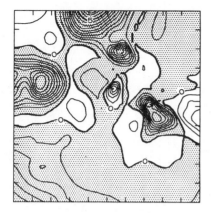

FIGURE 5.106 Product map of grain-size standard deviation multiplied by grain-size skewness of sediments collected in North Carolina estuary. Original size parameters are standardized as shown in Figure 5.103. Positive areas (shaded) indicate high correlation between map surfaces; negative areas are regions of inverse correlation between maps.

Comparing Map Coefficients

A very simple procedure has been used to compare trend surfaces of different stratigraphic horizons. Because the predicted values of $\hat{Y}_i$ are based exclusively on a polynomial equation, the degree of similarity between the surfaces can be determined by comparing the polynomial coefficients (the β's) alone. Of course, the surfaces must be of the same degree, because the equations that are compared must contain the same number of terms. Also, we must use the same sample locations to compute all surfaces that are to be compared. These restrictions limit the utility of the method to such problems as a study of structural changes in a region as reflected in the similarity of trend surfaces on successive stratigraphic horizons. Another possible application is comparison of trend surfaces of different mineralogical or geochemical constituents, provided all variables are derived from the same set of samples and all surfaces are of the same degree.

The coefficients may be compared by any of several measures of similarity. The correlation coefficient, r, is appropriate for this purpose, and may be defined as

$$r = \frac{\text{cov } \beta_{12}}{\sqrt{\text{var } \beta_1 \text{ var } \beta_2}} \qquad (5.106)$$

which is derived directly from (2.24). However, instead of comparing observations of X_1 and X_2, we are comparing the coefficients (β's) of the equations for trend surface 1 and trend surface 2.

Another measure of similarity that has been used for this purpose is the taxonomic distance, d, defined as

$$d = \sqrt{\frac{\sum_{i=1}^{n} (\beta_{1i} - \beta_{2i})^2}{n}} \qquad (5.107)$$

That is, the taxonomic distance is the square root of the mean of the squared differences between equivalent coefficients of the two surfaces being compared. We will discuss this measure further in Chapter 6 when we consider the problem

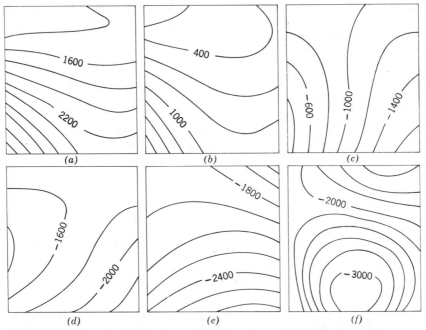

FIGURE 5.107   Third-degree trend surfaces of stratigraphic units in northwestern Kansas. Units are (a) Cretaceous, (b) Permian, (c) Pennsylvanian, (d) Mississippian, (e) Ordovician, and (f) Precambrian. (After Merriam and Harbaugh, 1964.)

of classification. It is no more efficient than the correlation coefficient, but it is sometimes easier to interpret because it is always positive and is not constrained to values less than 1.0 as is the correlation coefficient. Identical data sets give taxonomic distances of zero; increasingly dissimilar data sets have increasingly greater taxonomic distances between them.

Figure 5.107 shows a series of third-degree trend surfaces fit to successive formations in an area in northwestern Kansas. The correlation coefficient matrix in Table 5.24 gives the correlation between the coefficients of each surface. Table 5.25 is the matrix of taxonomic distances between the same surfaces. An exami-

TABLE 5.24   Matrix of Correlations between Coefficients of Third-Order Trend Surfaces (β_0 Coefficients not Included)

	(a)	(b)	(c)	(d)	(e)	(f)
(a) Cretaceous	1.000					
(b) Permian	0.684	1.000				
(c) Pennsylvanian	0.672	0.250	1.000			
(d) Mississippian	−0.547	−0.032	−0.036	1.000		
(e) Ordovician	−0.002	−0.010	0.521	0.256	1.000	
(f) Top Precambrian	−0.470	−0.486	−0.838	−0.248	−0.693	1.000

Source: Merriam and Sneath (1966).

TABLE 5.25 Matrix of Taxonomic Distances between Coefficients of Third-Order Trend Surfaces (β_0 Coefficients not Included)

	(a)	(b)	(c)	(d)	(e)	(f)
(a) Cretaceous	0.000					
(b) Permian	0.510	0.000				
(c) Pennsylvanian	0.565	0.691	0.000			
(d) Mississippian	1.316	0.975	1.108	0.000		
(e) Ordovician	0.943	0.789	0.629	0.936	0.000	
(f) Top Precambrian	2.096	1.839	2.296	2.057	2.205	0.000

Source: Merriam and Sneath (1966).

nation of the two matrices will show that the relative similarities indicated by the two measures are the same.

The trend-surface equation derived from one set of control points in a region may differ radically from an equation of the same degree derived from other points in the same region. However, the form of the two trend surfaces may be identical and both fit equally well. This results from the fact that the trend equation is a regression on the geographic coordinates of the sample localities, so any change in these is reflected in the coefficients. For this reason, comparisons between trend coefficients are valid only if the same sample points are used for all trends being compared. Further, the absolute values of the regression coefficients are sometimes subject to extreme fluctuations, especially in the higher powers, as a result of machine rounding and truncation. These factors restrict the applicability of this method of comparison of map surfaces.

SELECTED READINGS

Agterberg, F. P., 1967, Computer techniques in geology: *Earth-Science Reviews,* **3,** p. 47–77.

A general review article, including several examples in two-dimensional analysis.

Aitchison, J., and J. A. C. Brown, 1969, *The lognormal distribution, with special reference to its uses in economics:* Cambridge Univ. Press, Cambridge, 176 p.

Armstrong, Margaret, and Romain Jabin, 1981, Variogram models must be positive-definite: *Jour. Int'l. Assoc. Mathematical Geology,* **13,** no. 5, p. 455–459.

Bartels, C. P. A., and R. H. Ketellapper (eds.), 1979, *Exploratory and explanatory statistical analysis of spatial data:* Martinus Nijhoff Publishing, Boston, 268 p.

Tests of point distributions and spatial dependence are discussed in this collection of papers on "regional analysis."

Batschelet, Edward, 1965, Statistical methods for the analysis of problems in animal orientation and certain biological rhythms: *American Institute of Biological Sciences Monograph,* Washington, D.C., 57 p.

Bennett, R. J., 1979, *Spatial time series, analysis–forecasting–control:* Pion Ltd., London, 674 p.

An advanced treatment of temporal and spatial variation, developed from the viewpoint of classical time-series analysis.

Berry, B. J. L., and D. F. Marble (eds.), 1968, *Spatial analysis, a reader in statistical geography:* Prentice-Hall, Inc., Englewood Cliffs, N.J., 512 p.

The collection of definitive articles in this volume will be of interest to geologists as well as geographers. Many of these are from relatively obscure sources and may be difficult to find elsewhere.

Birkhoff, Garrett, and Lois Mansfield, 1974, Compatible triangular finite elements: *Jour. Mathematical Analysis and Applications,* **47,** no. 3, p. 531–553.

An advanced mathematical discussion of the principles behind contouring algorithms that utilize a triangular mesh.

Boon, J. D., III, D. A. Evans, and H. F. Hennigar, 1982, Spectral information from Fourier analysis of digitized quartz grain profiles: *Jour. Int'l. Assoc. Mathematical Geology,* **14,** no. 6, p. 589–605.

This article points out the importance of determining the true centroid in polar Fourier analysis, especially as applied to the study of grain shape.

Chayes, F., 1970, On deciding whether trend surfaces of progressively higher order are meaningful: *Geol. Soc. America Bulletin,* **81,** p. 1273–1278.

Cheeney, R. F., 1983, *Statistical methods in geology:* George Allen & Unwin Ltd., London, 169 p.

Chapters 8 and 9 of this text contain exceptionally well-written discussions of the analysis of two- and three-dimensional orientation data from the geological sciences.

Chorley, R. J., and P. Haggett, 1965, Trend-surface mapping in geographical research: *Trans. Inst. British Geographers,* Publ. No. 37, p. 47–67.

A clear discussion of trend-surface analysis, from a nonstatistical viewpoint. Most of the examples are geologic. Reprinted as article IV-7 in Berry and Marble (1968).

Clark, Isobel, 1979, *Practical geostatistics:* Applied Science Publishers, London, 129 p.

This compact tome is the most lucid discussion of regionalized variable theory and the practice of kriging that has been written in English.

Clark, M. W., 1981, Quantitative shape analysis: A review: *Jour. Int'l. Assoc. Mathematical Geology,* **13,** no. 4, p. 303–320.

An extensive review of particle shape analysis, with emphasis on Fourier techniques.

Cliff, A. D., and J. K. Ord, 1981, *Spatial processes, models and applications:* Pion Ltd., London, 266 p.

This advanced text emphasizes spatial autocorrelation and analysis of the patterns of points. Most of the applications are to discontinuous geographic variables.

Cochran, W. T. and others, 1967, What is the Fast Fourier Transform?: *Proc. Inst. Electrical and Electronics Engineers,* **55,** p. 1664–1674.

Cole, J. P., and C. A. M. King, 1968, *Quantitative geography:* John Wiley & Sons, Inc., New York, 692 p.

A valuable look at spatial analysis from the viewpoint of a geographer. Part 2, "Spatial distributions and relationships," is especially pertinent to material covered in this chapter.

Cote, L. J., J. O. Davis, W. Marks, R. J. McGough, E. Mehr, W. J. Pierson, Jr., J. F. Roper, G. Stephenson, and R. C. Vetter, 1960, The directional spectrum of a wind generated sea as determined from data obtained by the stereowave observation project: *New York Univ. Meteorological Papers,* **2,** no. 6, 88 p.

Dacey, M. F., 1967, Description of line patterns: *Northwestern Studies in Geography,* **13,** p. 277–287.

One of the few discussions of patterns formed by lines in a plane.

Dahlberg, E. C., 1975, Relative effectiveness of geologists and computers in mapping potential hydrocarbon exploration targets: *Jour. Int'l. Assoc. Mathematical Geology,* **7,** no. 5/6, p. 373–394.

A revealing comparison between contour maps produced by geologists and those produced by computer.

David, Michel, 1977, *Geostatistical ore reserve estimation:* Elsevier Publ. Co., Amsterdam, 364 p.

An extensive discussion of the practical aspects of regionalized variable theory and the use of kriging for mine evaluation. The treatment is at an intermediate level. Several FORTRAN programs are included in the text.

Davis, J. C., 1976, Contouring algorithms, *in AUTOCARTO II—Proceedings of the International Symposium on Computer-assisted Cartography,* Sept. 21–25, 1975: U.S. Bureau of the Census/Amer. Congress on Survey and Mapping, p. 352–359.

An assessment of options in gridding algorithms and their relative performance.

Davis, J. C., and F. W. Preston, 1972, Optical processing—An alternative to digital computing, *in* Fenner, P. (ed.), Quantitative Geology: *Geol. Soc. America,* Spec. Paper 146, p. 49–68.

Donnelly, K. P., 1978, Simulations to determine the variance and edge effect of total nearest neighbor distance, *in* Hodder, I. (ed.), *Simulation studies in archaeology:* Cambridge Univ. Press, Cambridge, p. 91–95.

An important article describing the corrections necessary to compensate for edge effects in nearest-neighbor analysis.

Doveton, J. H., and A. J. Parsley, 1970, Experimental evaluation of trend surface distortions induced by inadequate data-point distributions: *Inst. Mining and Metallurgy, Trans. Sec. B,* p. B197–B208.

The discussion of the influence of data-point distribution on trend surfaces in this chapter is based in large part on these experiments. Figures 5.86 and 5.87 are adapted from this article.

Ehrlich, R., J. P. Brown, and J. M. Yarus, 1980, The origin of shape frequency distributions and the relationship between size and shape: *Jour. Sedimentary Petrology,* **50,** no. 2, p. 475–484.

One of a series of articles by these authors on the use of circular Fourier analysis in studies of grain shape.

Folk, R. L., 1968, *Petrology of sedimentary rocks:* Hemphill's, Austin, Texas, 184 p.

Discusses traditional measures of grain shape.

Gaile, G. L., and J. E. Burt, 1980, Directional statistics: Concepts and Techniques in Modern Geography No. 25, *Geo Abstracts,* Univ. East Anglia, Norwich, 39 p.

A compact monograph that summarizes many of the test procedures that are applied to directional data.

Getis, Arthur, and Barry Boots, 1978, *Models of spatial processes, an approach to the study of point, line and area patterns:* Cambridge Univ. Press, Cambridge, 198 p.

Gold, C. M., T. D. Charters, and J. Ramsden, 1977, Automated contour mapping using triangular element data structures and an interpolant over each irregular triangular domain: *Computer Graphics,* **11,** no. 2, p. 170–175.

Griffiths, J. C., 1962, Frequency distributions of some natural resource materials: Pennsylvania State Univ., *Mineral Industries Experiment Station Circular* 63, p. 174–198.

This and the following reference discuss the negative binomial distribution as a model for the occurrence of mineral deposits and oil fields.

Griffiths, J. C., 1966, Exploration for natural resources: *Jour. Operations Research Soc. America,* **14,** no. 2, p. 189–209.

Gumbel, E. J., J. A. Greenwood, and D. Durand, 1953, The circular normal distribution: Tables and theory: *Jour. American Statistical Society,* **48,** p. 131–152.

Haggett, P., A. D. Cliff, and A. Frey, 1977, *Locational analysis in human geography,* 2nd ed.: John Wiley & Sons, Inc., New York, 605 p.

An advanced book, containing techniques which may be unfamiliar to many geologists. Part 2, "Methods in locational analysis," discusses sampling, classification of regions, and testing of statistical hypotheses about spatial relationships.

Harbaugh, J. W., 1964, A computer method for four-variable trend analysis illustrated by a study of oil-gravity variations in southeastern Kansas: *Kansas Geological Survey Bull.* 171, 58 p.

Harbaugh, J. W., and D. F. Merriam, 1968, *Computer applications in stratigraphic analysis:* John Wiley & Sons, Inc., New York, 282 p.

Harbaugh, J. W., and F. W. Preston, 1966, Fourier series analysis in geology: *Computers and Computer Applications in Mining and Exploration,* School of Mines, Univ. Arizona, Tucson, p. R1–R46.

A discussion of the use of both double and single Fourier series for the fitting of "trends" in geologic data. Reprinted as article IV-8 in Berry and Marble (1968).

Hodder, Ian, and Clive Orton, 1979, *Spatial analysis in archaeology:* Cambridge Univ. Press, Cambridge, 270 p.

An inexpensive paperback edition of the 1976 original containing discussions of point distributions, trend surfaces, and map comparison methods used in archaeology.

IBM, 1965, Numerical surface techniques and contour map plotting: International Business Machines, Data Processing Applications, White Plains, N.Y., 35 p.

An introductory treatment of contouring.

James, W. R., 1966, FORTRAN IV program using double Fourier series for surface fitting of irregularly spaced data: *Kansas Geological Survey Computer Contribution* 5, 19 p.

A brief discussion of the least-squares method of fitting double Fourier surfaces. Figure 5.92 is adapted from this article.

Journel, A. G., and Ch. J. Huijbregts, 1978, *Mining geostatistics:* Academic Press, London, 600 p. [reprinted with corrections, 1981].

The most complete English-language discussion of regionalized variable theory, with emphasis on mine evaluation. The treatment is at an advanced level.

Kaesler, R. L., and J. A. Waters, 1972, Fourier analysis of the ostracode margin: *Geol. Soc. America Bulletin,* **83,** p. 1169–1178.

Kendall, M. G., and P. A. P. Moran, 1963, *Geometrical probability:* Hafner Publ. Co., New York, 125 p.

Chapter 2 is concerned with the distribution of points on a plane.

King, L. J., 1969, *Statistical analysis in geography:* Prentice-Hall, Inc., Englewood Cliffs, N.J., 288 p.

This graduate-level text covers many of the same topics as this book. Testing of point distributions is covered in Chapter 5, and trend-surface analysis in Chapter 6.

Koch, G. S., Jr., and R. F. Link, 1980, *Statistical analysis of geological data:* Dover Publications, Inc., New York, 850 p.

Trend-surface analysis is discussed in Chapter 9. Problems of systematic searching and the applications of response surfaces are given in Chapter 12.

Krige, D. G., 1966, Two-dimensional weighted moving average trend surfaces for ore valuation, *in* Proc. Symposium on Mathematical Statistics and Computer Applications in Ore Valuation, Mar. 7–8: *Jour. South African Inst. Mining and Metallurgy,* Johannesburg, p. 13–38.

A critique of trend-surface analysis and discussion of moving-average schemes for predicting ore values. The discussion that follows Krige's paper (p. 39–79) points out the areas of disagreement between two opposing camps.

Krumbein, W. C., 1939, Preferred orientation of pebbles in sedimentary deposits: *Jour. Geology,* **47,** p. 673–706.

Lambie, Fred, 1981, An analysis of the probability of hitting an arbitrary elliptical target with sets of parallel search lines: Terrasciences Inc., Unpub. Report, San Ramon, Calif., 17 p.

Lee, D. R. and G. T. Sallee, 1970, A method of measuring shape: *Geographical Review,* **60,** no. 4, p. 555–563.

Li, J. C. R., 1964, *Statistical inference,* v. 2: Edwards Bros. Inc., Ann Arbor, Mich., 575 p.

Chapter 30 discusses tests of curvilinear regressions, which are directly extendible to trend surfaces.

Mardia, K. V., 1972, *Statistics of directional data:* Academic Press Ltd., London, 357 p.

A complete treatment of statistical methods appropriate for both two- and three-dimensional orientation data. Many of the examples are taken from geology.

Matheron, G., 1963, Principles of geostatistics: *Economic Geology,* **58,** p. 1246–1266.

This review of the theory of regionalized variables and the application of kriging was written by the person who has contributed most heavily to this branch of applied statistics.

McCammon, R. B., 1977, Target intersection probabilities for parallel-line and continuous-grid types of search: *Jour. Int'l. Assoc. Mathematical Geology,* **9,** no. 4, p. 369–383.

Contains equations and graphs for determining the probabilities of hitting an elliptical target with a search pattern consisting of regularly spaced lines.

McCullagh, M. J., 1981, Creation of smooth contours over irregularly distributed data using local surface patches: *Geographical Analysis,* **13,** no. 1, p. 51–63.

This and the following article discuss calculation of a triangular mesh connecting irregularly spaced points, and the contouring of a surface using this mesh.

McCullagh, M. J., and C. G. Ross, 1980, Delaunay triangulation of a random data set for isarithmic mapping: *Cartographic Journal,* **17,** no. 2, p. 93–99.

Mendenhall, W., 1968, *Introduction to linear models and the design and analysis of experiments:* Wadsworth Publ. Co., Inc., Belmont, Calif., 465 p.

Chapter 10 describes the fitting of response surfaces to experimental designs.

Merriam, D. F., and J. W. Harbaugh, 1964, Trend-surface analysis of regional and residual components of geologic structure in Kansas: *Kansas Geological Survey Spec. Distribution Publ.* 11, 27 p.

Merriam, D. F., and P. H. A. Sneath, 1966, Quantitative comparison of contour maps: *Jour. Geophysical Research,* **71,** p. 1105–1115.

The method of map comparison by correlation of trend surface coefficients is described in this article, using data from Merriam and Harbaugh (1964). Tables 5.24 and 5.25 are taken from this article.

Miles, R. E., 1964, Random polygons determined by lines in a plane, I and II: *Proceedings of the National Academy of Sciences,* **52,** p. 901–907, 1157–1160.

Mills, F. C., 1955, *Statistical methods,* 3rd ed.: Holt, Rinehart, and Winston, New York, 842 p.

Moellering, Harold, and J. N. Rayner, 1979, Measurement of shape in geography and cartography: Ohio State Univ., Report of the Numerical Cartography Laboratory, NSF Grant No. SOC77-11318, 109 p.

An extensive review of shape analysis, including polar Fourier methods.

Olea, R. A., 1975, Optimum mapping techniques using regionalized variable theory: *Kansas Geological Survey Series on Spatial Analysis,* no. 2, 137 p.

This monograph contains a complete derivation of the equations used to perform punctual and universal kriging, as used to make contour maps.

Olea, R. A., 1982, Optimization of the High Plains aquifer observation network, Kansas: *Kansas Geological Survey Ground Water Series,* no. 7, 73 p.

Rayner, J. N., 1971, *An introduction to spectral analysis:* Pion Ltd., London, 174 p.

One- and two-dimensional Fourier analysis, written from the viewpoint of a geographer. The discussion of the Fast Fourier Transform is especially lucid.

Ripley, B. D., 1981, *Spatial statistics:* John Wiley & Sons, New York, 252 p.

A thorough review of spatial analysis. Trend-surface analysis and quadrat analysis are discussed at length. Ripley has extensively reanalyzed several of the data sets from the first edition of this text, including the magnetite crystal data listed in Table 5.4.

Robinson, J. E., 1982, *Computer applications in petroleum geology:* Hutchinson Ross Publ. Co., New York, 164 p.

An elementary text emphasizing petroleum data file handling and mapping of subsurface information. Chapters 7 and 8 cover contour maps, trend surfaces, and spatial filtering.

Robinson, J. E., H. A. K. Charlesworth, and M. J. Ellis, 1969, Structural analysis using spatial filtering in interior plains of south-central Alberta: *Bull. American Assoc. Petroleum Geologists,* **53,** p. 2341–2367.

A description of two-dimensional Fourier analysis and filtering of structural data.

Rogers, A., 1974, *Statistical analysis of spatial dispersion, the quadrat method:* Pion Ltd., London, 164 p.

This is an advanced text on quadrat methods for analysis of point patterns. The examples are from cultural geography.

Sampson, R. J., 1975, The SURFACE II graphics system, *in* Davis, J. C., and M. J. McCullagh (eds.), *Display and analysis of spatial data:* Wiley Interscience, London, p. 244–266.

Silk, John, 1979, Statistical Concepts in Geography: George Allen & Unwin Ltd., London, 276 p.

A very readable introductory text for geographers, this book contains chapters on point and area patterns, and on sampling in a plane. Available as a paperback.

Singer, D. A., and F. E. Wickman, 1969, Probability tables for locating elliptical targets with square, rectangular, and hexagonal point-nets: Pennsylvania State Univ., *Mineral Sciences Experiment Station Spec. Publ.* 1–69, 100 p.

Stephens, M. A., 1967, Tests for the dispersion and for the modal vector of a distribution on a sphere: *Biometrika*, **54**, p. 211–223.

Stephens, M. A., 1969, Tests for randomness of directions against two circular alternatives: *Jour. American Statistical Association*, **64**, no. 325, p. 280–289.

Tipper, J. C., 1979, Surface modelling techniques: *Kansas Geological Survey Series on Spatial Analysis*, no. 4, 108 p.

A survey of computer-aided design methods, such as spline fitting, that have potential applications in the Earth sciences. Example applications are taken from paleontology and petroleum reservoir modelling.

Unwin, D., 1981, *Introductory spatial analysis:* Methuen & Co., Ltd., London, 212 p.

A compact but thorough discussion of the distribution of points, lines, and areas on maps. Also included are chapters on contouring and map comparison.

Uspensky, J. V., 1937, *Introduction to mathematical probability:* McGraw-Hill, Inc., New York, 411 p.

Walters, R. F., 1969, Contouring by machine: A user's guide: *Bull. American Assoc. Petroleum Geologists*, **53**, p. 2324–2340.

A dated, but readable discussion of contour mapping.

Watson, G. S., 1970, Orientation statistics in the Earth sciences: *Bull. Geological Institute of Uppsala*, **2**, no. 9, p. 73–89.

This article is a compact survey of the applications of directional statistics to data from geology and geography. Many of the examples are included in Watson (1983).

Watson, G. S., 1971, Trend-surface analysis: *Jour. Int'l. Assoc. Mathematical Geology*, **3**, no. 3, p. 215–226.

Watson, G. S., 1983, *Statistics on spheres:* John Wiley & Sons, Inc., New York, 238 p.

This is an advanced treatment of the statistics of point distributions on spheres. The subject includes directional data, since the points may represent the tips of vectors touching the unit sphere. Issued as a paperback, the book unfortunately contains numerous typographical errors.

Woodcock, N. H., 1977, Specification of fabric shapes using an eigenvalue method: *Geol. Soc. America Bulletin*, **88**, p. 1231–1236.

A classification of petrofabric diagrams based on ratios of their eigenvalues.

Wrigley, N. (ed.), 1979, *Statistical applications in the spatial sciences:* Pion Ltd., London, 310 p.

A collection of papers covering a wide variety of geographic topics. Part 2, on environmental science applications, is of special interest.

Analysis of Multivariate Data

In previous chapters, we have considered the analysis of data consisting of only a single variable measured on each specimen or observational unit. In Chapters 4 and 5 we also considered the influence of the temporal or geographic coordinates of the sample points. We will now examine techniques for the analysis of *multivariate data,* in which each observational unit is characterized by several variables. Multivariate methods allow us to consider changes in several properties simultaneously. Examples of data appropriate for multivariate analysis abound in geology. They include chemical analyses, where the variables may be percentage compositions or parts per million of trace elements; measures on streams, such as discharge, suspended sediment load, depth, dissolved solids, pH, and oxygen content; and paleontologic variables, perhaps a large number of measurements made on specimens of an organism. Dozens of other examples quickly spring to mind. Some are simple extensions of problems we have considered previously; others are entirely new classes of problems.

Multivariate methods are extremely powerful, for they allow the researcher to manipulate more variables than he can assimilate by himself. However, they are complicated, both in their theoretical structure and in their operational methodology. For many of the procedures, statistical theory and tests have been worked out only for the most restricted set of assumptions. The nature and behavior of the tests under more relaxed, general assumptions (such as those necessary for most real-world problems) are inadequately known. In fact, some of the procedures we will consider have no theoretical statistical basis at all, and tests of significance have yet to be devised. Nevertheless, these methods seem to hold the most promise for fruitful returns in geological investigations. Most of the problems in geology involve complex and interacting forces, which are impossible to isolate and study individually. Often a meaningful decision as to the relative worth of one of a number of possible variables cannot be made. The best course of action frequently is to examine as many facets of a problem as possible, and sort out, a posteriori, the major factors. The methods discussed in this chapter can be a significant help.

Multiple Regression

The first topic we will consider in our final chapter is actually a familiar subject under a new and more general guise. This is multiple regression and includes polynomial curve fitting (discussed in Chapter 4) and trend-surface analysis (discussed in Chapter 5). However, we will now remove the restrictions that limited us to considerations of change as a function of distance or spatial coordinates. Any observed variable can be considered to be a function of any other variable measured on the same samples. In Chapter 4, we considered changes in moisture content that occurred with changes in depth in the sediment. We could equally well have measured the montmorillonite content of the sediment in the core and examined changes in water content that may accompany changes in montmorillonite percentage. In fact, we could have measured several variables, perhaps organic content, mean grain size, and bulk density, and could have examined differences in water content associated with changes in each or all of these variables. In a sense, variables may be considered as spatial coordinates, and we can envision changes occurring "along" a dimension defined by a variable such as mineral content. Casting variables as dimensions is nothing new; we perform this every time we plot two variables against one another, because we are substituting spatial scales in the plot for the original scales on which the variables were measured. Such interchangeability is explicit in references to "p-dimensional space" which abound in the literature of multivariate analysis. Just as trend surfaces are a generalization of curve-fitting procedures to two-dimensional space, multiple regression is a further generalization to many-dimensional space.

We will not consider multiple regression in great detail, because the theoretical and computational essentials have been presented in earlier chapters. You will recall from Chapter 4 that polynomial regressions (having one independent variable) can be represented in a model equation of the general form

$$Y = \beta_0 + \beta_1 X_1 + \beta_2 X_1^2 + \cdots + \beta_m X_1^m + \epsilon \tag{6.1}$$

The model states than an observation Y is equal to a constant term plus a series of powers of an independent variable, plus a random error. A least-squares solution to a linear equation of this type can be found by solving a set of normal equations for the β coefficients. These can be expressed in matrix form as

$$[\Sigma X] \cdot [\beta] = [\Sigma Y] \tag{6.2}$$

with a solution

$$[\beta] = [\Sigma X]^{-1} \cdot [\Sigma Y] \tag{6.3}$$

where $[\Sigma Y]$ is a column matrix of the sum of squares and cross-products of Y with $X_1, X_1^2, \ldots, X_1^m$; $[\Sigma X]$ is a matrix of sums of squares and cross-products of the $X_1, X_1^2, \ldots, X_1^m$ powers; and $[\beta]$ is a column matrix of the unknown coefficients. In Chapter 4, we found the entries in the various matrices by labeling rows and columns and cross-multiplying.

Although we regarded this problem as involving only one independent variable (or two, in the case of trend-surface analysis discussed in Chapter 5), it can be regarded as containing m independent variables. This can readily be seen if we rewrite the model equation as

$$Y = \beta_0 + \beta_1 X_1 + \beta_2 X_2 + \cdots + \beta_m X_m + \epsilon \qquad (6.4)$$

and define the variables as $X_1 = X_1$, $X_2 = X_1^2$, $X_3 = X_1^3$, and so forth. Thus, the regression procedures we have considered up to this point have simply involved the definition of the independent variables in a specific manner. The relationships between the various categories of regression analysis were shown in Figure 5.89.

A regression of any m independent variables upon a dependent variable can be expressed as in (6.4). The normal equations that will yield a least-squares solution can be found by appropriate labeling of the rows and columns of the matrix equation and cross-multiplying to find the entries in the body of the matrix. For three independent variables, we obtain

$$
\begin{array}{cccc}
X_0 & X_1 & X_2 & X_3
\end{array}
\qquad\qquad\qquad Y
$$

$$
\begin{matrix}
X_0 \\ X_1 \\ X_2 \\ X_3
\end{matrix}
\begin{bmatrix}
\ & \ & \ & \ \\
\ & \ & \ & \ \\
\ & \ & \ & \ \\
\ & \ & \ & \
\end{bmatrix}
\begin{bmatrix}
b_0 \\ b_1 \\ b_2 \\ b_3
\end{bmatrix}
=
\begin{bmatrix}
\ \\ \ \\ \ \\ \
\end{bmatrix}
$$

where again X_0 is a dummy variable equal to 1 for every observation. The matrix equation, after cross-multiplication, is

$$
\begin{bmatrix}
n & \Sigma X_1 & \Sigma X_2 & \Sigma X_3 \\
\Sigma X_1 & \Sigma X_1^2 & \Sigma X_1 X_2 & \Sigma X_1 X_3 \\
\Sigma X_2 & \Sigma X_2 X_1 & \Sigma X_2^2 & \Sigma X_2 X_3 \\
\Sigma X_3 & \Sigma X_3 X_1 & \Sigma X_3 X_2 & \Sigma X_3^2
\end{bmatrix}
\begin{bmatrix}
b_0 \\ b_1 \\ b_2 \\ b_3
\end{bmatrix}
=
\begin{bmatrix}
\Sigma Y \\ \Sigma X_1 Y \\ \Sigma X_2 Y \\ \Sigma X_3 Y
\end{bmatrix}
\qquad (6.5)
$$

The β's in the regression model are estimated by b's, the sample *partial regression coefficients*. They are called partial regression coefficients because each gives the rate of change (or slope) in the dependent variable for a unit change in that particular independent variable, *provided* all other independent variables are held constant. Some statistics books emphasize this point by using the notation

$$Y = b_0 + b_{1.23} X_1 + b_{2.13} X_2 + b_{3.12} X_3 + \epsilon$$

The coefficient $b_{1.23}$, for example, is read "the regression coefficient of variable 1 on Y, as variables 2 and 3 remain constant." In general, these coefficients will differ from the *total regression coefficients* which are the simple regressions of each individual X variable on the Y variable. We ordinarily expect multiple regression to account for more of the total variation in Y than will any of the total regression coefficients. This is because multiple regression considers all possible interactions within combinations of variables as well as the variables themselves.

We will consider a problem in geomorphology to illustrate a typical application of multiple regression. For this study, a well-dissected area of relatively homogeneous geology was selected in eastern Kentucky. The study region contains many drainage basins of differing sizes; from these, all third-order basins were chosen, and several variables were measured on each. The order of a basin is defined by the number of successive levels of junctions on its stream from the stream's sources to the point where it joins another stream of equal or higher order. Thus, a third-order basin has two levels of junctions within its boundaries. Basin size, however, may be defined by many alternative methods. One of these is basin magnitude, which essentially is a count of the number of sources in the basin. A collection of basins of specified order may contain many different magnitudes. The relationship between magnitude and order of basins is shown in Figure 6.1.

On the collection of third-order basins, the following seven variables were measured:

1. Elevation of the basin outlet, in feet.
2. Relief of the basin, in feet.
3. Basin area, in square miles.
4. Total length of the stream in the basin, in miles.
5. Drainage density, defined as total length of stream in basin/basin area.
6. Basin shape, measured as the ratio of inscribed to circumscribed circles.
7. Basin magnitude, defined by the number of sources.

Our problem is to determine the influence of the first six variables on the seventh. Multiple regression, using basin magnitude as the dependent variable, is an appropriate technique. From the regression, the influence of all of the variables on basin magnitude can be assessed. Measurements of these variables for 50 third-order basins in the study region are given in Table 6.1, which is taken from a study by Krumbein and Shreve (1970). The significance of the linear relationship

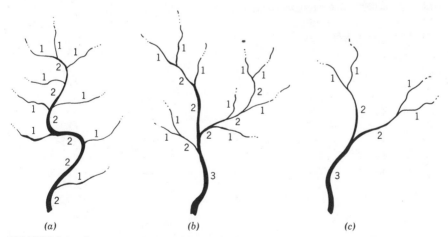

FIGURE 6.1   Contrast between stream magnitude and stream order. (a) Tenth-magnitude stream of second order. (b) Tenth-magnitude stream of third order. (c) Fourth-magnitude stream of third order. Magnitude is based on number of joining streams, order is based on succession of joining.

TABLE 6.1 Seven Geomorphic Variables Measured on Third-Order Basins in Kentucky; Variables Are (Y) Basin Magnitude, (X_1) Elevation of Basin Outlet, (X_2) Basin Relief, (X_3) Basin Area, (X_4) Total Stream Length in the Basin, (X_5) Drainage Density (Total Stream Length Divided by Basin Area), and (X_6) Ratio of Largest Inscribed Circle to Smallest Circumscribed Circle

Y	X_1 (ft)	X_2 (ft)	X_3 (mile^2)	X_4 (mile)	X_5	X_6
14	720	570	7	154	2,200	61
6	670	610	3	80	2,667	62
5	860	550	11	84	763	62
7	870	610	11	122	1,110	63
11	730	570	14	185	1,321	52
14	690	590	12	200	1,667	50
12	880	640	11	170	1,545	41
18	760	690	28	340	1,215	57
6	820	600	5	100	2,000	41
5	720	480	3	80	2,667	60
17	670	670	19	290	1,526	51
5	660	600	5	90	1,800	53
22	830	660	18	260	1,444	57
7	780	620	17	111	652	57
15	750	740	15	184	1,227	67
17	770	630	21	227	1,080	59
5	750	570	4	60	1,500	55
18	750	580	20	259	1,295	39
14	740	760	9	62	689	64
21	750	740	6	95	1,583	53
22	750	760	11	105	954	64
23	740	770	32	350	1,094	55
28	940	510	21	232	1,105	52
42	700	600	23	266	1,156	34
22	810	580	44	390	886	29
10	920	500	13	142	1,092	65
11	920	490	12	145	1,208	72
12	790	605	33	253	766	59
13	860	550	23	241	1,048	76
31	860	630	87	702	807	55
18	880	520	37	288	778	51
13	780	460	17	162	953	40
4	720	440	8	67	838	60
5	780	300	3	52	1,733	57
9	700	460	10	121	1,210	50
13	680	520	26	220	846	41
10	820	520	8	123	1,537	51
13	710	520	24	238	992	41
13	800	440	19	231	1,216	51
11	700	510	16	178	1,113	76
12	675	570	18	168	933	42

TABLE 6.1 (*Continued*)

Y	X_1 (ft)	X_2 (ft)	X_3 (mile²)	X_4 (mile)	X_5	X_6
4	740	510	8	65	812	49
17	740	520	31	334	1,078	67
9	770	600	21	184	876	47
8	820	520	11	136	1,237	56
13	850	490	22	233	1,059	74
22	820	629	34	410	1,206	39
10	820	510	11	149	1,354	60
19	680	640	46	348	757	55
27	660	789	55	382	695	38

Source: Krumbein and Shreve (1970).

can be tested by analysis-of-variance methods presented in Chapter 4. Table 4.12, for example, outlines the ANOVA for simple linear regression which may be expanded to multiple regression by changing the various degrees of freedom to account for additional variables. The modified ANOVA is shown in Table 6.2. The completed ANOVA for multiple regression on basin magnitude is shown in Table 6.3. The regression coefficients are also shown.

In multiple-regression problems, we usually are interested in the relative effectiveness of the independent variables as predictors of the dependent variable. We cannot determine this from a direct examination of the regression coefficients, however, because their magnitudes are dependent upon the magnitudes of the variables themselves. This is apparent in trend-surface analysis, where coefficients of higher orders almost invariably decrease in absolute size, even though higher orders may make greater contributions to the trend than lower orders. This results from the fact that a variable, raised to a power as it is in high orders, is much larger in magnitude than the original variable. The higher order regression coefficients become correspondingly smaller.

Fortunately, it is easy to standardize the partial regression coefficients by converting them to units of standard deviation (Li, 1964, p. 136). The standard partial regression coefficients, B_k, are found by

$$B_k = b_k \frac{s_k}{s_y} \tag{6.6}$$

TABLE 6.2 ANOVA for Multiple Regression with m Independent Variables

Source	Sum of Squares	Degrees of Freedom	Mean Squares	F test
Regression	SS_R	m	MS_R	MS_R/MS_D
Deviation	SS_D	$n - m - 1$	MS_D	
Total	SS_T	$n - 1$		

TABLE 6.3 Completed ANOVA for the Significance of Regression of Six Geomorphic Variables on Basin Magnitude[a]

Source	Sum of Squares	Degrees of Freedom	Mean Squares	F test
Regression	1,800.70	6	300.12	11.38
Deviation	1,134.12	43	26.37	
Total	2,934.82	49		

[a] Regression equation:

$$Y = -2.24 + 0.01X_1 + 0.02X_2 - 23.28X_3 + 6.26X_4 - 0.20X_5 - 11.66X_6$$

Coefficient of multiple regression: $R = 0.78$.

where s_k is the standard deviation of variable X_k and s_y is the standard deviation of Y. Because the standard partial regression coefficients are all expressed in units of standard deviation, they may be compared directly with each other and the most effective variables determined.

To compute the matrix of sums of squares and products necessary in the normal equation set, we found the diagonal entries ΣX_k^2. It is a simple matter to convert these sums of squares to corrected sums of squares, SS_k, and then to the standard deviations necessary to compute the partial correlation coefficients. However, it is possible to solve the normal equations in a manner that will yield the standardized partial regression coefficients directly, and gain an important computational advantage in the process.

The major sources of error in multiple regression come in the creation of the entries in the $[\Sigma X]$ matrix and during the inversion process. The sums of squares of the X_k variables may become so large that significant digits are lost by truncation. If the entries in the $[\Sigma X]$ matrix differ greatly in their magnitudes, an additional loss of digits may occur during inversion, especially if high correlations exist among the variables. Some computer programs may be capable of retaining only one or two significant digits in the coefficients, and with certain data sets retention may even be worse. Studies have shown that calculations using double-precision arithmetic may not be sufficient to overcome this problem. However, a few simple modifications in our computational procedure will gain us two to six significant digits during computation and greatly increase the accuracy of the computed regression (Longley, 1967, pp. 821–827).

The most obvious step that can be taken is to convert all observations to deviations from the mean. This reduces the absolute magnitude of variables and centers them about a common mean of zero. As an inevitable consequence, the coefficient b_0 will become zero, so the matrix equation can be reduced by one row and one column. This simple step may gain several significant digits. However, we also may reduce the size of entries in the matrix still further by converting them all to correlations. This is equivalent to expressing the original variables in the standard normal form of zero mean and unit standard deviation. The matrix equation for regression then has the form

$$[r_{xx}] \cdot [B] = [r_{xy}]$$ (6.7)

which can be solved by the operation

$$[B] = [r_{xx}]^{-1} \cdot [r_{xy}]$$ (6.8)

where $[r_{xy}]$ represents the column vector of correlations between Y and the X_k independent variables. The $m \times m$ matrix of correlations between the X_k variables is represented by $[r_{xx}]$. For example, the normal equation for three independent variables has the form

$$\begin{bmatrix} 1 & r_{12} & r_{13} \\ r_{21} & 1 & r_{23} \\ r_{31} & r_{32} & 1 \end{bmatrix} \cdot \begin{bmatrix} B_1 \\ B_2 \\ B_3 \end{bmatrix} = \begin{bmatrix} r_{x_1y} \\ r_{x_2y} \\ r_{x_3y} \end{bmatrix}$$ (6.9)

Note that the equation has one less row and column than the equivalent equation (6.5).

Computing the regression equation in standardized form has the disadvantage that the correlation matrix must be created first, increasing the computational effort. In order to preserve accuracy, the correlations must be calculated by the definitional equation rather than with the computational form given in (2.24). This is because (2.24) involves squaring the quantities ΣX_j and ΣX_k. If these sums are large, the squares may be inaccurate because of truncation. This problem is avoided if the means are subtracted from each observation prior to calculation of the sums of squares. The sums of squares are then found by (2.16) and (2.19). This process requires that the data be handled twice—first to calculate the means, and then to subtract out this quantity during calculations. Although this involves a significant increase in labor if computations are performed by hand, the additional effort is trivial on a digital computer. Also, the resulting coefficients must be "unstandardized" if they are to be used in a predictive equation with raw data. However, these disadvantages are more than offset by the increased stability and accuracy of the matrix solution, and the standardized coefficients provide a way of assessing the importance of individual variables in the regression. Partial regression coefficients can be derived from the standardized partial regression coefficients by the transformation

$$b_k = B_k \frac{s_y}{s_k}$$ (6.10)

The constant term, b_0, can be found by

$$b_0 = \overline{Y} - b_1\overline{X}_1 - b_2\overline{X}_2 - \cdots - b_m\overline{X}_m$$ (6.11)

Although the various sums of squares change if the data are standardized or the correlation form of the matrix equation is used, the ratios of the sums of squares remain the same. Therefore, tests of significance based on standardized regression are identical to those based on an unstandardized regression. Quantities such as the coefficient of multiple correlation (R) and percentage of goodness-of-fit $(100\% \cdot R^2)$ also remain unchanged.

The data given in Table 6.4 are reservoir performance figures for a gas-drive oil field in Arkansas. The dependent variable is the estimated oil in place in the reservoir, calculated by a materials-balance model. The materials-balance equation is basically a relationship between oil production, gas production, and pressure. It also involves assumptions about the reservoir volume and the initial volumes of oil, gas, and water. The independent variables are the time after completion of the field, the reservoir pressure, cumulative oil production, and the cumulative gas-to-oil ratio. Because of the relationships implicit between the independent variables and the dependent variable in the materials-balance equation, we expect extremely high intercorrelations. Indeed, if the materials-balance model is appropriate and our assumptions about the initial size and state of the reservoir are correct, the correlation will be perfect. Failure of the independent variables to completely explain the estimated oil in place may be attributed to errors in the initial assumptions or to factors overlooked in the materials-balance approach.

This data set has several characteristics that make it difficult to analyze. Because of the differences in size of the variables, the elements in the cross-product matrix will be radically different in their magnitudes. The data form a multivariate time series. In common with other series of this type, such as economic growth and employment curves, the variables are highly correlated. It will be difficult to retain sufficient digits in the calculation of the matrices, or to preserve accuracy during the inversion process. It may be instructive to compute the regression coefficients using both the matrix $[\Sigma X]$ and the matrix $[r_{xx}]$. The standardized partial regression coefficients should be converted to ordinary form using equations (6.10) and (6.11)

TABLE 6.4 Reservoir Performance Data for Gas-Drive Oil Field in Arkansas: $Y =$ Estimated Oil in Place, $X_1 =$ Time after Field Completion, $X_2 =$ Reservoir Pressure, $X_3 =$ Cumulative Oil Production, $X_4 =$ Cumulative Gas/Oil Ratio

Y ($\times 10^3$ bbl)	X_1 (month)	X_2 (psi)	X_3 ($\times 10^2$ bbl)	X_4 (ft^3/bbl)
110,273	1	3,520	0	760
111,105	4	3,125	28,183	853
114,992	8	2,910	46,536	906
119,437	12	2,785	60,302	939
118,961	16	2,650	73,604	960
116,968	20	2,505	87,513	990
119,663	24	2,425	98,738	1,018
117,514	28	2,290	112,597	1,070
117,292	32	2,125	126,192	1,200
114,776	36	1.950	139,981	1,310
113,969	40	1,785	153,219	1,440
111,881	44	1,670	161,327	1,500
114,455	48	1,601	173,485	1,516
116,196	52	1,537	185,832	1,520

for comparison purposes. The differences that you will see are due to rounding error when using matrix $[\Sigma X]$.

Although the standardized partial regression coefficients provide a guide to the most effective variables in the regression, they are not an infallible index to the "best possible" regression equation. Suppose you examine the regression equation and decide two variables are contributing a negligible amount to the regression and can be discarded. When one of the variables is omitted and the regression is recalculated, the goodness-of-fit and the regression equation, of course, change. Now suppose you decide to discard the second variable; again the regression changes. But the change might be quite different from the change that would occur if the first discarded variable were still in the regression. This occurs because the interaction effects of the two discarded variables with other variables cannot be assessed without recomputing the regression. If we want to search through a large set of variables and "weed out" those which are not helpful in the problem, we must do more than simply examine the partial regression coefficients.

Increasing the number of independent variables in the regression equation will always increase the SS_R (except in the situation where a new variable is completely correlated with a previous variable). However, the increase may not be significant. The loss of degrees of freedom for deviations may offset the reduction in SS_D, and actually increase the mean squares due to deviation. If this happens, the F ratio for the significance of the regression will decrease, and the addition of another variable has actually detracted from the regression. To determine the very best possible regression (in the sense of having the most significant F ratio) all possible combinations of the variables would have to be examined. This is possible if we are dealing with few variables, but the number of possible variable combinations is equal to 2^m, and the computational effort is formidable if m is large. Other procedures are available which yield an optimum regression with much less effort. These include schemes such as the backward elimination procedure, the forward selection procedure, stepwise regression, and stagewise regression. These methods may not find identical regression equations in a large selection of possible variables, but all will produce approximately equivalent results. A consideration of each is beyond the scope of this book; we will be content with a brief description of one of the techniques. These methods are well described in some of the texts listed in the Selected Readings at the end of the chapter, especially in Draper and Smith (1981) and in Marascuilo and Levin (1983).

The backward elimination procedure consists of computing a regression including all possible variables and selecting the least significant variable. The selection proceeds by examining the standardized partial regression coefficients for the smallest value, and then recomputing the regression omitting that variable. The significance of the deleted variable is tested by an analysis of variance similar to that outlined in Table 4.16. If the variable is not making a significant contribution to the regression, it is permanently discarded. The standardized partial regression coefficients of the reduced equation are then examined and the process is repeated. At each step, the regression equation is reduced by one variable, until all remaining variables are significant.

It is an instructive exercise to examine the collection of seven variables measured on river basins (Table 6.1) and see if any can be discarded without significantly affecting the multiple regression. We can find a minimal set of regressions by examining the standardized partial regression coefficients, deleting the smallest of these, and recomputing the regression. Repeatedly running a multiple-regression program obviously is less efficient than using a stepwise computer program, but it has the advantage that every step in the process can be examined closely. When you are confident that you understand the elimination process and the changes that occur in the regression coefficients, you may turn to a more automated procedure.

Although multiple regression is "multivariate" in the sense that more than one variable is measured on each observational unit, it really is a univariate technique, because we are concerned only with the variance of one variable, Y. Behavior of the independent variables, the X's, is not subject to analysis.

The next topic we will consider is discriminant function analysis, which involves *identification* or the placing of objects into predefined groups. The discrimination between two alternative groups is a process that is computationally intermediate between univariate procedures and true multivariate methods in which many variables are considered simultaneously. Two groups, each characterized by a set of multiple variables, can be discriminated by solving a set of simultaneous equations almost identical to those involved in multiple regression. The right-hand vector of the matrix equation, however, does not contain powers and cross-products of a single dependent variable, but rather differences between the multivariate means of the two groups.

Tests of discriminant functions involve multivariate extensions of simple univariate statistical tests of equality. These will be considered next, followed by a discussion of multivariate classification, or the sorting of objects into homogeneous groups. We will then consider eigenvector techniques, including principal components and factor analysis. The final topics will include multivariate extensions of discriminant analysis and multiple regression.

This list of topics certainly is not all-inclusive. However, the subjects selected have been chosen because they have found special utility in the Earth sciences. They include a wide variety of computational techniques and encompass many fundamental concepts. An understanding of the theory and operational procedures involved in these methods should provide sufficient background to evaluate other multivariate techniques as well.

Discriminant Functions

One of the most widely used multivariate procedures in Earth science is the discriminant function. We will consider it at length for two reasons: it is a powerful statistical tool, and discrimination can be regarded as either a univariate problem related to multiple regression or as a multivariate problem related to the statistical tests we have just discussed. Discriminant functions therefore provide an additional link between univariate and multivariate statistics.

First, however, we must define the process of *discrimination,* and carefully distinguish it from the related process of classification. Suppose we have collected two suites of shale samples of known freshwater and saltwater origin; we may have determined their origin from an examination of their fossil content. A number of geochemical variables are measured on the samples, including the content of vanadium, boron, iron, and so forth. The problem is to find the linear combination of these variables which produces the maximum difference between the two previously defined groups. If we find a function that produces a significant difference, we can use it to allocate new samples of unknown origin to one of the two original groups. In other words, new shale samples, not containing diagnostic fossils, can then be categorized as marine or freshwater on the basis of the linear discriminant function of their geochemical components. [This problem was considered by Potter, Shimp, and Witters (1963).]

Classification can be illustrated with a similar example. Suppose we have obtained a large, heterogeneous collection of shale samples, each of which has been geochemically analyzed. On the basis of the measured variables, can the samples be separated into groups (or *clusters,* as they are commonly called) that are both relatively homogeneous and distinct from other groups? The process by which this can be done has been highly developed by numerical taxonomists, and will be considered in the next section. There are several obvious differences between these procedures and discriminant functions. A classification is internally based; that is, it does not depend on a priori knowledge about relations between samples as does a discriminant function. The number of groups in a discriminant function is set prior to the analysis, while in contrast the number of clusters that will emerge from a classification scheme cannot ordinarily be predetermined. Similarly, each original sample is defined as belonging to a specific group in a discriminant analysis. In most classification procedures, a sample is free to enter any cluster that emerges. Other differences will become apparent as we examine these two procedures. The result of a cluster analysis of shales would be a classification of the samples into several groups. It would then be up to us to interpret the geological meaning (if any) of the groups so found.

A simple linear discriminant function transforms an original set of measurements on a sample into a single discriminant score. That score, or transformed variable, represents the sample's position along a line defined by the linear discriminant function. We can therefore think of the discriminant function as a way of collapsing a multivariate problem down into a problem which involves only one variable.

Discriminant function analysis consists of finding a transform which gives the minimum ratio of the difference between a pair of group multivariate means to the multivariate variance within the two groups. If we regard our two groups as consisting of two clusters of points in multivariate space, we must search for the one orientation along which the two clusters have the greatest separation while simultaneously each cluster has the least inflation. This can be graphically shown for two-dimensional cases, as in Figure 6.2. An adequate separation between groups A and B cannot be made using either variable X_1 or X_2. However, it is possible to

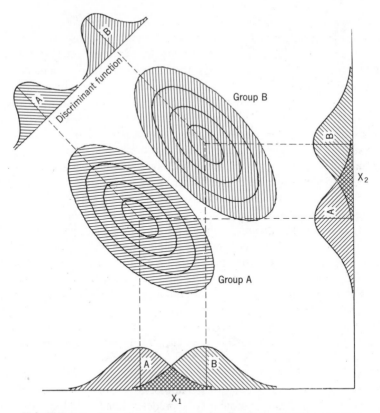

FIGURE 6.2 Plot of two bivariate distributions, showing overlap between groups A and B along both variables X_1 and X_2. Groups can be distinguished by projecting members of the two groups onto the discriminant function line.

find an orientation along which the two clusters are separated the most and inflated the least. The coordinates of this axis of orientation are the linear discriminant function.

One method that can be used to find the discriminant function is *regression;* however, the dependent variable consists of the differences between the multivariate means of the two groups. In matrix notation, we must solve an equation of the form

$$[s_p^2] \cdot [\lambda] = [D] \qquad (6.12)$$

where $[s_p^2]$ is an $m \times m$ matrix of pooled variances and covariances of the m variables. The coefficients of the discriminant equation are represented by a column vector of the unknown lambdas. Lowercase lambdas (λ) are used by convention to represent the coefficients of the discriminant function. These are exactly the same as the betas (β) used (also by convention) in regression equations. They should not be confused with lambdas used to represent eigenvalues in principal components or factor analyses.

The right-hand side of the equation consists of the column vector of m differences between the means of the two groups. You will recall from Chapter 3 that such an equation can be solved by inversion and multiplication, as

$$[\lambda] = [s_p^2]^{-1} \cdot [D] \tag{6.13}$$

To compute the discriminant function, we must determine the various entries in the matrix equation. The mean differences are found simply by

$$D_j = \overline{A}_j - \overline{B}_j = \frac{\sum_{i=1}^{n_a} A_{ij}}{n_a} - \frac{\sum_{i=1}^{n_b} B_{ij}}{n_b} \tag{6.14}$$

In this notation, A_{ij} is the ith observation on variable j in group A; $\overline{A}_j$ is the mean of variable j in group A, or the average of n_a observations. The same conventions apply to group B. The multivariate means of groups A and B can be regarded as forming two vectors. The difference between these multivariate means therefore also forms a vector

$$[D_j] = [\overline{A}_j] - [\overline{B}_j]$$

or, in expanded form,

$$\begin{bmatrix} D_1 \\ D_2 \\ \cdot \\ \cdot \\ \cdot \\ D_m \end{bmatrix} = \begin{bmatrix} \overline{A}_1 \\ \overline{A}_2 \\ \cdot \\ \cdot \\ \cdot \\ \overline{A}_m \end{bmatrix} - \begin{bmatrix} \overline{B}_1 \\ \overline{B}_2 \\ \cdot \\ \cdot \\ \cdot \\ \overline{B}_m \end{bmatrix}$$

To construct the matrix of pooled variances and covariances, we must compute a matrix of sums of squares and cross-products of all variables in group A and a similar matrix for group B. For example, considering only group A,

$$SPA_{jk} = \sum_{i=1}^{n_a} (A_{ij}A_{ik}) - \frac{\sum_{i=1}^{n_a} A_{ij} \sum_{i=1}^{n_a} A_{ik}}{n_a}$$

Here, A_{ij} denotes the ith observation of variable j in group A as before, and A_{ik} denotes the ith observation of variable k in the same group. Of course, this quantity will be the sum of squares of variable k whenever $j = k$. Similarly, a matrix of sums of squares and cross-products can be found for group B:

$$SPB_{jk} = \sum_{i=1}^{n_b} (B_{ij}B_{ik}) - \frac{\sum_{i=1}^{n_b} B_{ij} \sum_{i=1}^{n_b} B_{ik}}{n_b}$$

We will denote the sums of products matrix from group A as [SPA] and that from group B as [SPB]. The matrix of pooled variance can now be found as

$$[s_p^2] = \frac{[SPA] + [SPB]}{n_a + n_b - 2} \tag{6.15}$$

TABLE 6.5 Measurements on Median Grain Size and Sorting Coefficient for Two Suites of Sand Samples Collected on Texas Beach and Offshore

Group A, Beach Sands		Group B, Offshore Sands			
Median Grain Size	Sorting Coefficient	Median Grain Size	Sorting Coefficient	Median Grain Size	Sorting Coefficient
0.333	1.08	0.339	1.12	0.342	1.24
0.340	1.08	0.346	1.12	0.331	1.25
0.338	1.09	0.350	1.12	0.336	1.25
0.333	1.10	0.352	1.13	0.341	1.25
0.323	1.13	0.341	1.15	0.334	1.26
0.327	1.12	0.347	1.15	0.337	1.27
0.329	1.13	0.337	1.16	0.339	1.27
0.331	1.13	0.343	1.16	0.330	1.28
0.336	1.12	0.340	1.17	0.334	1.28
0.333	1.14	0.346	1.17	0.332	1.29
0.341	1.14	0.349	1.17	0.330	1.31
0.328	1.15	0.339	1.18	0.334	1.31
0.336	1.15	0.342	1.18	0.340	1.21
0.327	1.16	0.346	1.18		
0.329	1.16	0.351	1.18		
0.330	1.16	0.340	1.19		
0.323	1.17	0.344	1.19		
0.328	1.17	0.333	1.20		
0.332	1.17	0.337	1.20		
0.331	1.18	0.339	1.20		
0.326	1.18	0.342	1.20		
0.333	1.18	0.339	1.21		
0.330	1.19	0.340	1.21		
0.336	1.19	0.341	1.21		
0.327	1.20	0.335	1.22		
0.324	1.21	0.337	1.22		
0.332	1.21	0.340	1.22		
0.322	1.22	0.343	1.22		
0.329	1.22	0.334	1.22		
0.325	1.24	0.348	1.22		
0.328	1.26	0.337	1.22		
0.322	1.27	0.342	1.23		
0.318	1.22	0.334	1.24		
0.330	1.17	0.340	1.24		

You will note that this equation for pooled variance is exactly the same as that used in the T^2 test of the equality of multivariate means. Although the amount of mathematical manipulation that must be performed to calculate the coefficients of a discriminant function appears large, it actually is less formidable than it seems at first glance. To demonstrate, we can calculate a discriminant function between the two groups of data listed in Table 6.5. Group A consists of modern beach sands; the two variables are median grain size and sorting coefficient. Group B contains samples collected offshore. The same two measurements provide the variables. A scatter diagram or plot of the original observations is shown as Figure 6.3. Although the two clusters of points overlap, it is apparent that a line of division can be placed between the two clusters such that most observations from group A are on one side and most observations from group B are on the other.

Table 6.6 contains the calculations necessary to find the two vectors of multivariate means and the two matrices of sums of squares and products. From these, the matrix of pooled variances is calculated. We now have all of the entries necessary to solve the discriminant function:

$$
\begin{array}{cccc}
[s_p^2]^{-1} & \cdot & [D] & = & [\lambda]
\end{array}
$$

$$
\begin{bmatrix} 59{,}112.280 & 4312.646 \\ 4312.646 & 747.132 \end{bmatrix} \cdot \begin{bmatrix} -0.010 \\ -0.043 \end{bmatrix} = \begin{bmatrix} -783.63 \\ -75.62 \end{bmatrix}
$$

The set of λ coefficients we have found are entries in the discriminant function equation of the form

$$
R = \lambda_1 \psi_1 + \lambda_2 \psi_2 + \cdots + \lambda_m \psi_m \tag{6.16}
$$

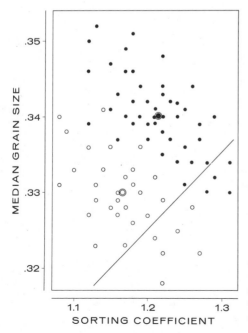

FIGURE 6.3 Plot of median grain size versus sorting coefficient for samples of modern sands. Samples indicated by open circles are beach sands, those indicated by solid dots are offshore sands. Larger symbols indicate bivariate means of the two groups. Line represents discriminant function.

TABLE 6.6 Matrices Necessary to Compute Discriminant Function between Two Groups of Observations Given in Table 6.5

Vector mean of group *A*

[0.330   1.167]

Vector mean of group *B*

[0.340   1.210]

Vector of mean differences (*A* − *B*)

[−0.010   −0.043]

Corrected sums of squares matrix *A*

$$\begin{bmatrix} 0.00092 & -0.00489 \\ -0.00489 & 0.07566 \end{bmatrix}$$

Corrected sums of squares matrix *B*

$$\begin{bmatrix} 0.00138 & -0.00844 \\ -0.00844 & 0.10700 \end{bmatrix}$$

Pooled variance—covariance matrix

$$\begin{bmatrix} 0.00003 & -0.00017 \\ -0.00017 & 0.00231 \end{bmatrix}$$

Inverse of pooled variance—covariance matrix

$$\begin{bmatrix} 59,112.280 & 4312.646 \\ 4312.646 & 747.132 \end{bmatrix}$$

This is a linear function; that is, all the terms are added together to yield a single number, the discriminant score. In a two-dimensional example, we can plot the discriminant function as a line on the scatter diagram of the two original variables. It is a line through the plot whose slope, α, is

$$\alpha = \frac{\lambda_2}{\lambda_1} \tag{6.17}$$

This line is shown on Figure 6.3.

Substitution of the midpoint between the two group means in this equation yields the discriminant index, R_0. That is, for each value of ψ_j in (6.16), we insert the terms

$$\psi_j = \frac{\overline{A}_j + \overline{B}_j}{2} \tag{6.18}$$

In our example,

$$R_0 = \lambda_1\psi_1 + \lambda_2\psi_2$$
$$= -783.63(0.335) - 75.62(1.189)$$
$$= -352.22$$

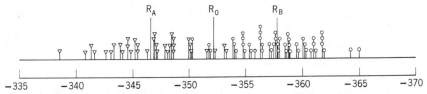

FIGURE 6.4 Projection of samples from Table 6.5 onto discriminant function line shown in Figure 6.3. R_A is projection of bivariate mean of beach sands, R_B is projection of bivariate mean of offshore sands, and R_0 is discriminant index.

The discriminant index, R_0, is the point along the discriminant function line which is exactly halfway between the center of group A and the center of group B. Next, we may substitute the multivariate mean of group A into the equation to obtain R_A (that is, we set $\psi_j = \overline{A}_j$) and the mean of group B to obtain R_B ($\psi_j = \overline{B}_j$). These define the centers of the two original groups along the discriminant function.

For group A,

$$R_A = \lambda_1\overline{A}_1 + \lambda_2\overline{A}_2$$
$$= -783.63(0.330) - 75.62(1.167)$$
$$= -346.64$$

and for group B,

$$R_B = \lambda_1\overline{B}_1 + \lambda_2\overline{B}_2$$
$$= -783.63(0.340) - 75.62(1.210)$$
$$= -357.81$$

The three points may be plotted as in Figure 6.4. In fact, every observation in the analysis can be entered into the equation and its position along the discriminant function located. This has been done on the same diagram; note that a few members of group A are located on the group B side of R_0 and a few members of group B are located on the group A side. These points are misclassified by the discriminant function.

Tests of Significance

If we are willing to make some assumptions about the nature of the data used in the discriminant function, we can test the significance of the separation between the two groups. Five basic assumptions about the data are necessary: (a) the observations in each group are randomly chosen, (b) the probability of an unknown observation belonging to either group is equal, (c) variables are normally distributed within each group, (d) the variance–covariance matrices of the groups are equal in size, and (e) none of the observations used to calculate the function were misclassified. Of these, the most difficult to justify are b–d. Fortunately, the function is not seriously affected by limited departures from normality or by limited inequality of variances. Justification of (b) must depend upon a priori assessment of the relative abundance of the groups under examination. If the assumption of equal abundance seems unjustified, a different assumption may be made, which will shift the position of R_0 [see Anderson (1958, Chapter 12) for an extensive discussion of alternative decision rules for discrimination].

The first step in a test of the significance of a discriminant function is to measure the separation or distinctness of the two groups. This can be done by computing the distance between the centroids, or multivariate means, of the groups. The measure of distance is derived directly from univariate statistics. We can obtain a measure of the difference between the means of two univariate samples, $\overline{X}_1$ and $\overline{X}_2$, by simply subtracting one from the other. However, this difference is expressed in the same units as the original observations, and usually it is more convenient if the difference is in standardized, or Z-score, form. By simply dividing the difference by the pooled standard deviation, we obtain a *standardized difference*.

$$d = \frac{\overline{X}_1 - \overline{X}_2}{s_p} \qquad (6.19)$$

When both sides of eq. (6.19) are squared, the denominator is the pooled variance of the two samples, s_p^2.

$$d^2 = \frac{(\overline{X}_1 - \overline{X}_2)^2}{s_p^2} \qquad (6.20)$$

Suppose that instead of a single variable, two variables are measured on each observation in the two groups. The difference between the bivariate means of the two groups can be expressed as the ordinary Euclidean or straight-line distance between them. Denoting the two groups as A and B,

$$\text{Euclidean distance} = \sqrt{(\overline{A}_1 - \overline{B}_2)^2 + (\overline{A}_2 - \overline{B}_2)^2} \qquad (6.21)$$

In general, if m variables are measured on each observation, the straight-line distance between the multivariate means of the two groups is:

$$\text{Euclidean distance} = \sqrt{\sum_{i=1}^{m} (\overline{A}_i - \overline{B}_i)^2} \qquad (6.22)$$

The square of the Euclidean distance is $\sum_{i=1}^{m} (\overline{A}_i - \overline{B}_i)^2$; you can verify that this is the same as the matrix product

$$\text{Euclidean distance}^2 = [\overline{A}_i - \overline{B}_i]'[\overline{A}_i - \overline{B}_i] \qquad (6.23)$$

The Euclidean distance and its square, unfortunately, are expressed as hodgepodges of the original units of measurement. To be interpretable, they must be standardized. Comparison with eq. (6.19) suggests that standardization must involve division by the multivariate equivalent of the variance, which is the variance–covariance matrix $[s_p^2]$. Of course, division is not a defined operation in matrix algebra, but we can accomplish the same end by multiplying by the inverse. Multiplying eq. (6.23) by the inverse of the variance–covariance matrix yields the *standardized squared distance*,

$$D^2 = [\overline{A}_i - \overline{B}_i]'[s_p^2]^{-1}[\overline{A}_i - \overline{B}_i] \qquad (6.24)$$

This measure of difference between the means of two multivariate groups is called "*Mahalanobis' distance*." Substituting quantities from Table 6.6 into eq. (6.24), we obtain

$$D^2 = [-0.010 \quad -0.043] \begin{bmatrix} 59,112.280 & 4312.646 \\ 4312.646 & 747.132 \end{bmatrix} \begin{bmatrix} -0.010 \\ -0.043 \end{bmatrix}$$

$$= 11.15$$

Interestingly, we can obtain exactly the same distance measure (within rounding error) by substituting the vector of mean differences into the discriminant function equation itself:

$$D^2 = (-783.63)(-0.010) + (-75.62)(-0.043)$$

$$= 11.17$$

Mahalanobis' distance is graphically shown on Figure 6.4, where it is equal to the distance between R_A and R_B.

The significance of Mahalanobis' distance can be tested using a multivariate equivalent of the t test of the equality of two means, called Hotelling's T^2 test. We will discuss this test more extensively in the next section. Here, we simply note that it has the form

$$T^2 = \frac{n_a n_b}{n_a + n_b} D^2 \tag{6.25}$$

and can be transformed to an F test. The test of multivariate equality, using this more familiar statistic, is

$$F = \left(\frac{n_a + n_b - m - 1}{(n_a + n_b - 2)m} \right) \left(\frac{n_a n_b}{n_a + n_b} \right) D^2 \tag{6.26}$$

with m and $(n_a + n_b - m - 1)$ degrees of freedom. The null hypothesis tested by this statistic is that the two multivariate means are equal, or that the distance between them is zero. That is,

$$H_0 : [D_i] = 0$$

against

$$H_1 : [D_i] > 0$$

The appropriateness of this as a test of a discriminant function should be apparent. If the means of the two groups are very close together, it will be difficult to tell them apart, especially if both groups have large variances. In contrast, if the two means are well separated and scatter about the means is small, discrimination will be relatively easy. As an exercise, it may be instructive to calculate the significance of the discriminant function for the example we have just worked.

Not all of the variables we have included in the discriminant function will be equally useful in distinguishing one group from another. We may wish to isolate those variables that are not especially helpful and eliminate them from future

TABLE 6.7 Trace-Element Analyses of Stream Sediments Collected in Two Areas in Sweden: Group A Samples Taken from Streams Draining Old Mining District; Group B Samples Taken from Streams Draining Area Believed to Be Barren; Group C Samples Taken from Streams in Unprospected Areas—These Are to Be Classified as Potentially Productive or Nonproductive

	Element												
	Ti^a	Mn^b	Ag^c	Ba^a	Co^d	Cr^a	Cu^c	Ni^a	Pb^a	Sr^a	V^a	Zn^a	Au^c
Group A— Productive Area	7,280	1,300	30.0	720	30	150	73	50	70	60	70	190	0.02
	10,300	1,200	0.7	1,280	20	160	25	50	70	90	50	50	0.02
	6,500	700	1.0	1,070	20	200	48	70	100	210	50	170	0.01
	7,000	1,500	0.7	760	30	160	70	40	110	240	40	250	0.01
	5,100	1,000	0.5	740	20	140	39	50	80	50	60	130	0.02
	10,600	2,100	0.3	980	30	50	25	30	70	150	160	110	0.01
	14,200	2,000	0.2	690	30	70	25	50	60	160	70	180	0.01
	9,700	900	0.2	680	35	70	38	30	70	80	110	250	0.01
	2,300	1,500	0.2	710	5	110	50	20	70	80	30	120	0.01
	12,100	6,300	0.1	1,520	30	30	24	30	80	320	160	190	0.02
	3,000	1,100	0.2	510	5	30	15	30	30	240	30	50	0.02
	7,500	2,400	0.7	690	30	30	31	10	100	210	40	280	0.03
	7,800	1,800	4.0	730	55	40	24	30	20	90	320	90	0.01
	6,900	1,500	1.0	326	30	50	25	10	90	70	200	70	0.04
	11,200	3,100	1.5	660	50	40	20	40	50	140	280	90	0.01
	5,200	1,400	0.8	680	35	50	42	20	50	30	150	150	0.01
	5,100	1,500	0.9	700	25	60	67	40	80	40	190	90	0.01
	10,600	2,900	0.4	1,640	25	20	21	30	30	320	90	200	0.01
	11,500	3,200	0.7	710	30	30	15	20	20	260	270	180	0.01
	7,100	1,800	0.9	490	75	50	8	10	30	80	180	100	0.02

Group												
Group B—Nonproductive Area	4,820	500	0.1	160	20	70	30	10	0	720	140	200	0.01
	3,040	500	0.2	150	20	30	82	10	20	1,580	160	70	0.01
	890	600	0.1	50	10	10	61	10	0	340	40	50	0.02
	2,100	500	0.1	100	15	30	77	10	0	650	90	80	0.02
	5,060	700	0.3	140	20	50	154	20	0	1,240	140	80	0.01
	1,980	700	0.1	80	15	20	63	20	0	720	80	110	0.00
	3,220	600	0.2	160	20	30	45	20	10	1,100	120	60	0.01
	3,280	800	0.2	90	15	10	40	30	20	1,480	70	40	0.00
	2,020	700	0.1	80	15	20	104	20	0	420	80	70	0.00
	4,600	700	0.3	160	20	60	48	10	20	780	150	50	0.02
	3,100	500	0.2	100	15	30	65	10	20	710	100	40	0.01
	3,020	600	0.2	90	15	10	69	0	30	1,310	110	30	0.02
	1,860	500	0.1	70	10	20	63	0	10	480	80	50	0.00
	2,800	700	0.1	110	15	20	58	10	20	730	120	80	0.01
	1,040	1,600	0.1	20	5	10	37	0	10	140	30	80	0.01
	4,640	800	0.3	220	15	20	121	20	20	1,200	210	160	0.00
	4,990	900	0.3	190	20	40	59	20	30	480	230	120	0.02
	2,830	800	0.2	120	15	20	40	10	20	690	140	60	0.00
	4,500	700	0.2	140	20	30	82	20	10	710	170	70	0.00
	2,900	600	0.1	80	15	10	99	0	0	760	80	90	0.01
Group C—Unprospected Areas	4,260	800	0.3	180	20	60	128	30	30	460	110	80	0.02
	6,500	1,200	0.5	380	30	40	72	50	20	320	90	160	0.01
	12,200	5,200	1.5	630	25	80	39	40	90	210	200	180	0.01
	1,080	1,600	0.2	80	5	10	102	0	10	160	30	80	0.00
	3,820	500	0.2	170	25	40	60	20	10	1,100	160	40	0.02
	1,020	2,400	0.1	20	0	10	28	0	0	1,320	20	60	0.00

[a] Determined to nearest 10 ppm.
[b] Determined to nearest 100 ppm.
[c] Determined to fraction of ppm.
[d] Determined to nearest 5 ppm.

analyses. Selecting the most effective set of discriminators for discriminant function analysis would seem to be analogous to selecting the most efficient predictors in multiple regression. The problem, however, is more complicated because the "dependent" or predicted variable in a discriminant function is composed of differences between two sets of the same variables that are used as "independent" predictors of the discrimination. Unlike regression, where the sums of squares of Y do not change as different variables X_i are added to the equation, the sums of squares of the differences between groups A and B do change as variables are added or deleted.

Some idea of the effectiveness of the variables as discriminators can be gained by computing the *standardized differences,*

$$D_i = \frac{\bar{A}_i - \bar{B}_i}{s_{pi}} \tag{6.27}$$

This is simply the difference between the means of the two groups A and B for variable i, divided by the pooled standard deviation of variable i. Since the measure does not consider interactions between variables, it is useful only as a general guide to discriminating power. Stepwise discriminant analysis programs may use standardized differences in choosing the order in which variables are added to the discriminant function. However, the significance of different combinations of variables can be tested only by computing the various functions and determining the relative amounts of separation the different equations produce between the two groups.

Discriminant function analysis provides a natural transition between two major classes of multivariate statistical techniques. On one hand, it is closely related to multiple regression and trend-surface analysis. On the other, it can be expressed as an eigenvalue problem, related to principal components, factor analysis, and similar multivariate methods. There are advantages to the use of eigenvectors in calculating the discriminant function, because it allows us to simultaneously discriminate between more than two groups. However, we will delay a consideration of this topic until we examine the basic elements of eigenvector analysis and some of the simpler eigenvector techniques.

The following problem is an example of the use of discriminant analysis in geology. Its solution may help gain a better understanding of the procedure. A government survey group in northern Sweden is prospecting for heavy metal deposits in densely forested mountains. Airborne magnetometer surveys have proved to be of limited value, so a geochemical prospecting approach is being evaluated, based on stream-water analyses. Seven variables have been selected and two suites of measurements performed. Group A consists of measurements on streams draining areas containing active mines or proven mineral deposits. Group B consists of similar measurements on streams draining areas that have been heavily prospected without results. Data are listed in Table 6.7. From these, calculate the discriminant function between productive and nonproductive regions. Determine if the difference between the two groups is significant, and investigate the relative importance of the variables utilized. For the purposes of this exercise, we will assume that the parent populations of the two groups are multivariate normal. Table 6.7 also contains a set of measurements made on streams draining areas not known to have been prospected. On the basis of the discriminant function, can any be selected as likely areas for prospecting?

Multivariate Extensions of Elementary Statistics

In Chapter 2, we considered some simple geologic problems that could be examined by elementary statistical methods. We will begin our consideration of multivariate methods in geology with some direct extensions of these simple tests. You will recall that the variation measured in most naturally occurring phenomena could be described by the normal distribution. This is a reflection of the central limit theorem, which states that observations which are the sums of many independently operating processes tend to be normally distributed as the number of effects becomes large. It is this tendency which allows us to use the normal probability distribution as a basis for statistical tests and provides the starting point for the development of other distributions such as the t, F, and χ^2. The concept of the normal distribution can be extended to include situations in which observational units consist of many variables.

Suppose we collect rocks from an area and measure a set of properties on each specimen. The measurements may include determinations of chemical or mineralogical constituents, specific gravity, magnetic susceptibility, radioactivity, or any of an almost endless list of possible variables. We can regard the set of measurements made on an individual rock as defining a vector $[X] = [X_1, X_2, \ldots, X_m]$, where there are m measured characteristics or variables. If samples, represented by vectors $[X]$, are randomly selected from a population that is the result of many independently acting processes, the observed vectors will tend to be multivariate normally distributed. Considered individually, each variate is normally distributed and characterized by a mean μ_k and a variance σ_k^2. The *joint probability distribution* is a p-dimensional equivalent of the normal distribution, having a vector mean $[\mu] = [\mu_1, \mu_2, \ldots, \mu_m]$ and a variance generalized into the form of a diagonal matrix:

$$[\Sigma^2] = \begin{bmatrix} \sigma_1^2 & 0 & 0 & 0 & \ldots & 0 \\ 0 & \sigma_2^2 & 0 & 0 & \ldots & 0 \\ \vdots & \vdots & \vdots & \vdots & & \vdots \\ 0 & 0 & 0 & 0 & \ldots & \sigma_m^2 \end{bmatrix}$$

In addition to these obvious extensions of the normal distribution to the multivariate case, the multivariate normal distribution has an important additional characteristic. This is the covariance, cov_{jk}, which occupies all of the off-diagonal positions of the matrix $[\Sigma^2]$. Thus, in the multivariate normal distribution, the mean is generalized into a vector and the variance into a matrix of variances and covariances. In the simple case of $m = 2$, the probability distribution forms a three-dimensional bell curve (Fig. 2.13), shown as a "contour map" in Figure 6.5. Although the distributions of variables X_1 and X_2 are shown along their respective axes, the essential characteristics of the joint probability distribution are better

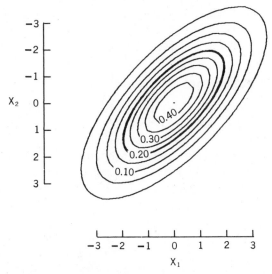

FIGURE 6.5 Contour map of bivariate normal probability distribution. See Figure 2.13 for perspective diagram of same distribution.

shown by the major and minor axes of the ellipsoid. Many of the multivariate procedures we will discuss are concerned with the relative orientations of these major and minor axes.

One of the simplest tests we considered in Chapter 2 was a t test of the probability that a random sample of n observations had been drawn from a normal population with a specified mean μ and an unknown variance σ^2. The test, given in (2.32), can be rewritten in the form

$$t = \frac{(\bar{X} - \mu)\sqrt{n}}{\sqrt{s^2}} \tag{6.28}$$

An obvious generalization of this test to the multivariate case is the substitution of a vector of sample means for $\bar{X}$, a vector of population means for μ, and a variance–covariance matrix for s^2. We have defined the vector of population means as $[\mu]$, so a vector of sample means can be designated as $[\bar{X}]$. Similarly, $[\Sigma^2]$ is the matrix of population variances and covariances, so $[s^2]$ may represent the matrix of sample variances and covariances. Both $[\bar{X}]$ and $[\mu]$ are taken to be column vectors, although equivalent equations may be written in which these are assumed to be row vectors. The two vectors may be subtracted to give a column vector of differences between the sample means and the population means. That is,

$$[\bar{X} - \mu] = [\bar{X}] - [\mu]$$

Substituting these quantities directly into (6.28) gives

$$t = \frac{[\bar{X} - \mu]\sqrt{n}}{\sqrt{[s^2]}}$$

Unfortunately, there is no equally obvious way of solving this equation so it yields a single value of t. We must reduce the vectors and the matrix to single numbers if we are to apply this test. If we were to multiply the column vector $[\overline{X} - \mu]$ by a row vector having the same number of elements, the result would be a single number. We will therefore define an arbitrary row vector $[A]$ whose transpose is a column vector $[A]'$. Multiplication of the column vector of differences $[\overline{X} - \mu]$ by the row vector $[A]$ gives a single number, and premultiplication of $[s^2]$ by $[A]$ and postmultiplication by $[A]'$ also yields a single number. That is, our test has become

$$t = \frac{[A]\,[\overline{X} - \mu]\sqrt{n}}{[A]\sqrt{[s^2]}[A]'}$$

However, we have also changed what we are testing, from a null hypothesis of

$$H_0 : [\mu_1] = [\mu_0]$$

to

$$H_0^* : [A] \cdot [\mu_1] = [A] \cdot [\mu_0]$$

The original hypothesis H_0 is true only if the new hypothesis H_0^* holds for all possible values of $[A]$. It is sufficient, however, to test only the maximum possible value of the test statistic, because if H_0^* is rejected for any value of $[A]$, the hypothesis H_0 also is rejected. With a bit of mathematical manipulation, we can determine the conditions under which a maximum test statistic will result for any arbitrary vector $[A]$. This involves introduction of the constraint $[A] \cdot [s^2] \cdot [A]' = 1$ and expressing the equation in a form incorporating a determinant. In the process, we can eliminate the troublesome square roots by squaring the equation. This also squares the test value, which we will now refer to as T^2. When all operations are complete, we find that the test statistic can be expressed as

$$T^2 = n[\overline{X} - \mu]'[s^2]^{-1}[\overline{X} - \mu] \tag{6.29}$$

That is, the arbitrary vector $[A]$ is equal to the vector of differences between the means, $[\overline{X} - \mu]$. We must find the inverse of the variance–covariance matrix, premultiply this inverse by a row vector of differences $[\overline{X} - \mu]'$, and then postmultiply by a column vector of these same differences. The test statistic is a multivariate extension of the t statistic; as we mentioned earlier, it is called *Hotelling's T^2* in honor of the statistician who first formulated it. The critical values of T^2 can be determined by the relation

$$F = \frac{n - m}{m(n - 1)} T^2 \tag{6.30}$$

where n is the number of samples and m is the number of variables, allowing us to use conventional F tables rather than special tables of the T^2 distribution. A more complete discussion of this test is given in many texts, such as Morrison (1976).

Although the expression of this test in a form such as (6.29) is easy, computation of a test value for an actual data set may be very laborious. For example, suppose we have measured the content of four elements in seven lunar samples. We wish to test the hypothesis that these samples have been drawn from a population having the same mean as terrestrial basalts. Assume we take our values for the populations means from the *Handbook of Physical Constants* (Clark, 1966, p. 4). Hotelling's T^2 seems appropriate to test the hypothesis that the vector of lunar sample means is no different from the vector of basalt means given in this reference.

We first must compute the vector of four sample means and the 4×4 matrix of variances and covariances. The vector of differences between sample and population means $[\overline{X} - \mu]$ also must be computed. Next, we must find the inverse of the variance–covariance matrix, or $[s^2]^{-1}$. We then must perform two matrix multiplications, $[\overline{X} - \mu]' \cdot [s^2]^{-1} \cdot [\overline{X} - \mu]$, and multiply by n to produce T^2. From this description, you can appreciate that the computational effort becomes increasingly great as the number of variables grows larger.

The data for the seven lunar samples are listed in Table 6.8, with the "popu-

TABLE 6.8 Abundances of Four Elements in Seven Lunar Samples, and Mean Abundances of Same Elements in Terrestrial Basalts

	Lunar Samples (%)			
	Si	Al	Fe	Mg
1	19.4	5.9	14.7	5.0
2	21.5	4.0	15.7	3.7
3	19.2	4.0	15.4	4.3
4	18.4	5.4	15.2	3.4
5	20.6	6.2	13.2	5.5
6	19.8	5.7	14.8	2.8
7	18.7	6.0	13.8	4.6
MEANS	19.66	5.31	14.69	4.19
"Population" Means	22.10	7.40	10.10	4.00
Differences	−2.44	−2.09	4.59	0.19

Variance–Covariance Matrix

$$
\begin{bmatrix}
1.1795350 & -0.3076159 & 0.0593007 & 0.0792885 \\
-0.3076159 & 0.8680964 & -0.6830902 & 0.3019053 \\
0.0593007 & -0.6830902 & 0.8014425 & -0.5469023 \\
0.0792885 & 0.3019053 & -0.5469023 & 0.8914289
\end{bmatrix}
$$

Inverse of Variance–Covariance Matrix

$$
\begin{bmatrix}
1.0614130 & 0.9946304 & 0.8169364 & 0.0699353 \\
0.9946304 & 5.2086160 & 5.3354270 & 1.4208490 \\
0.8169364 & 5.3354270 & 7.6584370 & 2.8189000 \\
0.0699353 & 1.4208490 & 2.8189000 & 2.3637960
\end{bmatrix}
$$

$T^2 = 11.93$

Source: Wanke and others (1970).

lation'' vector from Clark. Intermediate values in the computation of T^2 also are given, with the final test value of T^2. Compute the equivalent F statistic, which has m and $(n - m)$ degrees of freedom, and test the null hypothesis that the means of the lunar samples are equal to the mean of the population of terrestrial basalts.

We have dwelled on the T^2 test against a known mean, not because this specific test has greater utility in geology than other multivariate tests, but to illustrate the close relationship between conventional statistics and multivariate statistics. Multivariate equivalents can be formulated directly from most univariate tests with the proper expansion of the basic assumptions. However, the transition from ordinary algebra to matrix algebra often obscures the underlying similarity between the two applications. Although we usually regard multivariate methods as an extension of univariate statistics, the relationship can more generally be considered the reverse; univariate, or ordinary, statistical analysis is a special subset of the general area of multivariate analysis.

In the remaining discussion in this section, we will consider multivariate tests which are p-dimensional equivalents of some of the tests we considered in Chapter 2. However, we will not point out the details of the extrapolation from the univariate to the general case as we have done with the T^2 test. These derivations can be found in many texts on multivariate statistics, some of which are listed in the Selected Readings at the end of this chapter.

Equality of Two Vector Means

The test we have just considered is a one-sample test against a specified population mean vector. Suppose instead we have collected two independent random samples and we wish to test the equivalency of their mean vectors. We assume that the two samples are drawn from multivariate normal populations, both having the same unknown variance–covariance matrix $[\Sigma^2]$. We wish to test the null hypothesis

$$H_0 : [\mu_1] = [\mu_2]$$

against

$$H_1 : [\mu_1] \neq [\mu_2]$$

The null hypothesis states that the mean vector of the parent population of the first sample is the same as the mean vector of the parent population from which the second sample was drawn.

The test we must use is a multivariate equivalent of (2.33). In that two-sample t test, we used a pooled estimate of the population variance based on both samples. Accordingly, we must compute a pooled estimate, $[s_p^2]$, of the common variance–covariance matrix from our two multivariate samples. This is done by calculating a matrix of sums of squares and products for each sample. We can use the terminology of Chapter 2 and denote the matrix of sums of squares and cross-products of sample 1 as $[SP_1]$; similarly, the matrix from sample 2 is $[SP_2]$. The pooled estimate of the variance–covariance matrix is

$$[s_p^2] = \frac{1}{n_1 + n_2 - 2} ([SP_1] + [SP_2]) \tag{6.31}$$

We must next find the difference between the two mean vectors, $[\overline{X}_1]$ and $[\overline{X}_2]$, or $[\overline{X}_1 - \overline{X}_2]$. Our T^2 test has the form

$$T^2 = \frac{n_1 n_2}{n_1 + n_2} [\overline{X}_1 - \overline{X}_2]' \cdot [s_p^2]^{-1} \cdot [\overline{X}_1 - \overline{X}_2] \qquad (6.32)$$

The significance of the T^2 test statistic can be determined by the F transformation:

$$F = \frac{n_1 + n_2 - m - 1}{(n_1 + n_2 - 2)m} T^2 \qquad (6.33)$$

which has m and $(n_1 + n_2 - m - 1)$ degrees of freedom (Morrison, 1976).

TABLE 6.9   Abundance of Four Elements in Seven Samples of Basalts from the Pacific Region

	Pacific Samples (%)			
	Si	Al	Fe	Mg
1	22.5	9.6	6.6	3.4
2	22.1	8.4	7.8	3.6
3	25.9	8.7	4.8	4.0
4	23.5	8.1	5.0	5.2
5	21.7	10.0	8.2	4.9
6	21.9	8.2	9.3	4.9
7	23.7	7.2	9.5	3.3
MEANS	23.04	8.60	7.31	4.19
Difference Between Means in Tables 6.8 and 6.9	−3.39	−3.29	7.37	0.00

Variance–Covariance Matrix

$$\begin{bmatrix} 2.1828820 & -0.4399923 & -1.7223760 & -0.2409477 \\ -0.4399923 & 0.8966687 & -0.4199982 & 0.1266680 \\ -1.7223760 & -0.4199982 & 3.6547640 & -0.2480939 \\ -0.2409477 & 0.1266680 & -0.2480939 & 0.6380959 \end{bmatrix}$$

Pooled Variance–Covariance Matrix for Tables 6.8 and 6.9

$$\begin{bmatrix} 1.6812080 & -0.3738041 & -0.8315379 & -0.0808296 \\ -0.3738041 & 0.8823825 & -0.5515442 & 0.2142866 \\ -0.8315379 & -0.5515442 & 2.2281030 & -0.3974981 \\ -0.0808296 & 0.2142866 & -0.3974981 & 0.7647624 \end{bmatrix}$$

Inverse of Pooled Variance–Covariance Matrix

$$\begin{bmatrix} 1.1213670 & 0.8356105 & 0.6665248 & 0.2308193 \\ 0.8356105 & 1.9991410 & 0.7963902 & -0.0579050 \\ 0.6665248 & 0.7963902 & 0.9561165 & 0.3442555 \\ 0.2308193 & -0.0579050 & 0.3442555 & 1.5271490 \end{bmatrix}$$

$T^2 = 116.10$

Table 6.9 contains seven analyses for the same four elements as in Table 6.8. These analyses, however, are for oceanic basalts from the Pacific Ocean. We will test the hypothesis that the mean vectors of lunar samples and Pacific samples are the same, assuming both samples are randomly drawn from multivariate normal populations with the same variance–covariance matrix. Necessary computations are also given in Table 6.9. Is the null hypothesis of equality of multivariate means accepted at the 5% ($\alpha = 0.05$) significance level?

Equality of Variance–Covariance Matrices

An underlying assumption in the two preceding tests is that the samples are drawn from populations having the same variance–covariance matrix. This is the multivariate equivalent of the assumption of equal population variances necessary to perform t tests of means. In practice, an assumption of equality may be unwarranted, because samples which exhibit a high mean often will also have a large variance. You will recall from Chapter 4 that such behavior is characteristic of many geologic variables such as mine assay values and trace element concentrations. Equality of variance–covariance matrices may be checked by the following "test of generalized variances" which is a multivariate equivalent of the F test (Morrison, 1976).

Suppose we have k groups of observations, and have measured m variables on each sample. For each group a variance–covariance matrix $[s_i^2]$ may be computed. We wish to test the null hypothesis

$$H_0 : [\Sigma_1^2] = [\Sigma_2^2] = \cdots = [\Sigma_k^2]$$

against the alternative

$$H_1 : [\Sigma_i^2] \neq [\Sigma_j^2]$$

The null hypothesis states that all k population variance–covariance matrices are the same. The alternative is that at least two are different. Each variance–covariance matrix $[s_i^2]$ is an estimate of a population matrix $[\Sigma_i^2]$. If the parent populations of the k groups are identical, the sample estimates may be combined to form a pooled estimate of the population variance–covariance matrix. The pooled estimate is created by

$$[s_p^2] = \sum_{i=1}^{k} \frac{(n_i - 1)[s_i^2]}{(\sum_{i=1}^{k} n_i) - k} \tag{6.34}$$

where n_i is the number of samples in the ith group and Σn_i indicates the grand total number of all samples in all k groups. This equation is algebraically equivalent to (6.31) when $k = 2$, as $[s_i^2] = [SP_i]/(n_i - 1)$.

From the pooled estimate of the population variance–covariance matrix, a test statistic M can be computed:

$$M = [(\sum_{i=1}^{k} n_i) - k] \ln |s_p^2| - \sum_{i=1}^{k} [(n_i - 1) \ln |s_i^2|] \tag{6.35}$$

The test is based on the difference between the logarithm of the determinant of the pooled variance–covariance matrix and the average of the logarithms of the determinants of the sample variance–covariance matrices. If all the sample matrices are the same, this difference will be very small. As the variances and covariances of the samples deviate more and more from one another, the test statistic will increase. Tables of critical values of M are not widely available, so the transformation

$$C^{-1} = 1 - \frac{2m^2 + 3m - 1}{6(m + 1)(k - 1)} \left(\sum_{i=1}^{k} \frac{1}{n_i - 1} - \frac{1}{(\sum_{i=1}^{k} n_i) - k} \right) \quad (6.36)$$

can be used to convert M to an approximate χ^2 statistic:

$$\chi^2 \approx MC^{-1} \quad (6.37)$$

The approximate χ^2 value has degrees of freedom equal to $v = (1/2)(k - 1)$ $m(m + 1)$. If all the samples contain the same number of observations, n, (6.36) can be simplified to

$$C^{-1} = 1 - \frac{(2m^2 + 3m - 1)(k + 1)}{6(m + 1)k(n - 1)} \quad (6.38)$$

The χ^2 approximation is good if k and m do not exceed about 5 and each variance–covariance estimate is based on at least 20 observations. Using this test, determine the equivalency of the variance–covariance matrices of the two samples given in Tables 6.8 and 6.9. To compute the test statistic, you must find the determinants of the three 4×4 matrices, $[s_1^2]$, $[s_2^2]$, and $[s_p^2]$. Once the necessary determinants have been found, the test statistic can easily be calculated. For our purposes, we will assume that the sample sizes of the two data collections are sufficiently large so that the χ^2 approximation is accurate.

To illustrate the process of hypothesis testing using multivariate statistics, we will work through the following problem. Note that the number of samples is just sufficient for some of the approximations to be strictly valid; we will consider them to be adequate for the purposes of this demonstration.

In a local area in eastern Kansas, all potable water is obtained from wells. Some of these wells draw from the alluvial fill in stream valleys, while others tap a limestone aquifer that also is the source of numerous springs in the region. Residents prefer to obtain water from the alluvium, as they feel it is of better quality. However, the water resources of the alluvium are limited, and it would be desirable for some users to obtain their supplies from the limestone aquifer.

In an attempt to demonstrate that the two sources are equivalent in quality, a state agency sampled wells that tapped each source. The samples were analyzed for chemical compounds that affect the quality of water. Some of the data from these analyses are given in Table 6.10. The table also gives the variance–covariance matrices, inverses, and determinants for the two data sets and for the pooled data. From these, we can test the equivalence of the two vector means. We will assume that the samples have been drawn randomly from multivariate normal populations.

We must first test the assumption that the variance–covariance matrices for the two samples are equivalent using the test statistic M given in (6.35):

$$M = (20 + 20 - 2) \ln 2.0725 \cdot 10^8$$
$$- (19 \ln 1.8838 \cdot 10^8 + 19 \ln 2.2582 \cdot 10^8)$$
$$= 0.1835$$

The transformation factor C^{-1} must also be calculated to allow use of the χ^2 approximation:

$$C^{-1} = 1 - \frac{2 \cdot 5^2 + 3 \cdot 5 - 1}{6(5 + 1)(2 - 1)} \left(\frac{1}{19} + \frac{1}{19} - \frac{1}{40 - 2} \right)$$
$$= 0.861$$

The χ^2 statistic is 0.158, with degrees of freedom equal to

$$v = \frac{1}{2} (2 - 1)(5)(5 + 1) = 15$$

The critical value of χ^2 for $v = 15$ and 5% level of significance is 25. The computed statistic is far below this value, so we may conclude that there is nothing in our samples which suggests that the variance–covariance structures of the parent populations are different. We may proceed to the test of equality of multivariate means using the T^2 test of (6.32):

$$T^2 = \frac{20 \cdot 20}{20 + 20} [1.566] = 15.66$$

The value in brackets represents the product of the matrix multiplications specified in the equation. The T^2 statistic may be converted to an F statistic by (6.33):

$$F = \frac{20 + 20 - 5 - 1}{(20 + 20 - 2)5} 15.66 = 2.80$$

Degrees of freedom are

$$v_1 = 5$$
$$v_2 = 20 + 20 - 5 - 1 = 34$$

The critical value for F with 5 and 34 degrees of freedom at the 5% ($\alpha = 0.05$) level of significance is 2.49. Our computed test statistic just exceeds this critical value, so we conclude that our samples do, indeed, indicate a difference in the means of the two populations. In other words, there is a statistically significant difference in composition of water from the two aquifers. This simple test will not pinpoint those chemical variables responsible for this difference, but it does substantiate the natives' contention that they can tell a difference in the water!

Multivariate techniques equivalent to analysis of variance procedures discussed

TABLE 6.10 Cation Composition of Water Samples Collected from Wells in an Area of Eastern Kansas: X_1 = Silica, X_2 = Iron, X_3 = Magnesium, X_4 = Sodium + Potassium, X_5 = Calcium

		Cation (%)		
X_1	X_2	X_3	X_4	X_5
A. Samples from Wells Drilled into Alluvium				
15.9	20.0	27.8	41.8	26.1
16.5	14.1	38.3	20.7	21.9
9.6	15.6	10.0	32.3	44.5
11.5	15.2	34.8	21.4	29.4
12.0	21.3	38.9	34.6	7.3
13.1	26.1	59.0	23.9	20.6
12.7	15.9	38.8	21.3	29.3
15.8	15.9	52.7	13.0	2.2
11.6	19.8	30.8	17.9	24.4
11.2	17.2	22.3	41.0	28.0
8.2	19.5	31.6	26.6	33.6
10.8	16.8	39.2	20.9	7.7
10.8	8.2	29.1	27.1	33.1
15.5	15.5	35.3	29.6	62.6
10.2	6.1	32.6	35.0	60.3
9.6	18.1	30.6	19.2	11.6
12.2	15.1	32.3	42.6	7.4
8.9	18.0	40.3	42.9	25.3
12.3	6.4	37.3	35.5	18.3
12.7	16.8	27.6	50.9	7.5
B. Samples from Wells Drilled into Limestone Aquifer				
13.2	17.5	23.4	38.5	36.3
13.9	11.2	34.8	18.0	30.4
7.1	13.5	7.8	27.0	52.7
9.6	13.2	31.3	17.3	36.5
10.6	19.0	35.8	32.5	18.0
10.2	23.4	56.8	19.6	27.9
10.0	14.7	35.1	18.1	37.7
13.1	13.1	47.3	10.1	4.6
9.9	18.4	26.6	14.1	34.9
9.8	14.8	17.2	37.4	34.4
6.0	16.6	26.1	21.1	40.8
8.2	14.6	34.9	17.7	15.8
8.3	6.6	27.1	21.6	43.5
13.4	13.9	33.3	24.9	70.4
8.0	4.6	30.0	29.4	70.5
7.4	16.1	28.2	17.2	17.5
9.4	13.6	28.3	37.4	16.6
6.8	15.6	36.6	37.8	30.7
10.3	4.7	33.9	33.2	27.7
10.0	14.0	24.2	45.7	18.5

TABLE 6.10 (*Continued*)

VECTOR MEANS OF GROUP A

12.055	16.08	34.465	29.91	24.835

VECTOR MEANS OF GROUP B

9.76	13.955	30.935	25.93	33.27

Variance–Covariance Matrix A, Determinant = 2.25822×10^8

$$
\begin{bmatrix}
5.6394620 & 0.7332828 & 8.6868000 & -2.9821260 & -5.5767340 \\
0.7332828 & 23.1732900 & 12.7656400 & -4.5592230 & -26.9460700 \\
8.6868000 & 12.7656400 & 103.3983000 & -42.3948000 & -62.3459900 \\
-2.9821260 & -4.5592230 & -42.3948000 & 106.9526000 & 13.1360100 \\
-5.5767340 & -26.9460700 & -62.3459900 & 13.1360100 & 286.8151000
\end{bmatrix}
$$

Variance–Covariance Matrix B, Determinant = 1.88381×10^8

$$
\begin{bmatrix}
5.1614800 & 0.5133812 & 7.3683410 & -1.4102910 & -3.4401890 \\
0.5133812 & 21.0247300 & 10.6948600 & -4.0895870 & -25.3971700 \\
7.3683410 & 10.6948600 & 102.8046000 & -38.5268000 & -58.1688200 \\
-1.4102910 & -4.0895870 & -38.5268000 & 98.8654500 & 7.2520690 \\
-3.4401890 & -25.3971700 & -58.1688200 & 7.2520690 & 290.8707000
\end{bmatrix}
$$

Pooled Variance–Covariance Matrix, Determinant = 2.07252×10^8

$$
\begin{bmatrix}
5.4004710 & 0.6233320 & 8.0275700 & -2.1962090 & -4.5084610 \\
0.6233320 & 22.0990100 & 11.7302500 & -4.3244050 & -26.1716200 \\
8.0275700 & 11.7302500 & 103.1014000 & -40.4608000 & -60.2574000 \\
-2.1962090 & -4.3244050 & -40.4608000 & 102.9090000 & 10.1940400 \\
-4.5084610 & -26.1716200 & -60.2574000 & 10.1940400 & 288.8429000
\end{bmatrix}
$$

Inverse of Pooled Variance–Covariance Matrix

$$
\begin{bmatrix}
0.2100476 & 0.0029156 & -0.0176276 & -0.0023202 & -5.27736\text{e}{-05} \\
0.0029156 & 0.0518892 & -0.0036462 & 0.0004157 & 0.0039718 \\
-0.0176276 & -0.0036462 & 0.0148257 & 0.0050709 & 0.0023084 \\
-0.0023202 & 0.0004157 & 0.0050709 & 0.0116147 & 0.0006494 \\
-5.27737\text{e}{-05} & 0.0039718 & 0.0023084 & 0.0006494 & 0.0042798
\end{bmatrix}
$$

in Chapter 2 are available. In general, these involve a comparison of two $m \times m$ matrices that are the multivariate equivalents of the among-group and within-group sums of squares tested in ordinary analysis of variance. The test statistic consists of the largest eigenvalue of the matrix resulting from the comparison. We will not consider these tests here, because their formulation is involved and their applications to geologic problems have been, so far, minimal. This is not, however, a reflection on their potential utility. Interested readers are referred to the book by Cooley and Lohnes (1971), which contains computer programs for multivariate analysis of variance and examples of their use, and to Chapter 5 of the textbook by Morrison

(1976). Koch and Link (1980) include a brief illustration of the application of multivariate analysis of variance to geochemical data.

Cluster Analysis

Classification is the placing of objects into more or less homogeneous groups, in a manner so that the relation between groups is revealed. This is the special forte of taxonomists, who attempt to deduce the lineage of living creatures from their characteristics and similarities. Taxonomy is highly subjective and dependent upon the skills of individual taxonomists, developed through years of experience. In this respect, the field is analogous in many ways to geology. As in geology, a group of researchers have become dissatisfied with the subjectivity and capriciousness of traditional methods, and have sought new techniques of classification which incorporate the massive data-handling capabilities of the computer. These workers call themselves numerical taxonomists, and are responsible for many of the advances made in numerical classification.

At the present time, numerical taxonomy is the center of a controversy among biologists, much like the acrimonious debate among psychologists that swirled around factor analysis in the 1930s and 1940s. As in that dispute, the techniques of numerical taxonomy have been overzealously promoted by some practitioners. In addition, it has been claimed that a numerically derived taxonomy will better represent the phylogeny of a group of organisms than will any other type of classification. This, of course, cannot be demonstrated. At the present time, the theoretical underpinnings of taxonomic methods such as cluster analysis are incomplete; little is known of the statistical properties of the various methods, and only limited tests of significance are available (Hartigan, 1975). Presumably a more complete statistical basis will be fashioned as it has been for factor analysis. Already many of the methods of numerical taxonomy are important in geologic research, especially in the classification of fossil invertebrates and the study of paleoenvironments.

Cluster analysis is the name given to a bewildering assortment of techniques designed to perform classification by assigning observations to groups so each group is more-or-less homogeneous and distinct from other groups. There is no analytical solution to this problem, which is common to all areas of classification, not just numerical taxonomy. Although there are alternative classifications of classification procedures (Sneath and Sokal, 1973), most may be grouped into four general types.

1. *Partitioning methods* operate on the multivariate observations themselves, or on projections of these observations onto planes of lower dimension. Basically, these methods cluster by finding regions in the space defined by the m variables that are poorly populated with observations, and that separate densely populated regions. Mathematical "partitions" are placed in the sparse regions, subdividing the variable space into discrete classes. Although the analysis is done in the m-dimensional space defined by the variables rather

than the n-dimensional space defined by the observations, it proceeds iteratively and may be extremely time-consuming (Switzer, 1970).

2. *Arbitrary origin methods* operate on the similarity between the observations and a set of arbitrary starting points. If n observations are to be classified into k groups, it is necessary to compute an asymmetric $n \times k$ matrix of similarities between the n samples and the k arbitrary points that serve as initial group centroids. The observation closest or most similar to a starting point is combined with it to form a cluster. Observations are iteratively added to the nearest cluster, whose centroid is then recalculated for the expanded cluster.

3. *Mutual similarity* procedures group together observations that have a common similarity to other observations. First an $n \times n$ matrix of similarities between all pairs of observations is calculated. Then the similarity between columns of this matrix is iteratively recomputed. Columns representing members of a single cluster will tend to have intercorrelations near $+1$, while having much lower correlations with nonmembers.

4. *Hierarchical clustering* joins the most similar observations, then successively connects the next most similar observations to these. First an $n \times n$ matrix of similarities between all pairs of observations is calculated. Those pairs having the highest similarities are then merged, and the matrix recomputed. This is done by averaging the similarities that the combined observations have with other observations. The process iterates until the similarity matrix is reduced to 2×2. The levels of similarity at which observations are merged is used to construct a dendrogram.

Hierarchical techniques are the most widely applied clustering techniques in the Earth sciences, probably because the development of these methods has been closely linked with the numerical taxonomy of fossil organisms. Because of their widespread use, we will consider them in some detail.

Suppose we have a collection of objects we wish to arrange into a hierarchical classification. In biology, these objects are referred to as "operational taxonomic units" or OTU's. On each object, we can make a series of measurements which constitutes our data set. If we have n objects and measure m characteristics, the data set forms an $n \times m$ matrix. Next, some measure of resemblance or similarity must be computed between every pair of objects. Several coefficients of resemblance have been used, including the correlation coefficient r_{ij} and a standardized m-space Euclidian distance, d_{ij}. The distance coefficient is computed by

$$d_{ij} = \sqrt{\frac{\sum_{k=1}^{m} (X_{ik} - X_{jk})^2}{m}} \qquad (6.39)$$

where X_{ik} denotes the kth variable measured on object i and X_{jk} is the kth variable measured on object j. In all, m variables are measured on each object, and d_{ij} is the distance between object i and object j. As you would expect, a low distance indicates the two objects are similar or "close together," whereas a large distance indicates dissimilarity. Commonly, the $n \times m$ raw data matrix is standardized prior to computing distance measurements. This ensures that each variable is weighted equally. Otherwise, the distance will be influenced most strongly by the variable which has the greatest magnitude. In some instances this may be desirable, but unwanted effects can creep in through injudicious choice of measurement units. As an extreme example, we might measure three perpendicular axes on a collection of pebbles. If we measure two of the axes in centimeters and the third in millimeters, the third axis will have proportionally ten times the influence on the distance coefficient as either of the other two variables.

Computation of a similarity measurement between all possible pairs of objects will result in an $n \times n$ symmetrical matrix. Any coefficient c_{ij} in the matrix gives the resemblance between objects i and j. The next step is to arrange the objects into a hierarchy so objects with the highest mutual similarity are placed together. Then groups or clusters of objects are associated with other groups which they most closely resemble, and so on until all of the objects have been placed into a complete classification scheme. Several clustering techniques have been developed; a consideration of all of the possible variations and their relative merits is beyond the scope of this book. Rather, we will discuss one simple clustering technique, called the weighted pair-group method with arithmetic averages, and then point out some useful modifications to this scheme.

Extensive discussions of hierarchical and other classification techniques are contained in the books by Tryon and Bailey (1970), Bijnen (1973), Jardine and Sibson (1971), Clifford and Stephenson (1973), Sneath and Sokal (1973), Anderberg (1973), and Hartigan (1975). The first two books discuss classification as developed by experimental psychologists. The next three approach clustering from the viewpoint of numerical taxonomy. The last two volumes take a more general approach, and contain computer programs.

Table 6.11 contains a full, symmetric matrix of the coefficients of similarity between six objects, identified A, B, . . . , F. The ''objects'' are thin sections of sandstones, and the variables are descriptors of the rock fabric, including measures of grain size and shape, pore size and shape, and packing. In this example, the similarity measure is the correlation coefficient.

The first step in clustering by a pair-group method is to find the mutually highest

TABLE 6.11   Matrix of Correlations between Six Sandstones, Using Variables Measured on Thin Sections

	A	B	C	D	E	F
A	1.00	**0.57**	0.12	−0.65	−0.62	−0.39
B	**0.57**	1.00	**0.46**	−0.79	−0.72	−0.72
C	0.12	0.46	1.00	−0.58	−0.61	−0.52
D	−0.65	−0.79	−0.58	1.00	**0.66**	**0.41**
E	−0.62	−0.72	−0.61	**0.66**	1.00	0.40
F	−0.39	−0.72	−0.52	0.41	0.40	1.00

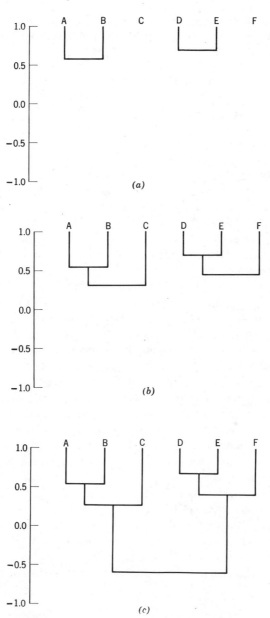

FIGURE 6.6 (a) Dendrogram with initial clusters. (b) Connection of remaining objects to clusters. (c) Final connection of two clusters, completing dendrogram.

TABLE 6.12 Matrix of Averaged Correlations between Two Clusters and Two Sandstones

	AB	C	DE	F
AB	1.00	**0.29**	−0.70	−0.55
C	**0.29**	1.00	−0.59	−0.52
DE	−0.70	−0.59	1.00	**0.41**
F	−0.55	−0.52	**0.41**	1.00

correlations in the matrix to form the centers of the clusters. The highest correlation in each column of the matrix in Table 6.11 is shown in boldface type. Objects *A* and *B* form mutually high pairs, because *A* most closely resembles *B*, and *B* most closely resembles *A*. However, *C* and *B* do not form a mutually high pair, because although *C* most closely resembles *B*, *B* resembles *A* more than it does *C*. To qualify as a mutually high pair, coefficients c_{ij} and c_{ji} must be the highest coefficients in their respective columns.

We can indicate the resemblance between our mutually high pairs in a diagram such as Figure 6.6a. Object *A* is connected to *B* at a level of 0.57, indicating the degree of their mutual similarity. In the same manner, *D* and *E* are connected. This is the first step in the construction of a *dendrogram,* or a tree diagram, which is the most common way of displaying the results of clustering.

Next, the similarity matrix must be recomputed, treating grouped or clustered elements as a single element. There are several methods for doing this. In the simple technique we are considering, new correlations between all clusters and unclustered objects are recalculated by simple arithmetic averaging. For example, the new correlation between cluster *AB* and object *C* is equal to the sum of the correlations of the elements common to both *AB* and *C*, divided by 2. Table 6.12 contains the results of these recalculations. The highest correlations in each column are shown in boldface type.

The clustering procedure is now repeated; mutually high pairs are sought out and clustered. In this cycle, object *C* joins cluster *AB* and object *F* joins cluster *DE* (Figure 6.6b). Then the process is continued until all clusters are joined together. The final matrix of similarities will be a 2 × 2 matrix between the last remaining two clusters, as shown in Table 6.13. This indicates that cluster *ABC* has a resemblance of −0.59 with cluster *DEF*. Our dendrogram can then be completed (Figure 6.6c).

TABLE 6.13 Matrix of Averaged Correlations between Final Two Clusters

	ABC	DEF
ABC	1.00	−0.59
DEF	−0.59	1.00

Clustering is an efficient way of displaying complex relationships among many objects. However, the process of averaging together members of a cluster and treating them as a single new object introduces distortion into the dendrogram. This distortion becomes increasingly apparent as successive levels of clusters are averaged together. We can evaluate the severity of this distortion by examining what numerical taxonomists call the matrix of cophenetic values. This is nothing more than the matrix of apparent correlations contained within the dendrogram. For example, the dendrogram in Figure 6.6 implies that the correlations between D, E, and F, on one hand, with A, B, and C, on the other, are all -0.59. Similarly, the correlations between F and D and also between F and E appear to be 0.41. Only the correlations between A and B and between D and E are correct. Table 6.14 contains the complete matrix of cophenetic values extracted from the dendrogram. We can obtain a visual impression of the degree of distortion in the dendrogram by plotting elements in the cophenetic value matrix against elements in the original correlation matrix (Figure 6.7). If the two matrices were identical, the plot would form a straight line. Deviations indicate distortions in the dendrogram; if a point falls above the line, the correlation expressed in the dendrogram is too high. Conversely, if a point falls below the line, averaging has resulted in a correlation which is lower than the true correlation. A numerical measure of the similarity between the two matrices can be found simply by computing the correlation between equivalent elements. Only one-half of the matrices, either above or below the diagonal, need be used, because both matrices are symmetrical about the diagonal. In our example, the correlation is $r = 0.98$.

The essential features of this particular method of cluster analysis can be summarized in list form:

1. The correlation coefficient is used as a similarity measure.
2. Highest similarities are clustered or linked first.
3. Two objects can be connected only if they have mutually highest correlations with each other.
4. After two objects are clustered, their correlations with all other objects are averaged.

An obvious modification to this scheme is to incorporate some other similarity measure. Although many measures have been proposed, only two are widely used:

TABLE 6.14 Matrix of Cophenetic Correlations Derived from Dendrogram in Figure 6.6

	A	B	C	D	E	F
A	1.00	0.57	0.29	−0.59	−0.59	−0.59
B	0.57	1.00	0.29	−0.59	−0.59	−0.59
C	0.29	0.29	1.00	−0.59	−0.59	−0.59
D	−0.59	−0.59	−0.59	1.00	0.66	0.41
E	−0.59	−0.59	−0.59	0.66	1.00	0.41
F	−0.59	−0.59	−0.59	0.41	0.41	1.00

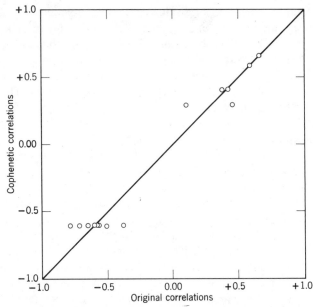

FIGURE 6.7 Plot of cophenetic values derived from dendrogram in Figure 6.6 against equivalent original correlations contained in Table 6.11. If dendrogram represented the structure of the correlation matrix exactly, all points would fall on diagonal line. Deviations from this line represent distortions in dendrogram.

the correlation coefficient and the distance coefficient. If the raw data are standardized prior to computing the similarity coefficient, correlations and distance coefficients can be directly transformed from one to another. Dendrograms constructed from the two measures generally are similar. However, the distance coefficient is not constrained within the range ± 1.0 as is the correlation coefficient, so it may produce more effective dendrograms if a few of the objects are very

TABLE 6.15 Similarity Measurements between Seven Objects; Upper Diagonal Half Contains Distance Coefficients, d_{ij} (in Parentheses); Lower Diagonal Half Contains Correlation Coefficients, r_{ij}

	A	B	C	D	E	F	G
A		(2.15)	(0.70)	(1.07)	(0.85)	(1.16)	(1.56)
B	−0.93		(1.53)	(1.14)	(1.38)	(1.01)	(2.83)
C	0.59	−0.44		(0.43)	(0.21)	(0.55)	(1.86)
D	−0.55	0.67	0.31		(0.29)	(0.22)	(2.04)
E	0.26	0.02	0.85	0.63		(0.41)	(2.02)
F	−0.79	0.94	−0.20	0.80	0.30		(2.05)
G	0.37	−0.64	−0.38	−0.90	−0.79	−0.82	

dissimilar from the other. Table 6.15 contains both distance and correlation coefficients for seven objects, in this instance, specimens of carbonate minerals. Variables include a number of physical properties measured on the minerals. The resulting dendrograms from each similarity matrix are shown in Figure 6.8. Although the general grouping is similar, two differences are apparent. The most obvious of these is the replacement of *B* in one of the central clusters by *D*, and the relegation of *B* to a position far down the hierarchical structure. It is worthwhile to examine the cause of this change.

Suppose we measure seven variables on each of three objects. These might be, for example, size measurements on three fossils, or chemical analyses of three rocks. If we plot each measurement as in Figure 6.9, we find that two of the objects

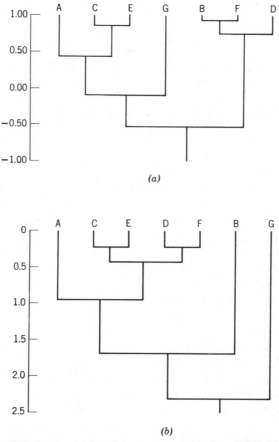

FIGURE 6.8 (a) Dendrogram constructed by weighted pair-group method, using arithmetic averaging of correlation coefficients. Original matrix given in Table 6.15. Cophenetic correlation coefficient = 0.77. (b) Dendrogram constructed by same method but using distance coefficients. Cophenetic correlation coefficient = 0.91.

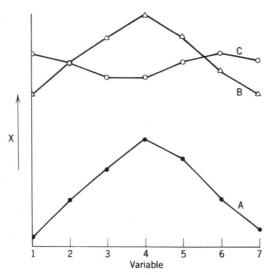

FIGURE 6.9 Plot of variables measured on three objects. Curves A and B are highly correlated but are separated by a large distance. Curves B and C are negatively correlated but are "close" in distance.

are similar in their relationships among the variables. These two will form a more or less parallel pattern, as do A and B in the diagram. The third may form a divergent pattern, but be much "closer" to the set of measurements of one of the other objects. In the illustration, A and B are highly correlated, or have a high linear relationship, but B and C have the smallest distance between them. If the variables are measurements of lengths on fossils, perhaps brachiopod shells, we would conclude that A and B are alike in their shapes, but B and C are nearly the same size. If the variables are the percentage of heavy elements in ore samples, we might conclude that samples A and B are compositionally alike, but A is dilute with respect to B. The ore concentrations of B and C are similar, but the chemical ratios are different.

The second characteristic in our list needs little explanation. The correlation coefficient indicates greatest similarity at high positive values, while the distance coefficient indicates greatest similarity by the smallest distance. Therefore, correlations must be linked or interconnected at high values, and distance coefficients must be linked at low values.

Our criterion for linking two objects to form a cluster requires that both have mutually the highest correlation with each other. Other criteria are possible. A simple method is known as single linkage clustering, which connects objects to clusters on the basis of the highest similarity between the object and any object in the cluster. Results of clustering the correlation matrix in Table 6.15 by single linkage are shown in Figure 6.10. Because objects are allowed to enter a cluster on the basis of their highest correlation with any object already in the cluster,

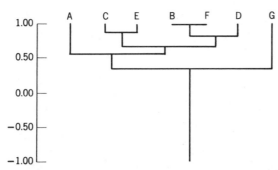

FIGURE 6.10 Dendrogram of correlation matrix in Table 6.15, clustered by single linkage method. Cophenetic correlation coefficient = 0.71.

linkage tends to occur at higher levels than in pair-group methods. In addition to the compression of the dendrogram, some connections are different. For example, cluster CE joins cluster BFD directly because of the high correlation between E and D. If correlations with C and E are averaged, the highest correlation is between CE and A.

This leads directly to the final characteristic, arithmetic averaging of the similarity measures of objects that have been clustered. In a single linkage method, no averaging is done at all. The methods illustrated in Figures 6.8a and 6.8b and in the initial example (Figure 6.6) are called averaged or weighted techniques, although the name perhaps should be equally weighted. In Figure 6.8a, C and E are joined at the onset of clustering. The correlations of the new cluster CE are found by combining the C and E rows and columns and dividing each entry by two. Next, object A enters the cluster and the correlations of the new cluster ACE are found by combining the CE cluster row and column with the A row and column and dividing by 2. That is, CE is treated as though it were a single object, when in fact it is two. The new object A has twice the influence on the correlations of the cluster ACE as do either E or C. Objects added late in the clustering procedure have a greater influence on the similarity matrix than do those linked earlier. Unweighted average or centroid methods attempt to avoid this by weighting each cluster proportionally to the number of objects within it during the averaging process. For example, having formed cluster CE, we could then link object A to form the new cluster ACE. However, the similarity measures of this new cluster would be found by summing the correlations of A with all elements except C and E, the correlations of C with all elements except A and E, and the correlations of E with all elements except A and C. That is, we would sum the correlations of all of the *original* elements in the cluster. Then, each sum is divided by three. This, in principle, gives each object in the cluster equal influence in the similarity characteristics of the whole cluster. This technique has the opposite drawback of weighted methods: late entries into a large cluster have almost no influence on the similarity measures of the cluster. Figure 6.11 shows a dendrogram of the correlation data in Table 6.15 clustered by unweighted averaging.

We can illustrate the effect of the four different linkage strategies by considering

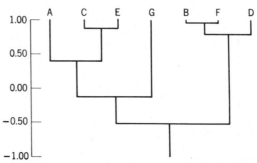

FIGURE 6.11 Dendrogram of correlation matrix in Table 6.15, clustered by unweighted averages method. Cophenetic correlation coefficient $= 0.72$.

a very simple clustering problem where only two variables are measured on each object. Then, all relationships between the objects can be shown in two dimensions, as in Figure 6.12. The distances between objects on the diagram are directly proportional to the degree of dissimilarity between them. Four of the objects, A through D, form a tight cluster. The dotted lines indicate the order in which the four have been joined together. A somewhat less similar object, E, also has been joined into the cluster. A sixth object, labelled F, is now being considered for possible inclusion in the growing cluster. The point M_1 is the centroid of points A through E, and M_2 is the average of object E and the previous cluster average.

Using a single linkage criterion, object F will be joined to the cluster if the distance CF is smaller than the distance to any other object in any other cluster.

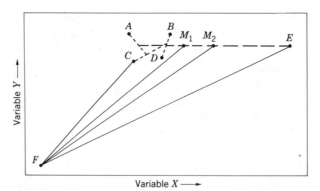

FIGURE 6.12 Diagram representing manner in which objects characterized by two variables, X and Y, enter a cluster. Objects A, B, C, and D form a cluster. Object E has joined this cluster, and object F is a candidate to join on the next iteration. M_1 is the centroid of objects A through E. M_2 is the average of object E and the last average of objects A through D.

In an unweighted average or centroid linkage, object F will join the cluster if the distance M_1F is smaller than the distance to the centroid of any other group. In weighted pair-group or average linkage, the candidate object F will join if the distance M_2F is smaller than the distance to the average of any other cluster. (Note that the point M_2 is half-way between the average of the cluster $ABCD$ and object E, which was admitted on the previous cycle.) Finally, with complete linkage, object F will join the cluster if the distance EF is less than the distance to the most distant point in any other cluster.

Faced with such a welter of alternative methods, all yielding slightly different results, a researcher may justifiably ask which is best. Unfortunately, there is no clear answer to this crucial question. Experience suggests that weighted pair-group methods tend to be superior to either single-linkage or unweighted average methods. Relative superiority is defined by the tendency to produce the highest cophenetic correlation coefficient, which is interpreted as indicating low distortion in the dendrogram. Cophenetic correlations below about 0.8 may indicate such severe distortion in the dendrogram at lower linkages that the diagram is misleading. Jardine and Sibson (1971), however, argued for single-linkage as the only procedure yielding entirely reproduceable results. Distance matrices usually cluster more successfully than do correlation matrices, in the sense that they yield higher cophenetic correlations. Distance matrices also seem to be less susceptible to drastic changes among different clustering methods. However, no statistical tests are available for hierarchical clustering, nor has any statistical theory been developed and applied. [For other methods of clustering, a certain amount of theoretical justification has been developed. For examples, see Switzer (1970) and Hartigan (1975).] Most researchers who use clustering methods experiment with a variety of similarity measures and clustering techniques, and they choose the combination that yields the most satisfactory results with their data.

A pragmatic consideration may dictate the choice of a clustering procedure. Most hierarchical techniques may require creation and manipulation of an inordinately large matrix of similarities if the number of objects is large. (In the fields of ecology and archeology, studies involving thousands of objects are not unusual.) Clustering procedures using a limited number of arbitrary cluster centers were devised to offset this computational difficulty. Probably the most widely used of these is the *k-means procedure* of McQueen (1967). Here, k points characterized by m variables are designated (either by the user or arbitrarily by the program) as initial "centroids" of clusters. A matrix of similarities between the k "centroids" and the n observations is calculated, and the closest or most similar observations are clustered with the nearest centroids. New centroids are then calculated and the process iterates exactly like a hierarchical procedure. In principle, the centroid will rapidly shift toward the true center of a growing cluster, as the influence of the true observations overwhelms that of the arbitrary starting point. The advantage of the k-means procedure is that only a $k \times m$ matrix of similarities is necessary, rather than an $m \times m$ matrix. If k is small (5 to 10) and n is large (1000 or more), the process may be faster than a hierarchical method by more than two orders of magnitude. The disadvantage of k-means is that a suboptimal clustering may result if the

TABLE 6.16 Ten Variables Measured as Ratios on Ten Species of Cambrian Trilobites Collected in Utah[a]

Species	X_1	X_2	X_3	X_4	X_5	X_6	X_7	X_8	X_9	X_{10}
Aphelaspis brachyphasis	0.208	0.250	0.542	0.237	0.875	0.292	0.284	0.925	0.343	0.373
A. haguei	0.318	0.318	0.545	0.428	1.000	0.318	0.296	0.796	0.444	0.537
A. subditus	0.174	0.304	0.391	0.375	0.913	0.304	0.297	0.946	0.405	0.486
Dicanthopyge convergens	0.259	0.370	0.370	0.859	0.852	0.333	0.500	0.591	0.591	0.818
D. quadrata	0.250	0.350	0.500	0.615	0.900	0.351	0.434	0.783	0.478	0.652
D. reductus	0.316	0.421	0.474	0.736	1.158	0.421	0.500	0.675	0.500	0.775
Prehousia alata	0.136	0.409	0.273	0.469	1.000	0.136	0.269	0.769	0.327	0.423
P. indenta	0.192	0.308	0.269	0.628	0.923	0.154	0.308	0.795	0.308	0.436
P. prima	0.261	0.261	0.261	0.545	0.956	0.261	0.296	0.833	0.333	0.407
A. longispina	0.259	0.370	0.556	0.444	0.852	0.296	0.372	0.824	0.431	0.706

[a]Key: C = correction for size of glabella = length of glabella; X_1 = length of border/C; X_2 = length of brim/C; X_3 = length of palpebral lobes/C; X_4 = width of glabella./C; X_5 = width of fixed cheek/C; X_6 = length of genal spine/length of free cheek; D = correction for size of pygidium = width of pygidium; X_7 = width of pygidial axis/D; X_8 = width of pleural axes/D; X_9 = length of pygidial axis/D; X_{10} = length of pygidium/D.

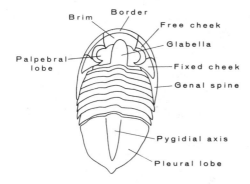

Brim
Border
Free cheek
Glabella
Palpebral
lobe
Fixed cheek
Genal spine
Pygidial axis
Pleural lobe

FIGURE 6.13 An opisthoparian trilobite, showing anatomical parts measured to create variables listed in Table 6.16.

arbitrary starting points do not fall within divergent clusters. This may lead to premature merger of the centroids and failure to detect outlying clusters.

The many considerations that enter into the choice of a clustering procedure introduce a certain subjectivity into a process designed to promote objectivity, but the cophenetic correlation provides a measure of guidance. The great benefits of cluster analysis are that it provides a relatively simple and direct way to classify objects, and it presents results in a manner that is both familiar and easy to understand.

As an exercise in clustering, we can examine a collection of data measured on Cambrian trilobites collected in the Great Basin region of western United States. The specimens have been assigned to three genera by conventional taxonomic procedures. Ten characters or variables have been measured on ten trilobites, each of which represents a separate species. Data are listed in Table 6.16. Commonly, trilobites are found disaggregated. To avoid the confusion that would result if a pygidium from a large individual were accidently associated with the cephelon of a small individual, all of the measurements have been converted to ratios. All measurements made on the glabella, or head, have been divided by the length of the glabella. Similarly, all measurements made on the pygidium, or tail, are divided by the width of the pygidium. The anatomical parts used in the variables are illustrated in Figure 6.13. Perform a cluster analysis on the trilobite data, after appropriate standardization, and see if numerical methods recover the same classification as that found by conventional taxonomy. Compute and utilize both correlation and distance as similarity measures. Which performs ''best'' in the sense of conforming to conventional taxonomy?

Introduction to Eigenvector Methods, Including Factor Analysis

A number of related computational procedures are loosely referred to as ''factor analysis''; they share the common objective of attempting to reveal a simple underlying structure that is presumed to exist within a set of multivariate observations. This structure is expressed in the pattern of variances and covariances between variables, and the similarities between observations. Factor methods all operate by extracting the eigenvalues and eigenvectors from a square matrix produced by multiplying a data matrix (or some transformation of a data matrix) by its transpose. The basic mathematical operation is exactly the same as that demonstrated in Chapter 3. Upon this basic objective and methodology there is superimposed an array of embellishments and variations, often of an arbitrary nature, which has resulted in a bewildering assortment of computational techniques. Often the basic similarities among these methods are obscured by the complicated mathematical notation and terminology used by the various practitioners.

Factor analysis was originally developed by experimental psychologists in the 1930s, and much factor terminology has meaning only within the context of this field. Indeed, the very name "factor" alludes to hypothetical mental attributes, referred to as "factors of the mind." Sociologists and biometricians also have contributed to the richness of the jargon of factor analysis, helping to create a controversial and poorly understood methodology that extends the beguiling promise of instant insight to the researcher faced with more data than comprehension.

Dozens of methodological variants of factor analysis have been developed. We will examine some of the most widely used of these, but will not dwell deeply on the philosophical or mathematical arguments that accompany them. Rather, we will explore some of the mathematical relationships that exist between a data matrix, its matrices of cross-products, and their eigenvalues and eigenvectors.

Factor methods can be divided into two broad classes, called *R-mode* and *Q-mode* techniques. The first is concerned with interrelations between variables, and operates by extracting eigenvalues and eigenvectors from a covariance or correlation matrix. The second is concerned with the relationships between objects, often as an attempt to discern patterns or groupings within their arrangement in multivariate space. Most *Q*-mode analyses proceed by extracting the eigenvalues and eigenvectors from a matrix of similarities between all possible pairs of objects. *R*-mode techniques are statistical procedures, in the sense that the data are regarded as samples taken from a much larger population, and the results pertain to the general properties or behavior of the variables. Because *Q*-mode methods focus on the similarities between *individuals* in the data set, they are not usually amenable to statistical analysis.

The first basic step in either *R*-mode or *Q*-mode factor analysis is to convert the data matrix into a square, symmetric matrix that expresses either the degrees of interrelationships between the variables, or between the objects on which these variables are measured. This is done by pre- or post-multiplying the data matrix by its transpose. In the simplest case, when a matrix of raw data $[X]$ consisting of n rows of observations and m columns of variables is pre-multiplied by its transpose $[X]'$, the result is a square matrix $[R]$, which is $m \times m$.

$$[R] = [X]'[X]$$

The elements of $[R]$ will consist of the raw sums of squares and cross-products of the m variables. That is,

$$r_{jk} = \sum_{i=1}^{n} x_{ij} x_{ik}$$

where j and k are two columns of the data matrix. If the data are standardized so each variable has a mean of zero and a standard deviation of one, the matrix $[R]$ will contain the correlation coefficients between the m variables.

Instead, the data matrix $[X]$ may be post-multiplied by its transpose $[X]'$, to yield the square symmetric matrix $[Q]$, which has n rows and n columns.

$$[Q] = [X][X]'$$

If $[X]$ contains raw observations, $[Q]$ contains the squares and cross-products for all pairs of objects, summed across variables. In effect, the elements of $[Q]$ are

$$q_{il} = \sum_{j=1}^{m} x_{ij} x_{lj}$$

where i and l are two rows of the data matrix. In most investigations, we have many more objects than variables, so $[Q]$ may be much larger than $[R]$, even though both are created from the same original data matrix $[X]$.

Most geological applications of factor analysis have then proceeded to extract eigenvalues and eigenvectors from either $[R]$ or $[Q]$. However, it is obvious that there must be a close link between the two, since both are generated from the same data set. This link was established in the early days of factor analysis, but its implications have been overlooked until very recent times. In part, this is because the psychologists and sociologists who were responsible for most of the pioneering work in factor analysis were concerned exclusively with R-mode studies. Not until biologists and geologists became interested in factor analysis were Q-mode techniques widely used, and most of these were direct adaptations of R-mode methods in which $[Q]$ was simply substituted for $[R]$. Since $[Q]$ often may be an enormous matrix, these direct approaches were not feasible until large, general-purpose computers became available in the late 1950s. Unfortunately, the basic relationship between $[R]$ and $[Q]$ was overlooked, and the methods of Q-mode factor analysis often were unduly complex (for example, see Imbrie and Purdy, 1962; Ondrick and Srivastava, 1970). Recent authors have exploited the duality between the R and Q modes, achieving a great simplification in the computations of Q-mode factor analysis (Jöreskog, Klovan, and Reyment, 1976; David, Dagbert, and Beauchemin, 1977; Zhou, Chang, and Davis, 1983).

Eckart-Young Theorem

The critical interrelationships between a data matrix and the eigenvalues and eigenvectors of its two cross-product matrices are expressed in the *Eckart-Young theorem*, first given by these two authors in their classic article in the first volume of *Psychometrika*, which appeared in 1936. The Eckart-Young theorem is the cornerstone of several multivariate techniques, including factor analysis. It states that for any real matrix $[X]$, two orthogonal matrices $[V]$ and $[U]$ can be found for which the product $[V]'[X][U]$ is a real diagonal matrix $[\Lambda]$ with no negative elements. A proof of the theorem is provided by Johnson (1963); we will examine the consequences of this theorem as it pertains to factor analysis, adopting a numerical example originally devised by Burt (1937).

TABLE 6.17 Measurements, in Millimeters, Made on Four Specimens of Goniatite Ammonoids, Representing Species of the Genus *Manticoceras*

	Umbilical Diameter	Height of Whorl	Width of Whorl
	X_1	X_2	X_3
Species *A*	4	27	18
Species *B*	12	25	12
Species *C*	10	23	16
Species *D*	14	21	14
	$\bar{X}_1 = 10$	$\bar{X}_2 = 24$	$\bar{X}_3 = 15$

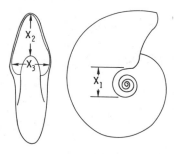

The measurements in Table 6.17 are typical of those that might be obtained in a simple geological investigation. We will pretend they are measurements made by a paleontologist on the shells of four goniatite ammonoid specimens, representing species of the genus *Manticoceras*. The variables include diameter of the umbilicus or exposed part of the inner whorls, height of the outer whorl at the peristome or shell opening, and width of the outer whorl at the peristome. The mean of each variable may be subtracted from each observation to simplify calculations. The resulting data matrix is

$$[X] = \begin{bmatrix} -6 & 3 & 3 \\ 2 & 1 & -3 \\ 0 & -1 & 1 \\ 4 & -3 & -1 \end{bmatrix}$$

Rearranging the matrices in the Eckart-Young theorem shows that the data matrix can be viewed as the product of three other matrices

$$[X] = [V][\Lambda][U]' \qquad (6.40)$$

where $[V]$ is an $n \times r$ matrix whose columns are *orthonormal*. This means that $[V]'[V] = [I]$, where $[I]$ is $r \times r$. Likewise, $[U]$ is an $m \times r$ matrix whose columns are orthonormal, so $[U]'[U] = [I]$, where $[I]$ is also $r \times r$. $[\Lambda]$ is an $r \times r$ square matrix containing r positive elements along the diagonal. These are called the *singular values* of $[X]$; all off-diagonal elements of $[\Lambda]$ are zero.

The minor product matrix $[R] = [X]'[X]$ is of size $m \times m$ and has r nonzero eigenvalues and $m - r$ eigenvalues that are equal to zero. The nonzero eigenvalues are equal to the square of the singular values in matrix $[\Lambda]$. That is,

$$[\Lambda]^2 = [I][\lambda]'$$

or equivalently (6.41)

$$[\Lambda] = [I][\sqrt{\lambda}]'$$

where $[\lambda]$ is a vector containing the r nonzero eigenvalues of $[R]$. The major product matrix $[Q] = [X][X]'$ is of size $n \times n$ but also has only r nonzero eigenvalues. These are identical to the eigenvalues extracted from $[R]$, except there are additional eigenvalues that are all equal to zero if n, the number of objects, is larger than m, the number of variables.

Furthermore, the columns of the matrix $[U]$ contain the eigenvectors of $[R]$ that are associated with each eigenvector λ. The columns of $[V]$ contain the eigenvectors from $[Q]$. Since the eigenvalues from both $[R]$ and $[Q]$ are identical, there must be a relationship between the two sets of eigenvectors $[U]$ and $[V]$. This relationship is

$$[V] = [X][U][\Lambda]^{-1}$$

or (6.42)

$$[U] = [X]'[V][\Lambda]^{-1}$$

In a factor analytic context, the vector formed by multiplying an eigenvector by its corresponding singular value is referred to as a *factor*. Recall that eigenvectors are calculated so that the sum of their squared elements is equal to 1.0. That is, the eigenvectors are of unit length. If they are multiplied by their corresponding singular values (or square roots of their eigenvalues), they are scaled so their lengths are proportional to the magnitudes of their singular values. The individual elements of a factor are called *loadings* and relate the factor to the original variables. In matrix notation, the R-mode factors are

$$[A^R] = [U][\Lambda]$$ (6.43)

The loadings represent the proportion or weighting that must be assigned to each variable in order to project the objects onto the factor axes as scores. They also represent the correlations of the individual variables with the factors. The corresponding equation for the factors in Q-mode analysis is

$$[A^Q] = [V][\Lambda]$$ (6.44)

and the loadings are the proportions of each individual object necessary to project the variables onto the factor axes.

R-mode scores are found by multiplying the data by the factor loadings, or

$$[S^R] = [X][A^R]$$ (6.45)

which project the n individual objects onto the factor axes. For a specific observation i,

$$s_{ik} = \sum a_{mk} x_{mi}$$

or

$$s_{ik} = a_{1k} x_{1i} + a_{2k} x_{2i} + a_{3k} x_{3i} + \cdots + a_{mk} x_{mi}$$

where s_{ik} is the score of the ith observation on the kth factor, x_{mi} is the value of variable m measured on object i, and a_{mk} is the loading of variable m on factor k. In turn, a_{mk} is the product of element m of the kth eigenvector, times the square root of the kth eigenvalue.

In a similar manner, Q-mode scores are found by multiplying the transpose of the data matrix by the Q-mode factor loadings:

$$[S^Q] = [X]'[A^Q] \tag{6.46}$$

This equation will project the m variables onto the factor axes.

Some algebraic rearrangements will demonstrate the relationship between factor loadings and scores in R and Q modes. Equation (6.43) defines R-mode factor loadings as

$$[A^R] = [U][\Lambda]$$

and the Eckart-Young theorem defines $[U]$ as

$$[U] = [X]'[V][\Lambda]^{-1}$$

Multiplying both sides by $[\Lambda]$,

$$[U][\Lambda] = [X]'[V][\Lambda]^{-1}[\Lambda]$$

$$[A^R] = [X]'[V]$$

Q-mode scores are defined by eq. (6.46) as

$$[S^Q] = [X]'[A^Q]$$

and the Q-mode factor loadings $[A^Q]$ are defined as

$$[A^Q] = [V][\Lambda]$$

Substituting,

$$[S^Q] = [X]'[V][\Lambda]$$

But, $[X]'[V] = [A^R]$, so

$$[S^Q] = [A^R][\Lambda] \tag{6.47}$$

Similar manipulations will show that

$$[S^R] = [A^Q][\Lambda] \tag{6.48}$$

So, Q-mode scores are proportional R-mode loadings, and vice-versa. The constant of proportionality is equal to $[\Lambda]$, the singular values. Equivalent expressions are

$$[A^R] = [S^Q][\Lambda]^{-1} \tag{6.49}$$

and

$$[A^Q] = [S^R][\Lambda]^{-1} \tag{6.50}$$

This means that if we perform an R-mode factor analysis, we can also automatically perform a Q-mode analysis, since both the Q-mode loadings and scores can be obtained from the R-mode solution.

We can illustrate these relationships using measurements on ammonoid shells introduced earlier. The R-mode, or minor product matrix, is obtained by premultiplying the data matrix by its transpose:

$$[X]'[X] = [R]$$

$$
\begin{bmatrix}
-6 & 2 & 0 & 4 \\
3 & 1 & -1 & -3 \\
3 & -3 & 1 & -1
\end{bmatrix}
\cdot
\begin{bmatrix}
-6 & 3 & 3 \\
2 & 1 & -3 \\
0 & -1 & 1 \\
4 & -3 & -1
\end{bmatrix}
=
\begin{bmatrix}
56 & -28 & -28 \\
-28 & 20 & 8 \\
-28 & 8 & 20
\end{bmatrix}
$$

The eigenvalues of $[R]$ are $\lambda_1 = 84$, $\lambda_2 = 12$, $\lambda_3 = 0$. Since the final eigenvalue is zero, the matrix $[\Lambda]^2$ has a rank of only two rather than three. That is,

$$
[\Lambda]^2 =
\begin{bmatrix}
84 & 0 \\
0 & 12
\end{bmatrix}
$$

so

$$
[\Lambda] =
\begin{bmatrix}
9.165 & 0.0 \\
0.0 & 3.464
\end{bmatrix}
$$

The eigenvectors of $[R]$ are

$$
[U] =
\begin{bmatrix}
0.8165 & 0.0 & 0.0 \\
-0.4082 & 0.7071 & 0.0 \\
-0.4082 & -0.7071 & 0.0
\end{bmatrix}
$$

Because the final eigenvalue is zero, the last column of $[U]$ disappears, leaving the 3×2 matrix

$$
[U] =
\begin{bmatrix}
0.8165 & 0.0 \\
-0.4082 & 0.7071 \\
-0.4082 & -0.7071
\end{bmatrix}
$$

The matrix of R-mode factor loadings $[A^R]$ is given by eq. (6.43):

$$
\begin{bmatrix}
0.8165 & 0.0 \\
-0.4082 & 0.7071 \\
-0.4082 & -0.7071
\end{bmatrix}
\cdot
\begin{bmatrix}
9.165 & 0.0 \\
0.0 & 3.464
\end{bmatrix}
=
\begin{bmatrix}
7.4832 & 0.0 \\
-3.7412 & 2.4494 \\
-3.7412 & -2.4494
\end{bmatrix}
$$

We can now project the four specimens onto the R-mode factor axes by computing their factor scores by eq. (6.45).

$$
\begin{bmatrix} -6 & 3 & 3 \\ 2 & 1 & -3 \\ 0 & -1 & 1 \\ 4 & -3 & -1 \end{bmatrix} \cdot \begin{bmatrix} 7.4832 & 0.0 \\ -3.7412 & 2.4494 \\ -3.7412 & -2.4494 \end{bmatrix} = \begin{bmatrix} -67.3 & 0.0 \\ 22.4 & 9.8 \\ 0.0 & -4.9 \\ 44.9 & -4.9 \end{bmatrix}
$$

The scores can be graphically displayed by plotting them in the space defined by the orthogonal factor axes. Figure 6.14 shows the four ammonoid specimens plotted on the first and second factors.

A Q-mode factor analysis begins by post-multiplying the data matrix by its transpose:

$$[X][X]' = [Q]$$

$$
\begin{bmatrix} -6 & 3 & 3 \\ 2 & 1 & -3 \\ 0 & -1 & 1 \\ 4 & -3 & -1 \end{bmatrix} \cdot \begin{bmatrix} -6 & 2 & 0 & 4 \\ 3 & 1 & -1 & -3 \\ 3 & -3 & 1 & -1 \end{bmatrix} = \begin{bmatrix} 54 & -18 & 0 & -36 \\ -18 & 14 & -4 & 8 \\ 0 & -4 & 2 & 2 \\ -36 & 8 & 2 & 26 \end{bmatrix}
$$

The matrix $[V]$ can be converted into the matrix of Q-mode factor loadings by eq. (6.44):

$$[V][\Lambda] = [A^Q]$$

$$
\begin{bmatrix} -0.8018 & 0.0 \\ 0.2673 & 0.8165 \\ 0.0 & -0.4082 \\ 0.5345 & -0.4082 \end{bmatrix} \cdot \begin{bmatrix} 9.165 & 0.0 \\ 0.0 & 3.464 \end{bmatrix} = \begin{bmatrix} -7.3485 & 0.0 \\ 2.4498 & 2.8284 \\ 0.0 & -1.4140 \\ 4.8987 & -1.4140 \end{bmatrix}
$$

Q-mode scores are the projections of variables onto the factor axes and are found by multiplying the transpose of the data matrix by the factor loadings.

$$[X]'[A^Q] = [S^Q]$$

$$
\begin{bmatrix} -6 & 2 & 0 & 4 \\ 3 & 1 & -1 & -3 \\ 3 & -3 & 1 & -1 \end{bmatrix} \cdot \begin{bmatrix} -7.3485 & 0.0 \\ 2.4498 & 2.8284 \\ 0.0 & -1.4140 \\ 4.8987 & -1.4140 \end{bmatrix} = \begin{bmatrix} 68.6 & 0.0 \\ -34.3 & 8.5 \\ -34.3 & -8.5 \end{bmatrix}
$$

Figure 6.15 is a plot of the three variables measured on the ammonoid specimens projected onto the plane defined by the first two Q-mode factors.

We may now confirm that Q-mode scores are proportional to R-mode loadings using eq. (6.47):

$$[A^R][\Lambda] = [S^Q]$$

$$
\begin{bmatrix} 7.4832 & 0.0 \\ -3.7412 & 2.4494 \\ -3.7412 & -2.4494 \end{bmatrix} \cdot \begin{bmatrix} 9.165 & 0.0 \\ 0.0 & 3.464 \end{bmatrix} = \begin{bmatrix} 68.6 & 0.0 \\ -34.3 & 8.5 \\ -34.3 & -8.5 \end{bmatrix}
$$

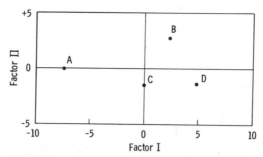

FIGURE 6.14 *R*-mode factor scores of four amminoid species. Horizontal axis is Factor I. Vertical axis is Factor II.

Finally, we may demonstrate the Eckart-Young theorem by recreating the 4 ×
3 data matrix [*X*] from its orthonormal parts:

$$[X] = [V][\Lambda][U]'$$

$$= \begin{bmatrix} -0.8018 & 0.0 \\ 0.2673 & 0.8165 \\ 0.0 & -0.4082 \\ 0.5345 & -0.4082 \end{bmatrix} \cdot \begin{bmatrix} 9.165 & 0.0 \\ 0.0 & 3.464 \end{bmatrix} \cdot \begin{bmatrix} 0.8165 & -0.4082 & -0.4082 \\ 0.0 & 0.7071 & -0.7071 \end{bmatrix}$$

$$= \begin{bmatrix} -6 & 3 & 3 \\ 2 & 1 & -3 \\ 0 & -1 & 1 \\ 4 & -3 & -1 \end{bmatrix}$$

In this simple numerical example, we have taken a multivariate set of observations
and have resolved it into a smaller number of factors. We have also shown that
R-mode solutions are equivalent to those obtained by *Q*-mode analysis. Both of
these are critical points and will be referred to repeatedly in the following discus-

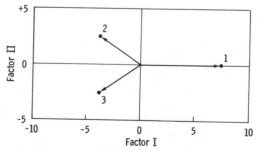

FIGURE 6.15 *Q*-mode factor scores of three
variables measured on amminoid species. Horizontal axis is Factor I. Vertical axis is Factor II.

sions where we will investigate some of the elaborations that are placed upon the relatively simple structure we have just examined.

As noted at the beginning of this section, "factor analysis" is a catchall term including a multitude of techniques that involve the extraction of eigenvalues and eigenvectors from a matrix of the cross-products of a data set. Factor analysis is also used (more correctly) in a strict sense to mean a statistical procedure by which a data matrix is decomposed into a prescribed number of uncorrelated factors and a residual set of "unique," random variations. Other important eigenvalue methods include principal components analysis (PCA), correspondence analysis, and in the Q mode, principal vector and principal coordinates analysis.

The characteristics of the various eigenvalue procedures will be illustrated with an artificial data set similar to the one published by Cooley and Lohnes (1971, pp. 133–136). We are all familiar with the problem of specifying the "size" of something. Should we measure a length, a width, an area, a volume, or some ratio of these? How is the concept of "size" to be separated from that of "shape"? To investigate these questions, we have created a set of 25 objects in the form of

TABLE 6.18   Measurements on 25 Blocks Having Randomly Generated Dimensions[a]

	X_1	X_2	X_3	X_4	X_5	X_6	X_7
a	3.760	3.660	0.540	5.275	9.768	13.741	4.782
b	8.590	4.990	1.340	10.022	7.500	10.162	2.130
c	6.220	6.140	4.520	9.842	2.175	2.732	1.089
d	7.570	7.280	7.070	12.662	1.791	2.101	0.822
e	9.030	7.080	2.590	11.762	4.539	6.217	1.276
f	5.510	3.980	1.300	6.924	5.326	7.304	2.403
g	3.270	0.620	0.440	3.357	7.629	8.838	8.389
h	8.740	7.000	3.310	11.675	3.529	4.757	1.119
i	9.640	9.490	1.030	13.567	13.133	18.519	2.354
j	9.730	1.330	1.000	9.871	9.871	11.064	3.704
k	8.590	2.980	1.170	9.170	7.851	9.909	2.616
l	7.120	5.490	3.680	9.716	2.642	3.430	1.189
m	4.690	3.010	2.170	5.983	2.760	3.554	2.013
n	5.510	1.340	1.270	5.808	4.566	5.382	3.427
o	1.660	1.610	1.570	2.799	1.783	2.087	3.716
p	5.900	5.760	1.550	8.388	5.395	7.497	1.973
q	9.840	9.270	1.510	13.604	9.017	12.668	1.745
r	8.390	4.920	2.540	10.053	3.956	5.237	1.432
s	4.940	4.380	1.030	6.678	6.494	9.059	2.807
t	7.230	2.300	1.770	7.790	4.393	5.374	2.274
u	9.460	7.310	1.040	11.999	11.579	16.182	2.415
v	9.550	5.350	4.250	11.742	2.766	3.509	1.054
w	4.940	4.520	4.500	8.067	1.793	2.103	1.292
x	8.210	3.080	2.420	9.097	3.753	4.657	1.719
y	9.410	6.440	5.110	12.495	2.446	3.103	0.914

[a]See text for listing of variables.

rectangular blocks. The three dimensions of the blocks were chosen randomly and range up to 10 units. All shapes and sizes are equally probable in the resulting collection, from a cube less than one unit on a side, through rodlike prisms and flat plates, to a cube $10 \times 10 \times 10$. A variety of measurements have been made on each of the blocks, and these constitute our variables. These are

$X_1 =$ long axis

$X_2 =$ intermediate axis

$X_3 =$ short axis

$X_4 =$ longest diagonal

$X_5 =$ ratio $\dfrac{\text{radius of smallest circumscribed sphere}}{\text{radius of largest inscribed sphere}}$

$X_6 =$ ratio $\dfrac{\text{long axis } + \text{ intermediate axis}}{\text{short axis}}$

$X_7 =$ ratio $\dfrac{\text{surface area}}{\text{volume}}$

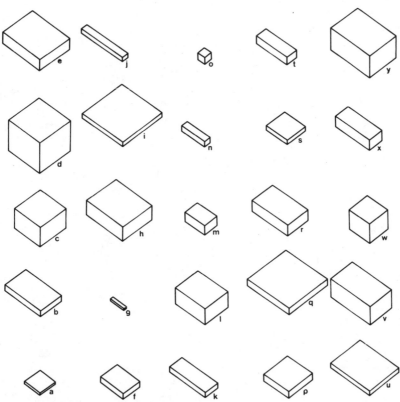

FIGURE 6.16 Twenty-five blocks with random lengths, widths, and heights.

Table 6.18 contains the 25 observations for the seven variables. The collection of blocks is shown in Figure 6.16. Note that this data set has some interesting properties; it should possess at most only three independent dimensions, as variables X_4 through X_7 have been created by various combinations of the length, width, and height measurements. Also, the data contain a certain amount of induced correlation caused by the inherent nature of the variables. The long axis of each block, by definition, must be longer than the intermediate axis, which in turn must be longer than the short axis. This means that if, for example, length and width are plotted against each other, the scatter of points is confined to the lower diagonal half of the diagram (Fig. 6.17). This induces a positive correlation that is significantly greater ($r = 0.58$) than that expected from two independent variables.

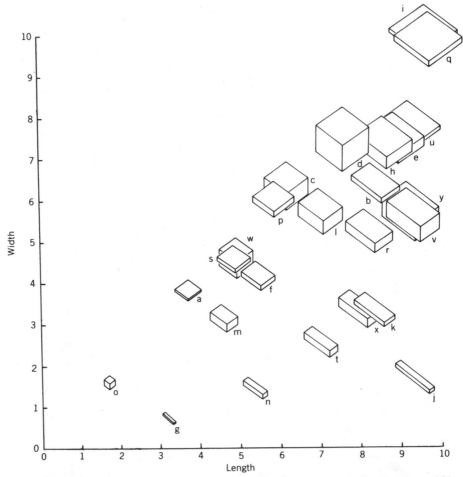

FIGURE 6.17   Cross-plot of lengths versus widths of random blocks. Because width must be less than length, all blocks plot in lower diagonal half of diagram. This induces a correlation of $r = 0.58$ between length and width. Length and width are given in arbitrary units.

Of course, we ordinarily would not collect measurements that we know to be dependent upon one another. Unfortunately, it happens more often than not that geologic variables are interrelated; compositional variables contain induced correlations because they are parts of a whole, and taxonomic measurements may be interrelated because of the effect of size. In this artificial example, we know what interdependencies exist, and we can expect to see certain consequences of these interdependencies in the outcomes of our analyses. This may help us understand the results that we obtain from analyses of real data, where the possibilities of similar interdependencies cannot be established before the fact.

Principal Components Analysis

The first major procedure that will be considered in this section is called principal components analysis, or PCA. Principal components are nothing more than the eigenvectors of a variance–covariance or a correlation matrix. By themselves they may provide significant insight into the structure of the matrix, and they often may be interpreted in much the same manner as factors. Many factor-analytic schemes employ principal components as starting points for the analysis. For this reason, and because their derivation and interpretation are more straightforward, we will begin with a discussion of principal components. Also, as we have noted, geologists have been rather confused in their use of terminology; most of the published studies that geologists have called "factor analyses" actually are principal components analyses. The "factors" cited in these articles properly should be called compo-

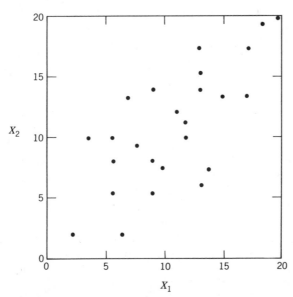

FIGURE 6.18 Scatter diagram of bivariate data from Table 6.19.

TABLE 6.19 Bivariate Observations with Variance of $X_1 = 20.3$, Variance of $X_2 = 24.1$, and Covariance 15.6

X_1	X_2	X_1	X_2
3	2	12	10
4	10	12	11
6	5	13	6
6	8	13	14
6	10	13	15
7	2	13	17
7	13	14	7
8	9	15	13
9	5	17	13
9	8	17	17
9	14	18	19
10	7	20	20
11	12		

nents. Some authors have found the confusion so great that they have resorted to labelling factor analysis as "*true* factor analysis" to distinguish it from components analysis (Jöreskog, Klovan, and Reyment, 1976).

Suppose we measure two variables on a collection of objects, perhaps lengths and widths of brachiopod shells, and obtain the data shown plotted in Figure 6.18 and listed in Table 6.19. The variance of X_1 is 20.3, the variance of X_2 is 24.1, and the covariance between the two is 15.6. These constitute the elements of the variance–covariance matrix, which is

$$[s^2] = \begin{bmatrix} 20.3 & 15.6 \\ 15.6 & 24.1 \end{bmatrix}$$

You will recall from Chapter 3 that a matrix can be expressed in geometric form as a series of vectors in multidimensional space. We regard each row of the matrix as giving the coordinates of the end-point of the vector which represents that row. A 2×2 matrix can be plotted on a flat diagram, as in Figure 6.19. Furthermore, these vectors can be considered to define arbitrary axes of an m-dimensional ellipsoid. The eigenvectors of the matrix yield the orientations of the principal axes of the ellipsoid, and the eigenvalues represent the lengths of each of the successive principal semiaxes. In Chapter 3, we interpreted arbitrary matrices in this manner, but it is obvious that variance–covariance matrices can be interpreted geometrically with equal facility. The PCA method is concerned with finding these axes and measuring their magnitudes.

If we measure m variables on a collection of objects, we can compute an $m \times m$ matrix of variances and covariances, $[s^2]$. From $[s^2]$, we can extract m eigenvalues and m eigenvectors. Because a variance–covariance matrix is symmetrical, the m eigenvectors will be mutually orthogonal.

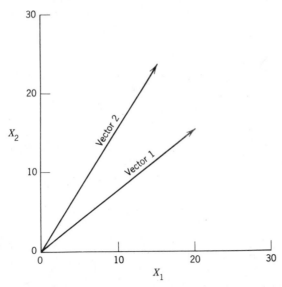

FIGURE 6.19 Vector representation of variances and covariances in 2 × 2 matrix. Vector 1 represents row 1 of matrix; vector 2 represents row 2.

We can compute the eigenvalues and eigenvectors of our 2 × 2 matrix $[s^2]$ and plot the vectors. The first eigenvector, expressed in standardized form is

$$I = \begin{bmatrix} 0.66 \\ 0.75 \end{bmatrix}$$

which means that the vector corresponding to the first principal axis of the ellipsoid slopes 0.66 units in s_1^2 for every 0.75 units in s_2^2. The first eigenvalue is 37.9, which we can plot as the length of the principal semiaxis. The second eigenvector is

$$II = \begin{bmatrix} 0.75 \\ -0.66 \end{bmatrix}$$

which you will recognize as being at right angles to the first. The second eigenvalue is 6.5, which we can use as the length of the second principal semiaxis. These geometric relationships are shown in Figure 6.20. Keep in mind that we have plotted vectors from a variance–covariance matrix, so the measurements on the diagram are given in the same units as the variances, or in this instance, units of length^2.

We can define the total variance in a data set as the sum of the individual variances. Because the variances are located along the diagonal of the variance–covariance matrix, this is equivalent to finding the trace of the matrix. In our example, the total variance is 20.3 + 24.1 = 44.4. The first variable contributes 20.3/44.4 or about 46% of the total variance, and the second contributes the remainder, about 54%. You will recall from Chapter 3 that the sum of the eigenvalues of a matrix is equal to the trace of the matrix, so the total of our two

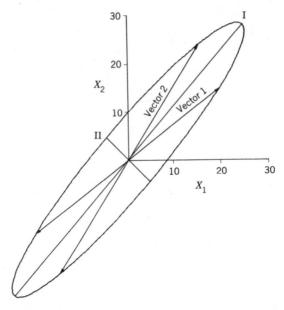

FIGURE 6.20 Ellipse defined by variances and covariances of data in Table 6.19, shown plotted in Figure 6.18. First principal component accounts for 86% of total variance, second principal component for 14%.

TABLE 6.20 Principal Components Scores of Data from Table 6.19, Calculated by Projecting Original Data onto Principal Axes; Variance of $Y_1 = 37.9$, Variance of $Y_2 = 6.5$

Y_1	Y_2	Y_1	Y_2
3.49	0.92	15.44	2.35
10.14	-3.64	16.19	1.69
7.72	1.18	13.11	5.75
9.97	-0.81	19.10	0.45
11.46	-2.14	19.85	-0.22
6.14	3.91	21.35	-1.54
14.37	-3.38	14.52	5.84
12.04	0.02	19.68	2.60
9.71	3.42	21.00	4.10
11.96	1.43	24.00	1.45
16.45	-2.45	26.16	0.87
11.87	2.84	28.23	1.70
16.28	0.28		

eigenvalues is $37.9 + 6.5 = 44.4$. Since these eigenvalues represent the lengths of the two principal semiaxes, the axes also represent the total variance of the data set, and each accounts for an amount of the total variance equal to the eigenvalue divided by the trace. The first principal axis contains $37.9/44.4$ or about 86% of the total variance, whereas axis II represents only 14%. In other words, if we measure the variation in our data set along the first principal axis, we can represent four-fifths of the total variation in the observations. It inevitably happens that at least one of the principal axes will be more efficient (in terms of accounting for total variance) than any of the original variables. On the other hand, at least one of the axes must be less efficient than any of the original variables.

If we make a transformation of the form $Y_1 = \alpha_1 X_1 + \alpha_2 X_2$, where the α's are the elements of the first eigenvector, we create a new set of data which will have a variance exactly equal to the first eigenvalue, or 37.9. A similar transformation $Y_2 = \beta_1 X_1 + \beta_2 X_2$, where the β's are elements of the second eigenvector, will yield a series of data points having a variance equal to the second eigenvalue, or only 6.5. Because the two new variables are measured along axes at right angles to each other, the correlation between the two is zero. Table 6.20 contains the data from Table 6.19 transformed in this manner.

In the table, each original observation has been converted to what is called a *principal component score* by projecting it onto the principal axes. This was done by

$$Y_{1i} = 0.66X_{1i} + 0.75X_{2i}$$

which projects the ith observation onto the first principal axis, multiplying the observed values of X_1 and X_2 by the corresponding elements of the first eigenvector. The same observation can be projected onto the second principal axis by

$$Y_{2i} = 0.75X_{1i} - 0.66X_{2i}$$

The elements of the eigenvectors that are used to compute the scores of observations are called *principal component loadings*. They are simply coefficients of the linear equation which the eigenvector defines. In writings of factor analysts, references are made to the "loading of variable A onto Factor I." This is just a way of referring to the coefficient in eigenvector I which corresponds to variable A. This operation in matrix form can be represented as

$$[X][U] = [S^R]$$

where $[S^R]$ is the $n \times m$ matrix of principal components scores, $[X]$ is the $n \times m$ matrix of original observations, and $[U]$ is a square matrix containing the successive eigenvectors in m columns of m elements, each corresponding to an original variable. In our example,

$$\begin{bmatrix} 3 & 2 \\ 4 & 10 \\ 6 & 5 \\ \vdots & \vdots \\ 20 & 20 \end{bmatrix} \cdot \begin{bmatrix} 0.66 & 0.75 \\ 0.75 & -0.66 \end{bmatrix} = \begin{bmatrix} 3.49 & 0.92 \\ 10.14 & -3.64 \\ 7.72 & 1.18 \\ \vdots & \vdots \\ 28.23 & 1.70 \end{bmatrix}$$

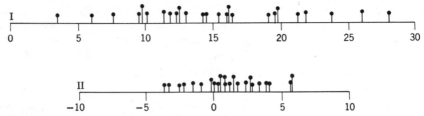

FIGURE 6.21   Plot of projections of data points in Table 6.19 onto their principal components. Variance along component I is 37.9, variance along component II is only 6.5.

Let us look again at our original data set. We have found the eigenvectors of the covariance matrix and determined that eigenvector I accounts for 86% of the total variance. Suppose we decide that it is imperative that we reduce our system to only one variable. This could be done by discarding either variable X_1 or X_2, but this will cause a loss of 46% or 54% of the total variance, depending upon which variable we retain. If, however, we convert our observations to scores on the first principal axis, we lose only 14% of the variation in our data set in the process.

Figure 6.21 shows the scores or projection of data points onto the principal axes.

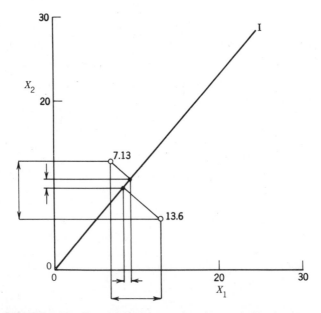

FIGURE 6.22   Two original data points (open circles ○) projected onto the first principal component. Projection of these scores (solid dots ●) back onto original variables results in smaller variance. Loss of variance occurs because second principal component is not utilized.

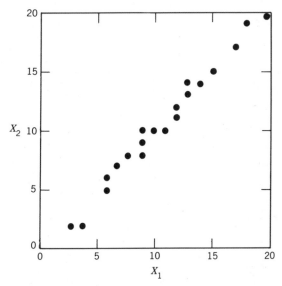

FIGURE 6.23 Data from Table 6.19 ranked and plotted.

It should be emphasized that the variance along the first principal axis will be greater than the variance along any other possible straight line drawn through the data points. However, the variance is not as great as the sum of the variances along the two lines X_1 and X_2, and if the second eigenvector is eliminated, an unavoidable loss of variance results. We can see how this occurs if we project the vector scores Y_{1i} back onto the X_1 and X_2 axes. Although some points may shift farther from their mean, the overall effect is to reduce the variance in both variables (Fig. 6.22).

Suppose we take the data in Table 6.19 and rank each observation within the two variables. Then the data will plot as in Figure 6.23. Ranking causes the variables to become highly correlated, and this is reflected in the covariance which now is 21.9. Because we are using the same observations, the variances remain unchanged. If we extract the eigenvectors and eigenvalues of this new variance–covariance matrix, we discover that the eigenvectors are almost the same:

$$I = \begin{bmatrix} 0.68 \\ 0.74 \end{bmatrix}, \qquad II = \begin{bmatrix} 0.74 \\ -0.68 \end{bmatrix}$$

However, the two new eigenvalues are radically different. Eigenvalue I is 44.2, so the first principal axis now accounts for 44.2/44.4 or over 99% of the total variance in the data. The second principal axis is so short that it is almost impossible to plot on our diagram (Fig. 6.24). Obviously, we could discard the second principal component and lose very little of the variance in the data set. If we do this, we are representing our original bivariate data by a single new variable (defined by the first principal component), and we have reduced the dimensionality of our data from two to one.

Rather than ranking the observations, we might randomize them as in Figure

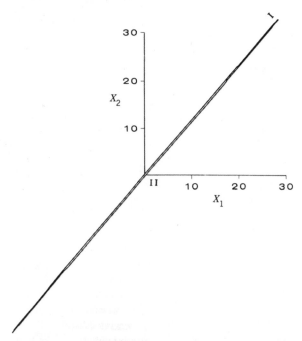

FIGURE 6.24 Ellipse defined by variance–covariance vectors of ranked data in Figure 6.23. The first principal component of the extremely elongated ellipse accounts for over 99% of the total variance.

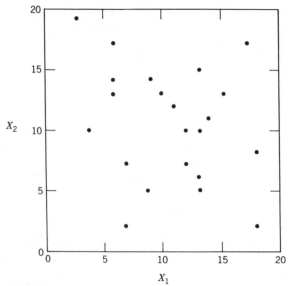

FIGURE 6.25 Data from Table 6.19 randomized and replotted.

6.25. Randomization destroys all correlation between the two variables, resulting in a covariance that is essentially zero. The variances, of course, remain the same because we are using the same data and have simply rearranged their order. If we extract the eigenvalues and eigenvectors of this variance–covariance matrix, we will obtain the two vectors

$$\text{I} = \begin{bmatrix} -0.22 \\ 0.98 \end{bmatrix}, \qquad \text{II} = \begin{bmatrix} 0.98 \\ 0.22 \end{bmatrix}$$

The two eigenvalues are almost identical, the first being 24.3 and the second 20.1. What we have found are the two principal axes of an ellipse which is almost circular (Fig. 6.26). This accords with what we expect, because the correlation between the two original variables is nearly zero; hence the two original axes are nearly orthogonal. Because the two original variable axes are almost equal in magnitude, the axes define a near-circular ellipse. No other set of axes, even those found by PCA, will be significantly better than the two original variables. In this situation, there is no transformation of the original data which will allow us to reduce the number of variables without a significant loss of information. Of course, it is very unlikely that genuine data sets would exhibit zero correlation between all variables.

In this simple example we first computed a variance–covariance matrix whose elements reflect the original units of measurement. Provided all variables are expressed in the same or commensurate units, the principal components will properly reflect the relative importance of the different variables. However, principal components analysis is sensitive to the magnitudes of the measurements, so if lengths

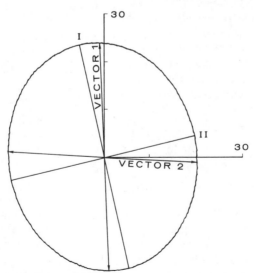

FIGURE 6.26 Near-circular ellipse defined by variance–covariance vectors of randomized data in Figure 6.15. The first principal component accounts for little more variance than either of the original variables.

of the brachiopod shells had been given in centimeters while their widths had been given in millimeters, the width variable would have exerted ten times more influence than the length variable on the outcome.

An obvious way around this difficulty is to standardize all variables so they have means of 0.0 and variances of 1.0. Then, the elements of the variance–covariance matrix will consist of correlations, and the principal components will be produced in dimensionless form. Standardization tends to inflate variables whose variance is small, and reduce the influence of variables whose variance is large. This may be undesirable but unavoidable if the original variables are expressed in different, incompatible units. Le Maitre (1982, pp. 108–110) gives an interesting discussion of the effects of standardization on principal components computed from geochemical and petrographic variables.

In most geologic studies, the relative magnitudes of the variables are important, so we should strive to work with the original variables and the variance–covariance matrix. In circumstances where we must standardize the variables in order to make them compatible, we must remember that a property that seems relatively insignificant may exert a strong influence on the analysis. Also, if the components are calculated from the correlation matrix, scores must be calculated from standardized, and not raw, variables.

We can examine the effects of principal components analysis on a larger data set, using the measurements on 25 randomly generated blocks listed in Table 6.18. The matrix of variances and covariances between the seven variables is given in Table 6.21, with the eigenvalues and eigenvectors of the matrix. The eigenvectors are the coordinates of the principal components axes of the block data.

The first two eigenvectors account for 93% of the variance in the data set. The first, accounting for 60% of the total variance, heavily weights the contributions of variations X_5 and X_6. Both of these are ratios which contain the short axis measurement (X_3) in the denominator. Therefore, we infer that principal component I measures differences in the thickness of blocks. An examination of Figure 6.27, which is a plot of the blocks at positions corresponding to their scores on the first two principal component axes, shows that this is the case; very flat blocks are placed far to the right, whereas equidimensional blocks are placed on the extreme left. A separation of rod- and plate-shaped blocks is not apparent along the first principal component, as either form may have very abbreviated short axes. We may conclude that the first component reflects the height of a block relative to its overall size.

The second eigenvector is heavily weighted on all three axes and on the length of the major diagonal. We may interpret this as a general reflection of size, an interpretation borne out by Figure 6.27. Along the second principal component, the blocks are sorted according to their size, with the smallest at the bottom and the largest at the top.

Interpreting the meanings (if any) of the principal component loadings is sometimes called "reification." Possibly some analysts feel the use of this term makes this subjective process more respectable. Plotting the principal components loadings as bar graphs (Fig. 6.28) may aid interpretation. Remember from Chapter 3 that

TABLE 6.21 Variance–Covariance Matrix of Seven Variables Measured on 25 Blocks; Only the Lower Half of the Symmetric Matrix is Shown

	Variable						
Variable	X_1	X_2	X_3	X_4	X_5	X_6	X_7
X_1	5.400						
X_2	3.260	5.846					
X_3	0.7785	1.465	2.774				
X_4	6.391	6.083	2.204	9.107			
X_5	2.155	1.312	−3.839	1.610	10.710		
X_6	3.035	2.877	−5.167	2.782	14.770	20.780	
X_7	−1.996	−2.370	−1.740	−3.283	2.252	2.622	2.594

Matrix of Eigenvectors

	Eigenvector						
Variable	I	II	III	IV	V	VI	VII
X_1	0.164	0.422	0.645	−0.090	0.225	0.415	−0.385
X_2	0.142	0.447	−0.713	−0.050	0.395	0.066	−0.329
X_3	−0.173	0.257	−0.130	0.629	−0.607	0.280	−0.211
X_4	0.170	0.650	0.146	0.212	0.033	−0.403	0.565
X_5	0.546	−0.135	0.105	0.165	−0.161	−0.596	−0.513
X_6	0.768	−0.133	−0.149	−0.062	−0.207	0.465	0.327
X_7	0.073	−0.313	0.065	0.719	0.596	0.107	0.092

Eigenvalues

34.490	19.000	2.540	0.810	0.340	0.033	0.003

Percentage of Total Variance Contributed by Each Eigenvalue

60.290	33.210	4.440	1.410	0.600	0.060	0.004

eigenvectors are standardized so the squares of the elements sum to one. Therefore, a loading reflects only the relative importance of a variable within a principal component, and does not reflect the importance of the component itself.

Although PCA produces a result in general agreement with our expectations, the separation of shapes is not definitive in this experiment. The third principal component accounts for only 4% of the variance of the data set and is essentially comprised of the longest versus the intermediate axis. However, it is adequate, when used in conjunction with the first two components, to completely separate platelike blocks from rod-shaped blocks. This result is not unexpected, because all of the variables in the experiment were created from three independent variables, the lengths of axes. Although two components are sufficient to express most of the variation in the data, a third independent component is necessary to recapture all of the essential details. In this example, we can see that PCA may be a powerful tool to determine the true number of linearly independent vectors that exist in a matrix. Therefore, it can measure the redundancy in the original set of variables.

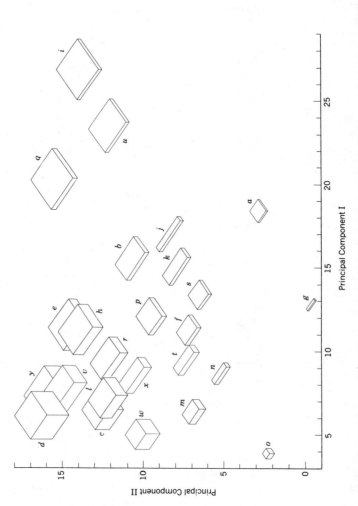

FIGURE 6.27 Principal component scores of block data plotted on first two principal components. Horizontal axis is component I, vertical axis is component II. Blocks shown plotted at their respective positions on the two components.

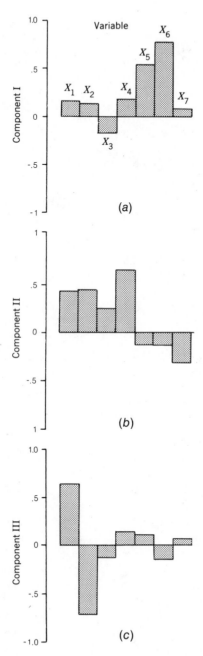

FIGURE 6.28 Plot of principal component loadings for first three components of random block data. (a) Component I; (b) Component II; (c) Component III.

TABLE 6.22   Fifty Grain-Size Analyses of Bottom Sediment Collected in Barataria Bay, Louisiana; Samples are Classified into Five Types

Type[a]	φ Categories, Analyses (wt %)						
	1–2	2–3	3–4	4–5	5–6	6–7	7–8
I	0.6	70.2	29.2	0.0	0.0	0.0	0.0
	1.0	69.9	29.1	0.0	0.0	0.0	0.0
	0.8	73.7	25.5	0.0	0.0	0.0	0.0
	0.9	75.3	23.8	0.0	0.0	0.0	0.0
	0.6	62.5	36.9	0.0	0.0	0.0	0.0
	1.1	68.8	30.1	0.0	0.0	0.0	0.0
	0.8	10.2	79.2	9.8	0.0	0.0	0.0
	1.0	16.3	73.8	8.9	0.0	0.0	0.0
	1.8	35.7	61.9	0.6	0.0	0.0	0.0
II	9.5	15.8	59.0	8.4	0.9	0.9	1.4
	2.4	14.5	53.9	12.2	5.5	1.6	2.5
	2.2	38.8	42.2	7.9	1.4	1.8	1.0
	1.7	30.4	44.5	11.2	3.0	1.9	2.9
	0.0	40.0	32.5	3.8	4.5	6.5	2.7
	0.0	37.0	45.4	7.3	3.8	3.3	3.8
	0.3	15.6	54.1	21.3	4.1	2.6	2.0
	0.3	24.4	56.0	15.1	4.2	0.0	0.0
	10.5	29.2	37.3	15.1	4.2	3.7	0.0
	0.3	13.3	63.5	14.2	4.0	3.4	1.3
	1.2	26.9	54.7	11.0	3.9	2.3	0.0
III	0.4	3.9	45.2	24.7	3.7	8.1	3.0
	0.0	13.8	39.3	15.4	9.1	4.5	6.4
	0.4	4.0	38.2	28.5	6.0	4.3	4.7
	1.9	11.5	49.5	22.4	5.7	4.5	2.0
	0.4	5.1	31.8	30.3	5.4	7.8	3.0
	0.5	5.9	32.2	32.7	4.9	5.4	2.7
	1.1	4.9	31.1	41.9	13.9	7.8	3.7
	7.9	8.5	21.0	19.9	8.9	5.9	6.3
	0.9	13.6	43.9	20.1	7.2	4.8	9.5
	2.9	15.5	37.0	30.3	5.1	1.9	2.2
	2.1	16.7	39.6	17.7	8.3	8.3	7.3
	0.3	20.6	55.4	16.6	6.2	6.1	5.5
IV	1.2	1.6	15.3	38.4	13.0	9.5	5.6
	2.3	7.9	23.9	25.5	9.2	7.9	7.7
	1.0	3.1	15.2	32.0	14.3	10.0	7.2
	0.0	11.5	28.4	19.1	7.3	7.8	4.8
	0.8	7.0	31.6	21.1	10.2	9.0	6.3
	0.5	2.1	14.0	37.2	19.9	11.4	6.1
	0.0	3.4	19.7	25.4	15.7	10.2	9.9
	1.4	1.9	14.4	40.2	8.5	8.4	7.1
	0.4	3.5	18.8	29.5	11.2	10.4	7.5
	0.8	6.3	18.2	28.0	9.1	9.7	9.9

TABLE 6.22   (*Continued*)

Type[a]	φ Categories, Analyses (wt %)						
	1–2	2–3	3–4	4–5	5–6	6–7	7–8
V	1.0	2.3	6.6	16.2	12.0	11.4	13.3
	3.2	3.9	10.5	24.1	14.2	15.4	13.5
	2.1	2.1	10.7	23.6	15.1	14.0	11.8
	4.4	8.1	8.9	19.9	12.0	11.4	10.8
	0.6	3.6	4.2	17.8	12.4	10.8	9.9
	0.5	4.1	9.8	27.9	13.5	13.5	7.4
	0.7	2.3	5.2	23.2	19.4	14.1	10.1
	3.4	1.6	4.4	18.0	14.7	15.3	15.1

Source: Modified from original data by Krumbein and Aberdeen (1937, Table 1).
[a]Key: I, beach and foreshore sands; II, silty channel sands; III, silty channel margin sands; IV, organic bottom silts; V, organic muds from lees of islands.

As an example of the application of PCA to a geologic problem, we will consider the data, taken from Krumbein and Aberdeen (1937, Table 1), given in Table 6.22. These are 50 grain-size analyses of bottom samples collected in Barataria Bay, Louisiana, which is a large embayment on the west side of the Mississippi delta. The bay includes a variety of depositional environments, characterized by a range of sediment types. The analyses are sieve and pipette separations made at 1-φ intervals. Raw data consist of the weight percent of sediment in each size fraction.

Variables are the relative fractions of each sample which fall within specified size limits. These are the same variables used to compute either graphic or moment statistics such as the mean, sorting, and skewness of sediment grain size. By PCA, we can examine the interactions between the various size fractions and find the most efficient linear combination of them, the term ''most efficient'' meaning the one that accounts for the greatest amount of total variance. We would expect the loadings on the first principal component to approximate in some way the mean, as this statistic traditionally is regarded as being the most efficient of all possible statistics.

The first step in our analysis is to compute the matrix of variances and covariances (Table 6.23). Standardization is not necessary in this problem, because the raw data are measured in the same units for all variables. It should be noted that the data constitute a nearly closed array (i.e., the variables sum to 100% for most observations), which raises troublesome theoretical questions concerning induced negative correlations. The covariance matrix is "overdetermined;" that is, it has more rows and columns than necessary. Obviously, if we know A, B, and C, and the total $(A + B + C)$, we have more information than we really need, and one of the three variables is superfluous. Of necessity, one of the eigenvalues of a matrix resulting from such data will be zero. In this example, not all of the observations sum to exactly 100%, as material smaller than 8 φ was discarded.

TABLE 6.23 Variance–Covariance Matrix of Grain-Size Measurements Made on Sediments from Barataria Bay, Louisiana; Only the Lower Half of the Symmetric Matrix is Shown[a]

	X_1	X_2	X_3	X_4	X_5	X_6	X_7
X_1	4.8443						
X_2	-2.6234	468.8480					
X_3	-0.0011	81.3941	353.1255				
X_4	-1.5449	-200.2109	-84.6165	130.2741			
X_5	-0.5972	-84.2597	-73.0435	44.7616	30.4350		
X_6	-0.3805	-71.2097	-66.5433	34.9927	23.7565	22.4189	
X_7	-0.0222	-57.8578	-56.1533	23.9136	19.3907	17.9388	17.9670

[a]Key: X_1 = 1–2ϕ, X_2 = 2–3ϕ, X_3 = 3–4ϕ, X_4 = 4–5ϕ, X_5 = 5–6ϕ, X_6 = 6–7ϕ, X_7 = 7–8ϕ.

Therefore, the final eigenvalue of the covariance matrix is very small but not zero, as it would be if all observations were closed.

The principal components or eigenvectors of the Barataria Bay data are given in Table 6.24. Note that the first two components alone account for over 90% of the variance in the data set. Figure 6.29 is a plot of variable loadings on the two

TABLE 6.24 Eigenvalues and Eigenvectors (Principal Components) of Covariance Matrix in Table 6.23[a]

Vector	Eigenvalue	Total Variance (%)	Total Variance (Cumulative %)
I	659.7759	64.18	64.19
II	318.4384	30.98	95.17
III	35.1959	3.42	98.59
IV	6.7528	0.66	99.25
V	3.8193	0.37	99.62
VI	2.3763	0.23	99.85
VII	1.5540	0.15	100.00

	Eigenvector						
Variable	I	II	III	IV	V	VI	VII
X_1	-0.0019	0.0039	-0.0689	-0.5829	0.7554	0.2793	0.0818
X_2	0.7710	-0.4777	0.3194	0.1885	0.1169	0.1581	0.0326
X_3	0.4167	0.8647	0.0531	0.2116	0.1123	0.1294	0.0421
X_4	-0.3907	0.0761	0.8844	0.0704	0.0490	0.2280	0.0028
X_5	-0.1895	-0.0794	-0.0775	0.6308	0.6255	-0.3240	-0.2401
X_6	-0.1618	-0.0813	-0.1629	0.3330	0.0526	0.2570	0.8723
X_7	-0.1308	-0.0735	-0.2750	0.2570	-0.0815	0.8107	-0.4146

[a]Key: X_1 = 1–2ϕ, X_2 = 2–3ϕ, X_3 = 3–4ϕ, X_4 = 4–5ϕ, X_5 = 5–6ϕ, X_6 = 6–7ϕ, X_7 = 7–8ϕ.

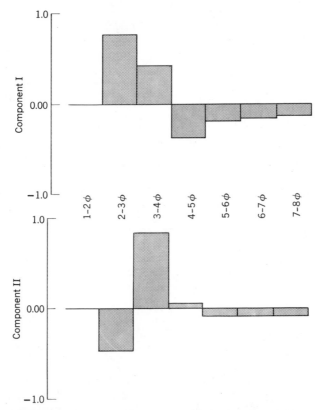

FIGURE 6.29 Loadings of variables on first two principal components of Barataria Bay data.

components. It is obvious from the plots that the first principal component essentially represents the relative proportion of fine and very fine sand in the sediment, or the sand/(silt + clay) ratio. The second component represents the ratio between fine and very fine sand, all other variables being weighted near zero. These two components alone are sufficient to account for almost all of the variance in the original data set, which suggests that very little of importance is contained within the subdivisions of the silt and clay-sized fractions. The differences between the sediments can be almost completely described by only two variables.

We can verify our analysis by computing the scores of the observations on the first two principal components. Figure 6.30 is a scatter diagram of scores for the first two principal components; the five different sediment categories are shown by symbols. Compare the separation between sediment types in this diagram with that shown in Figure 6.31, which is a plot of the median grain size versus sorting (quartile deviation). Perhaps even more interesting is Figure 6.32, which is a plot of percentage of sand versus the ratio of fine to very fine sand; these are variables

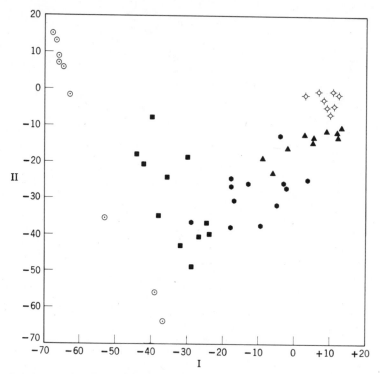

FIGURE 6.30 Barataria Bay sediment analyses plotted on first two principal components. Symbols correspond to five sediment types recognized by Krumbein and Aberdeen (1937): ⊙ I, beach and foreshore sands; ■ II, silty channel sands; ● III, silty channel margin sands; ▲ IV, organic bottom silts;-✧-V, organic muds from lees of islands.

suggested by the nature of the loadings on the first two principal components. Each of these diagrams is approximately equally efficient in terms of separating the five sediment types. However, the amount of experimental effort necessary to find the data plotted in Figure 6.32 is a fraction of that required to plot Figure 6.31. Rather than laborious sieve and pipette separation of samples into seven size intervals, only two simple sieving operations are necessary. In addition, results of the PCA suggest that sediments in the bay may profitably be regarded as mixtures of two populations, sand-sized material and silt–clay material. In this example, principal components not only suggest a new way of looking at the composition of these sediments, but also indicate a possible modification of experimental techniques that will result in considerable savings in effort with negligible loss of information. This experiment, with minor modifications, was given by Davis (1970). It is instructive to compare these results with those obtained by Klovan (1966) from a Q-mode factor analysis of the same data.

It may be interesting to examine the relative effectiveness of the mean, the first

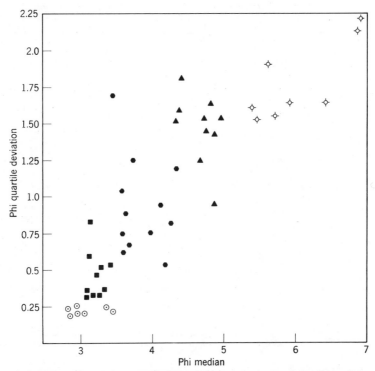

FIGURE 6.31   Plot of φ quartile deviation versus φ median of Barataria Bay sediments. Symbols correspond to those in Figure 6.30.

principal component, and the percentage of sand to distinguish between the five sediment types in the bay. This can be done by running a series of one-way analyses of variance using the five sediment types as groups. The ratio between the sums of squares among groups to total sum of squares is a measure of how tightly groups are clustered and separated from other groups. The variable which produces the highest ratio SS_A/SS_T is the most effective for characterizing the sediment types. Run the appropriate ANOVA's as outlined in Chapter 2, and determine which of the three variables is the most effective.

Another study that may shed light on the utility of the multiplicity of statistics and quasistatistics used to characterize sediments is to compute a variety of these measures and enter them as variables in a PCA. The analysis may select out interpretable combinations of these that are effective for characterizing sediments. Computational equations for numerous grain-size statistics are given in various reference books on sedimentary petrology, such as Folk (1980). These measures can be computed for the raw data given in Table 6.22. A principal components analysis of these new variables may be quite instructive. A similar study was done by Griffiths and Ondrick (1969, pp. 86–88).

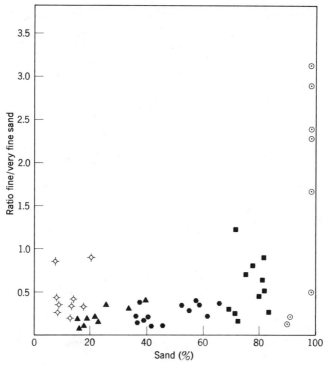

FIGURE 6.32   Plot of percentage of sand versus ratio of fine to very fine sand in Barataria Bay sediments. Symbols correspond to those in Figure 6.30.

R-Mode Factor Analysis

Principal components analysis consists of a linear transformation of m original variables to m new variables, where each new variable is a linear combination of the old. The process is performed in a fashion that requires that each new variable accounts for, successively, as much of the total variance as possible. When m new variables have been computed, all of the original variance will be accounted for. Nothing is said about probability, hypotheses, or testing, because PCA is not, strictly speaking, a statistical procedure. Rather, it is a mathematical manipulation. However, it assumes some of the characteristics of statistical procedures when decisions are made to discard some new variables or components as being inconsequentially small. Some statistical tests are available for checking the significance of discarded variables, but these are based on highly restrictive assumptions and are seldom applicable [Morrison (1976) gives a review]. Principal components analysis belongs to that category of techniques, including cluster analysis, in which utility is judged by performance and not by theoretical considerations.

Factor analysis is somewhat different, for it is commonly regarded as a statistical technique. It relies on a set of assumptions about the nature of the parent population from which the samples were drawn. These assumptions provide the rationale for the operations that are performed, and the manner in which results are interpreted. Some factor procedures even provide tests of significance (Lawley and Maxwell, 1971), although these are not widely used.

In factor analysis, the relationship within a set of m variables is regarded as reflecting the correlations of each of the variables with p mutually uncorrelated underlying factors. The usual assumption is that $p < m$. Variance in the m variables is therefore derived from variance in the p factors, but in addition a contribution is made by unique sources which independently affect the m original variables. Factor analysts refer to the p underlying factors as *common factors* and summarize the independent contribution as a *unique factor*. The factor model may be expressed as

$$X_j = \sum_{r=1}^{p} a_{jr} f_r + \epsilon_j \qquad (6.51)$$

where f_r is the rth common factor, p is the specified number of factors, and ϵ_j is random variation unique to the original variable X_j. Because there are m original variables X_j, there are m random variables ϵ_j; taken together, these constitute the unique factor. The coefficient a_{jr} is the loading of the jth variate on the rth factor. It corresponds to the loadings or weights on principal components.

We may assume that the X_j are multivariate normally distributed. The variances and covariances form an $m \times m$ matrix. From (6.51), we can determine that the diagonal elements of the matrix, the variances of the m variables, should be

$$s_{jj}^2 = \sum_{r=1}^{p} a_{jr}^2 + \text{var } \epsilon_{jj} \qquad (6.52)$$

and the off-diagonal elements, or covariances, should be

$$\text{cov}_{jk} = \sum_{r=1}^{p} a_{jr} a_{kr} \qquad (6.53)$$

The hypothesis underlying factor analysis may be expressed in matrix notation in the following way. The observed matrix of variances and covariances, which we will denote as $[s^2]$ is the product of an $m \times p$ matrix of factor loadings (which we will call $[A^R]$) multiplied by its transpose, plus an $m \times m$ diagonal matrix of unique variances, $[\text{var } \epsilon_{jj}]$:

$$[s^2] = [A^R] \cdot [A^R]' + [\text{var } \epsilon_{jj}] \qquad (6.54)$$

Multiplying an $m \times p$ matrix by its transpose will create an $m \times m$ matrix which, however, will have only p positive eigenvalues and associated eigenvectors. If $p = m$, the matrix $[\text{var } \epsilon_{jj}]$ will vanish and our problem is equivalent to PCA. In cases where $p < m$, we must estimate the matrix of parameters $[A^R]$, which are the loadings on the factors, and the unique variance, $[\text{var } \epsilon_{jj}]$. Note that the factor model requires that p, the number of factors, be known prior to analysis. This implies that the investigator has some insight into the probable nature of the factors, and can predict a suitable number of factors to be extracted. If p cannot be specified, the partition of variances between the common factors and the unique factor becomes indeterminate. This is a point sometimes overlooked by experimenters who wish to use factor analysis for "fishing expeditions." Stated in another way, p, the number of factors, $[A^R]$, the matrix of factor loadings, and the unique variances [var ϵ_{jj}] are all interrelated. They cannot all be estimated simultaneously, so it is necessary to introduce various constraints in order to find a unique solution. The simplest constraint is to assume some prior value for p, the number of factors. Unfortunately, in most geological problems the number of possible factors is not knowable in advance, and may even be an important objective in a study. Another approach is to assume some prior limit for either $[A^R][A^R]'$ or [var ϵ_{jj}] and to extract factors until this limit is reached.

We will examine two of the many factor analytic schemes that have been suggested. The principal components approach to factor analysis starts by extracting the eigenvalues and eigenvectors of the correlation matrix, and then discarding the less important of these. This does not lead to a "true" factor solution, but the mathematics are relatively straightforward and this is the approach used almost universally in the Earth sciences when factor analysis is employed. We will also take a brief look at the method of maximum likelihood, which does yield "true" factors. Unfortunately, the underlying mathematics are so involved that most authors dismiss them as "too complicated to be described."

Although the first method of factor analysis we will consider utilizes principal components, the computation of eigenvalues and eigenvectors is different in two respects. First, the eigenvalue operation is always performed on a standardized variance–covariance, or correlation, matrix. This assures that all variables are weighted equally, and also allows us to convert the principal component vectors into factors. Second, the eigenvectors, which are calculated in "normalized" form (see p. 135), are transformed so they define vectors whose lengths are proportional to the variation they represent. These transformed eigenvectors are the *factors* of the data set.

The conversion of normalized, or unit, eigenvectors into factors does not affect the directions of the vectors, only their lengths. It is done by multiplying every element in a normalized eigenvector by the corresponding singular value, or square root of the corresponding eigenvalue. The resultant factor is a vector that is weighted proportionally to the amount of total variance it represents.

The effect of standardization has been discussed in the preceding section on principal components analysis. Here we will demonstrate the effect using as an example the data plotted in Figure 6.33. The raw data have a variance–covariance matrix

$$\begin{bmatrix} 6.08 & 11.08 \\ 11.08 & 27.54 \end{bmatrix}$$

with eigenvectors and eigenvalues:

$$\text{eigenvector I} = \begin{bmatrix} 0.39 \\ 0.92 \end{bmatrix}, \quad \text{eigenvalue} = 32.23, \text{ or } 96\%$$

$$\text{eigenvector II} = \begin{bmatrix} 0.92 \\ -0.39 \end{bmatrix}, \quad \text{eigenvalue} = 1.39, \text{ or } 4\%$$

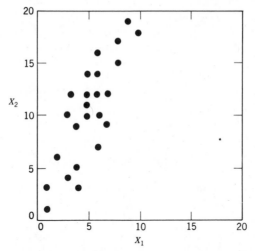

FIGURE 6.33 Data set to show effect of standardization. Raw data have means of $\bar{X}_1 = 5$ and $\bar{X}_2 = 10$.

If these data are standardized by subtracting the means and dividing by the standard deviations (Fig. 6.34), the standardized variance–covariance (or correlation) matrix is

$$\begin{bmatrix} 1.00 & 0.86 \\ 0.86 & 1.00 \end{bmatrix}$$

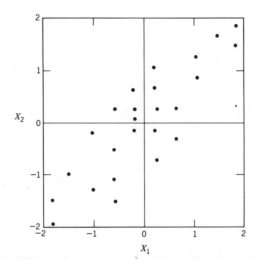

FIGURE 6.34 Data from Figure 6.33 standardized to have zero means and unit standard deviations. Note that both variables now cover same range.

with eigenvectors and eigenvalues:

$$\text{eigenvector I} = \begin{bmatrix} 0.707 \\ 0.707 \end{bmatrix}, \qquad \text{eigenvalue} = 1.86, \text{ or } 93\%$$

$$\text{eigenvector II} = \begin{bmatrix} -0.707 \\ 0.707 \end{bmatrix}, \qquad \text{eigenvalue} = 0.14, \text{ or } 7\%$$

The eigenvectors define the matrix $[U]$,

$$[U] = \begin{bmatrix} 0.707 & -0.707 \\ 0.707 & 0.707 \end{bmatrix}$$

and the eigenvalues, or rather their square roots, define the matrix of singular values:

$$[\Lambda] = \begin{bmatrix} \sqrt{1.86} & 0.0 \\ 0.0 & \sqrt{0.14} \end{bmatrix} = \begin{bmatrix} 1.36 & 0.0 \\ 0.0 & 0.37 \end{bmatrix}$$

The eigenvectors can be converted into factors by eq. (6.42),

$$[A^R] = [U][\Lambda]$$

Converting the first eigenvector into a factor gives

$$\text{factor I} = \begin{bmatrix} 0.707 \cdot \sqrt{1.86} \\ 0.707 \cdot \sqrt{1.86} \end{bmatrix} = \begin{bmatrix} 0.964 \\ 0.964 \end{bmatrix}$$

The elements of the factor are referred to as *factor loadings*. If we have performed the conversion correctly, the sum of the squares of the factor loadings should equal the eigenvalue:

$$(0.964)^2 + (0.964)^2 \approx 1.86$$

Factor II is equal to

$$\text{factor II} = \begin{bmatrix} -0.707 \cdot \sqrt{0.14} \\ 0.707 \cdot \sqrt{0.14} \end{bmatrix} = \begin{bmatrix} -0.264 \\ 0.264 \end{bmatrix}$$

Again, the sum of the squares of the factor loadings equals the eigenvalue:

$$(-0.264)^2 + (0.264)^2 \approx 0.14$$

The two factors are shown plotted in Figure 6.35. The orientations of the factors are the same as the original eigenvectors, but their lengths are equal to the square roots of the eigenvalues. The eigenvalues represent the proportion of the total variance accounted for by the eigenvectors. Now, the lengths of the factors represent the eigenvalues (or rather, the singular values or square roots of the eigenvalues), so the factors also represent the variances (or more correctly, the standard deviations). In fact, because of the way the eigenvectors were first normalized and then multiplied by the singular values, each factor loading is weighted proportionally to the square root of the amount of variance contributed by that variable to the factor. In our example, the first factor accounts for $1.86/2.00 = 93\%$ of the total

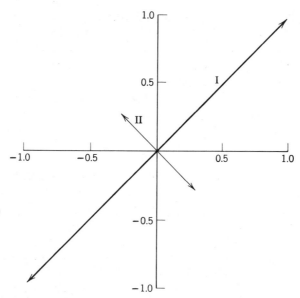

FIGURE 6.35 Plot of two factors extracted from bivariate data shown in Figure 6.34.

variance of our data. Of this, $0.964^2/1.86 = 50\%$ is derived from variable 1 and
$0.964^2/1.86 = 50\%$ is derived from variable 2. Similarly, factor II accounts for
7% of the total variance, deriving $-0.264^2/0.14 = 50\%$ from variable 1 and
$0.264^2/0.14 = 50\%$ from variable 2. One hundred percent of the variance of
variable 1 is accounted for in the two factors, as is 100% of the variance of variable
2. (The reciprocal nature of the influence of these variances on the two factors is
an inevitable consequence of working with a 2×2 matrix. This relationship is
not generally found in larger matrices.)

If we arrange the factor loadings in matrix form, we have the *factor matrix,*
$[A^R]$. The factor matrix of the data in Figure 6.34 is

$$\text{Variables} \begin{matrix} \\ 1 \\ 2 \end{matrix} \begin{matrix} \text{Factors} \\ \begin{matrix} \text{I} & \text{II} \end{matrix} \\ \begin{bmatrix} 0.964 & -0.264 \\ 0.964 & 0.264 \end{bmatrix} \end{matrix}$$

If we square the elements in the factor matrix and sum within each variable, the
totals are the amount of variance of each variable retained in the factors. That is,

$$\text{Variables} \begin{matrix} \\ 1 \\ 2 \end{matrix} \begin{matrix} \text{Factors} \\ \begin{matrix} \text{I} & \text{II} \end{matrix} \\ \begin{bmatrix} 0.964^2 & -0.264^2 \\ 0.964^2 & 0.264^2 \end{bmatrix} \end{matrix} = \begin{matrix} \text{Communalities} \\ \\ \begin{bmatrix} 1.00 \\ 1.00 \end{bmatrix} \end{matrix}$$

These sums are referred to as *communalities* and are symbolically represented as

h_j^2, where j refers to the jth variable. If we extract m factors from an $m \times m$ matrix of variances and covariances, the communalities are equal to the original variances. Because we are working with standardized variables, these communalities will be 1.00. However, if we extract fewer than m factors, the communalities will be less than the original variances, and can provide an index to the efficiency of our reduced set of factors. For example, if we were to retain only one factor from our 2×2 matrix, the communalities would be

$$h_1^2 = 0.964^2 = 0.93$$

$$h_2^2 = 0.964^2 = 0.93$$

That is, retaining only a single factor still accounts for 93% of the variance of variable 1 and 93% of the variance of variable 2.

Of course, the magnitude of the communalities is dependent upon the number of factors that are retained, bringing us up against one of the major problems of this approach to factor analysis. How many factors should be retained? There is, unfortunately, no simple answer and this question is one of the major sources of argument among factor analysts. Early experimental psychologists solved the problem in a very straightforward way; they extracted as many factors as their current ruling theory demanded. Another equally pragmatic approach is to extract only two or three factors, because this is the maximum number that can conveniently be displayed as scatter diagrams, and any number larger than this increases the dimensionality of the problem to the point where it again becomes difficult to grasp.

Some analysts recommend retaining all factors that have eigenvalues greater than one. That is, retain all factors that contain greater variance than the original standardized variables. In most instances, only a few factors will contain most of the variance in the data set, and this recommendation is useful. However, if the original variables are only weakly correlated or uncorrelated, half or more of the factors may have eigenvalues greater than one. You will be left not only with an inordinate number of factors, but may discover that none of them is interpretable anyway. If the factor theory is applicable to a given data set (i.e., the variances observed are the result of correlations between variables and underlying factors), a few factors should account for a very high percentage of the variance, and communalities will be high. If a large number of factors must be retained to account for much of the original variance, or if the communalities of the first few factors are low, the factor model probably is not appropriate.

Before continuing to the next step in factor analysis, which is rotation of the factor axes to "simple structure," let us apply what we have covered so far to an example. We will use the data in Table 6.18 and will retain two factors, because our intuition tells us two factors should be involved, one a size factor and the other a measure of shape. The matrix of standardized variances and covariances is given in Table 6.25. Table 6.26 contains the matrix of eigenvectors or principal components and their corresponding eigenvalues. We will retain the first two of these and convert them to factors. This is done by multiplying the normalized eigenvectors by the square roots of the corresponding eigenvalues to yield the factor loadings.

TABLE 6.25 Standardized Variances and Covariances (Correlations) between Seven Variables Measured on 25 Blocks Listed in Table 6.18; Only the Lower Half of the Symmetric Matrix is Shown

	X_1	X_2	X_3	X_4	X_5	X_6	X_7
X_1	1.000						
X_2	0.580	1.000					
X_3	0.201	0.364	1.000				
X_4	0.911	0.834	0.439	1.000			
X_5	0.283	0.166	−0.704	0.163	1.000		
X_6	0.287	0.261	−0.681	0.202	0.990	1.000	
X_7	−0.533	−0.609	−0.649	−0.676	0.427	0.357	1.000

These actually form an $m \times p$ factor loading matrix $[A^R]$, but we show it here as the transpose $[A^R]'$ for the sake of compactness:

$$\begin{matrix} \text{factor I} \\ \text{factor II} \end{matrix} \begin{bmatrix} 0.747 & 0.795 & 0.710 & 0.910 & -0.235 & -0.178 & -0.886 \\ 0.491 & 0.373 & -0.596 & 0.389 & 0.963 & 0.971 & 0.218 \end{bmatrix}$$

The communality of each variable is

$$h_j^2 = \quad [0.798 \quad 0.771 \quad 0.860 \quad 0.979 \quad 0.983 \quad 0.976 \quad 0.833]$$

The remaining $(1.00 - h_j^2)$ variance of variable j is the unique component associated

TABLE 6.26 Eigenvalues and Matrix of Eigenvectors of Matrix in Table 6.25; Only the Two Eigenvectors with Eigenvalues Larger than 1.000 Will Be Retained as Factors

Vector	Eigenvalue	Total Variance (%)	Total Variance (Cumulative %)
I	3.3946	48.4949	48.4949
II	2.8055	40.0783	88.5731
III	0.4373	6.2473	94.8204
IV	0.2779	3.9707	98.7911
V	0.0810	1.1565	99.9476
VI	0.0034	0.0487	99.9963
VII	0.0003	0.0037	100.0000

Variable	\multicolumn{7}{c}{Eigenvector}
	I	II	III	IV	V	VI	VII
X_1	0.4053	−0.2929	−0.6674	0.0888	−0.2267	0.4098	−0.2782
X_2	0.4316	−0.2224	0.6980	−0.0338	−0.4366	0.1443	−0.2540
X_3	0.3854	0.3559	0.1477	0.6276	0.5121	0.1875	−0.1081
X_4	0.4939	−0.2323	−0.1186	0.2103	−0.1054	−0.5878	0.5359
X_5	−0.1277	−0.5751	0.0294	0.1108	0.3890	−0.4232	−0.5562
X_6	−0.0968	−0.5800	0.1743	−0.0061	0.3549	0.5003	0.4975
X_7	−0.4809	−0.1303	0.0176	0.7353	−0.4553	0.0332	0.0489

TABLE 6.27 Reproduced Correlation Matrix for Two Factors Extracted from Random Block Data; Residual Correlation Matrix Is Shown Below, and Contains Correlations between Variables Unaccounted for by the Two Factors; Only the Lower Half of Each Symmetric Matrix is Shown

	X_1	X_2	X_3	X_4	X_5	X_6	X_7
Reproduced Correlation Matrix							
X_1	0.7983						
X_2	0.7766	0.7711					
X_3	0.2379	0.3426	0.8596				
X_4	0.8704	0.8685	0.4143	0.9794			
X_5	0.2969	0.1718	−0.7413	0.1606	0.9833		
X_6	0.3434	0.2201	−0.7057	0.2157	0.9778	0.9756	
X_7	−0.5546	−0.6233	−0.7594	−0.7214	0.4187	0.3701	0.8328
Residual Correlation Matrix							
X_1	0.2017						
X_2	−0.1963	0.2289					
X_3	−0.0367	0.0212	0.1404				
X_4	0.0409	−0.0348	0.0243	0.0206			
X_5	−0.0135	−0.0060	0.0371	0.0024	0.0167		
X_6	−0.0569	0.0409	0.0252	−0.0134	0.0124	0.0244	
X_7	0.0214	0.0146	0.1105	0.0459	0.0085	−0.0129	0.1672

with that variable. Magnitudes of the unique components are

$$[\text{var } \epsilon_{jj}] = [0.202 \quad 0.229 \quad 0.140 \quad 0.021 \quad 0.017 \quad 0.024 \quad 0.167]$$

If m factors were retained from a set of m variables, the original covariance matrix $[s^2]$ could be recreated by multiplying together all possible pairs of factor loadings and summing across the factors. Of course, when $p < m$ factors are retained, the original covariance matrix cannot be reproduced exactly. For variables j and k, the *reproduced covariance* $\hat{s}_{jk}^2$ is

$$\hat{s}_{jk}^2 = a_{j1}a_{k1} + a_{j2}a_{k2} + \cdots + a_{jp}a_{kp} \tag{6.55}$$

where a_{j1}, for example, is the loading of the jth variable on factor I. Denoting the matrix of factor loadings as $[A^R]$, this is equivalent to

$$[\hat{s}^2] = [A^R] \cdot [A^R]'$$

if the factor loadings are considered to constitute column vectors as we have done. The *residual standardized variance–covariance matrix* (or residual correlation matrix, as it also is called) can be found by subtraction:

$$[s_{\text{res}}^2] = [s^2] - [\hat{s}^2] \tag{6.56}$$

The reproduced and residual covariance matrices for our example are given in Table 6.27. The residual matrix is a measure of the inability of two factors to account for all of the variability in the original data set.

Factor Rotation

Although factor analysis may reduce the dimensionality of a problem to manageable size, the meaning of the factors may be difficult to deduce. Under factor theory, this may be the result of the fact that positions of the p orthogonal factor axes in m space are constrained by $m - p$ unnecessary axes, which also must be placed orthogonally through the sample space. However, we need only p factor axes to explain our data. If we ''chop off'' the extraneous orthogonal axes, it seems possible to further rotate the factors and perhaps find a better position for them. This we can do by a variety of rotational procedures. The particular technique we will employ is called *Kaiser's varimax* scheme, which has as its objective the moving of each factor axis to positions so that projections from each variable onto the factor axes are either near the extremities or near the origin. The method operates by adjusting the factor loadings so they are either near ± 1 or near zero. For each factor, there will be a few significantly high loadings and many insignificant loadings. Interpretation, in terms of original variables, is thus made easier. However, in certain instances rigid rotation of the factor axes will not improve the analysis, and may even confuse the results further. This may indicate that the factors are oblique, or intercorrelated, or it may imply that the factor model is inappropriate.

The varimax criterion involves maximization of the variance of the loadings on the factors. We may define the variance s_k^2 of the loadings on the kth factor as

$$s_k^2 = \frac{p \sum_{j=1}^{m} (a_{jp}^2/h_j^2)^2 - (\sum_{j=1}^{m} a_{jp}^2/h_j^2)^2}{p^2} \tag{6.57}$$

where, as before, p is the number of factors, m is the number of original variables, a_{jp} is the loading of variable j on factor p, and h_j^2 is the communality of the jth variable. The quantity we wish to maximize is

$$V = \sum_{k=1}^{p} s_k^2 \tag{6.58}$$

The variance is calculated from the factor loadings, a_{jp}, which are corrected by dividing each by its communality, h_j^2. In other words, only the common part of the variance of each variable is considered, removing the constraint imposed by the $m - p$ components necessary to account for all of the variance of each variable. Maximizing the variance implies maximizing the range of the loadings, which tends to produce either extreme (positive or negative) or near-zero loadings, satisfying the purpose of factor rotation.

No simple analytical scheme exists whereby the varimax criterion may be maximized. Rather, rotation of the factor axes must be performed iteratively, a sort of trial-and-error process. The factor axes are adjusted two at a time, holding the other axes stationary. After all axes have been adjusted, the process is repeated until the increase in variance of the loadings with each iteration drops below some specified cutoff level.

The varimax rotation process can best be illustrated with an example. We will attempt to ''clean up'' the factors produced from our artificial random block data set by rotating the two factors we retained. In Figure 6.36, the loadings of the seven original variables on factor I are plotted against the loadings on factor II. Connecting the plotted points with the origin yields a diagram in which the factor loadings are shown as vectors. The orientation of the vectors with respect to the factor axes reflects their degree of correlation with the factors. The length of each vector is proportional to the communality of the variable which the vector represents. If the two factors that are plotted completely account for all of the variation in an original variable, that variable has a communality of one and will lie somewhere along a circle of unit radius in the diagram. In the example, all communalities are high, so the vectors representing the seven original variables all extend to near the unit circle.

Varimax rotation changes the factor loadings so the original variables have either a high positive or high negative correlation (near ± 1.0) with a factor, or a correlation near zero. Figure 6.37 shows the positions of the variables on the factor axes after rotation. Note that the positions of the variables with respect to each other are not altered by rotation; only their relation to the factor axes is changed. Also note that the lengths of the vectors remain unchanged.

Similarly, the relationships among the observations themselves are not changed by rotation, although the position of individual objects within the space defined by the factor axes is altered. Figure 6.38 is a plot of factor scores, similar to the plot of principal component scores given in Figure 6.27. Note that two sets of factor axes are shown on the diagram, one for unrotated factor scores and the other

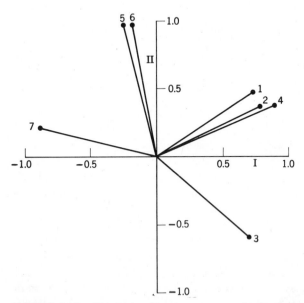

FIGURE 6.36 Plot of loadings on two raw factors extracted from measurements on 25 random blocks. Data for seven variables are listed in Table 6.18.

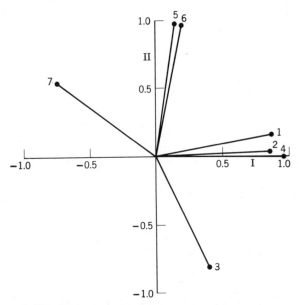

FIGURE 6.37 Plot of loadings on two factors rotated
by the varimax criterion. Variables are those measured
on 25 random blocks listed in Table 6.18.

for scores after varimax rotation. The first factor does indeed seem to reflect overall
size of the blocks, as smaller blocks are located on the left and larger blocks are
placed on the right. The second factor separates equidimensional shapes at the top,
with plates and rods lower on the second factor. In this instance, varimax rotation
does not seem to contribute to our ability to interpret the factors. Nor is the pattern
of factor scores much different from that obtained by principal components analysis,
although the relative importance of the first and second factor axes are reversed
compared to the principal component axes.

Plotting of scores on factors (either rotated or unrotated) is more complex than
the plotting of principal component scores. Principal components are linear transformations, so we were able to plot PCA scores simply by projecting our original
observations onto the principal axes. In factor analysis, however, the scores represent estimates of the contributions of various factors to each original observation.
Because the factors themselves are estimated from these same data, the computation
of factor scores is a somewhat circular process, and the results are not unique. One
of the clearer discussions of this problem is given by Morrison (1976); also see
the procedure given by Harman (1967, pp. 348–350). In psychometrics, the factors
themselves are usually the items of interest, and factor scores are not often needed.
Consequently, the calculation of factor scores has received comparatively little
attention. The factor scores may be an important part of a geological factor study,
however, and it is essential that we be able to compute them.

We will refer to our original data as the $n \times m$ matrix $[X]$ in which the n rows

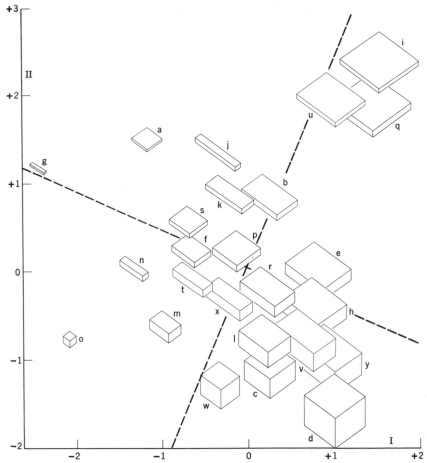

FIGURE 6.38 Plot of factor scores on first two factors of random block data. Horizontal axis is factor I, vertical axis is factor II. Blocks shown plotted at their respective positions on the two factors. Varimax-rotated factor axes shown as dashed lines.

are our observations and the m columns are the variables. Following the method used in PCA, it seems as though we could compute a matrix of factor scores, $[S^R]$, by multiplying this data matrix by the matrix of factor loadings, $[A^R]$. That is, by performing the operation

$$[X] \cdot [A^R] = [S^R]$$

If we retain p factors, the loading matrix $[A^R]$ will be $m \times p$ and the matrix of scores, $[S^R]$, will be $n \times p$. However, you will recall that the original data represent not only the factors, but also unique variation (6.51). Therefore, the matrix of scores computed in this manner will reflect in part the covariance structure of the original m variables as well as the structure of the p factors. The influence of the unique part of the original variables must, in effect, be divided out of these scores to obtain true factor scores. This may be done by multiplying the equation by the inverse of the covariance matrix:

$$[X] \cdot [s^2]^{-1} \cdot [A^R] = [\hat{S}^R] \tag{6.59}$$

The inverted covariance matrix is $m \times m$, the matrix of factor loadings is $m \times p$, so the matrix of "true" factor scores, $[\hat{S}^R]$, is $n \times p$, which is dimensionally correct. This operation will return factor scores which are free of the unique component present in each of the original observations.

Although computationally direct, this method of finding the factor scores is not used in practice. The matrix $[s^2]$ may be very large, especially in the Q-mode analyses which we will consider later, and inversion may be difficult because of its size. However, by using an algebraic identity, we can invert a $p \times p$ matrix of covariances derived from the factors and obtain the same result. Ordinarily, p is much smaller than m, simplifying the calculations although the number of matrix operations is increased.

We first compute the matrix $[S]$ by multiplying the matrix of factor loadings by its transpose:

$$[A^R]' \cdot [A^R] = [S] \tag{6.60}$$

The transpose of the $m \times p$ factor loading matrix is $p \times m$, so the result of this multiplication is a square $p \times p$ matrix. This smaller matrix is then inverted and multiplied by the factor loading matrix to yield a matrix of *score coefficients*, $[B]$:

$$[A^R] \cdot [S]^{-1} = [B] \tag{6.61}$$

These score coefficients can then be used to compute the true factor scores by matrix multiplication:

$$[X] \cdot [B] = [\hat{S}^R] \tag{6.62}$$

This series of operations in (6.60)–(6.62) can be expanded and expressed entirely in terms of the factor loading matrix, $[A^R]$:

$$[X] \cdot [B] = [\hat{S}^R]$$
$$[X] \cdot [A^R] \cdot [S]^{-1} = [\hat{S}^R]$$
$$[X] \cdot [A^R] \cdot ([A^R]' \cdot [A^R])^{-1} = [\hat{S}^R] \tag{6.63}$$

The same procedure is used to produce factor scores on unrotated or varimax rotated factor axes. Note that the data matrix $[X]$ contains standardized variables, and not raw variables.

The problem of specifying p, the number of factors to be retained, becomes critical at this point. The number of factors affects the magnitudes of the reproduced and residual correlation matrices, the communalities, and the loadings on the unique component. The factor loadings themselves are not affected. That is, if $p = 2$ factors are extracted from a data set, the loadings on factor I and factor II are not altered by the extraction of a third factor. However, if we extract and rotate two factors, the loadings may be radically different from those obtained if we extract and rotate three factors from the same data. The factors obtained when $p = 2$ are unconstrained during rotation. The same two factors are not as free to rotate if $p = 3$ because of the constraint of the third orthogonal axis that must also be accommodated in the m-dimensional space defined by the variables.

The varimax rotational scheme preserves the orthogonality of the factor axes. Even though the factors no longer coincide with the principal axes of the variance–covariance ellipsoid, they are at right angles to one another and hence are uncorrelated. A host of rotational schemes exists in which the requirement of orthogonality is relaxed and the factor axes may be oblique to one another. In some instances, the resulting oblique factors are more readily interpretable, because more extreme loadings may be obtained on the factors. However, some philosophical difficulties exist with oblique rotation schemes. For one, the factor model assumes that the observed matrix of variances and covariances results from correlations between the m variables and p mutually uncorrelated factors. Relaxing the restriction on orthogonality introduces intercorrelations between factors which would seem to be a violation of the original set of assumptions. If the factors themselves are correlated, relationships between factors and original variables are much more complex than the model assumes, because interactions exist between pairs of variables and pairs of factors. The presence of intercorrelation also brings up the nagging suspicion that perhaps the oblique factors are themselves nothing more than the result of correlations with some "superfactor" hidden still farther from direct observation.

Factor analysis was originally devised to explain the interrelationships in large numbers of variables by the presence of a few factors. The original applications were accompanied by theory which specified the expected nature of the factors and thus allowed their interpretation. When, however, factor analysis is applied to problems in areas where no theory of structure exists, is it necessary to deduce the meaning of the factors. At times this is not possible, either because no pattern emerges in the factor loadings or because the theoretical framework of the problem is too poorly developed for adequate understanding. However, rather than admit defeat, factor analysts have devised nonorthogonal rotation schemes that allow them to express the factors in terms of the original variables. Thus the analyst has come full circle from variables to factors for reduction in the size of the problem, back to variables for interpretation of the factors.

This should not be taken as implying that oblique methods are useless; significant results have been obtained in certain problems using these techniques. However, if ordinary orthogonal factor methods fail to yield interpretable results, most novice practitioners should resign themselves to the position that the problem is intractable using the factor approach, or that too little is known of causal relationships in the problem to allow interpretation. Many oblique factor studies of geologic data have led to trivial results, recapturing original variables now labeled as "factors" and yielding no more insight than could be gained from a careful inspection of the original correlation matrix. Oblique solutions introduce one more subjective decision into a process already arbitrary and probably should be avoided by all but the expert. Those interested should refer to Chapter 15 of Harman (1967) and to the earlier work by Thurstone (1947), especially his Chapter 15.

We will now briefly consider an alternative method for R-mode factor analysis, the maximum likelihood procedure developed by Lawley (1940) and subsequently modified by many workers. It avoids some, but not all, of the problems that beset other factor techniques. The maximum likelihood method does this by making certain initial presumptions about the nature of the factors and the unique variance. The factors are assumed to be normally distributed with means of zero and variances of one. The elements of the matrix of unique variances are also assumed to be normally distributed with a mean of zero and a variance [var ϵ_{jj}]. All of the factors and the elements of the unique variance are further assumed to be independent. Finally, the observed matrix of variances and covariances $[s^2]$ is assumed to be adequate for estimation of $[\Sigma]$, the unobservable matrix of variances and covariances between the factors.

Deriving the maximum likelihood estimators of the factor loadings requires mathematical gyrations of exceeding complexity; in common with most authors, we will forego tracing their development. Interested readers are referred to Lawley and Maxwell (1971) and Jöreskog (1977); an especially readable, short description is contained in Morrison (1976). We will content ourselves with examining the computational steps involved, as these are implemented in several commonly available libraries of computer programs.

The maximum likelihood procedure begins with the same model equation as other forms of factor analysis,

$$[s^2] = [A^R] \cdot [A^R]' + [\text{var } \epsilon_{jj}]$$

However, the maximum likelihood estimates of the factor loadings must be developed iteratively. To clarify the steps, we now introduce the notation $[_i a_{jr}]$, which means the ith iteration in the estimation of the loading of variable j on factor r. Similarly, $[_i \text{var } \epsilon_{jj \cdot r}]$ is the ith iteration in the approximation of the unique variance of variable j left after extraction of the rth factor.

The starting estimates of the loadings on the first factor, $[_0 a_{j1}]$, are based on the elements of the first eigenvector extracted from the observed matrix of variances and covariances, $[s^2]$. The elements of the eigenvector are scaled so the sum of their squares is equal to the first eigenvalue. A starting approximation of the specific variance is made as

$$[_0 \text{var } \epsilon_{jj \cdot 1}] = \text{diag}([s^2] - [_0 a_{j1}][_0 a_{j1}]') \tag{6.64}$$

(The operator *diag* means that only the diagonal elements of the matrix are retained.)

Next, we form the matrix

$$[_0 \text{var } \epsilon_{jj \cdot 1}]^{-1/2} ([s^2] - [_0 \text{var } \epsilon_{jj \cdot 1}])[_0 \text{var } \epsilon_{jj \cdot 1}]^{-1/2} \tag{6.65}$$

and extract its first eigenvalue and eigenvector. The eigenvector is again scaled so the sum of its squared elements is equal to the eigenvalue; this we designate $[_1a_{j1}]$. The estimate of the factor loading matrix at the end of the first iteration is

$$[_1A_1^R] = [_0\text{var } \epsilon_{jj\cdot1}]^{1/2} [_1a_{j1}] \tag{6.66}$$

The unique variance as estimated at the end of the first iteration is

$$[_1\text{var } \epsilon_{jj\cdot1}] = \text{diag}([s^2] - [_1A_1^R][_1A_1^R]') \tag{6.67}$$

This is analogous to the initial estimate of the unique variance given in eq. (6.64). The process is repeated again from that point, using the new estimate of $[\text{var } \epsilon_{jj\cdot1}]$. The iterations continue until $[_iA_1^R]$ and $[_{i+1}A_1^R]$ differ by no more than a trivial amount. The column vector $[_iA_1^R]$ is the maximum likelihood estimate of the loadings on the first factor.

The hypothesis that the data contain only a single factor, so that

$$[s^2] = [A_1^R][A_1^R]' + [\text{var } \epsilon_{jj\cdot1}]$$

can be tested by a χ^2 procedure described by Morrison (1976). If the hypothesis is rejected, additional factors must be estimated. These are found in an iterative process similar to that used to find the first factor, except that the process begins with the residual matrix $[s_{\text{res}}^2]$.

$$[s_{\text{res}}^2] = [s^2] - [_iA_1^R][_iA_1^R]'$$

The first two eigenvalues and eigenvectors are extracted from this matrix. The eigenvector $[_0a_{j2}]$ is the initial approximation of the second factor. This initial approximation is combined with the vector $[_iA_1^R]$ to form the $m \times 2$ matrix

$$[_0A_2^R] = [_iA_1^R {}_0a_{j2}]$$

This initial estimate of the two-factor loading matrix is used to create an equivalent of eq. (6.64). The process that was used to estimate the loadings on the first factor is now used to estimate the loadings on the second factor. After a stable estimate has been found, it is tested and if significant, the process is repeated to extract a third factor, then a fourth and so on. Eventually, all of the systematic sources of variation will be extracted and the factoring process can be stopped.

The extensions of the maximum likelihood factor process are similar to those of factor analysis based on principal components. The factors may be rotated to simple structure or even to oblique positions in the search for meaning. The problem of specifying p in advance of analysis has been avoided, and the factors are free of the bias inherent in factors extracted by simpler procedures. Unfortunately, the fundamental criticisms of factor analysis remain. In those fields with well-developed theories of causality, factor analysis may be especially useful. In geology, today's strongly held truths tend to be tomorrow's discredited concepts, and factor interpretations probably will fare no better. The skeptically minded are invited to read the critique of factor analysis in geology by Temple (1978) and the more developed criticism of the use of factor analysis in hydrology by Matalas and Reiher (1967).

Q-Mode Factor Analysis

We now turn to Q-mode factor analysis, where attention is devoted exclusively to interpretation of the inter-object relationships in a data set, rather than to the inter-variable (or covariance) relationships explored with R-mode factor analysis. The fact that the two are really equivalent has escaped most investigators, and has led to the creation of Q-mode procedures that are extremely cumbersome and computationally extravagant.

The first step in Q-mode analysis is to create an $n \times n$ matrix of similarities between samples. The correlation coefficient, however, may be considered inappropriate as a measure of similarity between samples because it requires calculation of "variances" across variables. Such a measure is at best obscure.

The most widely used similarity measure in Q-mode factor analysis is the cosine θ coefficient of proportional similarity,

$$\text{cosine } \theta_{ij} = \frac{\sum_{k=1}^{m} x_{ik} x_{jk}}{\sqrt{\sum_{k=1}^{m} x_{ik}^2 \sum_{k=1}^{m} x_{jk}^2}} \tag{6.68}$$

This expresses the similarity between object i and object j by regarding each as a vector defined in m-dimensional space. The cosine θ coefficient is the cosine of the angle between the two vectors. Note that the equation is very similar in form to the correlation coefficient (eq. 2.24); if the variables are standardized to have means of zero and standard deviations of one, the two measures are numerically identical.

Cosine θ ranges from 1.0 for two objects whose vector representations coincide, to 0.0 for objects whose vectors are at $90°$. Since cosine θ measures only the angular similarity, it is sensitive only to the relative proportions of the variables and not to their absolute magnitudes. If, for example, measurements were made on two brachiopods which were identical in shape but not in size, the cosine θ similarity measure between them would be 1.0.

The $n \times n$ matrix of similarities may be generated most conveniently if cosine θ is calculated in two steps (Jöreskog, Klovan, and Reyment, 1976). First, every element in a row of the data matrix is divided by the square root of the sum of squares of the elements in that row,

$$w_{ik} = \frac{x_{ik}}{\sqrt{\sum_{k=1}^{m} x_{ik}^2}} \tag{6.69}$$

This standardizes the objects so that the squares of the variables measured on each object sum to one. Then, cosine θ is given by

$$\text{cosine } \theta_{ij} = \sum_{k=1}^{m} w_{ik} w_{jk} \tag{6.70}$$

In matrix notation, we first define an $n \times n$ diagonal matrix $[D]$, which contains the sums of the squares of each row along the diagonal, and zeros elsewhere. The standardization step is

$$[W] = [D]^{-1/2}[X] \tag{6.71}$$

TABLE 6.28 Cosine θ Proportional Similarities Between 25 Blocks Listed in Table 6.18; Only the Lower Half of the Symmetric 25 × 25 Matrix is Shown

	a	b	c	d	e	f	g	h	i	j	k	l	m	n	o	p	q	r	s	t	u	v	w	x	y
a	1.0000																								
b	0.8993	1.0000																							
c	0.5992	0.8554	1.0000																						
d	0.5096	0.7916	0.9920	1.0000																					
e	0.7406	0.9520	0.9658	0.9284	1.0000																				
f	0.9167	0.9959	0.8566	0.7917	0.9462	1.0000																			
g	0.9314	0.7963	0.5005	0.4152	0.6284	0.8276	1.0000																		
h	0.6782	0.9191	0.9848	0.9586	0.9951	0.8790	0.5708	1.0000																	
i	0.9458	0.9817	0.7881	0.7129	0.9021	0.9834	0.8130	0.8577	1.0000																
j	0.9137	0.9732	0.7591	0.6909	0.8790	0.9664	0.8628	0.8369	0.9472	1.0000															
k	0.9113	0.9935	0.8138	0.7477	0.9225	0.9877	0.8326	0.8850	0.9705	0.9929	1.0000														
l	0.6457	0.8963	0.9939	0.9766	0.9851	0.8927	0.5517	0.9965	0.8265	0.8154	0.8628	1.0000													
m	0.7873	0.9585	0.9529	0.9156	0.9830	0.9623	0.7258	0.9761	0.9042	0.9123	0.9423	0.9736	1.0000												
n	0.8868	0.9683	0.8183	0.7586	0.9074	0.9714	0.8724	0.8768	0.9259	0.9837	0.9824	0.8648	0.9531	1.0000											
o	0.7909	0.8171	0.7825	0.7385	0.7981	0.8579	0.8659	0.7860	0.7813	0.8098	0.8174	0.7943	0.8834	0.8924	1.0000										
p	0.8777	0.9887	0.8962	0.8392	0.9703	0.9924	0.7652	0.9449	0.9759	0.9332	0.9673	0.9251	0.9703	0.9442	0.8411	1.0000									
q	0.8766	0.9904	0.8826	0.8207	0.9664	0.9889	0.7448	0.9377	0.9820	0.9342	0.9687	0.9133	0.9561	0.9332	0.8003	0.9971	1.0000								
r	0.7393	0.9545	0.9603	0.9252	0.9960	0.9464	0.6462	0.9914	0.8934	0.8979	0.9335	0.9844	0.9888	0.9270	0.8081	0.9626	0.9573	1.0000							
s	0.9589	0.9828	0.7937	0.7193	0.9006	0.9915	0.8652	0.8579	0.9930	0.9598	0.9772	0.8319	0.9220	0.9538	0.8452	0.9772	0.9753	0.8972	1.0000						
t	0.8154	0.9758	0.8962	0.8495	0.9636	0.9685	0.7599	0.9456	0.9174	0.9624	0.9764	0.9359	0.9819	0.9801	0.8432	0.9602	0.9542	0.9794	0.9317	1.0000					
u	0.9460	0.9882	0.7900	0.7156	0.9061	0.9880	0.8244	0.8620	0.9984	0.9630	0.9819	0.8321	0.9129	0.9424	0.7893	0.9761	0.9816	0.9023	0.9943	0.9386	1.0000				
v	0.6095	0.8833	0.9833	0.9702	0.9759	0.8729	0.5244	0.9896	0.7987	0.8138	0.8561	0.9948	0.9639	0.8609	0.7651	0.9016	0.8917	0.9823	0.8041	0.9328	0.8091	1.0000			
w	0.5899	0.8396	0.9961	0.9943	0.9494	0.8438	0.5096	0.9715	0.7671	0.7526	0.8026	0.9869	0.9498	0.8182	0.8007	0.8812	0.8609	0.9487	0.7799	0.8906	0.7711	0.9776	1.0000		
x	0.7359	0.9503	0.9395	0.9059	0.9802	0.9394	0.6682	0.9747	0.8774	0.9166	0.9410	0.9704	0.9850	0.9446	0.8128	0.9450	0.9385	0.9939	0.8874	0.9900	0.8921	0.9769	0.9332	1.0000	
y	0.5749	0.8571	0.9925	0.9857	0.9676	0.8495	0.4835	0.9869	0.7739	0.7739	0.8227	0.9954	0.9520	0.8280	0.7547	0.8861	0.8740	0.9698	0.7784	0.9127	0.7814	0.9968	0.9877	0.9584	1.0000

TABLE 6.29 Eigenvalues and First Three Eigenvectors and Factor Loading Vectors of Cosine θ Matrix in Table 6.28. Eigenvalues VIII to XXV are Identically Zero

Vector	Eigenvalue	Total Variance (%)	Total Variance (Cumulative %)
I	22.2999	89.20	89.20
II	1.9907	7.96	97.16
III	0.4546	1.82	98.98
IV	0.2192	0.88	99.86
V	0.0324	0.13	99.99
VI	0.0029	0.01	100.00
VII	0.0004	0.00	100.00

Block	Eigenvector I	Eigenvector II	Eigenvector III	Factor I	Factor II	Factor III
a	0.1780	−0.3709	−0.0414	0.8407	−0.5234	−0.0414
b	0.2085	−0.0969	−0.1593	0.9845	−0.1368	−0.1593
c	0.1961	0.2574	0.0625	0.9261	0.3632	0.0625
d	0.1859	0.3260	0.1047	0.8780	0.4600	0.1047
e	0.2079	0.1155	−0.1101	0.9816	0.1629	−0.1101
f	0.2087	−0.1123	−0.0613	0.9858	−0.1585	−0.0613
g	0.1606	−0.4137	0.4273	0.7586	−0.5836	0.4273
h	0.2042	0.1804	−0.0635	0.9644	0.2545	−0.0635
i	0.2004	−0.1783	−0.2520	0.9462	−0.2516	−0.2520
j	0.1997	−0.1894	−0.0658	0.9431	−0.2672	−0.0658
k	0.2055	−0.1417	−0.1168	0.9706	−0.1999	−0.1168
l	0.2020	0.2126	0.0179	0.9537	0.2999	0.0179
m	0.2103	0.0604	0.1123	0.9933	0.0852	0.1123
n	0.2045	−0.1301	0.1498	0.9658	−0.1835	0.1498
o	0.1833	−0.0913	0.7009	0.8655	−0.1288	0.7009
p	0.2095	−0.0459	−0.1089	0.9893	−0.0648	−0.1089
q	0.2078	−0.0535	−0.2168	0.9814	−0.0755	−0.2168
r	0.2085	0.1095	−0.0637	0.9847	0.1545	−0.0637
s	0.2026	−0.1917	−0.0826	0.9566	−0.2705	−0.0826
t	0.2090	−0.0069	0.0094	0.9870	−0.0097	0.0094
u	0.2018	−0.1792	−0.2294	0.9531	−0.2528	−0.2294
v	0.1993	0.2327	0.0011	0.9411	0.3283	0.0011
w	0.1943	0.2603	0.1567	0.9176	0.3673	0.1567
x	0.2073	0.0925	−0.0103	0.9791	0.1305	−0.0103
y	0.1957	0.2697	0.0197	0.9243	0.3805	0.0197

and the similarity matrix [cosine θ] is

$$[\text{cosine } \theta] = [W][W]' \tag{6.72}$$
$$= [D]^{-1}[X][X]'[D]^{-1}$$

The fact that the $n \times n$ matrix [cosine θ] can have at most only m eigenvalues suggests that it should be possible to take advantage of the Eckart-Young theorem and to extract the eigenvectors from the $m \times m$ matrix $[W]'[W]$ rather than from the larger matrix [cosine θ], and then transform the R-mode scores into Q-mode loadings and vice versa. This topic is considered in greater detail in the next section. Under certain circumstances the reciprocal relationship between R- and Q-mode factors and scores holds exactly, but this depends upon the nature of any scaling that is performed on the data matrix $[X]$.

Table 6.28 contains the Q-mode cosine θ similarity matrix for the random block data, and Table 6.29 gives the eigenvalues and first three eigenvectors. As we would expect, the first few eigenvalues account for almost all of the variation among the blocks. Their eigenvectors can be converted into factor loadings by multiplying each element in a vector by the corresponding singular value (or square root of the corresponding eigenvalue). This is exactly the same procedure used in R-mode factor analysis. It scales the Q-mode factor axes so their lengths are proportional to the amount of variation between the objects that they contain. Table 6.29 also lists the Q-mode factors that correspond to the first three eigenvectors.

In Q-mode analysis, we plot the loadings rather than the factor scores if we wish to see relationships between the objects in our sample. Figure 6.39 is a plot of the first two Q-mode factors; the blocks are shown in positions representing their loadings on the factors. The arc on the diagram is part of a circle representing a communality of 1.00; if an object falls on the circle, the two factors account for all of its variability. Blocks that plot inside the circle are characterized by variability that is not represented by the two factors.

Figure 6.40 is a plot of the second and third Q-mode factors. This plot and that shown in Figure 6.39 together represent 99% of the variation in the blocks, which is exactly what we expect from our knowledge of how the blocks were originally created. Although Figure 6.39 shows a general progression from larger to smaller blocks, the distinction by shape is not as clear-cut as in R-mode factor analysis (Fig. 6.38). You will note that the second and third factors shown in Figure 6.40 produce patterns of loadings that are superior for classification purposes to the first factor, even through the first eigenvector is an order of magnitude larger than the second.

In R-mode factor analysis, the data are centered about zero by subtracting the means from each observation prior to factoring. This is not done in Q-mode factor analysis, with the result that the first Q-mode factor is simply a vector from the origin to the centroid of the set of objects. In effect, this first factor expresses "size," or the combined magnitudes of the variables measured on each object. Typically, all of the loadings on the first factor will be positive in sign and about equal in magnitude. Since the first factor reveals very little about the structure of

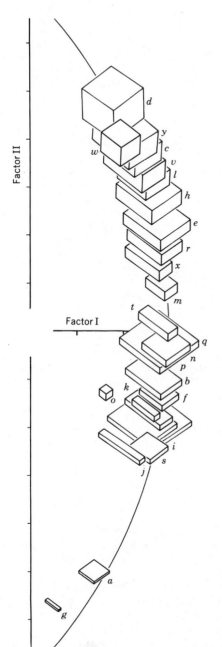

FIGURE 6.39 Plot of *Q*-mode factor loadings on the first two factor axes of block data, calculated from cosine θ similarity matrix (Table 6.28). Horizontal axis is Factor I, vertical axis is Factor II. Note that vertical axis has been shifted from origin. Circular arc represents communality of 1.00.

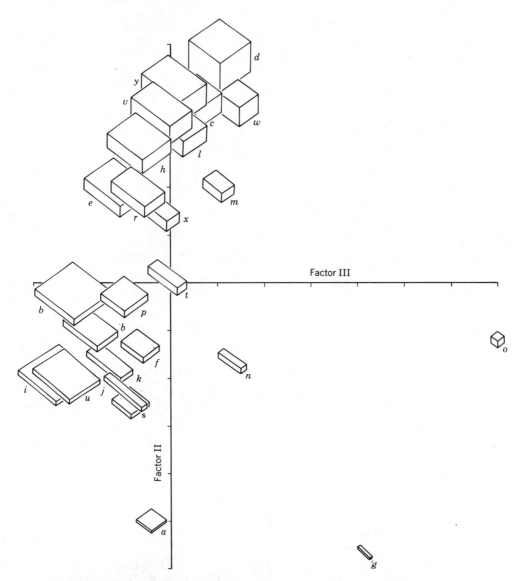

FIGURE 6.40 Plot of Q-mode factor loadings on second and third factor axes of block data, calculated from cosine θ similarity matrix (Table 6.28).

the data, usually it is regarded as a ''nuisance factor'' and discarded. The second and third factors, as in Figure 6.40, are considered to be much more revealing.

The geologic example we will consider is taken from igneous petrology. Table 6.30 contains data on the major chemical constituents of 20 rock samples taken from a complex and apparently differentiated igneous body. By Q-mode analysis, we hope to place each sample in its proper place within the differentiation series.

TABLE 6.30 Major Oxides Contained in 20 Samples Collected from an Igneous Complex; Sample Names Assigned by Petrographic Analysis

Sample Name	Sample Number	$X_1 = SiO_2$	$X_2 = Al_2O_3$	$X_3 = Fe_2O_3$	$X_4 = FeO$	$X_5 = MgO$	$X_6 = CaO$	$X_7 = Na_2O$	$X_8 = K_2O$
Syenite	1	61.7	15.1	2.0	2.3	3.7	4.6	4.4	4.5
Syenite	2	58.3	17.9	3.2	1.7	1.5	3.7	5.9	5.3
Syenite	3	51.2	17.6	3.5	4.3	3.2	4.5	5.7	4.4
Monzonite	4	54.4	14.3	3.3	4.1	6.1	7.7	3.4	4.2
Diorite	5	58.0	15.7	0.7	2.8	5.0	10.9	3.0	3.2
Diorite	6	46.9	15.9	2.9	10.0	7.0	9.6	2.7	0.7
Diorite	7	58.0	17.3	2.2	3.8	2.2	4.3	4.3	4.1
Quartz diorite	8	55.5	16.5	1.7	4.6	6.7	6.7	3.2	2.5
Gabbro	9	55.4	15.3	2.7	5.5	5.8	9.9	2.9	1.5
Gabbro	10	55.9	13.5	2.7	5.9	6.5	8.9	2.4	1.7
Norite	11	47.2	14.5	1.6	13.8	5.2	8.1	3.1	1.2
Norite	12	48.2	18.3	1.3	6.1	10.8	9.4	1.3	0.7
Hypersthene gabbro	13	44.8	18.8	2.2	4.7	11.3	14.6	0.9	0.1
Hypersthene gabbro	14	47.0	14.1	0.8	15.0	16.0	2.3	0.4	1.7
Syenite	15	59.8	17.3	3.6	1.6	1.2	3.8	5.0	5.1
Quartz syenite	16	66.2	16.2	2.0	0.2	0.8	1.3	6.5	5.8
Altered syenite[a]	17	50.0	9.9	3.5	5.0	11.9	8.3	2.4	5.0
Monzonite	18	57.4	18.5	3.7	2.1	1.7	6.8	4.5	3.7
Monzonite	19	59.8	15.8	3.8	3.3	2.2	3.9	3.0	4.4
Diabase	20	52.2	18.2	3.3	4.4	4.7	6.5	4.6	1.9

[a]Contains secondary minerals including diopside.

TABLE 6.31 Cosine θ Coefficients Between 20 Samples of Igneous Rock from a Differentiated Pluton

Sample	1	2	3	4	5	6	7	8	9	10	11	12	13	14	15	16	17	18	19	20
1	1.000																			
2	0.997	1.000																		
3	0.994	0.997	1.000																	
4	0.996	0.991	0.994	1.000																
5	0.993	0.988	0.989	0.997	1.000															
6	0.972	0.968	0.981	0.987	0.984	1.000														
7	0.998	0.999	0.998	0.995	0.992	0.977	1.000													
8	0.995	0.991	0.995	0.998	0.996	0.989	0.995	1.000												
9	0.991	0.986	0.990	0.998	0.998	0.992	0.991	0.998	1.000											
10	0.992	0.985	0.988	0.998	0.996	0.991	0.991	0.997	0.999	1.000										
11	0.966	0.961	0.975	0.978	0.974	0.996	0.972	0.981	0.984	0.984	1.000									
12	0.971	0.966	0.978	0.985	0.985	0.993	0.974	0.990	0.990	0.988	0.981	1.000								
13	0.948	0.943	0.958	0.970	0.973	0.984	0.951	0.973	0.978	0.973	0.965	0.993	1.000							
14	0.934	0.922	0.940	0.950	0.937	0.970	0.936	0.957	0.952	0.956	0.972	0.969	0.945	1.000						
15	0.998	1.000	0.996	0.992	0.989	0.967	0.999	0.991	0.987	0.986	0.960	0.965	0.941	0.921	1.000					
16	0.997	0.997	0.989	0.985	0.982	0.953	0.995	0.985	0.978	0.978	0.948	0.951	0.922	0.911	0.997	1.000				
17	0.979	0.967	0.973	0.990	0.984	0.980	0.973	0.987	0.987	0.990	0.970	0.982	0.969	0.965	0.968	0.961	1.000			
18	0.996	0.998	0.997	0.994	0.994	0.977	0.998	0.994	0.992	0.990	0.968	0.975	0.958	0.925	0.998	0.992	0.971	1.000		
19	0.999	0.998	0.995	0.995	0.991	0.973	0.999	0.994	0.990	0.991	0.968	0.970	0.946	0.934	0.999	0.996	0.975	0.997	1.000	
20	0.992	0.993	0.998	0.996	0.993	0.989	0.996	0.998	0.996	0.993	0.980	0.988	0.972	0.947	0.992	0.984	0.977	0.997	0.993	1.000

TABLE 6.32   First Five Factors Extracted from Cosine θ Matrix in Table 6.31

Sample	I	II	III	IV	V	Communality
1	0.9948	−0.0910	0.0242	0.0324	0.0069	0.9996
2	0.9918	−0.1223	0.0081	−0.0177	−0.0268	0.9997
3	0.9958	−0.0587	0.0085	−0.0457	−0.0344	0.9983
4	0.9989	−0.0126	−0.0070	0.0357	0.0178	0.9997
5	0.9963	−0.0191	−0.0596	0.0297	0.0353	0.9986
6	0.9904	0.1188	−0.0133	−0.0594	0.0309	0.9997
7	0.9959	−0.0838	0.0191	−0.0235	−0.0086	0.9998
8	0.9996	0.0010	−0.0017	0.0112	−0.0132	0.9996
9	0.9983	0.0204	−0.0336	0.0055	0.0391	0.9997
10	0.9978	0.0223	−0.0049	0.0291	0.0498	0.9994
11	0.9833	0.1202	0.0550	−0.0988	0.0746	0.9997
12	0.9890	0.1259	−0.0512	0.0008	−0.0538	0.9995
13	0.9721	0.1719	−0.1552	0.0066	−0.0365	0.9999
14	0.9561	0.2323	0.1691	0.0146	−0.0527	0.9997
15	0.9918	−0.1257	0.0102	−0.0084	−0.0137	0.9998
16	0.9844	−0.1665	0.0458	0.0203	−0.0113	0.9994
17	0.9866	0.0783	0.0214	0.1316	0.0259	0.9980
18	0.9950	−0.0870	−0.0367	−0.0275	−0.0089	0.9998
19	0.9945	−0.0946	0.0296	0.0035	0.0066	0.9989
20	0.9981	−0.0161	−0.0236	−0.0395	−0.0295	0.9995
Variance	98.124	1.148	0.349	0.204	0.116	
Cumulative variance	98.124	99.272	99.621	99.825	99.942	

The succession of samples within the sequence between end members can be shown by plotting the samples as loadings on pairs of factors. In this example, the communalities of all samples on the first two factors is almost 1.00, so the samples define vectors that plot as radii of a circle. The angles between different radii are a measure of the similarity between samples, as expressed by the cosine θ coefficient. The cosine θ similarity matrix between observations is given in Table 6.31, and the factor matrix is given in Table 6.32. Two factors were retained for rotation; rotated factor loadings are given in Table 6.33. These loadings are shown plotted in Figure 6.41. Note that the observations form a gradational sequence from samples classified as hypersthene gabbro at one extreme to quartz syenite at the other. Table 6.33 also contains the scores of the original variables (chemical constituents) on the two factors. It is obvious that both factors heavily weight the relative abundance of silica plus alumina, conforming to traditional interpretations of igneous rock sequences.

A few final remarks should be made about Q-mode factor analysis. A primary purpose of the technique is to identify end members—real or hypothetical objects

TABLE 6.33   Rotated Factor Loadings and Varimax Factor Scores

Sample	I	II	Communality
1	0.7851	0.6177	0.9980
2	0.8044	0.5929	0.9986
3	0.7636	0.6418	0.9950
4	0.7342	0.6774	0.9980
5	0.7368	0.6709	0.9929
6	0.6377	0.7671	0.9950
7	0.7809	0.6236	0.9988
8	0.7254	0.6878	0.9993
9	0.7111	0.7009	0.9970
10	0.7094	0.7020	0.9960
11	0.6316	0.7632	0.9814
12	0.6319	0.7712	0.9940
13	0.5879	0.7930	0.9745
14	0.5348	0.8259	0.9681
15	0.8068	0.5904	0.9995
16	0.8295	0.5556	0.9968
17	0.6628	0.7350	0.9796
18	0.7825	0.6207	0.9976
19	0.7873	0.6148	0.9979
20	0.7360	0.6744	0.9965
Variance	52.311	46.962	
Cumulative variance	52.311	99.272	

Varimax Factor Score Matrix

Variable	Factor I	Factor II
X_1	70.2648	5.6766
X_2	14.5830	8.1431
X_3	4.5006	-1.0267
X_4	-16.2185	24.5371
X_5	-19.4934	28.8944
X_6	-6.5178	16.8625
X_7	11.4400	-6.9660
X_8	11.1204	-7.2130

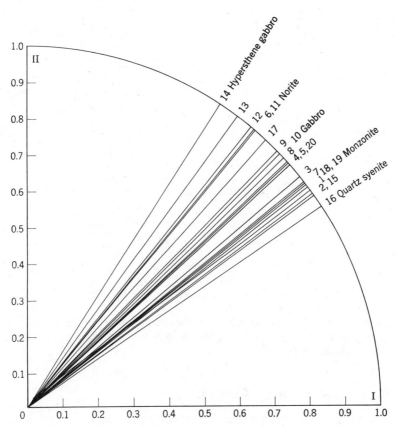

FIGURE 6.41 Plot of factor loadings from Q-mode analysis of 20 igneous rock analyses.

having extreme properties. Often intermediate objects can be regarded as mixtures of the two end members, just as the igneous rocks shown in Figure 6.41 could be interpreted as being from a gradational sequence between sialic and femic extremes. Q-mode factor analysis also is used for classification, an objective that is essentially the same as cluster analysis, yet is much more costly in terms of computing time. If the object of an analysis is to search through samples for groups or clusters, either cluster analysis or R-mode factor analysis may provide more efficient means for doing so. If significant factors can be extracted by R-mode methods, scatter diagrams of factor scores or cluster diagrams usually will reveal the relationships between samples. As an example, Figure 6.42 is a dendrogram constructed by weighted pair-group averaging of the matrix of correlations between the igneous rock data from Table 6.30. The relative arrangement of samples is almost exactly the same as that obtained by Q-mode analysis.

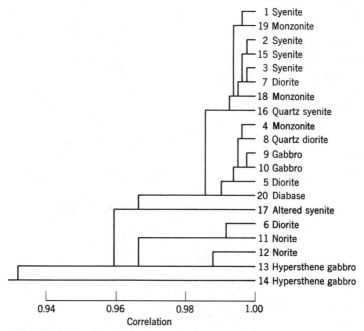

FIGURE 6.42   Cluster analysis of chemical analyses of igneous rocks. The clustering shows essentially the same arrangement as Q-mode factor analysis.

Principal Coordinates Analysis

Principal coordinates analysis is a widely used Q-mode factor technique. Most applications have been in quantitative biology and paleontology, although the method has also been used in petrology. Principal coordinates analysis was popularized by Gower (1966), and is extensively discussed in Reyment, Blackith, and Campbell (1984) and Jöreskog, Klovan, and Reyment (1976). The objective of the procedure is the same as that of Q-mode factor analysis: to determine if a collection of multivariate observations represents a sample from a single population or a mixture from several different populations.

In principal coordinates, the first step is to calculate an $n \times n$ matrix of similarities in the form of distances between the n objects in the data. Any of a number of distance measures can be used, including the Euclidean distance:

$$D_{ij} = \sum_{k=1}^{m} (x_{ik} - x_{jk})^2 \tag{6.73}$$

Perhaps the most widely used distance measure is the *Gower distance,*

$$G_{ij} = \sum_{k=1}^{m} \left(1 - \frac{|x_{ik} - x_{jk}|}{\text{Range}_k}\right) \bigg/ m \tag{6.74}$$

To compute the Gower distance between object i and object j, the absolute difference between them for variable k is divided by the range of variable k. This gives a number in the interval 0.0 to 1.0, with a small number representing close similarity and a number approaching 1.0 representing maximum dissimilarity. To make the similarity measure behave in a manner analogous to the correlation coefficient, this quantity is subtracted from 1.0. The calculations are repeated for all variables k measured on objects i and j, then summed and divided by the number of variables to give the quantity, G_{ij}.

In computing the Gower distance, no assumptions are made about the nature of the data; observations may be nominal or ordinal, or of higher rank. In fact, the data matrix may consist of mixtures of different types of numbers, such as counts of the number of plates in the calyxes of crinoids, the lengths of their arms, and the ratios of calyx heights to diameters. The similarity measures for all possible pairs of objects are assembled in an $n \times n$ *association matrix*, [A]. This matrix

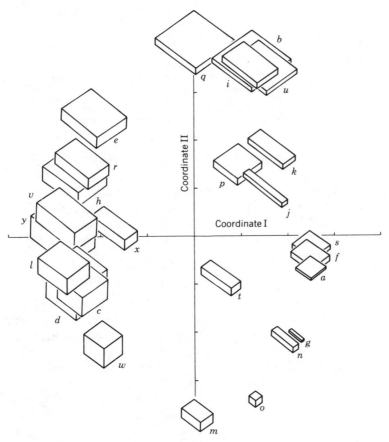

FIGURE 6.43 Plot of first two principal coordinates of random block data. Blocks are shown at positions corresponding to their loadings on the principal coordinates.

TABLE 6.34 Matrix of Similarities between 25 Random Blocks; Upper Diagonal Half of formed by Subtraction of Row and Column Totals and Addition of Grand Mean

	a	b	c	d	e	f	g	h	i	j	k	l
	1.0000	0.2262	−0.7412	−0.8077	−0.8564	0.9841	0.8343	−0.9455	0.3172	0.4510	0.4431	−0.8790
a	1.0720	1.0000	−0.7339	−0.6850	0.1049	0.2151	−0.0963	−0.2045	0.8481	0.6541	0.8157	−0.6104
b	0.2415	0.9586	1.0000	0.9852	0.5418	−0.7002	−0.6039	0.7751	−0.5441	−0.8599	−0.8943	0.9693
c	−0.7098	−0.7593	0.9906	1.0000	0.5617	−0.7892	−0.6494	0.7944	−0.5669	−0.7646	−0.8084	0.9811
d	−0.7810	−0.7151	0.9711	0.9813	1.0000	−0.7909	−0.9036	0.9460	0.1543	−0.4760	−0.3594	0.6787
e	−0.8551	0.0495	0.5024	0.5177	0.9306	1.0000	0.7957	−0.8875	0.3480	0.3475	0.3649	−0.8397
f	1.0546	0.2289	−0.6704	−0.7640	−0.7910	1.0690	1.0000	−0.8896	−0.1708	0.4622	0.3298	−0.7092
g	0.9071	−0.0803	−0.5719	−0.6220	−0.9016	0.8669	1.0734	1.0000	−0.1405	−0.6391	−0.5781	0.8794
h	−0.9390	−0.2548	0.7408	0.7555	0.8817	−0.8826	−0.8825	0.9408	1.0000	0.2692	0.4780	−0.5221
i	0.3470	0.8211	−0.5550	−0.5825	0.1134	0.3762	−0.1403	−0.1763	0.9876	1.0000	0.9666	−0.7675
j	0.4833	0.6296	−0.8683	−0.7777	−0.5144	0.3782	0.4952	−0.6724	0.2593	0.9925	1.0000	−0.7905
k	0.4697	0.7855	−0.9084	−0.8271	−0.4036	0.3900	0.3571	−0.6171	0.4623	0.9534	0.9811	1.0000
l	−0.8590	−0.6471	0.9487	0.9558	0.6280	−0.8211	−0.6885	0.8338	−0.5443	−0.7872	−0.8159	0.9680
m	0.3111	−0.8425	0.3595	0.2940	−0.5675	0.3131	0.5830	−0.2869	−0.7371	−0.3018	−0.4739	0.1880
n	0.7655	−0.0741	−0.5688	−0.5674	−0.9086	0.6940	1.0072	−0.8617	−0.2898	0.6344	0.4691	−0.6365
o	0.6957	−0.5691	−0.0561	−0.1427	−0.7883	0.6882	0.9193	−0.6164	−0.0041	−0.1701		−0.2410
p	0.4746	0.2308	−0.1474	−0.2901	0.0296	0.5791	0.0752	−0.1165	0.6894	−0.3168	−0.1592	−0.2761
q	−0.1584	0.6667	−0.1923	−0.2124	0.5752	−0.0982	−0.5417	0.3158	0.8304	−0.0631	0.1482	−0.1169
r	−0.9751	−0.0728	0.5287	0.6020	0.8124	−0.9549	−0.8468	0.8404	−0.2192	−0.3431	−0.3066	0.6971
s	1.0461	0.2712	−0.6559	−0.7527	−0.7369	1.0598	0.8016	−0.8458	0.4692	0.3434	0.3738	−0.8145
t	0.1256	0.1052	−0.4431	−0.3302	−0.4639	0.0198	0.4518	−0.4368	−0.3946	0.7529	0.6067	−0.3409
u	0.4113	0.9005	−0.6983	−0.7017	−0.0028	0.4195	−0.0412	−0.3031	0.9610	0.4535	0.6391	−0.6653
v	−0.9803	−0.3818	0.7456	0.8145	0.6970	−0.9766	−0.7733	0.8360	−0.4826	−0.4848	−0.5086	0.8588
w	−0.4782	−0.9064	0.9361	0.9146	0.1804	−0.4646	−0.2833	0.4838	−0.7217	−0.7670	−0.8664	0.8408
x	−0.7142	−0.0972	0.2293	0.3639	0.3156	−0.7764	−0.3890	0.3961	−0.5164	0.1573	0.0741	0.4088
y	−0.9549	−0.4889	0.8523	0.8964	0.7001	−0.9377	−0.7756	0.8652	−0.4918	−0.6322	−0.6533	0.9264

will be symmetrical and have 1's down the diagonal, with values between 0.0 and 1.0 elsewhere.

Each row of the matrix $[A]$ is summed and divided by n to give the row averages; each column of $[A]$ is also summed and divided by n to give column averages. These are designated $\bar{a}_{j \cdot}$ and $\bar{a}_{\cdot k}$, respectively. The grand total of either the row or column totals, designated $\bar{\bar{a}}_{\cdot \cdot}$, is also found. The elements a_{jk} are now transformed to give a new matrix $[Q]$ by the operation

$$q_{jk} = a_{jk} + \bar{\bar{a}}_{\cdot \cdot} - (\bar{a}_{j \cdot} + \bar{a}_{\cdot k}) \qquad (6.75)$$

We may regard the n objects as being located in the m-dimensional space defined by the variables. The transformation in eq. (6.75) moves the origin of the coordinate system describing the m dimensions so it coincides with the centroid of the cloud of points. The operation also closes the data set, as all rows and columns now sum to zero, so one eigenvalue of $[Q]$ is forced to become zero. This tends to increase the relative magnitudes of the first few eigenvalues.

Next, the eigenvalues and eigenvectors are extracted from $[Q]$; these are the principal coordinates of the data set. The relative importance of each coordinate may be assessed simply by calculating the percentage of the trace of $[Q]$ contained in each successive eigenvalue. Usually, only the first few coordinates are of interest, as these hopefully will account for most of the differences between the observations.

Matrix Contains Gower Distances, Lower Diagonal Half Contains Gower Distances Trans-

m	n	o	p	q	r	s	t	u	v	w	x	y	
0.2417	0.7021	0.6164	0.4201	-0.1729	-0.9702	0.9750	0.0980	0.3836	-0.9852	-0.5220	-0.7157	-0.9656	a
-0.8552	-0.0808	-0.5916	0.2330	0.7089	-0.0112	0.2568	0.1343	0.9295	-0.3299	-0.8934	-0.0419	-0.4429	b
0.3308	-0.5915	-0.0947	-0.1612	-0.1661	0.5743	-0.6863	-0.4300	-0.6854	0.7814	0.9330	0.2685	0.8823	c
0.2700	-0.5855	-0.1766	-0.2993	-0.1815	0.6522	-0.7785	-0.3126	-0.6841	0.8549	0.9162	0.4077	0.9311	d
-0.5661	-0.9013	-0.7969	0.0458	0.6314	0.8881	-0.7372	-0.4208	0.0402	0.7629	0.2073	0.3848	0.7600	e
0.2453	0.6321	0.6104	0.5261	-0.1112	-0.9484	0.9903	-0.0064	0.3933	-0.9799	-0.5069	-0.7764	-0.9469	f
0.5130	0.9431	0.8394	0.0199	-0.5568	-0.8426	0.7299	0.4235	-0.0696	-0.7789	-0.3278	-0.3912	-0.7870	g
-0.2907	-0.8595	-0.6301	-0.1054	0.3669	0.9109	-0.8513	-0.3988	-0.2652	0.8967	0.5056	0.4622	0.9200	h
-0.7642	-0.3110	-0.5023	0.6771	0.8581	-0.1720	0.4404	-0.3800	0.9755	-0.4453	-0.7232	-0.4756	-0.4603	i
-0.3314	0.6108	-0.0437	-0.3316	-0.0378	-0.2985	0.3120	0.7650	0.4656	-0.4499	-0.7711	0.1956	-0.6032	j
-0.4977	0.4512	-0.2039	-0.1683	0.1792	-0.2563	0.3482	0.6245	0.6569	-0.4680	-0.8647	0.1181	-0.6185	k
0.1706	-0.6479	-0.2683	-0.2786	-0.0794	0.7540	-0.8336	-0.3165	-0.6410	0.9060	0.8490	0.4593	0.9677	l
1.0000	0.5012	0.8785	-0.1803	-0.8831	-0.4139	0.1718	0.0896	-0.7714	-0.1077	0.6221	-0.1589	-0.0154	m
1.0667	1.0000	0.7526	-0.2895	-0.6827	-0.6937	0.5425	0.6872	-0.1568	-0.6164	-0.2941	-0.0945	-0.6701	n
0.5620	1.0548	1.0000	0.0295	-0.7698	-0.7373	0.5440	0.1489	-0.4812	-0.5195	0.2127	-0.4203	-0.4458	o
0.9551	0.8232	1.0865	1.0000	0.5796	-0.4083	0.6087	-0.8063	0.5398	-0.5221	-0.2617	-0.8887	-0.4085	p
-0.1285	-0.2437	0.0912	1.0370	1.0000	0.3030	-0.0128	-0.5204	0.7763	0.0179	-0.4825	-0.2003	0.0061	q
-0.8713	-0.6769	-0.7481	0.5765	0.9569	1.0000	-0.9431	-0.0169	0.2108	0.9366	0.3190	0.7579	0.8824	r
-0.4215	-0.7073	-0.7350	-0.4308	0.2406	0.9181	1.0000	-0.1066	0.4668	-0.9853	-0.5184	-0.8367	-0.9415	s
0.2402	0.6050	0.6224	0.6623	0.0008	-0.9490	1.0701	1.0000	-0.1713	-0.0178	-0.2727	0.6199	-0.1728	t
0.1146	0.7062	0.1838	-0.7962	-0.5503	-0.0662	-0.0799	0.9833	1.0000	-0.4969	-0.8292	-0.3781	-0.5469	u
-0.7464	-0.1378	-0.4463	0.5500	0.7465	-0.2601	0.4935	-0.1880	0.9834	1.0000	0.6137	0.7614	0.9819	v
-0.1055	-0.6202	-0.5074	-0.5348	-0.0348	0.8646	-0.9813	-0.0573	-0.5364	0.9377	1.0000	0.2032	0.7235	w
0.6632	-0.2589	0.2637	-0.2354	-0.4963	0.2858	-0.4756	-0.2733	-0.8297	0.5904	1.0155	1.0000	0.6345	x
-0.1601	-0.1016	-0.4115	-0.9048	-0.2563	0.6825	-0.8361	0.5770	-0.4209	0.6958	0.1765	0.9310	1.0000	y
-0.0074	-0.6680	-0.4279	-0.4153	-0.0407	0.8162	-0.9317	-0.2065	-0.5805	0.9254	0.7060	0.5747	0.9494	

TABLE 6.35 Eigenvalues Associated with First Seven Coordinates Extracted from Block Data; Column 1 Contains the Successive Eigenvalues, Column 2 the Percent of Total Variation Accounted for by Each, and Column 3 the Cumulative Variation Accounted for

	1	2	3
I	13.3598	53.5758	53.5758
II	6.9122	27.7197	81.2954
III	4.2627	17.0943	98.3897
IV	0.3291	1.3200	99.7097
V	0.0682	0.2735	99.9832
VI	0.0042	0.0168	100.0000
VII	0.0000	0.0000	100.0000

The final step is to plot the individual loadings of the principal coordinates; this is done by cross-plotting the n elements (each corresponding to an object) of the eigenvectors.

We will use the artificial block data to illustrate principal coordinates analysis, so the results may be compared to those obtained from other eigenvector techniques. Table 6.34 contains the 25×25 matrix of similarities between the individual blocks, calculated using the Gower distance measure. The upper diagonal half contains Gower distance measures as defined by eq. (6.74) and taken from matrix $[A]$; the lower diagonal half contains similarity measures after transformation by subtracting the row and column mean from every element, then adding the grand mean, as defined in eq. (6.75). It is from this matrix $[Q]$ that the eigenvalues and eigenvectors are extracted.

TABLE 6.36 First Two Principal Coordinates of Block Data; Each Element Corresponds to a Specific Block

	Principal Coordinate	
Block	I	II
a	0.2685	−0.0847
b	0.1318	0.3113
c	−0.2405	−0.1235
d	−0.2503	−0.1110
e	−0.2071	0.2086
f	0.2606	−0.0745
g	0.2250	−0.2210
h	−0.2499	0.1019
i	0.1249	0.3108
j	0.1880	0.0683
k	0.1897	0.1463
l	−0.2609	−0.0727
m	0.0105	−0.3822
n	0.2005	−0.2333
o	0.1266	−0.3410
p	0.0978	0.1222
q	−0.0050	0.3500
r	−0.2304	0.1256
s	0.2573	−0.0431
t	0.0758	−0.1090
u	0.1532	0.3030
v	−0.2572	0.0085
w	−0.1942	−0.2370
x	−0.1491	−0.0076
y	−0.2657	−0.0159

Table 6.35 gives the first seven eigenvalues extracted from matrix $[Q]$. Note that the seventh and all succeeding eigenvalues are zero. In fact, the first two eigenvalues account for 81% of the total variation in the block data, and the third eigenvalue accounts for an additional 17%, or essentially all of the remaining variation. (Recall that the data were generated from only three independent variables. The small amount of variation not accounted for by the first, second, and third principal coordinates can be ascribed to various rounding errors in the calculations.)

The first two principal coordinates, consisting of the elements of eigenvectors I and II, are listed in Table 6.36. Each element corresponds to an individual observation. These loadings are plotted in Figure 6.43. Compare the pattern of the principal coordinates solution to that obtained by Q-mode factor analysis (Fig. 6.39). Note that the fact that the diagonal elements of $[Q]$ may not be 1.00 means that it is not possible to plot a unit circle on the diagram to represent the communality.

Correspondence Analysis

Factor analysis is designed for interval or ratio data—that is, measurements made on a continuous numerical scale. It is not appropriate for enumerative data, such as counts of the numbers of fossils of different types that are present in samples, or the number of fractures of different orientations that occur within mining blocks. Such nominal or ordinal observations may be all that are available, and in some instances it seems desirable to process these using an eigenvalue technique similar to factor analysis.

Problems in which count data are all that are available are very common in the social sciences. Surveys by questionnaire, for example, result in responses that can only be tallied in categories. Consequently, most of the research on the use of eigenvalue methods for analyzing such data has been done by sociologists and statisticians working on sociological problems. These data may be summarized conveniently in the form of contingency tables; the first work of the type we are considering was applied to such summaries and was known as "contingency table analysis" (Hirschfeld, 1935; Fisher, 1940). More recently, Benzécri and others (1980) have written extensively on the subject, and Benzécri's term "correspondence analysis" has come to be widely used. His work provides the basis for many applications in geology, such as those of Teil (1975), Teil and Cheminée (1975), and David, Dagbert, and Beauchemin (1977). In these geological applications, however, the methods of Benzécri and his predecessors have been extensively modified. Hill (1974) discusses the history of correspondence analysis, and the interrelationships between the work of various authors. A detailed discussion of correspondence analysis and its extensions is contained in the monograph by Lebart, Morineau, and Warwick (1984).

Correspondence analysis proceeds by operating on a matrix derived from a contingency table which has been transformed so that the elements of the table can be regarded as conditional probabilities. Because of the nature of the transformation (actually a form of scaling), relationships between rows and columns of the transformed table are the same as those within the original data matrix. This means that the Eckart-Young theorem holds exactly, and R- and Q-mode solutions are equivalent.

The raw data matrix $[X]$ has n rows that represent observations and m columns of variables. The elements themselves are tallies. In a problem from paleoecology, for example, the columns might consist of different microfossil species, the rows could be chip samples taken from different stratigraphic intervals within a well, and the entries in the table would be counts of the number of individual specimens of each species of microfossil recovered from the samples. The total number of individuals is simply the sum of all of the elements in the data matrix, or

$$\sum_{i=1}^{n} \sum_{j=1}^{m} x_{ij}$$

The sum of the rows

$$\sum_{j=1}^{m} x_{ij}$$

is the number of microfossils of all types that are found in each sample, and the sum of the columns

$$\sum_{i=1}^{n} x_{ij}$$

is the number of microfossils of each species that are recovered from all samples. The tallies can be converted to percentages of the total, which may be regarded as probabilities:

$$P_{ij} = \frac{x_{ij}}{\sum_{i=1}^{n} \sum_{j=1}^{m} x_{ij}} \tag{6.76}$$

These P_{ij} values can be thought of as the *joint probabilities* that specific microfossil species will be found in specific chip samples. The row totals divided by the grand total give the *marginal probabilities*

$$P_{i\cdot} = \frac{\sum_{j=1}^{m} x_{ij}}{\sum_{i=1}^{n} \sum_{j=1}^{m} x_{ij}} \tag{6.77}$$

which are the probabilities that specific chip samples will contain microfossils, without considering what the fossils might be. The column totals, treated in the same manner, give the marginal probabilities

$$P_{\cdot j} = \frac{\sum_{i=1}^{n} x_{ij}}{\sum_{i=1}^{n} \sum_{j=1}^{m} x_{ij}} \tag{6.78}$$

which are the probabilities that specific microfossils will occur, regardless of which sample they may be in. If the joint probabilities are divided by their corresponding marginal probabilities, the results are the *conditional probabilities*

$$P_{(i|j)} = \frac{P_{ij}}{P_{\cdot j}}$$

and

$$P_{(j|i)} = \frac{P_{ij}}{P_{i.}}$$

The first of these conditional probabilities describes the situation where we are *given* that we will find a microfossil of species j, and we want the probability that it will come from chip sample i. The second conditional probability, based on the row totals, gives the probability that the microfossil we find will be of species j, when we are given that we will find it in the ith sample.

In Chapter 2 we noted that in a contingency table the observations could be expressed as proportions of the total number of observations. Then, if the rows and columns of the table are independent, these observations should be approximately equal to the products of the marginal probabilities of their respective rows and columns. If two variables j and k are closely related, however, all the expected values which occur in the jth and kth columns should be very similar. This suggests that it should be possible to express the degree of similarity by computing a cross-product that would involve the observed and expected probabilities for all rows in the two columns being compared. Such a measure is used in correspondence analysis and is a form of the correlation coefficient between two variables (Kendall and Stuart, 1967). It is given by

$$r_{jk} = \sum_{i=1}^{n} \left(\frac{P_{ij} - P_{i.}P_{.j}}{\sqrt{P_{i.}P_{.j}}} \right) \left(\frac{P_{ik} - P_{i.}P_{.k}}{\sqrt{P_{i.}P_{.k}}} \right) \tag{6.80}$$

P_{ij} is the "observed" probability in row i and column j of the body of the contingency table, and $P_{i.}P_{.j}$ is the "expected" probability computed as the product of the marginal probabilities. Expressed in the terms used in Chapter 2, this becomes

$$r_{jk} = \sum_{i=1}^{n} \left(\frac{O_{ij} - E_{ij}}{\sqrt{E_{ij}}} \right) \left(\frac{O_{ik} - E_{ik}}{\sqrt{E_{ik}}} \right) \tag{6.81}$$

The relationship between this expression and the χ^2 statistic applied to contingency tables becomes clearer if we square one of the terms:

$$\left[\frac{O_{ij} - E_{ij}}{\sqrt{E_{ij}}} \right]^2 = \frac{(O_{ij} - E_{ij})^2}{E_{ij}}$$

We then see that the similarity measure used in correspondence analysis can be regarded as the product of two χ^2 values. This leads to the expression "χ^2 distance," which is sometimes applied to this measure (Lebart, Morineau, and Warwick, 1984). If the similarity measure r_{jk} is computed between all pairs of columns j and k, it will form a square $m \times m$ matrix. The eigenvalues and eigenvectors can be extracted from this matrix; these are the principal axes of correspondence analysis. Note that expressing all of the elements in the contingency table as a proportion of the total insures that the sum of the columns (and of the rows) is equal to 1.00. Therefore, we have closed the data set, and one eigenvalue must be zero. This means the dimensionality of our problem is guaranteed to be reduced from m to $m - 1$, and hopefully much less.

Rather than use eq. (6.80) directly, an alternative formulation for the similarity is given in eq. (6.82), which will yield the same set of eigenvectors. This is

$$r_{ik} = \sum_{i=1}^{n} \frac{P_{ij}P_{ik}}{P_{i\cdot}\sqrt{P_{\cdot j}P_{\cdot k}}} \qquad (6.82)$$

The last eigenvalue that results from the use of eq. (6.80) will be trivial and exactly equal to 0.0. Because the data are not centered around zero prior to factoring when using eq. (6.82), the factor solution will contain an initial trivial eigenvalue that is identically equal to 1.0. The computations involved in eq. (6.82) are easier to express in matrix form. First, we denote the original $n \times m$ data matrix as $[X]$. Elements of $[X]$ are converted to joint probabilities by dividing every element in the matrix by the grand total, which is the scalar $\Sigma\Sigma x_{ij}$, to give a matrix $[B]$:

$$[B] = \frac{1}{\Sigma\Sigma x_{ij}} [X] \qquad (6.83)$$

Then, we define an $m \times m$ square matrix $[M]$, which contains the column totals of $[B]$ arranged in order along the diagonal, and with zeros in all off-diagonal positions. We also define another square matrix $[N]$, which is $n \times n$ and contains the row totals of $[B]$ along the diagonal and zeros elsewhere. These two matrices contain the column and row marginal probabilities and are used to transform $[B]$.

$$[W] = [N]^{-1/2}[B][M]^{-1/2} \qquad (6.84)$$

(Since we are dealing with diagonal matrices, the operations $[N]^{-1/2}$ and $[M]^{-1/2}$ are equivalent to replacing each diagonal element n_{ii} and m_{jj} with $1/\sqrt{n_{ii}}$ and $1/\sqrt{m_{jj}}$. The off-diagonal zero elements in each matrix, of course, remain zero.)

The matrix $[W]$ will be $n \times m$, with a transformed element w_{ij} corresponding to every original element x_{ij}. The matrix of cross-products of the columns is simply

$$[R] = [W]'[W] \qquad (6.85)$$

Similarly, the matrix of cross-products of the rows is

$$[Q] = [W][W]' \qquad (6.86)$$

The eigenvalues of $[R]$ and $[Q]$ will be identical, except that $[Q]$ will have $(n - m)$ additional eigenvalues, each of which will be zero. The eigenvectors of $[R]$ can be converted to correspondence factor loadings by multiplying each vector by its corresponding singular value, which is the square root of the corresponding eigenvalue. That is,

$$R\text{-mode loadings} = \sqrt{\lambda} \cdot R\text{-mode eigenvector}$$

In the matrix notation we have used earlier, the singular values of $[R]$ can be thought of as occurring along the diagonal of an $m \times m$ matrix $[\Lambda]$, whose off-diagonal elements are all zero. The eigenvectors of $[R]$ form the columns of an

TABLE 6.37 Counts of Conodont Tests Recovered from 10-kg Samples of Rock; Columns are Conodont Varieties; Rows are Stratigraphic Units that are Members in a Section of Missourian Age in Eastern Kansas; Megacyclothem Classifications are Outside Shale (O), Shoal Limestone (S), Upper Limestone (U), Middle Limestone (M), "Phantom Black Shale" (P), Black Shale (B)

			Counts of Conodonts										
N	Class	Rock Unit	Adetognathus	Ozarkodina	Aethotaxis	Idiognathodus delicatus	I. elegantulus	Magnilaterella	Hindeodella	Idioprioniodus	Gondolella	Others	TOTAL
			A	B	C	D	E	F	G	H	I	J	
1.	M	South Bend Ls.	13	10	0	0	37	0	0	0	0	0	60
2.	O	Rock Lake Sh.	0	0	0	0	11	0	0	0	0	0	11
3.	U	Stoner Ls.	4	2	1	51	26	1	0	0	0	0	85
4.	B	Eudora Sh.	0	7	1	207	350	0	0	34	14	3	606
5.	M	Captain Creek Ls.	8	28	6	0	60	0	0	0	0	0	102
6.	O	Vilas Sh.	145	20	5	0	10	0	0	0	0	0	180
7.	U	Spring Hill Ls.	5	134	8	0	353	1	0	4	0	0	505
8.	P	Hickory Creek Ls.	20	60	0	0	920	0	0	0	0	0	100
9.	M	Merriam Ls.	115	255	10	0	1140	0	0	0	0	0	1520
10.	S	Bonner Springs Sh.	1	0	0	0	3	0	0	0	0	0	4
11.	S	Farley Ls.	31	21	7	0	4	1	0	0	0	0	61
12.	—	Island Creek Sh.	100	5	1	0	5	0	0	0	0	0	110
13.	U	Argentine Ls.	0	39	0	0	80	0	1	0	0	0	121
14.	P	Quindaro Sh.	10	70	0	0	538	0	0	5	0	0	623
15.	M	Frisbee Ls.	3	78	5	0	450	0	0	3	0	0	539
16.	O	Lane Sh.	0	0	0	0	28	0	0	0	0	0	28
17.	U	Raytown Ls.	38	20	3	100	267	3	0	25	0	0	456
18.	B	Muncie Creek Sh.	15	8	0	243	515	0	10	85	55	13	946
19.	M	Paola Ls.	10	130	10	200	900	0	0	50	0	0	1300
20.	O	Chanute Sh.	117	20	0	63	57	0	0	7	0	0	264
		TOTAL	258	389	31	367	1929	4	5	82	32	7	3104

TABLE 6.38 χ^2 Similarity Matrix, Eigenvalues, and First Two Eigenvectors Calculated for Conodont Abundance Data; Also Given are R- and Q-Mode Correspondence Factor Loadings for First Two Factors

χ^2 SIMILARITY MATRIX

	A	B	C	D	E	F	G	H	I	J
A	.3843	.0037	.0257	-.0273	-.1136	.0056	-.0079	-.0239	-.0204	-.0098
B	.0037	.0568	.0196	-.0645	.0088	.0015	-.0076	-.0292	-.0268	-.0129
C	.0257	.0196	.0216	-.0119	-.0117	.0063	-.0026	-.0066	-.0070	-.0034
D	-.0273	-.0645	-.0119	.1655	-.0477	.0090	.0150	.0620	.0486	.0233
E	-.1136	.0088	-.0117	-.0477	.0592	-.0066	-.0052	.0159	-.0136	-.0066
F	.0056	.0015	.0063	.0090	-.0066	.0075	-.0010	.0006	-.0024	-.0011
G	-.0079	-.0076	-.0026	.0150	-.0052	-.0010	.0091	.0129	.0179	.0088
H	-.0239	-.0292	-.0066	.0620	.0159	.0006	.0129	.0365	.0330	.0160
I	-.0204	-.0268	-.0070	.0486	-.0136	-.0024	.0179	.0330	.0430	.0210
J	-.0098	-.0129	-.0034	.0233	-.0066	-.0011	.0088	.0160	.0210	.0102

Vector	Eigenvalue	Total Similarity (%)	Total Similarity (Cumulative %)
1	.4262	53.7003	53.7003
2	.2634	33.1837	86.8841
3	.0468	5.8936	92.7776
4	.0385	4.8532	97.6308
5	.0101	1.2691	98.8999
6	.0044	.5488	99.4487
7	.0036	.4523	99.9010
8	.0008	.0988	99.9997
9	.0000	.0003	100.0000
10	.0000	.0000	100.0000

Eigenvector

Conodont	I	II
A	.9467	.0653
B	.0357	-.3489
C	.0756	-.0647
D	-.0945	.7561
E	-.2761	-.2769
F	.0166	.0287
G	-.0248	.1011
H	-.0735	.3242
I	-.0657	.2923
J	-.0316	.1409

Correspondence Axis Loadings

R Mode

Conodont	I	II
A	2.2655	.1229
B	.0715	-.5492
C	.6038	-.4064
D	-.1938	1.2193
E	-.2195	-.1730
F	.4087	.5546
G	-.4512	1.4456
H	-.3037	1.0531
I	-.4768	1.6680
J	-.4761	1.6703

Q Mode

Unit	I	II
1	.5628	-.3344
2	-.3362	-.3372
3	-.0968	1.3119
4	-.3338	.7964
5	.1589	-.5199
6	2.8147	.0333
7	-.1593	-.5114
8	-.2333	-.3696
9	.0349	-.4195
10	.6154	-.1930
11	1.8068	-.3260
12	3.1445	.1537
13	-.1850	-.5511
14	-.2260	-.3911
15	-.2395	-.4309
16	-.3362	-.3372
17	.0168	.4110
18	-.3055	.8712
19	-.2515	.0998
20	1.3905	.5737

$m \times m$ matrix, $[U]$. The matrix equation used to determine the R-mode loadings is then

$$[A^R] = [U][\Lambda] \tag{6.87}$$

The scores of each of the n observations on the m correspondence factors is simply

$$[S^R] = [W][A^R] \tag{6.88}$$

These scores can be plotted along axes defined by the R-mode correspondence factors in the same way as principal component or factor scores.

If, instead of extracting eigenvalues from $[R]$ we extract them from $[Q]$, we can calculate Q-mode correspondence loadings and scores. Again, loadings are found by multiplying the elements of the eigenvectors by the square roots of the associated eigenvalues,

$$[A^Q] = [V][\Lambda] \tag{6.89}$$

where $[V]$ is the $n \times n$ matrix whose columns contain the n eigenvectors of $[Q]$. Q-mode scores are

$$[S^Q] = [W]'[A^Q] \tag{6.90}$$

Because of the Eckart-Young theorem and the fact that the scaling of the original data matrix affected both rows and columns of the original data matrix $[X]$ in the same manner, there is a direct relationship between the R- and Q-mode solutions. This relationship is

$$[A^Q] = [W][A^R][\Lambda]^{-1}$$
$$= [S^R][\Lambda]^{-1} \tag{6.91}$$

In other words, the Q-mode correspondence loadings are equal to the R-mode correspondence scores, divided by the appropriate singular values. Thus, we can obtain a Q-mode solution by solving an R-mode problem, which is ordinarily a great computational advantage as matrix $[R]$ is usually much smaller than matrix $[Q]$.

Furthermore, we can plot both our samples and our variables in the same space, using the same axes. This can be done by converting the R-mode loadings and the Q-mode loadings so they both have the same metric. Scaling of the loadings is performed by

$$[\hat{A}^R] = [M]^{1/2} [A^R]$$

and   (6.92)

$$[\hat{A}^Q] = [N]^{1/2} [A^Q]$$

We will now use a geological data set to examine the "classical" application of correspondence analysis, which is the interpretation of enumerated data. Table 6.37 contains counts of the number of conodonts extracted from 10-kg composite samples of rock collected from each stratigraphic unit in a sequence in eastern Kansas. The rocks are Missourian in age and represent four megacyclothems, or

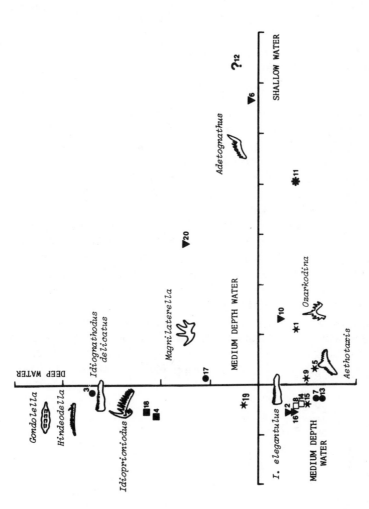

FIGURE 6.44 Plot of correspondence factor loadings for conodont abundance data given in Table 6.37. Approximate depth ranges for conodont types are indicated. Megacyclothem classifications of stratigraphic units include outside shale (▼), shoal limestone (✳), Upper limestone (●), middle limestone (✳), "phantom black shale" (□), and black shale (■).

repetitions of lithologies, that may reflect cyclic changes in the depth of seawater. Each unit has been classified as part of an idealized megacyclothem; the classifications are indicated in the table. Paleoecologists have suggested that conodonts were associated with specific depth zones, as are some modern pelagic organisms. If both lithologies and conodont assemblages were responses to changes in sea level, correspondence analysis should reveal patterns in conodont abundances that would be similar to the changes in lithology.

Since there are 10 species of conodonts and 20 stratigraphic units, it is most convenient to construct our matrix of similarities between variables. Table 6.38 gives the matrix of χ^2 similarity measures, its eigenvalues, and the most significant eigenvectors. Also given are the R- and Q-mode loadings on the correspondence axes. These are plotted in Figure 6.44. The megacyclothem categories for each stratigraphic unit are indicated. Both the samples (rock units) and variables (conodont species) can be plotted in the same space.

Figure 6.45 shows the relative depth ranges of the conodonts collected from the Missourian stratigraphic sequence. Note that the R-mode loadings plotted on Figure 6.44, which represent the conodont types, do seem to reflect the water depths at which these organisms lived. Conodonts from the shallowest marine environments appear on the positive end of factor I, those from the deepest water appear on the positive end of factor II, and those from environments of intermediate depth are found near the origin. By examining the positions of the Q-mode factor loadings on the same diagram, we can infer the water depths at which the various rock units were deposited. The classification of the various stratigraphic units into the megacyclothem pattern, indicated on Table 6.37, follows the use of Heckel and

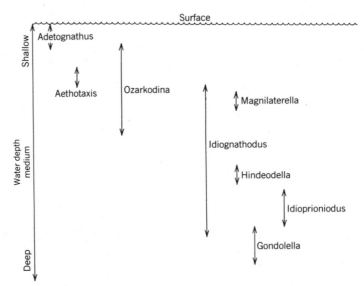

FIGURE 6.45   Relative depth ranges for conodonts collected from Missourian stratigraphic sequence.

Baesemann (1975). We can note that deltaic "outside shales" such as the Vilas, Bonner Springs, and Chanute shales do fall in the shallow-water part of the factor diagram, as does the Farley Limestone, which is classified as a shoal-water deposit. Phosphatic black shales such as the Eudora and Muncie Creek shales fall in the deep-water part of the diagram, with such fine-grained "upper limestones" as the Stoner and Raytown. Most rock units, however, plot near the origin, including the so-called "phantom black shales," which seem to be no different than most other rock types.

Correspondence analysis of these conodont data gives some insight into the nature of cyclic sedimentation in this particular sequence. The extremes of the marine environment do seem to yield distinctive lithologies that occur at characteristic positions within megacyclothems. Most lithologies, however, do not fit into a definitive pattern. In particular, the "phantom black shales" that presumably should have deep-water characteristics, seem indistinguishable from other rock types that originated at intermediate depths.

Application to Continuous Variables

In geology, as well as in other areas, correspondence analysis has been applied to continuous data rather than to enumerated counts (see Teil, 1975, and David, Dagbert, and Beauchemin, 1977, for applications). This poses some conceptual problems, because the transformed observations cannot be regarded as probabilities, although some authors refer to them in this way (David, Dagbert, and Beauchemin, 1977). Since in general the grand total (as well as the row totals) will consist of a mixture of various types of measurements, the transformation process is strongly dependent upon units of measurements. In spite of this philosophical objection, correspondence analysis is routinely applied to sets of interval and ratio data. In these applications, the transformation process can be regarded as no more than an arbitrary procedure designed to close the data set, and to insure that rows and columns of the data matrix are scaled in equivalent manners when computing either $[R]$ or $[Q]$. When used in this manner, the similarity measure often is referred to as the "profile distance"; it reflects the relative magnitude of the variables rather than their absolute values (Zhou, Chang, and Davis, 1983).

We can use the artificial data on 25 randomly created blocks to study the behavior of correspondence analysis when applied to measured variables. The original data are listed in Table 6.18; Table 6.39 is the matrix of similarities calculated using eq. (6.80). The eigenvalues of the matrix are also given, as are the eigenvectors, the R-mode loadings, and the Q-mode loadings on the first two correspondence axes. Figure 6.46 is a plot of the first two R- and Q-mode loadings, shown on the same diagram. From this presentation it is comparatively easy to see not only the similarities between the individual observations, but also the relative contributions of each of the original variables to the correspondence axes.

Hill (1974) has suggested a more appropriate mechanism for handling continuous data with correspondence analysis. If the interval or ratio data scales are divided into discrete categories and the number of measurements falling into each category

TABLE 6.39 Similarity Matrix, Eigenvalues, and First Two Eigenvectors Calculated for Random Block Data; Also Listed are the R- and Q-Mode Correspondence Factor Loadings for the First Two Factors

χ^2 SIMILARITY MATRIX

	X_1	X_2	X_3	X_4	X_5	X_6	X_7
X_1	.0094	.0023	.0089	.0088	−.0093	−.0118	−.0110
X_2	.0023	.0168	.0142	.0096	−.0135	−.0126	−.0184
X_3	.0089	.0142	.0376	.0174	−.0264	−.0318	−.0118
X_4	.0088	.0096	.0174	.0123	−.0153	−.0178	−.0160
X_5	−.0093	−.0135	−.0264	−.0153	.0218	.0253	.0148
X_6	−.0118	−.0126	−.0318	−.0178	.0253	.0307	.0134
X_7	−.0110	−.0184	−.0118	−.0160	.0148	.0134	.0428

Vector	Eigenvalue	Total Similarity %	Total Similarity (Cumulative %)
1	.1213	70.8332	70.8332
2	.0359	20.9437	91.7769
3	.0110	6.4431	98.2201
4	.0029	1.6891	99.9091
5	.0001	.0837	99.9928
6	.0000	.0072	100.0000
7	.0000	.0000	100.0000

Eigenvector

Variable	I	II
X_1	.1922	−.0538
X_2	.2783	−.2148
X_3	.4961	.4085
X_4	.3109	−.0447
X_5	−.4126	−.1458
X_6	−.4710	−.2967
X_7	−.3883	.8203

Correspondence Axis Loadings

R Mode			Q Mode		
Variable	I	II	Block	I	II
X_1	.1555	−.0237	a	−.5814	.0088
X_2	.2747	−.1153	b	−.1290	−.1257
X_3	.6980	.3126	c	.4855	.0899
X_4	.2219	−.0173	d	.6570	.1459
X_5	−.3809	−.0732	e	.1998	−.0955
X_6	−.3793	−.1299	f	−.1507	−.0214
X_7	−.5467	.6280	g	−.8040	.5665
			h	.3164	−.0467
			i	−.2322	−.2435
			j	−.3215	−.0086

TABLE 6.39 (*Continued*)

	Correspondence Axis Loadings				
	R Mode			Q Mode	
Variable	I	II	Block	I	II
			k	−.2154	−.0698
			l	.3782	.0457
			m	.1067	.1565
			n	−.2370	.2185
			o	−.2046	.7453
			p	−.0421	−.0861
			q	−.0520	−.2084
			r	.1873	−.0313
			s	−.2654	−.0455
			t	−.0217	.0662
			u	−.2393	−.2097
			v	.4181	.0395
			w	.5015	.2009
			x	.1477	.0399
			y	.5008	.0539

are counted, the data are reduced to ordinal rank. In effect, each continuous variable is replaced with a number of discrete variables. This decreases the information content of the data set, but considering the inexactitude of many geologic observations, this loss may not be significant.

As an experiment, the block data may be reduced to ordinal rank by dividing the range of each variable into a suitable number of discrete intervals (such as "low," "medium," "high" values), and noting into which of the categories each observation falls. The raw data matrix, therefore, consists of an array of ones and zeros (Table 6.40). These are then converted into joint probabilities of occurrence, transformed, and the similarity matrices [R] or [Q] calculated. Note that [Q] is still 25 × 25, but [R] is now expanded to 21 × 21. The successive eigenvalues are smaller and drop off more slowly than do the eigenvalues from a similarity matrix calculated from metric data. The first two eigenvalues, for example, account for only 43% of the trace of the similarity matrix. All eigenvalues after the fourteenth are identically zero. Even though the first few correspondence factor axes do not seem to be as efficient as those calculated from Table 6.39, a plot of the loadings on the correspondence axes shows patterns that are at least as meaningful as those obtained from metric data. Figure 6.47 shows the R- and Q-mode loadings for the first two correspondence axes. Compare this result with that obtained from metric data (Fig. 6.46), keeping in mind that the information content in the ordinal data is much lower.

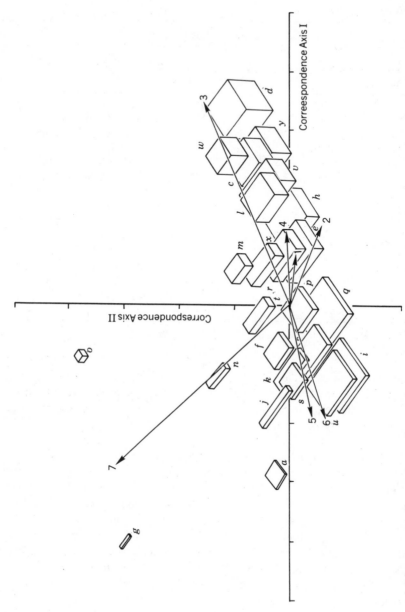

FIGURE 6.46 Plot of *R*- and *Q*-mode loadings on first two correspondence axes from block data. Horizontal axis is correspondence axis I; vertical axis is correspondence axis II. Blocks and variables shown plotted at their respective loadings on the two axes.

TABLE 6.40 Block Data with Each Original Variable Categorized into Three Ranges (L = Low, M = Medium, H = High); This Produces a 21 × 25 Matrix of Ordinal Data

Block	X₁ L 1	X₁ M 2	X₁ H 3	X₂ L 4	X₂ M 5	X₂ N 6	X₃ L 7	X₃ M 8	X₃ H 9	X₄ L 10	X₄ M 11	X₄ H 12	X₅ L 13	X₅ M 14	X₅ H 15	X₆ L 16	X₆ M 17	X₆ H 18	X₇ L 19	X₇ M 20	X₇ H 21
a	1	0	0	0	1	0	1	0	0	1	0	0	0	0	1	0	0	1	0	1	0
b	0	0	1	0	1	0	1	0	0	0	0	1	0	1	0	0	1	0	1	0	0
c	0	1	0	0	0	1	0	1	0	0	1	0	1	0	0	1	0	0	1	0	0
d	0	1	0	0	0	1	0	0	1	0	0	1	1	0	0	1	0	0	1	0	0
e	0	0	1	0	0	1	0	1	0	0	0	1	1	0	0	0	1	0	1	0	0
f	0	1	0	0	1	0	1	0	0	0	1	0	0	1	0	0	1	0	1	0	0
g	1	0	0	1	0	0	1	0	0	1	0	0	0	1	0	0	0	1	0	0	1
h	0	1	0	0	1	0	0	1	0	0	0	1	1	0	0	1	0	0	1	0	0
i	0	0	1	0	0	1	1	0	0	0	0	1	0	0	1	0	1	0	1	0	0
j	0	0	1	1	0	0	1	0	0	0	1	0	0	0	1	0	1	0	1	0	0
k	0	0	1	1	0	0	1	0	0	0	1	0	0	1	0	0	0	1	1	0	0
l	0	0	1	0	0	1	0	1	0	0	0	1	1	0	0	1	0	0	1	0	0
m	0	0	1	0	1	0	1	0	0	1	0	0	1	0	0	1	0	0	1	0	0
n	0	1	0	1	0	0	1	0	0	1	0	0	1	0	0	1	0	0	0	1	0
o	1	0	0	1	0	0	1	0	0	0	1	0	0	0	1	0	1	0	1	0	0
p	0	1	0	0	1	0	1	0	0	0	0	1	0	0	1	0	0	1	1	0	0
q	1	0	0	0	0	1	0	1	0	0	0	1	1	0	0	1	0	0	1	0	0
r	0	1	0	0	1	0	1	0	0	0	1	0	0	1	0	0	1	0	1	0	0
s	1	0	0	0	0	1	1	0	0	0	1	0	1	0	0	1	0	0	1	0	0
t	0	0	1	1	0	0	1	0	0	0	0	1	0	0	1	0	0	1	1	0	0
u	1	0	0	0	1	0	0	1	0	0	0	1	1	0	0	1	0	0	1	0	0
v	0	1	0	0	1	0	1	0	0	0	1	0	0	1	0	0	0	1	1	0	0
w	0	0	1	0	1	0	1	0	0	0	1	0	1	0	0	1	0	0	1	0	0
x	1	0	0	0	1	0	0	1	0	0	1	0	1	0	0	1	0	0	1	0	0
y	0	1	0	0	0	1	0	1	0	0	0	1	1	0	0	1	0	0	1	0	0

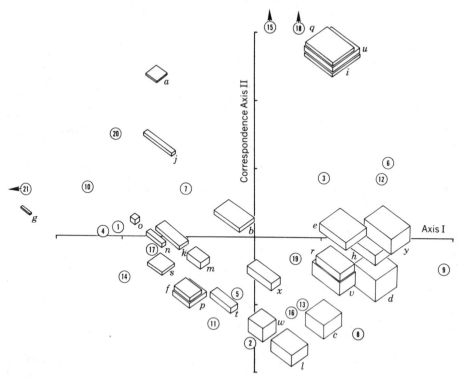

FIGURE 6.47 Plot of R- and Q-mode loadings on first two correspondence axes from block data expressed in ordinal classes. R-mode loadings are shown as circled numbers corresponding to those listed in Table 6.40. Horizontal axis is correspondence axis I; vertical axis is correspondence axis II. Variables 15, 18, and 21 plot outside the range of the diagram.

Simultaneous R- and Q-Mode Factor Analysis

Although the Eckart-Young theorem states that equivalent solutions can be obtained in either R- and Q-mode, in practice this may not be quite true. A plot of R-mode factor scores looks different than a plot of Q-mode factor loadings, as a glance at Figures 6.38 and 6.39 will attest. You will recall that an R-mode solution is derived from the symmetric minor product matrix $[W]'[W]$, while a Q-mode solution is derived from the major product $[W][W]'$. Unfortunately, the scaling procedures that are used to create $[W]$ from the original raw data $[X]$ are not the same in the two modes. For example, principal components involves transformation of each element of $[X]$ by dividing by the standard deviation of the columns, producing a scaled data matrix $[W]$. Q-mode factor analysis uses a standardization that includes dividing each element of $[X]$ by the square root of the sum of the squares of the rows, also producing a scaled data matrix $[W]$. However, the matrix $[W]$ in principal components analysis is *not* identical to the matrix $[W]$ produced as the first step in Q-mode factor analysis. The difference in scaling distorts the solution in one mode with respect to the other mode.

There are several ways around this problem. Obviously, if no scaling is done, the eigenvalues and eigenvectors of $[X]'[X]$ are the same as those from $[X][X]'$, except that one or the other may have additional zero eigenvalues. R-mode factor scores will be proportional to Q-mode factor loadings, and vice versa. In addition, the R-mode and Q-mode factor loadings both occur in the space defined by the same factor axes, so both can be plotted on the same diagram, as in Figure 6.48.

Unfortunately, using the raw cross-product matrix has distinct disadvantages. Since no scaling has been done, the analysis is very sensitive to the choice of measurement units, and results may simply reflect the average magnitudes of the variables rather than their variances and covariances. Although such a method is a mathematically simple way to simultaneously extract R- and Q-mode factors, it is almost never used in practice.

A second solution is to scale $[X]$ in a manner that treats rows and columns identically. This is done in correspondence analysis, where each element is divided by the product of the square roots of the row and column totals. Although this approach is eminently sensible when dealing with the data of contingency tables, it is less so when applied to measurement data.

A third alternative is to seek a way of scaling by rows that produces a meaningful measure of interrelation between the rows in the matrix $[W]$, and at the same time results in a meaningful measure of interrelation between the columns. This proves to be easier than might be supposed, and is the basis for at least two practical methods of simultaneous extraction of R- and Q-mode factors.

The elements of $[X]$ can be standardized by subtracting the column (variable) means and dividing by the square root of n, the number of observations. That is,

$$w_{ij} = \frac{x_{ij} - \bar{x}_{\cdot j}}{\sqrt{n}} \tag{6.93}$$

Then, the minor product matrix $[W]'[W]$ will contain the variances and covariances of the variables. At the same time, the major product matrix $[W][W]'$ is equivalent to the principal coordinates matrix $[Q]$ when the similarity between objects is defined by the Euclidean distance. That is,

$$q_{ij} = d_{ij} + \bar{\bar{d}}_{\cdot\cdot} - (\bar{d}_{i\cdot} + \bar{d}_{\cdot j}) \tag{6.94}$$

and d_{ij} is an element of the distance matrix $[D]$:

$$d_{ij} = \sum_{k=1}^{m} (x_{ik} - x_{jk}) \Big/ \sqrt{n} \tag{6.95}$$

Alternatively, we can standardize the elements of $[X]$ by subtracting the column (variable) means and dividing by the product of the column (variable) standard deviations and the square root of n:

$$w_{ik} = \frac{x_{ik} - \bar{x}_{\cdot k}}{s_k \sqrt{n}} \tag{6.96}$$

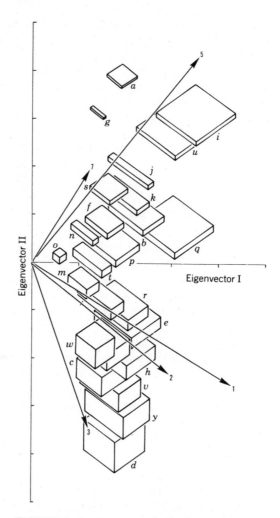

FIGURE 6.48   Plot of *R–Q* mode factor loadings on first two factor axes of block data, calculated from matrix of raw sums of squares and cross-products, $[X]'[X]$. Blocks are shown plotted at locations corresponding to *Q*-mode loadings. Variables are plotted as vectors defined by *R*-mode loadings. Variable X_4 plots outside diagram in same direction as variable X_2; variable X_6 plots outside diagram in same direction as variable X_5.

The minor product matrix $[W]'[W]$ will now contain the variances and covariances of the variables in standardized form, which of course are the correlations between the variables.

Again, the major product matrix $[W][W]'$ is equivalent to one version of the principal coordinates matrix $[Q]$. Now, however, the distance matrix $[D]$ contains Euclidean distances between the observations as defined by standardized variables, or

$$d_{ij} = \sum_{k=1}^{m} \left(\frac{x_{ik} - x_{jk}}{s_k} \right) \bigg/ \sqrt{n} \qquad (6.97)$$

To perform simultaneous R- and Q-mode factor analysis, we first compute the minor product matrix $[W]'[W]$ after scaling by either eq. (6.93) or (6.96). We then extract its eigenvectors and eigenvalues and compute the R-mode factor loadings by multiplying each element of an eigenvector by the corresponding singular value or square root of its eigenvalue. If the matrix of eigenvectors is denoted $[U]$, this is

$$[A^R] = [U][\Lambda] \qquad (6.98)$$

where again, $[\Lambda]$ is a diagonal matrix of the singular values of $[W]'[W]$. Next, we must find the Q-mode factor loadings, which can be found as the product of the scaled data matrix and the matrix of eigenvectors:

$$[A^Q] = [W][U] \qquad (6.99)$$

Of course, we can also complete factor scores as well. R-mode scores are found by multiplying the scaled data matrix by the R-mode factor loading matrix.

$$[S^R] = [W][A^R] \qquad (6.100)$$

The Q-mode scores are found by

$$[S^Q] = [W]'[A^Q] \qquad (6.101)$$
$$= [W]'[W][U]$$

Since the R-mode factor loadings $[A^R]$ give the coordinates of the variables as points in a "factor space," and the Q-mode factor loadings $[A^Q]$ give the coordinates of the objects as points in the same factor space, both sets of loadings can be plotted on the same factor diagrams. Variables that plot close together are very similar. The Eckart-Young theorem gives the relationship between the variables and the objects. Equation (6.40) can be rewritten as

$$[A^Q][\Lambda]^{-1/2}[A^R]' = [W] \qquad (6.102)$$

An element of $[W]$ is thus equal to

$$\frac{1}{\sqrt{\lambda_k}} \sum_{k=1}^{m} a_{ik}^Q a_{jk}^R = w_{ij} \qquad (6.103)$$

Observation w_{ij}, which is the scaled value of variable j observed on object i, can

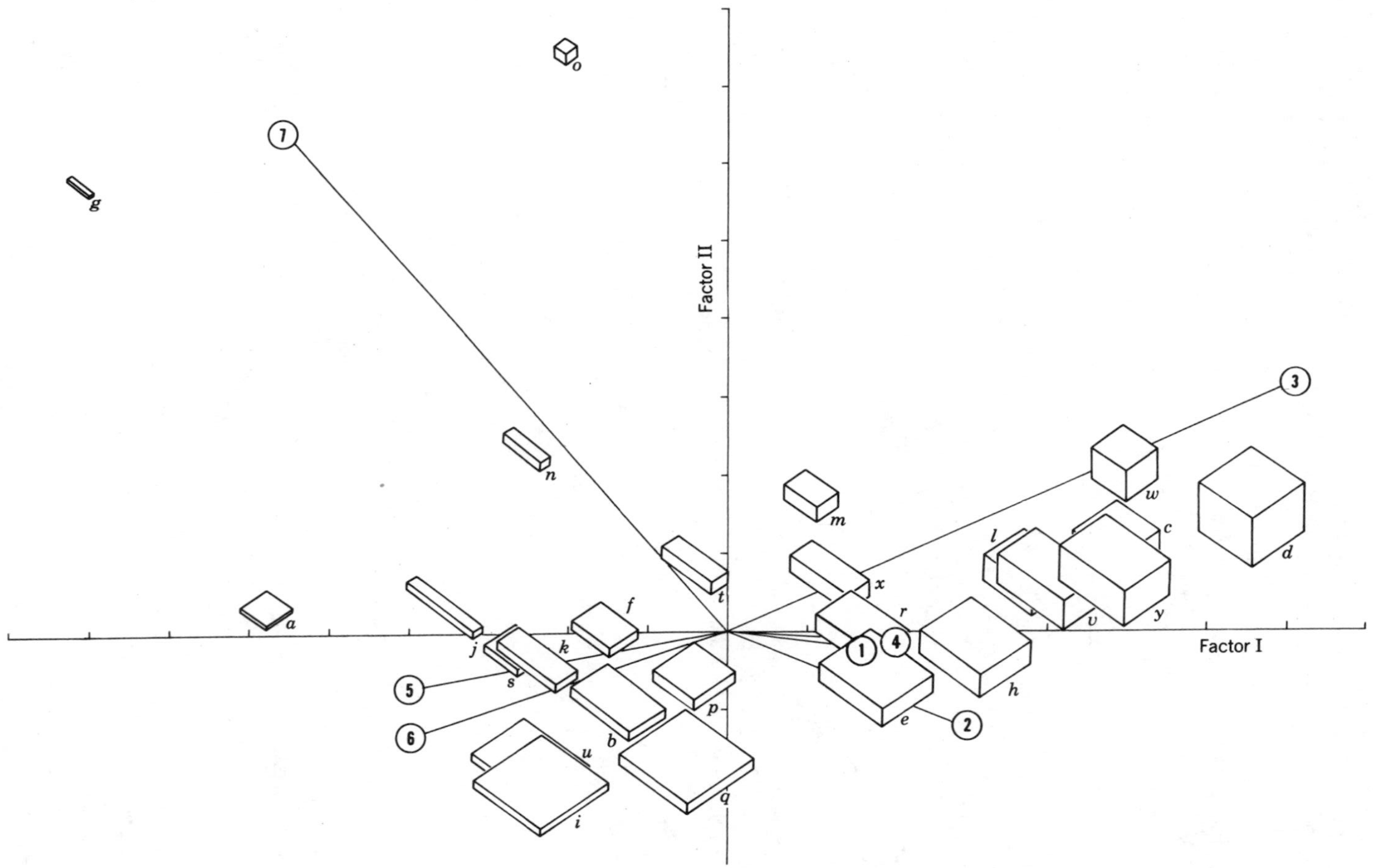

FIGURE 6.49 Plot of *R–Q* mode factor loadings on first two factor axes of block data, calculated from matrix of variances and covariances, $[W]'[W]$. Variables are plotted as vectors defined by *R*-mode loadings. Blocks are shown plotted at locations corresponding to *Q*-mode loadings.

TABLE 6.41 Content of Uranium, Thorium, and Potassium, and Airborne Radiometric
Intensity along a Traverse across Quartz Monzonite Intrusion near Berea, Virginia

	A	B	C	D
No.	AERO	U	TH	K
1	240	0.63	2.05	0.13
2	360	2.18	5.31	0.31
3	420	2.26	5.61	0.34
4	500	1.71	6.44	0.70
5	580	2.38	7.99	1.73
6	700	3.83	8.32	4.26
7	600	3.79	9.46	1.53
8	650	4.09	14.71	3.11
9	770	4.21	12.00	1.90
10	930	4.72	12.78	2.92
11	1020	6.24	16.31	2.29
12	1000	5.24	14.51	1.88
13	1000	4.73	15.79	4.64
14	1040	4.67	10.30	4.17
15	1150	5.08	13.11	3.97
16	1000	5.27	13.40	4.36
17	960	5.61	10.31	2.05
18	420	2.33	6.83	0.47
19	370	2.64	9.88	0.58
20	400	2.29	6.02	0.34
21	480	2.32	6.14	0.32
22	730	5.94	12.86	1.35

(A) Airborne radiometric measurements in counts per second; (B) Uranium in parts per million; (C)
Thorium in ppm; (D) Potassium in percent

After Sherman, Bunker, and Bush (1971).

be regarded as the product of an object loading vector $[a_i^Q]$ and a variable loading
vector $[a_f^R]$, multiplied by $1/\sqrt{\lambda_k}$. The magnitude of a vector product, you will
recall from Chapter 5, is inversely related to the distance between the ends of the
two vectors. Thus, the strength of the relationship between an object and a variable
in the factor diagram is directly expressed by the distance between the object point
and the variable point.

Although equivalent relationships exist between factor scores, they are not so
neatly expressed in terms of similarities. The best way of displaying the results of
simultaneous R- and Q-mode factor analysis is to plot the two sets of loadings on
the factor axes. This is done for the random block data in Figure 6.49. From an
R-mode perspective, the procedure is the same as principal components analysis,
so the critical matrices are the same as those given in Table 6.21.

The duality between principal components analysis and principal coordinates
analysis using the Euclidean distance was first pointed out by Gower (1966).

TABLE 6.42 Correlation Matrix, Eigenvalues, and R- and Q-Mode Factor Loadings of Composition and Radiometric Data from Berea, Virginia

(A) Correlation Matrix

	AERO	U	TH	K
AERO	1.00			
U	0.89	1.00		
TH	0.82	0.89	1.00	
K	0.82	0.67	0.69	1.00

(B) Eigenvalues

	Factors			
	I	II	III	IV
λ	3.39	0.39	0.15	0.06
%	84.81	9.76	3.87	1.55
Σ%	84.81	94.57	98.45	100.00

(C) R-Mode Loadings

	Factors			
	I	II	III	IV
AERO	0.96	0.05	-0.22	-0.16
U	0.94	-0.27	-0.13	0.17
TH	0.92	-0.25	0.28	-0.07
K	0.86	0.51	0.09	0.07

(D) Q-Mode Loadings

	Factors			
	I	II	III	IV
1	-0.75	0.12	-0.03	-0.01
2	-0.49	-0.02	-0.02	0.03
3	-0.45	-0.02	-0.04	0.00
4	-0.41	0.04	-0.00	-0.08
5	-0.22	0.09	0.03	-0.05
6	0.11	0.29	-0.00	0.12
7	-0.08	-0.05	0.00	0.04
8	0.21	0.00	0.23	0.03
9	0.12	-0.08	0.02	-0.03
10	0.31	0.00	-0.01	-0.03
11	0.51	-0.23	0.00	-0.01
12	0.35	-0.18	-0.03	-0.09
13	0.53	0.14	0.14	-0.04

TABLE 6.42 (*Continued*)

	Factors			
	I	II	III	IV
14	0.36	0.21	−0.11	−0.00
15	0.50	0.11	−0.08	−0.07
16	0.49	0.13	0.01	0.03
17	0.26	−0.09	−0.19	0.04
18	−0.40	−0.04	0.01	−0.00
19	−0.31	−0.11	0.15	0.01
20	−0.44	−0.03	−0.01	0.01
21	−0.41	−0.03	−0.05	−0.03
22	0.21	−0.26	−0.03	0.12

From Zhou, Chang, and Davis (1983).

However, the duality has not been exploited until recently, even though a number of factor analysis programs employ singular value decomposition, which is an implementation of the Eckart-Young theorem. A mathematical derivation of simultaneous *R*- and *Q*-mode factor analysis is given by Zhou, Chang, and Davis (1983), who also provide a number of geological examples of its application.

Table 6.41 contains measurements made along a profile across a small, highly

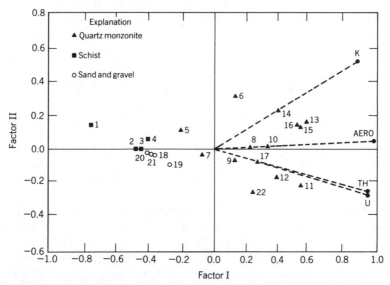

FIGURE 6.50 Plot of *R–Q* mode factor loadings on first two factor axes of Berea data, calculated from correlation matrix (Table 6.42). Dashed lines represent *R*-mode loadings; point symbols represent *Q*-mode loadings. From Zhou, Chang, and Davis (1983).

radioactive quartz monzonite pluton that has intruded a chlorite-actinolite schist near Berea, Virginia. The 22 auger samples on the profile were analyzed for the radioactive elements uranium, thorium, and potassium. Airborne radiometric measurements were also made at the same locations along the profile. The purpose of the study, originally conducted by Sherman, Bunker, and Bush (1971), was to relate the concentrations of the radioactive elements to the radiometric measurements.

The data were analyzed by Zhou, Chang, and Davis (1983) using the scaling given in eq. (6.96). Table 6.42 contains the resulting 4×4 matrix of correlations, the eigenvalues, and R- and Q-mode factor loadings. Figure 6.50 shows the first two R-mode and Q-mode loadings plotted in factor space. There is a clear distinction between samples from the pluton and those collected in the schist host rock and alluvial cover.

Multigroup Discriminant Functions

Multigroup discriminant analysis combines a rationale similar to that of analysis of variance with computational procedures related to those used in factor analysis. The problem is an extension of two-group discrimination, already discussed. As an example, suppose a paleontologist interested in sexual dimorphism of gastropods has collected specimens of modern whelks from several different localities. By examining the soft parts of their bodies, the investigator can distinguish males from females. Individuals can then be categorized as males from locality A, females from locality A, males from locality B, and so forth. Multivariate measurements made on the shells alone can be used in a discriminant analysis to find combinations of measurements that allow the various categories of sex and location to be distinguished. Hopefully, the distinctions between males and females will be greater than the distinctions between localities. The project might provide insight into the characteristics of gastropod shells that would permit classification of fossil shells according to their sex.

The analogy with analysis of variance comes from the way in which variances and covariances can be partitioned among categories or groups. You will recall from Chapter 2 that in one-way analysis of variance, the total sum of squares (SS_T) is equal to the sum of squares within the groups (SS_W) plus the sum of squares between the groups (SS_B). Exactly the same structure is invoked in discriminant analysis.

The notation of multigroup discriminant analysis is complicated because we must consider not only objects and variables, but also the groups in which the objects occur. Therefore, we must initially use a notation with three subscripts in which x_{ijk} denotes the ith variable measured on object j within group k. A complicating factor is that all groups do not necessarily contain the same number of objects, so we must denote the number of objects in the kth group as n_k. We will assume the observations are classed into g distinct groups. If we add all of the groups together, we find that there are a grand total of $N = \Sigma_{k=1}^{g} n_k$ objects in the entire data collection, each characterized by a set of variables.

The mean of the ith variable in the kth group is

$$\overline{X}_{i \cdot k} = \sum_{j=1}^{n_k} \frac{x_{ijk}}{n_k} \tag{6.104}$$

The grand mean of the ith variable is the average of all observations of variable i, regardless of the groups in which the observations are placed. The grand mean is equal to

$$\overline{\overline{X}}_{i \cdot \cdot} = \sum_{k=1}^{g} \sum_{j=1}^{n_k} \frac{x_{ijk}}{N} \tag{6.105}$$

The covariation between variable i and variable l for all objects, without regard to groups, is

$$s_{il} = \sum_{k=1}^{g} \sum_{j=1}^{n_k} (x_{ijk} - \overline{\overline{X}}_{i \cdot \cdot})(x_{ijk} - \overline{\overline{X}}_{l \cdot \cdot}) \tag{6.106}$$

If we compute this measure for all possible pairs of variables, they will form a $p \times p$ symmetric matrix $[S]$ that is referred to as the matrix of total sums of products. We also can compute a measure of covariation between variable i and variable l within the g groups by

$$w_{il} = \sum_{k=1}^{g} \sum_{j=1}^{n_k} (x_{ijk} - \overline{X}_{i \cdot k})(x_{ljk} - \overline{X}_{l \cdot k}) \tag{6.107}$$

Again, for all possible pairs of variables, this will form a $p \times p$ matrix $[W]$ that is the within-group sum of products. This quantity is equivalent to the sum of the matrices $[SPA]$ and $[SPB]$ used in simple two-group discriminant analysis. The final way in which we can express the variation is between groups:

$$b_{il} = \sum_{k=1}^{g} (\overline{X}_{i \cdot k} - \overline{\overline{X}}_{i \cdot \cdot})(\overline{X}_{l \cdot k} - \overline{\overline{X}}_{l \cdot \cdot}) \tag{6.108}$$

This also forms a $p \times p$ matrix $[B]$, which contains the between-group sums of products.

As in conventional analysis of variance, the within- and between-groups sums of products add to equal the total sums of products.

$$[S] = [B] + [W] \tag{6.109}$$

We would like for the ratio

$$\frac{[B]}{[W]}$$

to be as large as possible. You will recognize that this ratio is a multivariate analogue of the F ratio given by $F = MS_B/MS_W$, used to test the distinction between groups in an analysis of variance. If this ratio is large, the means of the groups are widely spread, while observations within groups are tightly clustered around the means. The problem of discriminant analysis is one of finding a set of linear weights for the variables that cause this ratio to be a maximum. If we refer to this set of weights as the vector $[A_1]$, discriminant analysis can be expressed as finding values for the elements of $[A_1]$ that cause the ratio

$$\frac{[A_1]'[B][A_1]}{[A_1]'[W][A_1]}$$

to be a maximum. Of course, we must place some constraints on $[A_1]$. In discriminant analysis, we usually specify that the denominator of this equation must be equal to one. That is,

$$[A_1]'[W][A_1] = 1$$

Under this constraint, the ratio will be a maximum when $[A_1]$ is the eigenvector corresponding to the largest eigenvalue of $[W]^{-1}[B]$. We can find a second set of linear weights $[A_2]$, which are the elements of the eigenvector corresponding to the second largest eigenvalue. A third set of weights also can be found, as well as a fourth set, and so on. In this manner, we can calculate a succession of discriminant functions along which the predefined groups are as distinct as possible. Because of the nature of eigenvectors, each is orthogonal to the others, and each is successively the most efficient discriminator possible. It is possible to compute a discriminant function for each positive eigenvalue. In general, the number of positive eigenvalues will be equal to the smaller of either $(g - 1)$ or p. Unfortunately, the matrix created by the operation $[W]^{-1}[B]$ is not symmetric, so its eigenvectors are not easily found. Some discriminant function programs compute the eigenvectors iteratively, using a process called singular value decomposition (Businger and Golub, 1969). Other programs first transform the matrix to symmetrical form, and then find a set of eigenvectors that in turn can be transformed into those required. The technique is described by Gnanadesikan (1977); critical steps are outlined by Maron (1982).

The observations used in the calculation of the discriminant function can be projected into the space defined by the discriminant axes. This is done by the matrix multiplication

$$[Z] = [A]'[X] \qquad (6.110)$$

where $[X]$ is the original $N \times p$ data matrix and $[A]$ is a $p \times t$ matrix whose columns consist of the t largest eigenvectors to be used as discriminant functions. The centroids of the g groups can be projected into the discriminant space by

$$[\bar{Z}] = [A]'[\bar{X}_k] \qquad (6.111)$$

where the matrix $[\bar{X}_k]$ is $g \times p$ and contains the means of all variables for each group. If we confine our attention to pairs of discriminant functions (usually the first and second), we can plot the observations and the group centroids as scatter diagrams. The data usually are scaled in some way prior to plotting. Some programs standardize by subtracting the grand mean from each observation and dividing by the standard deviation calculated over the entire data set. Others form the divisor by pooling the within-group standard deviations. Marascuilo and Levin (1983) provide an instructive comparison of the different approaches.

Obviously, an observation of unknown origin can be projected into the discriminant space simply by premultiplying it by the transpose of $[A]$. The group affinity of the new observation may be apparent from its position on the scatter diagrams, but it also is possible to compute a measure of its distance to the centroid of each group. The new observation is classified as belonging to the closest group.

To compute the generalized distances from a new observation x_i to each of the g group centroids, we must first determine all the differences $(x_i - \overline{X}_{i\cdot k})$, which can be arranged conveniently in a $g \times p$ matrix $[U]$. Then,

$$[D^2] = [U]'[A][A]'[U] \tag{6.112}$$

This will provide the generalized distances from the new observation to each of the g groups, measured in the t-dimensional discriminant space. Alternatively, we

TABLE 6.43 Chemical Analyses of Brines Recovered from Drill-Stem Tests of Three Carbonate Rock Units in Texas and Oklahoma

N		HCO3	SO4	CL	CA	MG	NA
		GROUP 1—ELLENBURGER DOLOMITE					
1		10.4	30.0	967.1	95.9	53.7	857.7
2		6.2	29.6	1174.9	111.7	43.9	1054.7
3		2.1	11.4	2387.1	348.3	119.3	1932.4
4		8.5	22.5	2186.1	339.6	73.6	1803.4
5		6.7	32.8	2015.5	287.6	75.1	1691.8
6		3.8	18.9	2175.8	340.4	63.8	1793.9
7		1.5	16.5	2367.0	412.0	95.8	1872.5
	$\overline{X} =$	5.6	23.1	1896.2	276.5	75.0	1572.3
		GROUP 2—GRAYBURG DOLOMITE					
8		25.6	0	134.7	12.7	7.1	134.7
9		12.0	104.6	3163.8	95.6	90.1	3093.9
10		9.0	104.0	1342.6	104.9	160.2	1190.1
11		13.7	103.3	2151.6	103.7	70.0	2054.6
12		16.6	92.3	905.1	91.5	50.9	871.4
13		14.1	80.1	554.8	118.9	62.3	472.4
	$\overline{X} =$	15.2	80.7	1375.4	87.9	73.4	1302.9
		GROUP 3—VIOLA LIMESTONE					
14		1.3	10.4	3399.5	532.3	235.6	2642.5
15		3.6	5.2	974.5	147.5	69.0	768.1
16		0.8	9.8	1430.2	295.7	118.4	1027.1
17		1.8	25.6	183.2	35.4	13.5	161.5
18		8.8	3.4	289.9	32.8	22.4	225.2
19		6.3	16.7	360.9	41.9	24.0	318.1
	$\overline{X} =$	3.8	11.9	1106.4	180.9	80.5	857.1
	$\overline{\overline{X}} =$	8.0	37.7	1482.3	186.8	76.2	1261.4

Source: Ostroff (1967).

can compute

$$[D^2] = [U]'[W]^{-1}[U] \tag{6.113}$$

which yields the generalized distances from the new observation to the centroids of each group measured in the original, p-dimensional space. The implications of these and other alternative definitions of similarity between an observation and the different group centroids are discussed at length by Gnanadesikan (1977), who also provides a method for drawing confidence regions around the group centroids.

Discriminant functions are useful for determining if several groups, presumed to be different, are in fact distinct. We will examine an application of this type.

Salt water is trapped in sedimentary rocks at the time they are formed in the marine environment. The chemical composition of the connate water is subsequently modified by ion exchange and other reactions, by mixing with other brines, and by dilution from infiltrating surface waters. Nonetheless, brines recovered during drill-stem tests of wells may have relict compositional characteristics that provide clues to the origin or depositional environment of their source rocks.

Table 6.43 contains brine analyses, reported in equivalent parts per million (EPM), for oil-field waters from three carbonate units in Texas and Oklahoma. Each sample has been collected from a different oil pool. Discriminant function analysis can be applied to these data to determine if they are distinctive. If they are, this suggests that the brine analyses might provide information about the nature of their original environment, since all three source rocks have approximately the same lithology and have undergone similar histories of burial.

Since there are three groups, only two discriminant functions can be calculated and only two positive eigenvalues will be found. The first of these, $\lambda_1 = 13.29$, accounts for 93.6% of the between-group variance and the second, $\lambda = 0.902$, accounts for the remaining 6.4%. Figure 6.51 shows two discriminant function

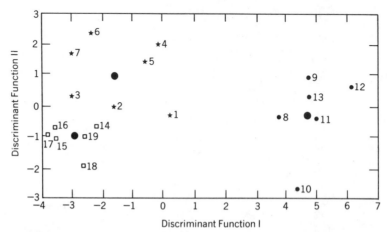

FIGURE 6.51   Plot of scores on first and second discriminant functions. Group 1, Ellenburger Dolomite ($\star$), Group 2, Grayburg Dolomite ($\bullet$), Group 3, Viola Limestone ($\square$). Large dots are group centroids.

axes, with the discriminant scores and centroids for the three groups plotted. The scores have been standardized so their grand means are zero and the standard deviations of the combined data are equal to one.

The first discriminant function clearly separates brines from the Grayburg Dolomite (Group 2) from those collected from the Ellenburger Dolomite (Group 1) and Viola Limestone (Group 3). Distinctions along the second discriminant function are less clear, with overlap between Groups 1 and 2, and between 2 and 3. However, when viewed together, both functions are adequate to completely separate the three groups. This encouraging result suggests that brines collected from formations having similar lithologies may retain unique, relatively homogeneous characteristics.

Canonical Correlation

We now turn to a multivariate technique that has the same computational basis as factor analysis, but which in its concept and objectives is closely related to multiple regression. You will recall that multiple regression is concerned with the relationship between a single dependent variable Y and a set of predictor variables $X_1, X_2, \ldots,$ X_m. An extension of this concern is the relationship(s) between a *set* of Y variables and a second set of X variables measured on the same objects. These relationships may be investigated by finding linear combinations of the X and Y variables that give the highest correlations between the two sets.

Such correlations are called *canonical correlations* and the linear combinations are called *canonical variables*. In effect, we convert the set of X's into a single new variable and the set of Y's into another single new variable, and then determine the correlation between these new variables. The conversion process is linear; that is, the original variables are each weighted and added together to yield the canonical variable. Applications of canonical correlation might include determining the relationship between a set of geochemical variables and a set of petrographic variables, or the relationship between the petrophysical responses on logs from wells and formation properties measured on core samples from the same wells.

Because all the variables are measured on the same samples, the observations form a data matrix whose dimensions are $n \times (p + q)$, where p represents the number of Y variables and q the number of X variables. (For computational convenience, the smaller of the two sets of variables is called Y, so $p \leq q$.) The matrix of variances and covariances, $[S]$, is $(p + q) \times (p + q)$ and can be thought of as composed of four parts: a $p \times p$ matrix $[S_{yy}]$ containing the variances and covariances of the Y variables, a $q \times q$ matrix $[S_{xx}]$ containing variances and covariances of the X variables, and the $p \times q$ matrix $[S_{xy}]$ (and its transpose $[S_{xy}]'$), which contains the covariances between the X's and Y's. That is,

$$[S] = \begin{bmatrix} S_{yy} & S_{xy} \\ S_{xy}' & S_{xx} \end{bmatrix}$$

Although the matrix $[S]$ may be thought of as being partitioned, it has the form of any other variance–covariance matrix. It is symmetrical around the diagonal, whose elements are the variances; the off-diagonal elements are the covariances.

We can denote the $n \times p$ part of the data matrix that contains the Y variables as $[Y]$ and the $n \times q$ part of the data matrix that contains the X variables as $[X]$. The $[Y]$ and $[X]$ matrices can be transformed by multiplication by arbitrary vectors, which results in new variables that are linear combinations of the old:

$$[A]'\,[Y]$$
$$[B]'\,[X]$$

where $[A]$ is a $1 \times p$ vector and $[B]$ is a $1 \times q$ vector. The variances of the two sets of transformed variables will be

$$[A]'\,[S_{yy}][A]$$
$$[B]'\,[S_{xx}][B]$$

The covariances between the transformed X and Y variables will be

$$[A]'\,[S_{xy}]'\,[B]$$

The objective of canonical correlation is to select elements of the two vectors $[A]$ and $[B]$ so the covariances are maximized, subject to the constraint that the variances are equal to one. If the variances are initially standardized to equal one, the covariances are simultaneously standardized and become the correlations between the variables. By using eigenvalue techniques, values of the vectors $[A]$ and $[B]$ can be found that have the desired properties. We are guaranteed that the canonical correlation will be greater than the largest correlation between any original X variable and any original Y variable—that is, greater than any element in the matrix $[S_{xy}]$. This is because we could immediately create linear combinations that would have a correlation this high by setting all elements of $[A]$ and $[B]$ to 0.0 except for those that correspond to the two highest correlated variables, which would be set to 1.0.

The equation that must be solved is very similar to the basic eigenvalue equation that occurs in principal components analysis:

$$\left|\,[\Lambda] - \lambda[I]\,\right| = [0] \tag{6.114}$$

Here, $[\Lambda]$ is a matrix that results from the multiplication of the various parts of the partitioned variance–covariance matrix $[S]$. The matrix multiplication yields a product matrix $[\Lambda]$ that is $q \times q$ and that represents a pooling of the variances in the two sets of variables. That is,

$$[\Lambda] = [S_{xx}]^{-1}[S_{xy}]'[S_{yy}]^{-1}[S_{xy}] \tag{6.115}$$

Unfortunately, the matrix $[\Lambda]$ is asymmetric, so heavy-handed computational methods may be necessary in order to find the determinant and to solve the equation.

The eigenvalue, λ, is numerically equal to the square of the correlation between the two canonical variables. Since the matrix $[\Lambda]$ is $q \times q$, it will have q distinct eigenvalues, each of which represents the correlation between a different pair of canonical variables. Successive eigenvalues will be of decreasing magnitudes, and each pair of canonical variables will be uncorrelated with all other pairs of canonical variables.

The vector $[B]$, used to transform $[X]$ into canonical variables, is found by determining the eigenvectors that correspond to the eigenvalues just found:

$$([\Lambda] - \lambda[I])[B] = [0]$$

or

$$([S_{xx}]^{-1}[S_{xy}]'[S_{yy}]^{-1}[S_{xy}] - \lambda[I])\,[B] = [0]$$

Recall that an eigenvector is calculated simply by substituting an eigenvalue into, and then solving, the set of q simultaneous equations. Once the transformation vector $[B]$ is found, the equivalent canonical transform for the Y's is found by

$$[A] = [S_{yy}]^{-1}[S_{xy}][B] \Big/ \sqrt{\lambda}$$

TABLE 6.44 (a) Partitioned Matrix of Standardized Variances and Covariances (Correlation Coefficients) for Block Data, (b) Eigenvalues and Corresponding Canonical Correlations, (c) Transformation Vectors to Convert Block Variables into Canonical Variables, for the Three Nonzero Canonical Correlations; [A] Vectors are Used to Convert Variables X_1-X_3 to Canonical Form, [B] Vectors are Used to Convert Variables X_4-X_7 to Canonical Form

1.0000	0.5802	0.2011	0.9112	0.2833	0.2865	−0.5331
0.5802	1.0000	0.3637	0.8337	0.1658	0.2610	−0.6087
0.2011	0.3637	1.0000	0.4385	−0.7041	−0.6805	−0.6488
0.9112	0.8337	0.4385	1.0000	0.1630	0.2022	−0.6755
0.2833	0.1658	−0.7041	0.1630	1.0000	0.9902	0.4272
0.2865	0.2610	−0.6805	0.2022	0.9902	1.0000	0.3571
−0.5331	−0.6087	−0.6488	−0.6755	0.4272	0.3571	1.0000

(a)

$$\lambda_1 = 0.9999 \qquad r_1 = 1.00$$
$$\lambda_2 = 0.8485 \qquad r_2 = 0.92$$
$$\lambda_3 = 0.6538 \qquad r_3 = 0.81$$
$$\lambda_4 = 0.0000 \qquad r_4 = 0.00$$

(b)

First Canonical Correlation		Second Canonical Correlation		Third Canonical Correlation	
Vector [A]	Vector [B]	Vector [A]	Vector [B]	Vector [A]	Vector [B]
0.3453	0.6453	−0.0594	−0.0415	0.0956	−0.0321
0.2945	−0.5731	0.1438	−0.6351	−0.0723	0.7468
0.1359	0.5027	−0.1559	0.7712	−0.0404	−0.6593
	0.0494		0.0137		−0.0826

(c)

Of course, there will be a vector $[A]$ and a vector $[B]$ corresponding to each λ. Each vector pair will transform $[X]$ and $[Y]$ into canonical variates; the correlation between these new variables will be $r = \sqrt{\lambda}$.

We can illustrate canonical correlation by turning once again to our data on random-sized blocks. The data fall naturally into two classes, because variables X_1, X_2, and X_3 are the fundamental dimensions of the blocks, while variables X_4 through X_7 are derived from these. We can therefore define the first three variables as forming the set of Y variables, and the next four variables as forming the set of X variables.

The standardized variance–covariance (or correlation) matrix of the block data, partitioned for canonical correlation, is given in Table 6.44. The matrix $[\Lambda]$ from which the eigenvalues must be extracted is also given in Table 6.44, as are its eigenvalues, the canonical correlations which they represent, and the $[A]$ and $[B]$ vectors corresponding to the largest canonical correlation. We can use the vector $[A]$ to transform the Y variables into scores or canonical variates, and the $[B]$ vector to transform the X variables into another set of scores or canonical variates. Figure 6.52 shows a cross-plot of the two sets of canonical variates from the block data. It is obvious that there is, indeed, a perfect linear relationship between the

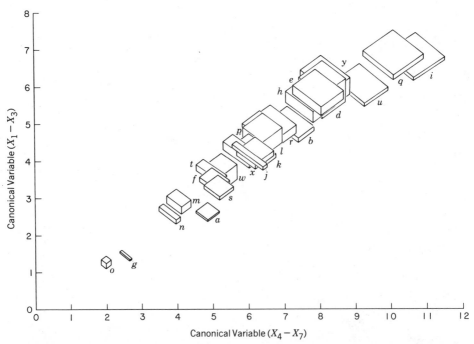

FIGURE 6.52   Cross-plot of the first pair of canonical variates of random block data. Canonical scores of X variables are measured along horizontal axis and canonical scores of Y variables are measured along vertical axis. Canonical correlation between first X and first Y canonical variables is $R = 1.00$.

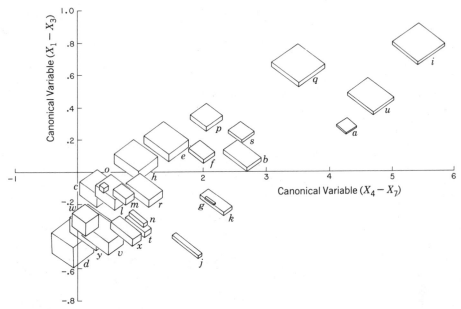

FIGURE 6.53 Cross-plot of the second pair of canonical variates of random block data. Canonical scores of X variables are measured along horizontal axis and canonical scores of Y variables are measured along vertical axis. Canonical correlation between second X and second Y canonical variables is $R = 0.85$.

two sets when placed in canonical form. Figure 6.53 is a similar cross-plot of the second pair of canonical variates. This diagram also exhibits a very strong correlation even though the transformation is quite different.

Comments made under the section on principal components about "reading" component loadings pertain equally to canonical transformations. The vectors $[A]$ and $[B]$ are weights used to transform the original variables $[Y]$ and $[X]$ into canonical variables. Under certain circumstances it may be possible to ascribe a physical meaning to a particular combination of weights. Canonical variables can then be discussed in the same manner as factors. However, there is no assurance that the pattern of weights will have any interpretable meaning, and the canonical correlations may simply reflect arbitrary mathematical combinations of the variables in the two sets. This is especially apt to be true if the canonical correlations between $[X]$ and $[Y]$ are weak. Marascuilo and Levin (1983), who have an exceptionally lucid chapter on canonical correlation, discuss alternative methods for interpreting canonical weights.

There are two commonly used statistical tests for canonical correlations. One checks for the presence of *any* significant canonical correlation among the q canonical relationships. This test will be significant if one *or more* of the pairs of canonical variables are correlated. The second test procedure checks only the significance of the largest canonical correlation.

The test of overall significance was defined by Bartlett (1938) as

$$L = (1 - \lambda_1)(1 - \lambda_2) \cdots (1 - \lambda_q) \tag{6.116}$$

The quantity

$$\chi^2 = \{-(n - 1) + \frac{1}{2}(p + q + 1)\} \ln L \tag{6.117}$$

is distributed as χ^2 with pq degrees of freedom. The null hypothesis is that all of the canonical correlations are equal to zero. If the test statistic falls in the critical region, at least one of the canonical correlations is significantly greater than zero.

A test of only the largest canonical correlation can be derived from Bartlett's test as

$$\chi^2 = \{-(n - 1) + \frac{1}{2}(p + q + 1)\} \ln(1 - \lambda_1) \tag{6.118}$$

where λ_1 is the largest eigenvalue. The test has approximately the degrees of freedom given by

$$df \approx p + q + 1 + \frac{1}{2}\{(p - 1)(q - 1)\}^{2/3} \tag{6.119}$$

The number of degrees of freedom calculated by eq. (6.119) should be rounded down to the nearest whole number.

TABLE 6.45 Log and Core Measurements Made on Pennsylvanian Limestones Encountered in a Well Drilled in Northwestern Kansas

γ	Δ_t	R_t	K	ϕ	S_o	S_w
3.1	64.0	28.8	0.1	3.9	28.2	53.8
3.4	69.0	25.1	0.4	7.0	17.2	55.6
3.4	65.0	38.0	0.1	6.1	24.6	54.2
2.8	62.0	15.1	0.4	6.2	19.3	63.0
2.5	56.0	58.9	0.1	5.9	15.3	73.0
2.3	56.0	61.7	0.3	4.7	14.9	61.6
2.3	54.0	129.0	0.2	6.2	29.0	37.1
2.6	60.0	110.0	1.6	12.7	26.7	34.6
6.0	97.0	5.2	0.0	3.0	0.0	96.6
5.2	67.0	18.2	18.0	18.9	26.4	32.3
3.9	82.0	26.9	6.5	18.4	19.0	48.3
4.7	80.0	12.9	2.5	17.9	16.8	48.0
5.1	77.0	12.0	0.1	12.3	11.7	72.6
5.0	79.0	11.0	0.0	10.4	0.0	91.4
6.1	81.0	70.8	0.0	5.2	0.0	97.8

γ = gamma ray intensity, in gamma ray units. Δ_t = sonic transit time in μsec/ft. R_t = microlaterolog resistivity, in ohm-meters. K = permeability, in millidarcies. ϕ = porosity, in percent. S_o = oil saturation, in percent of porosity. S_w = water saturation, in percent of porosity.

As a geological example, we will consider the data given in Table 6.45, which lists well log measurements made in a series of limestones of Pennsylvanian age encountered while drilling a well in northwestern Kansas. The logging tools measure the gamma ray intensity, sonic transmissivity, and electrical resistance of the interval of rock spanned by the tool. These measured properties reflect both the characteristics of the rock and those of the fluids in the pore space. The table also contains laboratory measurements made on cores taken from the same interval. These include the permeability, porosity, oil saturation, and water saturation.

In well log analysis, the logging tool measurements are transformed to yield estimates of porosity, oil saturation, and water saturation. These estimates are

TABLE 6.46 (a) Partitioned Matrix of Standardized Variances and Covariances (Correlation Coefficients) for Well Data in Table 6.45, (b) Pooled Variance Matrix $[\Lambda]$, (c) Eigenvalues of $[\Lambda]$ and Corresponding Canonical Correlations; $[A]$ Transformation Vector to Convert Log Measurements to Canonical Form, $[B]$ Transformation Vector to Convert Core Measurements to Canonical Form

$$
\begin{bmatrix}
1.0000 & 0.6459 & -0.2320 & 0.3031 & 0.1027 & 0.0881 & -0.1102 \\
0.6459 & 1.0000 & -0.0663 & 0.3063 & 0.4753 & 0.1899 & -0.2996 \\
-0.2320 & -0.0663 & 1.0000 & 0.2161 & 0.3118 & -0.0862 & -0.0789 \\
0.3031 & 0.3063 & 0.2161 & 1.0000 & 0.6334 & 0.0251 & -0.2150 \\
0.1027 & 0.4753 & 0.3118 & 0.6334 & 1.0000 & 0.2099 & -0.5134 \\
0.0881 & 0.1899 & -0.0862 & 0.0251 & 0.2099 & 1.0000 & -0.8639 \\
-0.1102 & -0.2996 & -0.0789 & -0.2150 & -0.5134 & -0.8639 & 1.0000
\end{bmatrix}
$$

(a)

$$
\begin{bmatrix}
0.1237 & -0.0891 & -0.0337 & 0.0406 \\
0.0769 & 0.4445 & 0.1144 & -0.2278 \\
-0.0750 & 0.0006 & 0.0762 & -0.0372 \\
-0.0962 & 0.0167 & 0.0620 & -0.0332
\end{bmatrix}
$$

(b)

$$\lambda_1 = 0.4010 \qquad r_1 = 0.63$$
$$\lambda_2 = 0.1697 \qquad r_2 = 0.41$$
$$\lambda_3 = 0.0405 \qquad r_3 = 0.20$$
$$\lambda_4 = 0.0000 \qquad r_4 = 0.00$$

(c)

First Canonical Correlation		Second Canonical Correlation		Third Canonical Correlation	
Vector $[A]$	Vector $[B]$	Vector $[A]$	Vector $[B]$	Vector $[A]$	Vector $[B]$
$\begin{bmatrix} 0.4960 \\ -0.8727 \\ -0.2869 \end{bmatrix}$	$\begin{bmatrix} 0.2953 \\ -0.9473 \\ -0.0573 \\ -0.1100 \end{bmatrix}$	$\begin{bmatrix} -0.6627 \\ 0.2195 \\ -0.3380 \end{bmatrix}$	$\begin{bmatrix} -0.6816 \\ 0.4287 \\ 0.3627 \\ 0.4692 \end{bmatrix}$	$\begin{bmatrix} -0.0589 \\ -0.2128 \\ 0.3301 \end{bmatrix}$	$\begin{bmatrix} -0.1225 \\ -0.0510 \\ -0.8266 \\ -0.5469 \end{bmatrix}$

(d)

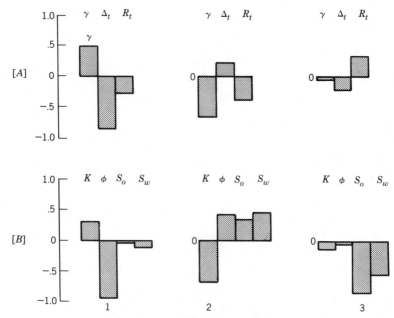

FIGURE 6.54 Canonical transformation vectors [A] and [B] calculated from log and core measurements made in well from northwestern Kansas. Elements of the transformation vectors are plotted as weights. Variables are identified in Table 6.45.

based on various combinations of the log responses, calibrated against known standards. This suggests that we should find a significant canonical correlation between logging tool responses and core measurements.

Table 6.46 gives the standardized variance–covariance matrix of the seven variables in Table 6.45. The partitioning of the matrix into the sub-matrices $[S_{yy}]$, $[S_{xx}]$, and $[S_{xy}]$ and its transpose are indicated. The matrix $[\Lambda]$ that results from the pooling of the variances and covariances in the two sets according to eq. (6.115) is also shown, as are the eigenvalues of $[\Lambda]$ and the corresponding canonical correlations. The [A] and [B] transformation vectors associated with the three nonzero eigenvalues given in Table 6.46 are shown in graphical form in Figure 6.54.

The first canonical correlation essentially reflects a relationship between gamma ray intensity and sonic transit time on one hand and porosity on the other. The signs of the weights suggest that a high gamma ray reading and low sonic value should indicate a low porosity. Shales give a relatively high gamma ray log response because of the presence of radioactive potassium-40 in clay minerals. The velocity of sound waves is much slower in the fluids that fill pores than in solid rock. Therefore, we expect that a combination of the gamma ray and sonic logs would be a good indicator of shales and other nonporous rock, and hence correlated with core porosity.

The second canonical correlation is more interesting, because it relates a combination of logs to core permeability, a notoriously difficult property to predict. The weights suggest that a combination of a low gamma ray response and low microresistivity should equate with a low permeability (and/or high porosity and fluid saturation). Unfortunately, the correlation is not statistically significant and may be merely a vagary of sampling. The third canonical correlation, which is very weak and not statistically significant, relates the microresistivity and sonic log responses to the relative saturations of oil and water.

SELECTED READINGS

Anderberg, M. R., 1973, *Cluster analysis for applications:* Academic Press, Inc., New York, 359 p.

Contains an excellent discussion of nonhierarchical clustering methods, and an appendix of FORTRAN programs.

Anderson, T. W., 1958, *An introduction to multivariate statistical analysis:* John Wiley & Sons, Inc., New York, 374 p.

An advanced treatment of multivariate statistics. The T^2 statistic is discussed in Chapter 5, and discriminant functions are discussed in Chapter 6, where it is called "classification." Principal components are discussed in Chapter 11 and briefly as a statistical method on p. 329–330.

Bartlett, M. S., 1938, Further aspects of the theory of multiple regression: *Proc. of the Cambridge Philosophical Soc.,* **34,** p. 33–40.

Benzécri, Jean-Paul, and others, 1980, *L'Analyse des données,* v. 2, *L'Analyse des correspondances:* Dunod, Paris, 628 p.

Bijnen, E. J., 1973, *Cluster analysis—Survey and evaluation of techniques:* Tilburg Univ. Press, Groningen, The Netherlands, 112 p.

A compact, very readable review of all types of clustering procedures.

Burt, C., 1937, Correlations between persons: *British Journal of Psychology, General Section,* **28,** p. 59–96.

This early article discusses the relationship between R- and Q-mode factor analysis. The hypothetical example used to demonstrate the Eckart-Young theorem in this chapter is adapted from Burt's illustration.

Businger, P. A., and G. H. Golub, 1969, Algorithm 358, singular value decomposition of a complex matrix: *Communications of the Assoc. for Computing Machinery,* **12,** p. 564–565.

Child, D., 1970, *The essentials of factor analysis:* Holt, Rinehart & Winston Ltd., London, 107 p.

A compact paperback survey of factor analysis and related methods. The treatment is non-mathematical and very readable.

Clark, S. P., Jr. (ed.), 1966, Handbook of Physical Constants: *Geol. Soc. America,* Mem. 97, 587 p.

Clifford, H. T., and W. Stephenson, 1973, *An introduction to numerical classification:* Academic Press, Inc., New York, 229 p.

A biologist's view of clustering and other classification procedures. Chapter 8 (on similarity matrices) and Chapter 13 (multivariate analysis) are especially pertinent.

Cooley, W. W., and P. R. Lohnes, 1971, *Multivariate data analysis:* John Wiley & Sons, Inc., New York, 364 p.

An operational discussion of, among other topics, multiple regression, discriminant functions, and factor analysis. Contains FORTRAN subroutines and flow charts for many procedures.

David, M., M. Dagbert, and Y. Beauchemin, 1977, Statistical analysis in geology: Correspondence analysis method: *Quart. Colorado Sch. Mines,* **72,** no. 1, 60 p.

The authors demonstrate the application of correspondence analysis to continuous variables and provide a justification based on probabilities. Contains a FORTRAN program.

Davis, J. C., 1970, Information contained in sediment-size analyses: *Jour. Int'l. Assoc. Mathematical Geology,* **2,** no. 2, p. 105–112.

The experiment on Barataria Bay sediments discussed in this chapter is taken from this article.

Draper, N. R., and H. Smith, 1981, *Applied regression analysis,* 2nd ed.: John Wiley & Sons, Inc., New York, 709 p.

A very thorough treatment of all aspects of linear regression. The comparison of sequential techniques in Chapter 6 is especially valuable.

Eckart, C., and B. Young, 1936, The approximation of one matrix by another of lower rank: *Psychometrika,* **1,** no. 3, p. 211–218.

The fundamental reference on the relationship between *R*- and *Q*-mode factor analysis.

Everitt, B. S., 1978, *Graphical techniques for multivariate data:* North-Holland, New York, 117 p.

Fisher, R. A., 1940, The precision of discriminant functions: *Annals of Eugenics,* **10,** p. 422–429.

Folk, R. L., 1980, *Petrology of sedimentary rocks,* 4th ed.: Hemphill's, Austin, Tex., 184 p.

Gnanadesikan, R., 1977, *Methods for statistical data analysis of multivariate observations:* John Wiley & Sons, Inc., New York, 311 p.

Graphical methods are emphasized in this text derived from work at Bell Telephone Laboratories. Clustering and other classification procedures are discussed in Chapter 4.

Gower, J. C., 1966, Some distance properties of latent root and vector methods used in multivariate analysis: *Biometrika,* **53,** nos. 3, 4, p. 325–338.

A discussion of the dual relationship between correspondence analysis and principal components analysis.

Griffiths, J. C., and C. W. Ondrick, 1969, Modelling the petrology of detrital sediments, *in* Merriam, D. F. (ed.), *Computer applications in the earth sciences:* Plenum Press, New York, p. 73–97.

The article is concerned with a process response model for the origin of detrital sediments. "Factors" exhibited by sediments are discussed on p. 76–84. The PCA experiment suggested in this chapter is described on p. 86–88.

Harman, H. H., 1967, *Modern factor analysis,* 2nd ed.: Univ. Chicago Press, Chicago, 474 p.

Although most examples in this text are psychometric, the treatment is computationally oriented and relatively free of the jargon of psychology. Emphasis is placed on computer implementations.

Harris, R. J., 1975, *A primer of multivariate statistics:* Academic Press, Inc., New York, 332 p.

Discussions of principal components (Chapter 6) and factor analysis (Chapter 7) in this book, although written in the context of the social and behavioral sciences, are equally applicable to geological data. Also covered are topics such as the effects of data transformation and factor rotation.

Hartigan, J. A., 1975, *Clustering algorithms:* John Wiley & Sons., Inc., New York, 351 p.

A discussion of clustering using a strictly algorithmic approach. Contains numerous FORTRAN subroutines and test data sets, drawn from a variety of disciplines.

Heckel, P. H., and J. F. Baesemann, 1975, Environmental interpretation of conodont distribution in Upper Pennsylvanian (Missourian) megacyclothems in eastern Kansas: *Bull. American Assoc. Petroleum Geologists,* **59,** no. 3, p. 486–509.

Hill, M. O., 1974, Correspondence analysis: A neglected multivariate method: *Jour. Royal Stat. Soc., Ser. C, Appl. Stat.,* **23,** no. 3, p. 340–354.

Hirschfeld, H. O., 1935, A connection between correlation and contingency: *Proc. of the Cambridge Philosophical Soc.,* **31,** p. 520–524.

Howarth, R. J. (ed.), 1983, *Statistics and data analysis in geochemical prospecting:* Elsevier Publ. Co., Amsterdam, 437 p.

Chapter 6 covers use of multivariate methods in geochemical exploration and lists sources of computer programs. A novel use of discriminant analysis in prospecting for tin is described in Chapter 9, and Chapters 10 and 11 review published applications of multivariate techniques.

Imbrie, J., and E. G. Purdy, 1962, Classification of modern Bahamian carbonate sediments, *in* Classification of Carbonate Rocks, a Symposium: *American Assoc. Petroleum Geologists,* Mem. 1, p. 253–272.

The basic reference in the geological literature on principal coordinates analysis.

Jardine, N., and R. Sibson, 1971, *Mathematical taxonomy:* John Wiley & Sons, Ltd., London, 286 p.

Clustering and classification discussed using set notation. The book contains a valuable glossary of the jargon of numerical taxonomy.

Johnson, R. M., 1963, On a theorem stated by Eckart and Young: *Psychometrika,* **28,** no. 3, p. 259–263.

A mathematical proof of the Eckart-Young theorem.

Jöreskog, K. G., 1977, Factor analysis by least-squares and maximum-likelihood methods, *in* Enslein, K., and others (eds.), *Statistical methods for digital computers,* Vol. 3: John Wiley & Sons, Inc., New York, p. 125–153.

A complete discussion, with flow charts and algorithms, for the method of "true" *R*-mode factor analysis described in this chapter.

Jöreskog, K. G., J. E. Klovan, and R. A. Reyment, 1976, *Geological factor analysis:* Elsevier Publ. Co., Amsterdam, 178 p.

Essential reading for those interested in applying factor analysis to geological problems. The authors do not shrink from expressing opinions about the relative merits of different methods.

Kendall, M. G., and A. Stuart, 1967, *Advanced theory of statistics,* 2nd ed., Vol. 2: Charles Griffin & Co., Ltd., London, 690 p.

Klovan, J. E., 1966, The use of factor analysis in determining depositional environments from grain-size distributions: *Jour. Sedimentary Petrology,* **36,** p. 115–125.

A *Q*-mode analysis of sediments from Barataria Bay is included.

Koch, G. S., Jr., and R. F. Link, 1980, *Statistical analysis of geological data:* Dover Publications, Inc., New York, 850 p.

Krumbein, W. C., and E. Aberdeen, 1937, The sediments of Barataria Bay: *Jour. Sedimentary Petrology,* **7,** p. 3–17.

Data in Table 6.22 were adapted from Table 1 of this article.

Krumbein, W. C., and R. L. Shreve, 1970, Some statistical properties of dendritic channel networks: Office of Naval Research, Tech. Rept. 13, ONR Task No. 389-150, 117 p. [available from Documents Clearinghouse, Arlington, VA, as document AD 705 625].

Results of a study by geology and geography students on the drainage network of an area in eastern Kentucky. The classes were sufficiently large so extensive data on operator error could be gathered, as well as a large quantity of data on the subject itself. An excellent example of the use of statistics in the classroom. Table 6.1 is adapted from this study.

Lawley, D. N., 1940, The estimation of factor loadings by the method of maximum likelihood: *Proc. Royal Soc. Edinburgh, Ser. A60,* p. 64–82.

Lawley, D. N., and A. E. Maxwell, 1971, *Factor analysis as a statistical method,* 2nd ed.: Butterworth & Co., Ltd., London, 153 p.

This short monograph describes factor analysis independently of its psychometric origins. Especially valuable are the worked examples with intermediate steps in the calculations.

Lebart, L., A. Morineau, and K. M. Warwick, 1984, *Multivariate descriptive statistical analysis:* John Wiley & Sons, Inc., New York, 231 p.

Translation of a French text which discusses correspondence analysis as developed by Benzécri.

Le Maitre, R. W., 1982, *Numerical petrology:* Elsevier Publ. Co., Amsterdam, 281 p.

This specialized text is particularly valuable for its discussions of the effects of closure on multivariate analyses and the application of these methods to geochemical and petrological data. Required reading for anyone analyzing constant-sum data.

Li, C. C., 1964, *Introduction to experimental statistics:* McGraw-Hill, Inc., New York, 460 p.

Chapter 30, "Multiple Measurements," contains a short but very readable discussion of D^2 and of discriminant functions.

Longley, J. W., 1967, An appraisal of least squares programs for the electronic computer from the point of view of the user: *Jour. American Statistical Assoc.,* **62,** no. 319, p. 819–841.

A discussion of the terrifying inaccuracies that can creep into multiple-regression programs through poor programming and maintenance of insufficient digits.

Lunneborg, C. E., and R. D. Abbott, 1983, *Elementary multivariate analysis for the behavioral sciences:* North-Holland, New York, 522 p.

Marascuilo, L. A., and J. R. Levin, 1983, *Multivariate statistics in the social sciences: A researcher's guide:* Brooks/Cole Publ. Co., Monterey, Calif., 530 p.

Provides detailed analyses of popular multivariate statistics program libraries with comparisons of output.

Maron, M. J., 1982, *Numerical analysis: A practical approach:* MacMillan Publ. Co., Inc., New York, 471 p.

Contains good descriptions of the mathematical operations involved in all kinds of mathematical procedures, including eigenvalue operations.

Marriott, F. H. C., 1974, *The interpretation of multiple observations:* Academic Press, Inc. Ltd., London, 117 p.

A very compact volume that provides a readable survey of multivariate statistics.

Matalas, N. C., and B. J. Reiher, 1967, Some comments on the use of factor analysis: *Water Resources Research,* **3,** p. 213–223.

McQueen, J., 1967, Some methods for classification and analysis of multivariate observations: *5th Berkeley Symposium on Mathematics, Statistics, and Probability,* **1,** p. 281–298.

Morrison, D. F., 1976, *Multivariate statistical methods,* 2nd ed.: McGraw-Hill, Inc., New York, 415 p.

One of the most lucid of the multivariate statistics textbooks, containing a very helpful section on the relation between the more familiar univariate methods and their multivariate extensions. Chapters 4, 6–9 cover topics discussed in this section. The treatment of factor analysis in Chapter 9 is especially straightforward.

Morrison, D. F., 1983, *Applied linear statistical methods:* Prentice-Hall, Inc., Englewood Cliffs, N.J., 562 p.

This text has an excellent discussion of two-group discriminant analysis.

Ondrick, C. W., and G. S. Srivastava, 1970, CORFAN-FORTRAN IV computer program for correlation, factor analysis (*R*- and *Q*-mode) and varimax rotation: *Kansas Geological Survey Computer Contribution 42,* 92 p.

Ostroff, A. G., 1967, Comparison of some formation water classification systems: *Bull. American Assoc. Petroleum Geologists,* **51,** no. 3, p. 404–416.

Potter, P. E., N. F. Shimp, and J. Witters, 1963, Trace elements in marine and fresh-water argillaceous sediments: *Geochimica et Cosmochimica Acta,* **27,** p. 669–694.

Reyment, R. A., R. E. Blackith, and N. A. Campbell, 1984, *Multivariate morphometrics,* 2nd ed.: Academic Press, Inc. Ltd., London, 233 p.

A thorough review of the application of multivariate statistical techniques to animal morphology, especially of fossils. Contains extensive citations to the literature and numerous summaries of published case studies.

Sherman, K. N., C. M. Bunker, and C. A. Bush, 1971, Correlation of uranium, thorium, and potassium with aeroradioactivity in the Berea area, Virginia: *Economic Geology,* **66,** p. 302–308.

Sneath, P. H. A., and R. R. Sokal, 1973, *Numerical taxonomy:* W. H. Freeman and Co., San Francisco, 573 p.

The definitive work on numerical taxonomy. Unfortunately, the geologist may find the jargon confusing and the long arguments for a numerically derived phylogeny irrelevant. It is, however, the easiest introduction to the biological literature on quantitative classification.

Switzer, P., 1970, Numerical classification, *in* Merriam, D. F. (ed.), *Geostatistics, a colloquium:* Plenum Press, New York, p. 31–43.

Teil, H., 1975, Correspondence factor analysis: An outline of its method: *Jour. Int'l. Assoc. Mathematical Geology,* **7,** no. 1, p. 3–12.

Teil, H., and J. L. Cheminée, 1975, Application of correspondence factor analysis to the study of major and trace elements in the Erta Ale Chain (Afar, Ethiopia): *Jour. Int'l. Assoc. Mathematical Geology,* **7,** no. 1, p. 13–30.

Temple, J. T., 1978, The use of factor analysis in geology: *Jour. Int'l. Assoc. Mathematical Geology,* **10,** no. 4, p. 379–387.

A strong criticism of factor analysis, particularly as used in geology and related areas.

Thurstone, L. L., 1947, *Multiple-factor analysis:* Univ. Chicago Press, Chicago, 535 p.

The classic work in factor analysis, in its original context as a psychometric tool. Written prior to the development of computers, much of the computational detail is now obsolete. However, the treatment of factor theory remains a fundamental reference.

Tryon, R. C., and D. E. Bailey, 1970, *Cluster analysis:* McGraw-Hill, Inc., New York, 347 p.

Clustering presented from the viewpoint of the experimental psychologist and sociologist. The techniques discussed are computationally distinct from those presented in this chapter; the "O-analysis" methods are the most similar.

Wanke, H., and others, 1970, Major and trace elements and cosmic-ray produced radioisotopes in lunar samples: *Science,* **167,** no. 3918, p. 523–525.

Zhou, D., T. Chang, and J. C. Davis, 1983, Dual extraction of *R*-mode and *Q*-mode factor solutions: *Jour. Int'l. Assoc. Mathematical Geology,* **15,** no. 5, p. 581–606.

How to Run the STAT Program

by Steve N. Yee

Note: In these instructions, the symbol ↵ indicates that the ENTER or RETURN key should be pressed.

Introduction

STAT is a computer program which performs some of the statistical procedures contained in this book. It can run on any personal computer using the PC-DOS® or MS-DOS® (Version 2.10) operating systems with 128K memory. The computer may be equipped with either a color or monochrome monitor and a DOS-compatible line printer.

These instructions assume that you are familiar with the basic operation of your computer and its attached peripherals; that is, you can turn everything on and off, can insert and remove diskettes, and understand the functions of the keys on the keyboard. This type of information is contained in your Guide to Operations manual. It also is assumed that you have some familiarity with the Disk Operating System manual supplied with your computer.

Back Up Your STAT Diskette

The diskette supplied with this book contains the STAT program and some example data sets. You should make a backup copy of the STAT diskette before proceeding, to prevent loss of the STAT program should the diskette become damaged or lost. See your Disk Operating System manual for instructions on how to make a backup copy of a diskette. After you have created a backup copy, store the original STAT diskette in a safe place and use your backup copy for all operations. In the following instructions, all references to the STAT diskette mean your backup copy.

Loading STAT

In addition to the STAT program and example data sets, the STAT diskette also contains a file that is used to reconfigure your DOS operating system. If your personal computer has a hard disk, you should start in the system or root directory;

621

your computer automatically places you in the root directory when first turned on, or you can reach the root directory by typing the Change Directory command:

CD\↵

(You can tell when you are in the root directory by the prompt, which will look like C:\>, which indicates the C or hard disk drive, and the \ or root directory.)
Insert the STAT diskette into the floppy disk drive (Drive A) and type

COPY A:CONFIG.SYS↵

This will copy the CONFIG.SYS file into the root directory on the hard disk. Next, remove the STAT diskette from Drive A and reboot your computer. This is done by pressing three keys *simultaneously*:

CNTRL

ALT

DEL

The computer will reboot from the hard disk, but will include the system reconfiguration information you have placed in the root directory.

You are now ready to load the STAT program onto the hard disk. You may wish to place STAT and related files in a subdirectory, rather than in the root directory. For example, you might wish to place the program in a subdirectory called TERRA. To do this, you should be in the root directory; use the Change Directory command

CD\↵

if necessary. Now type the Make Directory command

MD TERRA↵

This will create the TERRA subdirectory.

Next, access the new directory you have just created by typing the Change Directory command

CD\TERRA↵

Insert the STAT diskette into the floppy disk drive and type

A:LOAD↵

This will copy all of the files on the diskette into the TERRA subdirectory on your hard disk. In the future, when you want to run the STAT program you must first move to the TERRA subdirectory by typing

CD\TERRA↵

If your personal computer does not have a hard disk, the procedure for loading STAT is slightly different. First, you must create a backup DOS diskette; if you do not know how to do this, refer to the instructions in your Disk Operating System manual. The DOS operating system on the backup disk will be modified by instructions in the file CONFIG.SYS, but you can always restore DOS to its original configuration by using your original DOS diskette.

Insert the backup DOS diskette into Drive A. Insert the STAT diskette into Drive B. Type

COPY B:CONFIG.SYS A:

The computer will now copy file CONFIG.SYS onto the backup DOS diskette. Remove the STAT diskette from Drive B, but leave the backup DOS diskette in Drive A. Reboot your computer by pressing three keys *simultaneously*:

CNTRL

ALT

DEL

The computer will reboot from the DOS diskette in Drive A, and will include the system modifications you have incorporated on the DOS backup diskette. In the future, when you wish to run STAT, you must boot your system using this DOS backup diskette.

Running STAT

If your computer does not have a hard disk, insert the STAT diskette into Drive A. To run STAT, type

STAT↵

at the DOS level. When Drive A stops spinning and the disk drive indicator light goes off, remove the STAT diskette from Drive A. Insert into Drive A a diskette containing any data files you wish to analyze or to hold any output files you wish to create using STAT.

If your computer has a hard disk, run STAT by typing

STAT↵

at the DOS level. The title screen will appear, which lists the STAT version number and copyright notice. You are asked to enter the date, in the order month-day-year, followed by ↵ . Use 2-digit numbers separated by dashes. For example,

08-20-85↵

means August 20, 1985.

You next are asked to enter the time, as 2-digit numbers separated by a colon, in the order hours:minutes followed by ↵ . Use 24-hour time notation; for example

20:35↵

means 8:35 p.m. The date and time will appear on all printed output, so you can easily keep track of the printed material you may have generated during a STAT session. After you have entered the time as requested and pressed ◢ , you will go to the main menu of 5TAT options:

1 _____ DATA ENTRY
2 _____ ELEMENTARY STATISTICS
3 _____ MATRIX OPERATIONS
0 _____ EXIT STAT AND RETURN TO DOS

To choose the option you want, type the corresponding number and "RETURN" (◢).

Entering Numeric Data

All numerical data entered from the keyboard should be in free format. In other words, if a prompt requires that you provide more than one value, the numbers must be separated either by a blank space or a comma. When entering numerical data, use the top row of keys, not the numeric keypad.

Data Entry

All input data must have the structure of a matrix. The following table gives the row and column convention for different analyses:

Data	Rows	Columns
Univariate	Observations	1
Multivariate	Observations	Coordinate Values
One-Way Analysis of Variance	Replicates	Samples
Two-Way Analysis of Variance	Treatments	Samples

When the DATA ENTRY option is chosen, you are asked for the name of the file where the data are to be stored. Then you are prompted for the number of rows and columns in the data matrix. Finally, you are asked to enter each row in turn. The data are then stored in the designated file. Now and in all future sessions when 5TAT is run, this data set can be accessed by typing the name of the file. When data entry has been completed, press ◢ to return to the main menu of options.

Naming Files

5TAT file names must consist of at least one character and no more than eight characters. If you wish, you may use file name extensions of up to three characters, separated from the file name by a period. For example,

FILENAME.EXT

is a valid file name and extension. File names may be composed only of upper- and lowercase letters and numbers; no special characters may be used except for the period used to separate a file name from its extension.

Correcting Mistakes

If you type in an incorrect value but have not yet pressed the ⏎ key, you can correct your mistake by pressing the backspace key ← and retyping the correct value. When the ⏎ is pressed, STAT will automatically check for invalid characters (for example, a letter when a number is required). If an invalid character is found, a warning message will appear on the screen and you will be asked to re-enter the correct value. If you have entered a valid character but it is not the one you intended (for example, if you entered a data value of 2.32 when the correct value is 2.23), continue entering data until the file is complete. You can later edit the file using the DOS Line Edit command EDLIN. Refer to your Disk Operating System manual for instructions.

You may make a wrong choice from a menu, and find yourself in a part of the STAT program where you do not want to be. For example, you might accidentally type 3 ⏎ rather than 2 ⏎ at the main menu and go to the Matrix Operations option rather than to the Elementary Statistics option. You can easily back up to the previous menu by selecting Q ⏎ (for Quit), U ⏎ (for Up), or H ⏎ (for Help). You can then select the correct option.

If you are at a submenu within an option, typing Q ⏎, U ⏎, or H ⏎ will return you to the option menu. Typing Q ⏎, U ⏎, or H ⏎ at the option menu will return you to the main menu. If you type Q ⏎, U ⏎, or H ⏎, at the main menu, you will return to the prompt for date. You may enter a new date and press ⏎, which will move you to the time prompt. If you type Q ⏎, U ⏎, or H ⏎, you will return to the date prompt. In this way, you can change the date and time without leaving the STAT program. Typing Q ⏎, U ⏎, or H ⏎ at the date prompt will return you to DOS.

If you are entering a file name and make a mistake, do *not* use Q ⏎, U ⏎, or H ⏎ to go back to the previous menu, since the computer will assume these are file names. Instead, simply press ⏎ .

Printing Files

To print a file after returning to DOS, you must enter the following sequence of commands. First, *simultaneously* press the keys

CTRL

P

This will activate your printer. Then type the instruction

TYPE ⟨FILENAME⟩ ⏎

Note: In place of ⟨FILENAME⟩, insert the name of the STAT file you want printed.

Do not type the brackets. When the file has been printed, you can turn the printer off by again simultaneously depressing

CTRL

P

Elementary Statistics

When the ELEMENTARY STATISTICS option is chosen, you are presented with six possible selections:

1 —— UNIVARIATE STATISTICS AND ONE SAMPLE T TEST
2 —— BIVARIATE STATISTICS
3 —— LINEAR REGRESSION
4 —— ONE WAY ANALYSIS OF VARIANCE
5 —— TWO WAY ANALYSIS OF VARIANCE
0 —— RETURN TO MAIN MENU OF OPTIONS

Choose the desired option by typing its corresponding number, followed by ↵.

The following table gives the minimum and maximum limits on the size of the data matrix for each option:

Option	Number of Rows		Number of Columns	
	Minimum	Maximum	Minimum	Maximum
Univariate	2	No Limit	1	1
Bivariate	2	No Limit	2	2
Linear Regression	2	100	2	2
One-Way Analysis of Variance	2	No Limit	2	100
Two-Way Analysis of Variance	2	No Limit	2	100

The statistical output may take one of two forms. Results of analyses may be printed on the computer's monitor and/or written onto files. These files can be sent to a printer after returning to DOS. The prompts for selecting the desired types of output are:

OUTPUT RESULTS TO MONITOR – Y OR N
OUTPUT RESULTS TO FILE   – Y OR N

You are then asked for the name of the input data file. If you want the statistical results sent to a file, you will also be asked for the name of the output file.

The following list gives the various statistics which are computed by the different options of the ELEMENTARY STATISTICS program.

Option	Statistics
Univariate Statistics	Number of observations, sum, sum of squares, variance, standard deviation, mean, one sample t statistic
Bivariate Statistics	Number of observed pairs, sum of cross-products, covariance, correlation; for both X_1 and X_2: sum, sum of squares, mean, variance, standard deviation
Linear Regression	Terms in the normal equations (that is, n, the number of observed pairs; sum of X's; and sum of X's squared), sum of Y's, sum of cross-products of X and Y, the estimated parameters of the regression equation, the estimated values of $\hat{Y}$ based on the regression equation, the residuals, total sum of squares, sums of squares due to regression and deviation, goodness-of-fit, and correlation coefficient
One-Way ANOVA	One-Way ANOVA table
Two-Way ANOVA	Two-Way ANOVA table

Matrix Operations

When the MATRIX OPERATIONS option is selected from the main menu, you are presented with eight choices:

1 _____ PRINT OUT A MATRIX

2 _____ ADD TWO MATRICES

3 _____ MULTIPLY TWO MATRICES

4 _____ MULTIPLY A MATRIX BY A CONSTANT

5 _____ INVERT A MATRIX

6 _____ TRANSPOSE A MATRIX

7 _____ COMPUTE EIGENVALUES AND VECTORS OF SYMMETRIC MATRIX

0 _____ RETURN TO MAIN MENU OF OPTIONS

Select the desired option by typing in its corresponding number and a RETURN (   ). The maximum dimensions for any matrix in these operations are 50 rows by 50 columns.

Option 1 will print out any data matrix used or created by the STAT program. The output can be sent to the monitor or to a file for later printing. You are asked for the name of the input data file which contains the matrix to be printed. The matrix is then printed out in strips of five columns.

For options 2 to 7, you are asked to enter the file names of the input and output matrices. Option 4 requires that you also enter a constant. In option 7, the input matrix must be symmetric. The eigenvalues are stored in descending order as a column vector in the file named EIGENVAL.DAT which is automatically created by STAT. The corresponding eigenvectors are contained in the columns of the output matrix.

The same file name can be used for both an input and output matrix. If this is done, the output matrix will be written over the input matrix contained in the specified file.

For options **2** and **3**, two matrices are required for input. The two matrices must be in separate files. By judicious use of **2** through **7**, a long string of matrix operations can be performed with relative ease. In fact, it is possible to duplicate almost any of the multivariate statistical calculations described in this book by combinations of the matrix operations available in STAT.

Some Examples

Univariate Statistics

We will assume that the STAT diskette has been inserted into the floppy-disk drive of your computer, or that the STAT programs have been loaded onto your hard disk. In response to the DOS prompt, type

STAT↵

You will be prompted for the date and time. After you have provided this information, continue by pressing ↵. The main menu of options will appear. Select ELEMENTARY STATISTICS by typing

2↵

The Elementary Statistics menu of options will appear. Select UNIVARIATE STATISTICS AND ONE SAMPLE T TEST by typing

1↵

The program will now ask if you want your results to appear on the monitor. Type

Y↵

STAT will ask if you want your results to be loaded into a file. Type

N↵

STAT will ask for the name of your input data file. Five analyses of chromium in samples of an Upper Pennsylvanian shale from Kansas are listed in Table 2.1 and have been placed in a file called UNIV.DAT on the STAT disk. We will test the hypothesis that the mean chromium content of the population from which these samples have been taken is 200 ppm. Enter the name of the input data file by typing

UNIV.DAT↵

In response to the question, "DO YOU WANT TO COMPUTE ONE SAMPLE T STATISTIC?", answer "yes" by typing

Y↵

In response to the instruction "ENTER HYPOTHETICAL MEAN OF POPULA-TION," type

2OO↵

STAT will now produce a statistical analysis of the chromium analyses from Table 2.1. The one sample t-statistic for testing the hypothesis that $\mu = 200$ will be computed by STAT and is $t = 2.06$. Since there are $5 - 1 = 4$ degrees of freedom, and the critical value of $t_{.95,4}$ found from Table 2.11 is 2.13, we cannot reject the null hypothesis. This collection of analyses could have been drawn from a parent population of shales whose mean chromium content is 200 ppm. The t test is discussed on pages 59 to 66.

Bivariate Statistics

To continue, press ↵. You now see the ELEMENTARY STATISTICS menu of options again. Select the BIVARIATE STATISTICS option by typing

2↵

Again, choose to send results to the monitor and not to a file. In response to the instruction "ENTER NAME OF INPUT DATA FILE," type

BIV.DAT↵

This will load the data set listed in Table 2.5, giving the lengths and widths of shells of the brachiopod *Composita*. The measurements are in millimeters. STAT will compute the covariance and correlation between length and width and also provide all values for intermediate calculations. The first screen display shows the number of pairs of observations and the sums of squares and cross-products of the two variables. Press ↵. The screen will now display the means of the two variables, their standard deviations, variances, covariance, and correlation. Press ↵ after the second screen display to return to the ELEMENTARY STATISTICS menu of options. The calculation of covariance and correlation are described on pages 34–41.

Other Elementary Statistics Options

STAT provides three more options for elementary statistics; these are linear regression, one-way ANOVA, and two-way ANOVA. Example data sets for each procedure are included as files on the STAT diskette. Table 4.11, listing the moisture content in core samples through bottom-mud in a Louisiana estuary, is contained in the LINREG.DAT file. These eight observations provide a simple demonstration of linear regression, which is described on pages 176–186.

The one-way analysis of variance option can be demonstrated using data from Table 2.16, which gives the amount of carbonate cement, in percent $CaCO_3$, measured on five sandstone samples. Each of the sandstone samples has been broken into six replicates. The data are listed in the file named ANOVA1.DAT on the STAT diskette. One-way analysis of variance is discussed on pages 74–77.

Two-way analysis of variance can be demonstrated using data from Table 2.17, on directional permeabilities measured on core samples of the St. Peter Sandstone. The measurements, in millidarcies, can be classed into two treatments representing horizontal and vertical permeabilities, depending upon the orientation of the cores. The 20 measurements are contained in the file ANOVA2.DAT on the STAT diskette. Two-way analysis of variance is discussed in the text on pages 78–80.

Matrix Operations

A more extensive bivariate data set is provided in file PCA.DAT and can be used to try the various matrix operations in STAT. The data in PCA.DAT are taken from Table 6.19 and represent measurements of lengths and widths made on 20 brachiopod shells; these data are shown plotted in Figure 6.18. Also provided is a data file named ONES.DAT which consists of a 1 × 20 vector of 1's. We will demonstrate the matrix operations available in STAT by calculating the variance–covariance matrix of the data in Table 6.19. The variance–covariance matrix can be found by the equation

$$[S] = \frac{1}{n-1} ([X]'[X] - n[\overline{X}][\overline{X}]')$$

where

$$[S] = m \times m \text{ variance–covariance matrix}$$
$$[\underline{X}] = n \times m \text{ data matrix}$$
$$[\overline{X}] = m \times 1 \text{ vector of means}$$

You may wish to compare this matrix formulation with eq. (2.22) to confirm that the two are equivalent.

The first part of the calculation involves finding the mean of the data set. This can be done using the algebraic operation

$$[\overline{X}] = \frac{1}{n} [X]'[H]$$

where

$$[H] = n \times 1 \text{ vector of 1's}$$

Note that the matrix multiplication [X]'[H] has the effect of summing the observations in the data matrix.

From the main menu, select the MATRIX OPERATIONS option by typing

3↵

The MATRIX OPERATIONS menu of options will now appear. The first step we must perform is to transpose the data matrix. Select the transposition option by typing

6↵

STAT will present the equation

TRANSPOSE OF A = C

and you will be asked for the name of the file containing the matrix to be transposed. Type

PCA.DAT↵

STAT will now ask for the name of the file where the transpose is to be stored. Type

T.DAT↵

The transpose of the data matrix is now stored in file T.DAT. Press ↵ to return to the menu of matrix operations. We can now sum the data matrix by multiplying the transpose by the column vector of 1's. Type

3↵

The monitor now shows the matrix multiplication operation

A * B = C

The program will request the names of the files containing each of these three matrices. We want matrix A to contain the transpose of the data, so type

T.DAT↵

Matrix B should contain [H], the vector of 1's, so type

ONES.DAT↵

We will name the file where the product matrix is to be stored C1.DAT, so type

C1.DAT↵

The operation will be performed and the product of the matrix multiplication, which is a 1 × 2 column vector of the sums of the two variables, will be stored in file C1.DAT. Press ↵ to restore the menu of options.

To convert the sums into means, we must divide by n, or equivalently, multiply by $1/n$. This involves multiplying a matrix by a constant, which is option **4**. Type

4↵

The equation

CONST * A = C

appears, followed by the instruction to enter the value for the constant. Since $1/n = 1/20 = 0.04$, type

.04↵

In response to the request for the name of the file containing matrix A, type

C1.DAT⏎

We can place the means back into the same matrix occupied by the sums, since we will no longer need the sums. In response to the request for the name of the file where the matrix product is to be stored, type

C1.DAT⏎

Press ⏎ to restore the menu of matrix options.

The vector of means now must be transposed and cross-multiplied to obtain the final term in our equation. Transposition is selected by typing

6⏎

Matrix A should contain the vector of means. Type

C1.DAT⏎

The transpose will be placed in a file we will call C2.DAT. Type

C2.DAT⏎

Press ⏎ to restore the menu of matrix options.

Now we must multiply the vector of means by its transpose. From the menu of options, select matrix multiplication option **3**. Type

3⏎

Matrix A should contain the vector of means, which is in C1.DAT. Type

C1.DAT⏎

Matrix B should contain the transpose of the mean vector, which is in C2.DAT. Type

C2.DAT⏎

The results of the matrix multiplication will be the cross-products of the mean vector which we must store in a file we will call C3.DAT; type

C3.DAT⏎

Press ⏎ to return to the menu of matrix options.

The next steps consist of multiplying the matrix of cross-products of the mean vector by n, then subtracting this matrix from the cross-products of the data matrix. Again, we must multiply a matrix by a constant, so select this option by typing

4⏎

The required constant is n, or, carrying the sign of the operation, -25. Type

-25⏎

The matrix A should contain the cross-products of the means. Type

$$C3.DAT \dashv$$

The product of this operation must be stored in a file we will call C4.DAT. Type

$$C4.DAT \dashv$$

Press $\dashv$ to return to the menu of matrix options.

We now must find the cross-products of the original data matrix. We can do this by premultiplying the data matrix by its transpose. Type

$$3 \dashv$$

The matrix multiplication equation A * B = C again appears. Into matrix A we must place the transpose of the data, which are contained in the file T.DAT. Type

$$T.DAT \dashv$$

Matrix B must contain the original data which are in the file PCA.DAT. Type

$$PCA.DAT \dashv$$

The resulting cross-product matrix can be placed in a file we will call C5.DAT. Type

$$C5.DAT \dashv$$

After the operation is complete, return to the matrix options menu by pressing $\dashv$.

The next step in the calculation of the variance–covariance matrix is to add together the two parts of the equation

$$([X]'[X]) + (-n[X][X]')$$

From the menu, select the matrix addition option by typing

$$2 \dashv$$

Matrix A should contain $[X]'[X]$. Type

$$C5.DAT \dashv$$

Matrix B should contain the cross-products of the means. Type

$$C4.DAT \dashv$$

We must store their sum in another file. Type

$$C6.DAT \dashv$$

Press $\dashv$ to return to the matrix options menu.

The final step in the calculation consists of multiplying by $1/(n-1)$, which is $1/19$. Select option **4** by typing

$$4 \dashv$$

For the constant, enter the decimal equivalent of 1/19 by typing

.041666667↵

This value is to be multiplied by the matrix contained in file C6.DAT. Type

C6.DAT↵

This will produce the final result, our desired matrix of variances and covariances. Store the variance–covariance matrix in file C7.DAT by typing

C7.DAT↵

Press ↵ to return to the menu of matrix options.

The final step of the matrix operation is to display the variance–covariance matrix we have created. To print the matrix, select option **1**. Type

1↵

In response to the query from STAT

OUTPUT RESULTS TO MONITOR – Y or N

type

Y↵

to send results to the monitor. In response to the query

OUTPUT RESULTS TO FILE – Y or N

type

N↵

as we do not wish to store this output in a separate file. The variance–covariance matrix is stored in the file C7.DAT, so type

C7.DAT↵

The variance–covariance matrix will now appear on the monitor. It should be

	1	2
1	20.27666	15.58500
2	15.58500	24.06000

Return to the matrix options menu by pressing ↵.

This may seem a long, involved process to calculate two variances and a co-variance; but remember, exactly the same number of steps could have found all of the elements of a 20×20 matrix. In fact, with the matrix algebra operations contained in STAT, you should be able to perform *any* of the multivariate operations described in the textbook except for the calculation of eigenvalues and eigenvectors from asymmetric matrices.

Additional Programs

TERRASTAT©, an extensive library of programs for almost all of the procedures described in this book and keyed to appropriate parts of the text, is available on diskette for personal computers. The programs are written for the IBM-PC® and compatible computers that use the PC-DOS® or MS-DOS® operating systems. Information on TERRASTAT, including prices and computer compatibility, can be obtained from

TERRASCIENCES, INC.
7555 West Tenth Avenue
Lakewood, Colorado USA 80215

Index

Some terms listed in this index, such as *Mean,* occur repeatedly throughout the text. Only the page on which the definition of the term appears is cited.